Pericyclic Reactions

Pericyclic Reactions

A Mechanistic and Problem-Solving Approach

Sunil Kumar
Department of Chemistry
F.G.M. Govt. College
Haryana, India

Vinod Kumar
Department of Chemistry
Maharishi Markandeshwar University
Haryana, India

S.P. Singh
Department of Chemistry
Kurukshetra University, Kurukshetra
Haryana, India

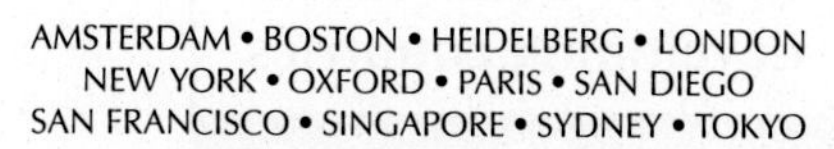

Academic Press is an imprint of Elsevier

Academic Press is an imprint of Elsevier
125 London Wall, London EC2Y 5AS, UK
525 B Street, Suite 1800, San Diego, CA 92101-4495, USA
225 Wyman Street, Waltham, MA 02451, USA
The Boulevard, Langford Lane, Kidlington, Oxford OX5 1GB, UK

Copyright © 2016 Elsevier Inc. All rights reserved.

No part of this publication may be reproduced or transmitted in any form or by any means, electronic or mechanical, including photocopying, recording, or any information storage and retrieval system, without permission in writing from the publisher. Details on how to seek permission, further information about the Publisher's permissions policies and our arrangements with organizations such as the Copyright Clearance Center and the Copyright Licensing Agency, can be found at our website: www.elsevier.com/permissions.

This book and the individual contributions contained in it are protected under copyright by the Publisher (other than as may be noted herein).

Notices
Knowledge and best practice in this field are constantly changing. As new research and experience broaden our understanding, changes in research methods, professional practices, or medical treatment may become necessary.

Practitioners and researchers must always rely on their own experience and knowledge in evaluating and using any information, methods, compounds, or experiments described herein. In using such information or methods they should be mindful of their own safety and the safety of others, including parties for whom they have a professional responsibility.

To the fullest extent of the law, neither the Publisher nor the authors, contributors, or editors, assume any liability for any injury and/or damage to persons or property as a matter of products liability, negligence or otherwise, or from any use or operation of any methods, products, instructions, or ideas contained in the material herein.

ISBN: 978-0-12-803640-2

British Library Cataloguing in Publication Data
A catalogue record for this book is available from the British Library

Library of Congress Cataloging-in-Publication Data
A catalog record for this book is available from the Library of Congress

For Information on all Academic press publications
visit our website at http://store.elsevier.com/

To Our Families

Sunil Kumar
Parents
Dr. Meenakshi, Ayush, Neerav

Vinod Kumar
Parents
Sushma, Mohit, Vignesh

S.P. Singh
Pushpa, Sunny, Romy
Preeti, Preety
Poorva, Uday, Adi, Veer

Contents

Preface

Ever since the appearance of the classic *The Conservation of Orbital Symmetry* by Woodward and Hoffmann in 1970, there has been a surge in the publication of many books and excellent review articles dealing with this topic. This was natural as after having established mechanisms of ionic and radical reactions, focus had shifted to uncover the mechanisms of the so-called "no-mechanism reactions." The uncovering of the fact that orbital symmetry is conserved in concerted reactions was a turning point in our understanding of organic reactions. It is now possible to predict the stereochemistry of such reactions by following the simple rule that stereochemical consequences of reactions initiated thermally will be opposite to those performed under photochemical conditions. Study of pericyclic reactions, as these are known today, is an integral part of our understanding of organic reaction mechanisms.

Despite the presence of many excellent books on this vibrant topic, there was an absence of a book that concentrates primarily on a problem-solving approach for understanding this topic. We had realized during our teaching career that the most effective way to learn a conceptual topic is through such an approach. This book is written to fill this important gap in the belief that it would be helpful to students to have problems pertaining to different types of pericyclic reactions compiled together in a single book.

The book opens with an introduction (Chapter 1), which, besides providing background information needed for appreciating different types of pericyclic reactions, outlines simple ways to analyze these reactions using orbital symmetry correlation diagram, frontier molecular orbital (FMO), and perturbation molecular orbital (PMO) methods. This chapter also has references to important published reviews and articles.

Electrocyclic, sigmatropic, and cycloaddition reactions are subsequently described in Chapters 2, 3, and 4, respectively. Chapter 5 is devoted to a study of cheletropic and 1,3-dipolar cycloaddition reactions as examples of concerted reactions. Many group transfer reactions and elimination reactions, including pyrolytic reactions, are included in Chapter 6. There are solved problems in each chapter that are designed for students to develop proficiency that can be acquired only by practice. These problems, about 450, provide sufficient breadth to be adequately comprehensive. Solutions to all these problems are provided in each chapter. Finally, in Chapter 7, we have compiled unworked problems whose solutions are provided separately in the Appendix. The aim behind introducing these unsolved problems is to let the students develop their own skills.

Assuming that a student has taken courses in organic chemistry that include reaction mechanisms and stereochemistry, the book is meant to be taught as a one-semester course to graduate and senior undergraduate students majoring in chemistry. One has to remember that a book designed for a one-semester course cannot include all the reactions reported in the literature; rather, only representative examples of each of various reaction types are given. A general index is included, which it is hoped will be of help to readers in searching for the types of reactions related to a particular problem.

We hope that our book will be well received by students and teachers. We encourage all those who read and use this book to contact us with any comments, suggestions, or corrections for future editions. Our email addresses are: chahal_chem@rediffmail.com, vinodbatan@gmail.com, and shivpsingh@rediffmail.com.

We thank our reviewers for carefully reading the manuscript and offering valuable suggestions. Finally, we thank the editorial staff of Elsevier for bringing the book to fruition.

July 2015

Sunil Kumar
Vinod Kumar
S.P. Singh

Chapter 1

Pericyclic Reactions and Molecular Orbital Symmetry

Chapter Outline

In organic chemistry, a large number of chemical reactions containing multiple bond(s) do not involve ionic or free radical intermediates and are remarkably insensitive to the presence or absence of solvents and catalysts. Many of these reactions are characterized by the making and breaking of two or more bonds in a *single concerted step* through the *cyclic transition state*, wherein all first-order bondings are changed. Such reactions are named as pericyclic reactions by Woodward and Hoffmann.

The word *concerted* means reactant bonds are broken and product bonds are formed synchronously, though not necessarily symmetrically without the involvement of an intermediate. The word *pericyclic* means the movement of electrons (π-electrons in most cases) in a cyclic manner or around the circle (i.e., *peri* = around, *cyclic* = circle or ring).

They are initiated by either *heat* (thermal initiation) or *light* (photo initiation) and are highly stereospecific in nature. The most remarkable observation about these reactions is that, very often, thermal and photochemical processes yield products with different stereochemistry. Most of these reactions are equilibrium processes in which direction of equilibrium depends on the enthalpy and entropy of the reacting species. Therefore, in general, three important points that should be considered while studying the

Pericyclic Reactions. http://dx.doi.org/10.1016/B978-0-12-803640-2.00001-4
Copyright © 2016 Elsevier Inc. All rights reserved.

pericyclic reactions are: involvement of π-electrons, type of activation energy required (thermal or light), and stereochemistry of the reaction.

There is a close relationship between the mode of energy supplied and stereochemistry for a pericyclic reaction, which can be exemplified by considering the simpler reactions shown in Scheme 1.1.

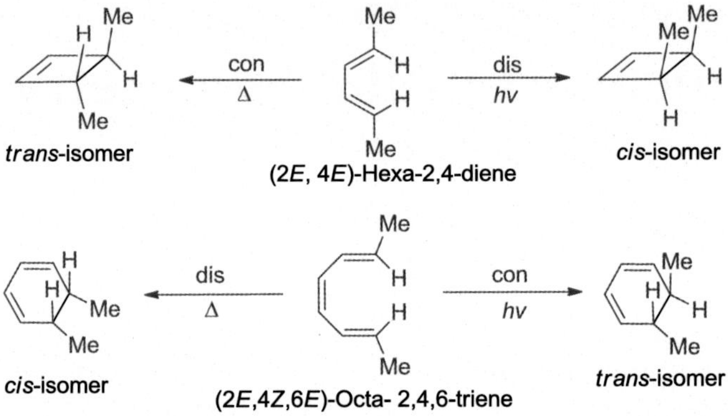

trans-isomer ← con, Δ — (2*E*, 4*E*)-Hexa-2,4-diene — dis, *hv* → *cis*-isomer

cis-isomer ← dis, Δ — (2*E*,4*Z*,6*E*)-Octa- 2,4,6-triene — con, *hv* → *trans*-isomer

SCHEME 1.1 Stereochemical changes in pericyclic reactions under thermal and photochemical conditions.

When heat energy is supplied to the starting material, then it gives one isomer, while light energy is responsible for generating the other isomer from the same starting material.

1.1 CLASSIFICATION OF PERICYCLIC REACTIONS

Pericyclic reactions are mainly classified into the four most common types of reactions as depicted in Scheme 1.2.

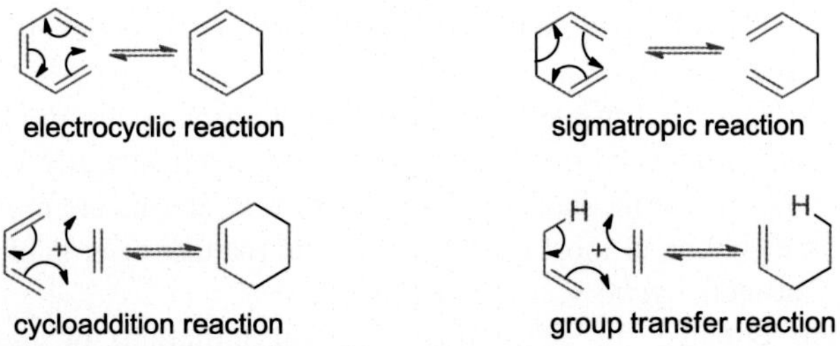

electrocyclic reaction

sigmatropic reaction

cycloaddition reaction

group transfer reaction

SCHEME 1.2 Common types of pericyclic reactions.

In an **electrocyclic reaction**, a cyclic system (ring closure) is formed through the formation of a σ-bond from an open-chain conjugated polyene system at the cost of a multiple bond and vice versa (ring opening). These reactions are unimolecular in nature as the rate of reactions depends upon the presence of one type of reactant species. Such reactions are reversible in nature, but the direction of the reaction is mainly controlled by thermodynamics. Most of the electrocyclic reactions are related to ring closing process instead of ring opening due to an interaction between the terminal carbon atoms forming a σ-bond (more stable) at the cost of a π-bond.

Sigmatropic rearrangements are the unimolecular isomerization reactions in which a σ-bond moves from one position to another over an unsaturated system. In such reactions, rearrangement of the π-bonds takes place to accommodate the new σ-bond, but the total number of π-bonds remains the same.

In **cycloaddition reactions**, two or more components containing π-electrons come together to form the cyclic system(s) through the formation of two or more new σ-bonds at the cost of overall two or more π-bonds, respectively, at least one from each component. Amongst the pericyclic reactions, cycloadditions are known as the most abundant, featureful, and valuable class of the chemical reactions. The reactions are known as intramolecular when cycloaddition occurs within the same molecule. The reversal of cycloaddition in the same manner is known as *cycloreversion*. There are some cycloaddition reactions that proceed through the stepwise ionic or free radical mechanism and thus are not considered as pericyclic reactions.

These reactions are further extended to cheletropic and 1,3-dipolar reactions, which shall be discussed in detail in Chapter 5.

Group transfer reactions involve the transfer of one or more atoms or groups from one component to another in a concerted manner. In these reactions, two components join together to form a single molecule through the formation of a σ-bond.

It is very important to note that in studying the pericyclic reactions, the curved arrows can be drawn in clockwise or anticlockwise direction (Scheme 1.3). The direction of arrows does not indicate the flow of electrons from one component or site to another as in the case of ionic reactions; rather, it indicates where to draw the new bonds.

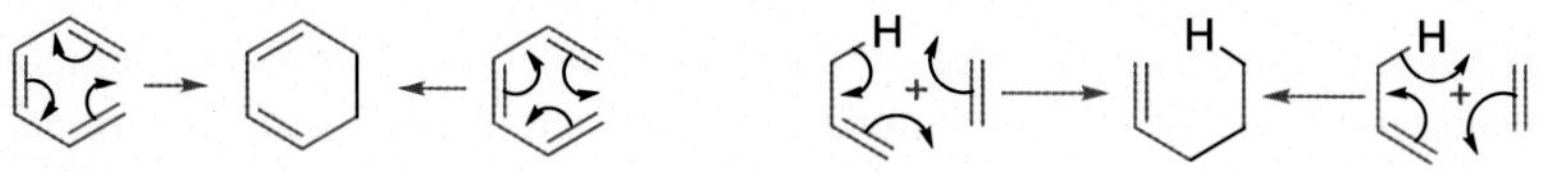

SCHEME 1.3 Clockwise and anticlockwise direction of the curved arrows in pericyclic reactions.

1.2 MOLECULAR ORBITALS OF ALKENES AND CONJUGATED POLYENE SYSTEMS

In order to understand and explain the results of the various pericyclic reactions on the basis of different theoretical models, a basic understanding of the molecular orbitals of the molecules, particularly those of alkenes and conjugated polyene systems and their symmetry properties, is required.

According to the molecular orbital theory, molecular orbitals (MOs) are formed by the linear combination of atomic orbitals (LCAO) and then filled by the electron pairs. In LCAO when two atomic orbitals of equivalent energy interact, they always yield two molecular orbitals, a bonding and a corresponding antibonding orbital. The bonding orbital possesses lower energy and more stability while antibonding possesses higher energy and less stability as compared to an isolated atomic orbital. Let us consider the simplest example of H_2 molecule formed by the combination of 1s atomic orbitals (Figure 1.1).

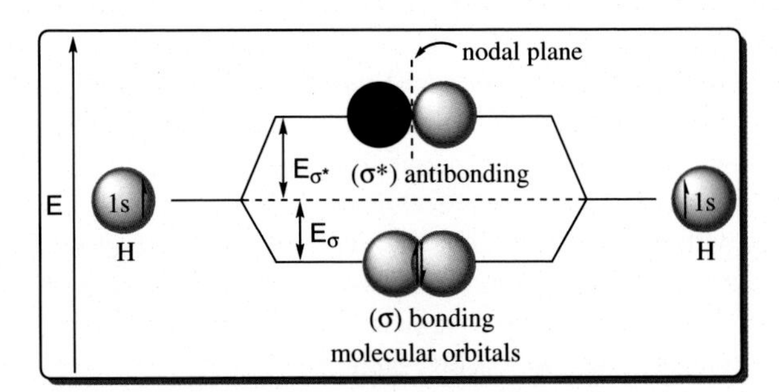

FIGURE 1.1 Formation of molecular orbitals in the case of an H_2 molecule.

The bonding molecular orbital is a result of positive (constructive) overlap, and hence electron density lies in the region between two nuclei. However, an antibonding molecular orbital is formed as a result of negative (destructive) overlap and, therefore, exhibits a nodal plane in the region between the two nuclei. The bonding and antibonding molecular orbitals exhibit unequal splitting pattern with respect to the atomic orbitals because a fully filled molecular orbital has higher energy due to interelectronic repulsion.

We now consider molecular orbital theory with reference to the simplest π-molecular system, ethene. As already discussed, the number of molecular orbitals formed is always equal to the number of atomic orbitals combining together. Similarly, in the case of an ethene molecule, sideways interaction between *p*-orbitals of the two individual carbon atoms results in the formation of the new π bonding and π* antibonding molecular orbitals that differ in energy (Figure 1.2). In the bonding orbital of ethene, there is a constructive overlap of two similar lobes of *p*-orbitals in the bonding region between the nuclei. However, in the case of an antibonding orbital, there is destructive overlap of two opposite lobes in the bonding region. Each *p*-orbital consists of two lobes with opposite phases of the wave function.

We ignore σ-bond skeleton in this treatment as sigma molecular orbitals remain unaffected during the course of a pericyclic reaction.

The conjugated polyenes constitute an important class of organic compounds exhibiting a variety of pericyclic reactions. On the basis of the

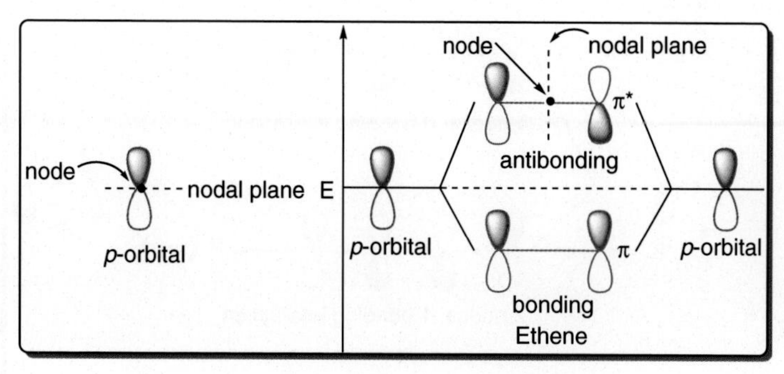

FIGURE 1.2 Formation of two molecular orbitals (π and π*) of ethene.

number of π-electrons, such compounds are classified into two categories bearing 4n or (4n + 2) π-electron systems. In order to construct the molecular orbitals for such polyene systems, let us consider buta-1,3-diene as the simplest example.

In the molecule of buta-1,3-diene, there are four *p*-orbitals located on four adjacent carbon atoms and hence this generates four new π-molecular orbitals on overlapping. The way to get these new π-molecular orbitals is the linear combination of two π-molecular orbitals of ethene according to the *perturbation molecular orbital* (*PMO*) *theory*. Like the combination of atomic orbitals, overlapping of the bonding (σ or π) or antibonding molecular orbitals (σ* or π*) of the reactants (here, ethene) results in the formation of the new molecular orbitals that are designated as Ψ_1, Ψ_2, etc. in the product (here, buta-1,3-diene).

According to PMO theory, linear combination always takes place between the two orbitals (two molecular orbitals or two atomic orbitals, or one atomic and one molecular orbital) having minimum energy difference. Thus, here we need to consider π−π and π*−π* interactions (constructive or destructive) instead of interactions between π and π* orbitals (Figure 1.3). In buta-1,3-diene, 4π-electrons are accommodated in the first two π-molecular orbitals, and the remaining two higher energy π-molecular orbitals will remain unoccupied in the ground state of the molecule.

The lowest energy orbital (represented as wave function Ψ_1) of buta-1,3-diene does not have any node and is the most stable due to the presence of three bonding interactions. However, the second molecular orbital Ψ_2 possesses one node, two bonding and one antibonding interactions, and would be less stable than Ψ_1. The Ψ_3 has two nodes and one bonding interaction. Due to the two antibonding interactions, Ψ_3 possesses overall antibonding character and thus energy of this orbital is more than the energy of Ψ_2. The Ψ_4 orbital is formed by the interaction between π* and π* of two ethene molecules. It bears three nodes and the highest energy.

Similarly, in the case of longer conjugated systems like a hexa-1,3,5-triene system, there are six *p*-orbitals on six adjacent carbon atoms, which can

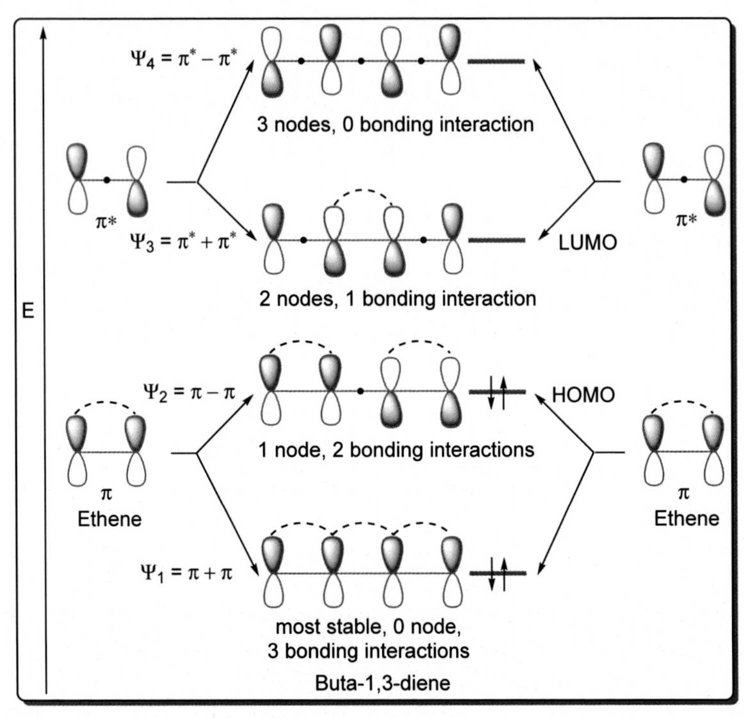

FIGURE 1.3 Formation of π-molecular orbitals in buta-1,3-diene.

generate six new π-molecular orbitals (Figure 1.4). In hexa-1,3,5-triene, 6π-electrons are accommodated in the first three bonding π-molecular orbitals (Ψ_1, Ψ_2, Ψ_3) and the remaining three higher energy antibonding π-molecular orbitals (Ψ_4, Ψ_5, Ψ_6) will remain unoccupied in the ground state.

On the basis of molecular orbital diagrams of ethene, buta-1,3-diene, and hexa-1,3,5-triene, the following points should be considered while constructing the molecular orbitals of the conjugated polyenes:

1. Consider only π-molecular orbitals and ignore σ-bond skeleton as sigma molecular orbitals remain unaffected during the course of a pericyclic reaction.
2. For a system containing n π-electrons (n = even), interaction of *p*-orbitals leads to the formation of n/2 π-bonding and n/2 π-antibonding molecular orbitals.
3. The bonding molecular orbitals are filled by the electrons, while antibonding orbitals remain vacant in the ground state of the molecule.
4. The lowest energy molecular orbital (for example, Ψ_1 in the case of buta-1,3-diene) always has no node, however, the next higher has one node and the second higher has two nodes and so on. Thus, the nth molecular orbital will have n − 1 nodes.

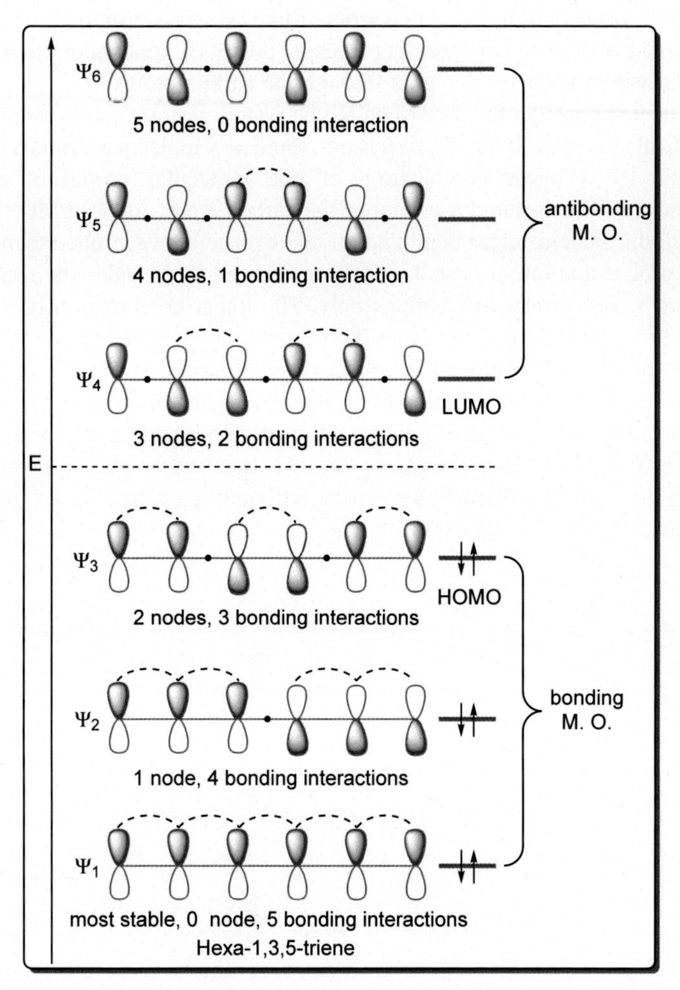

FIGURE 1.4 π-Molecular orbitals in a hexa-1,3,5-triene system.

5. It is important to note that the nodes are found at the most symmetric points in a molecular orbital. For example, in the case of Ψ_2 of buta-1,3-diene, a node is present at the center of C_2-C_3 bond, however, it will be incorrect if the node is present at the center of a C_1-C_2 bond or C_3-C_4 bond.

1.3 MOLECULAR ORBITALS OF CONJUGATED IONS OR RADICALS

The construction of molecular orbitals in the case of conjugated π-systems having an odd number of carbons can be made in a similar manner. Some important examples of this class include cation or anion or free radical of propenyl-, pentadienyl-, and heptatrienyl-like systems. Such systems, in addition to bonding and antibonding orbitals, possess a nonbonding molecular orbital in which nodal planes pass through the carbon atoms.

Let us first consider the case of an allylic system bearing cation or anion or free radical character. In an allylic system, three new molecular orbitals can be generated by a linear combination of one molecular orbital of ethene component and an isolated *p*-orbital of the carbon atom. As per PMO theory, in the allylic system linear combination takes place between one ethene molecular orbital and one *p*-orbital, and thus we need to consider the results of π–p and π^*–p orbital interactions only. The linear combination of π with *p*-orbital in a bonding manner (with the signs of the wave function of the two adjacent atomic orbitals matching) yields a new molecular orbital having least energy, i.e., Ψ_1, while in antibonding manner (with the signs of the wave function of the two adjacent atomic orbitals unmatched) this gives another new molecular orbital having more energy i.e., $\Psi_{2'}$. In a similar way, interaction of π^* with *p*-orbital in a bonding as well as antibonding manner yields two new molecular orbitals, one having low energy, i.e., $\Psi_{2''}$, and another having higher energy, i.e., Ψ_3 (Figure 1.5).

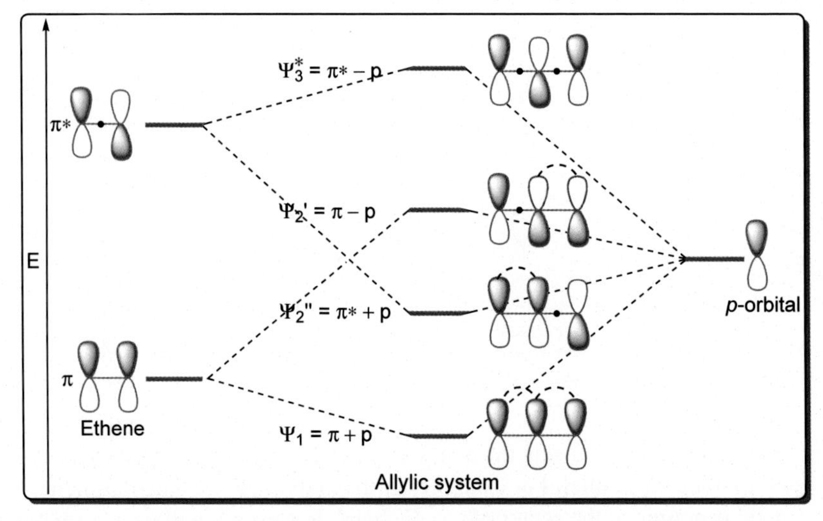

FIGURE 1.5 Mixing of *p*-orbital with molecular orbitals of ethene in an allylic system.

However, we cannot get four orbitals by using three orbitals. In fact, we do not get two separate orbitals $\Psi_{2'}$ and $\Psi_{2''}$ but something in between, namely Ψ_2. The orbital Ψ_2 can be created by adding $\Psi_{2'}$ and $\Psi_{2''}$ so that they cancel each other on C–2 and reinforce each other on C–1 and C–3. Thus Ψ_2 can be considered as a combination of $\Psi_{2'}$ and $\Psi_{2''}$, which is formed by mixing the *p*-orbital in an antibonding manner and with the π^*-orbital in a bonding manner. In case of Ψ_2, a nonbonding molecular orbital, a node is always present at the central carbon of the system. This means that there is no π-electron density at the central carbon atom. Moreover, the energy of a nonbonding molecular orbital is the same as the contributing atomic orbitals. Hence, there is no net stabilization as a result (Figure 1.6).

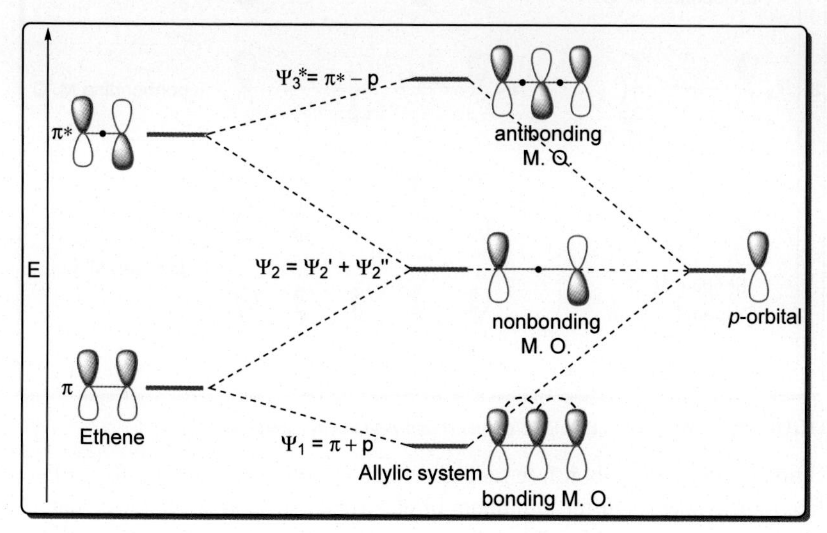

FIGURE 1.6 Mixing of *p*-orbital with molecular orbitals of ethene in an allylic system continued.

As illustrated in Figure 1.6, the following points need to be considered while constructing the molecular orbital diagram of a conjugated open-chain system having an odd number of carbon atoms.

1. In case of conjugated π-systems having an odd number of **n** carbon atoms, **n** number of molecular orbitals are present.
2. The system will have (n − 1)/2 bonding, (n − 1)/2 antibonding, and one nonbonding molecular orbital.
3. The nonbonding molecular orbital will be (n + 1)/2nd orbital and always lies between the bonding and antibonding molecular orbitals.
4. All nodal planes (n − 1) pass through the carbon atom(s) of the nonbonding molecular orbitals (Ψ_n).
5. All nodal planes pass between two carbon nuclei in case of odd Ψ_n (Ψ_1, Ψ_3, Ψ_5, so on) while one nodal plane passes through the central carbon atom and remaining nodal planes pass between two carbon atoms in case of even Ψ_n (Ψ_2, Ψ_4, Ψ_6, so on).

The molecular orbital diagrams for propenyl and pentadienyl systems are illustrated in Figure 1.7 in which the molecular orbitals for their corresponding

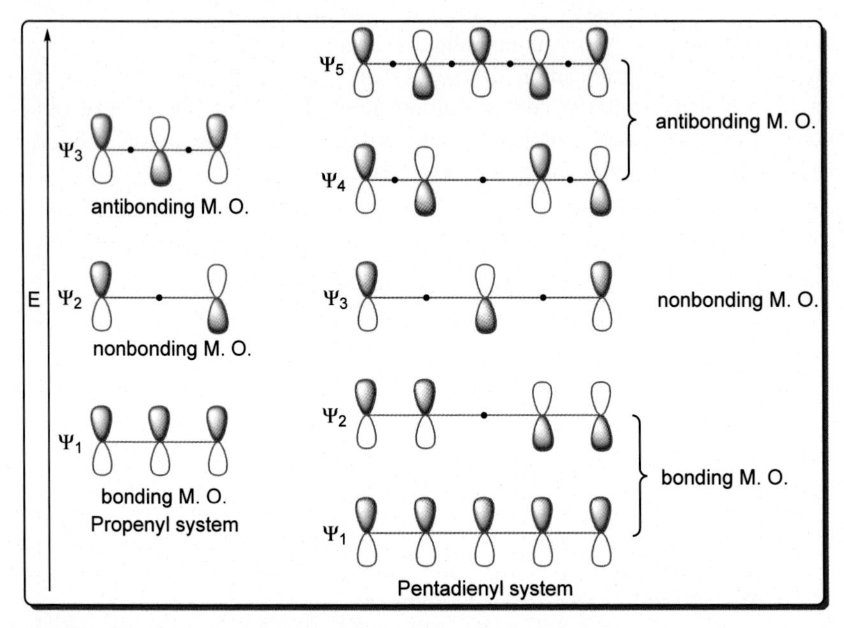

FIGURE 1.7 Molecular orbitals of propenyl and pentadienyl systems.

cation or anion or carbon free radical remain the same. The cation or anion or free radical species differ in number of electrons (electron occupancy) that are filled according to Aufbau's rule in their ground state as shown in Figure 1.8. Also, Hund's rule and the Pauli exclusion principle should be followed.

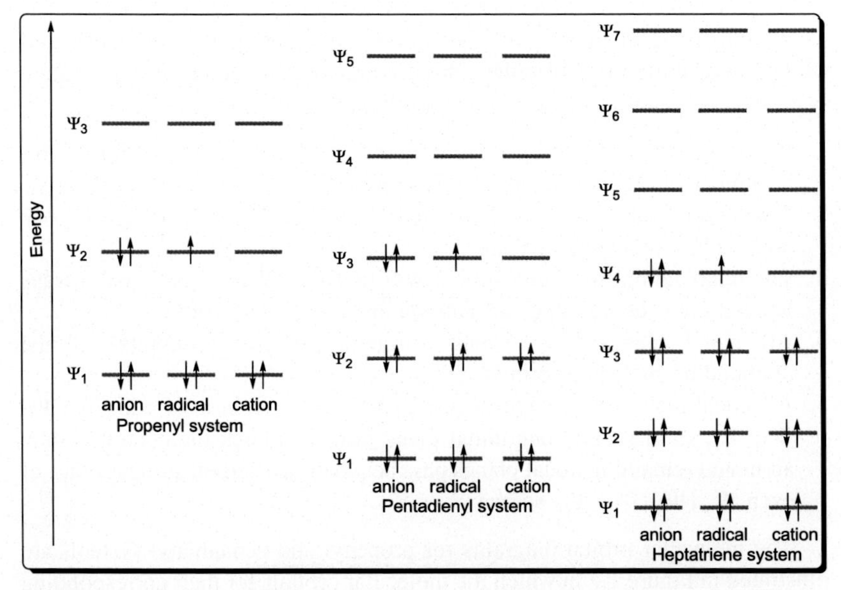

FIGURE 1.8 Electron occupancy diagram of propenyl, pentadienyl, and heptatriene systems.

1.4 SYMMETRY PROPERTIES OF π OR σ-MOLECULAR ORBITALS

There are two independent symmetry elements, viz., mirror plane, *m*, and twofold axis, C_2, that are used to characterize various molecular orbitals of alkenes or conjugated polyene systems.

1. Symmetry about a mirror plane (*m*) bisects the molecular orbital in such a way that lobes of the same color or sign are reflected, and, therefore, reflections on either side of the plane are identical. It is perpendicular to the plane of the atoms.
2. Symmetry about a twofold axis (C_2) passing at right angles in the same plane, and through the center of the framework of the atoms forming the molecular orbital is said to be present if the rotation of the molecule around the axis by 180° (360°/2) results in a molecular orbital identical with the original.

Let us examine symmetry properties of π-orbitals of ethene in the ground state and also in the excited state. The ground state (π) orbital is symmetric (S) with respect to the mirror plane, *m*, and antisymmetric (A) with respect to rotation axis, C_2. On the other hand, the antibonding orbital (π*) of ethene is antisymmetric with respect to *m* and symmetric with respect to the C_2 axis. However, the sigma orbital of a C—C covalent bond has a mirror plane symmetry, and since a rotation of 180° through its midpoint regenerates the same sigma orbital, it also has C_2 symmetry. A σ^* orbital is antisymmetric with respect to both *m* and C_2. The symmetry properties of these MOs (bonding or antibonding) are shown in Figures 1.9 and 1.10, and are summarized in Table 1.1.

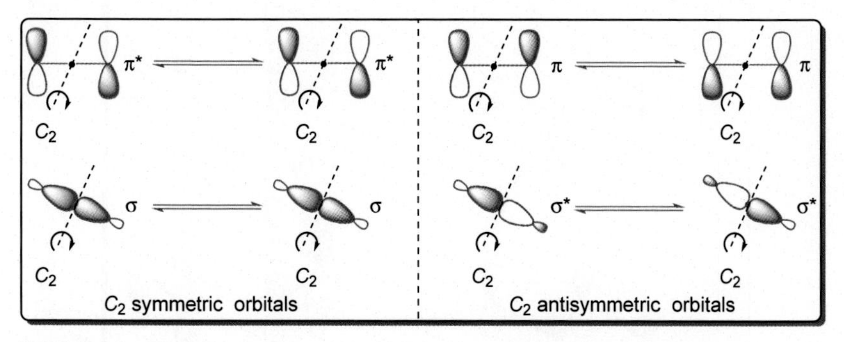

FIGURE 1.9 Twofold axis (C_2) symmetric and antisymmetric molecular orbitals.

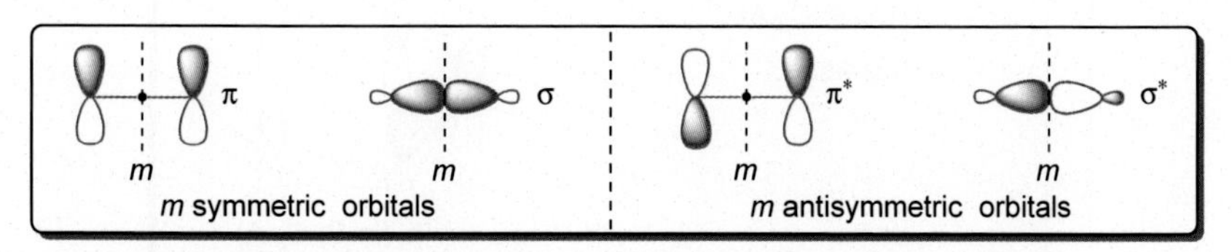

FIGURE 1.10 Mirror plane (*m*) symmetric and antisymmetric molecular orbitals.

TABLE 1.1 Symmetry properties of the σ and π-molecular orbitals; A = antisymmetric, S = symmetric.

Orbitals	*m*	C_2	Orbitals	*m*	C_2
π	S	A	σ	S	S
π*	A	S	σ*	A	A

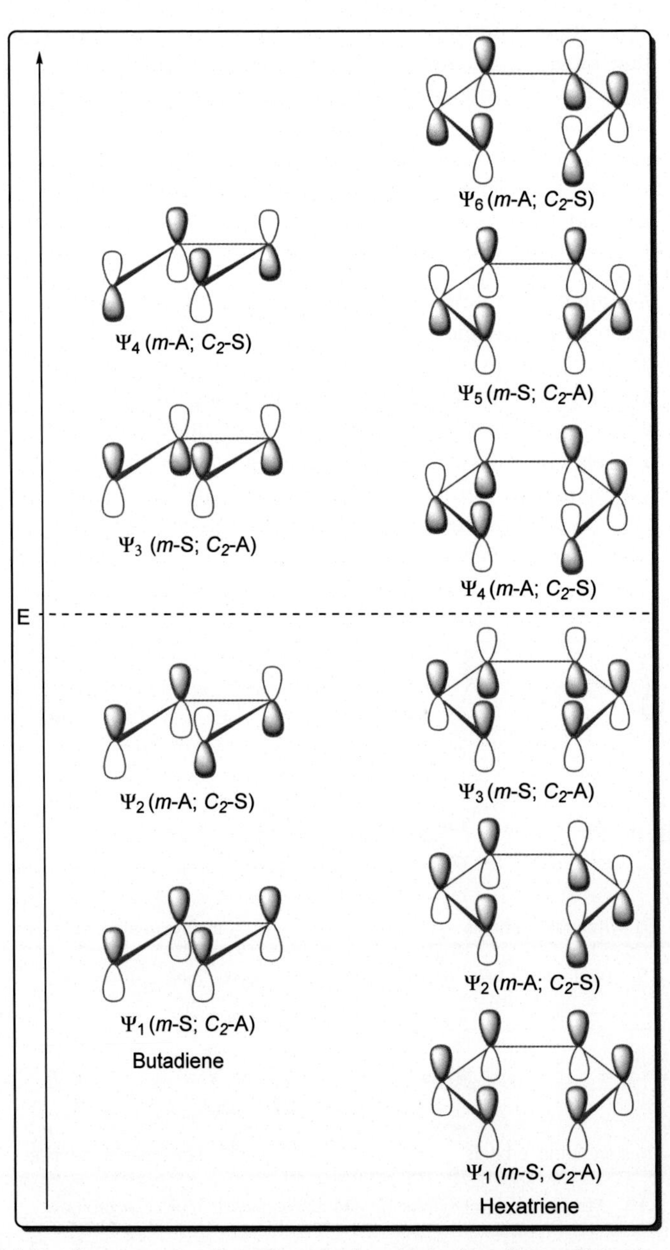

FIGURE 1.11 Symmetry properties of the molecular orbitals of butadiene and hexatriene systems.

A similar consideration leads to the following symmetry properties for the four π-molecular orbitals of butadiene and six π-molecular orbitals of hexatriene and are summarized in Figure 1.11.

In conclusion, for a linear conjugated π-system, the wave function Ψ_n will have n − 1 nodes. When n − 1 is zero or an even integer, Ψ_n will be symmetric with respect to mirror plane (*m*) and antisymmetric with respect to C_2. When n − 1 is an odd integer, Ψ_n will have the symmetry exactly reversed (Table 1.2).

TABLE 1.2 Symmetry elements in the orbital Ψ_n of a linear conjugated π-system.

Wave functions	Nodes (n − 1)	*m*	C_2
Ψ_{odd}	0 or Even integer	S	A
Ψ_{even}	Odd integer	A	S

1.5 ANALYSIS OF PERICYCLIC REACTIONS

Pericyclic reactions have been known for a long time, but it was in 1965 when Woodward and Hoffmann offered a reasonable explanation for them based on the principle of the "*Conservation of Orbital Symmetry.*" The principle states that orbital symmetry is conserved in the concerted reactions. Molecular orbitals in the reactant can only transform into those orbitals in the products that have the same *symmetry properties* with respect to the elements of symmetry preserved in the reaction. Even if symmetry is slightly disturbed in a reactant by a trivial substituent or by asymmetry of the molecule, a concerted reaction may still be analyzed by mixing the interacting orbitals according to quantum mechanical principles and following them through the reaction. The energy of the transition state of a symmetry allowed process will necessarily be lower than that of the alternative symmetry forbidden path, and even when this difference is small, a concerted reaction will take the path of least resistance, i.e., the symmetry allowed path, if that path is available.

Another explanation has been proposed by K. Fukuii on the basis of frontier molecular orbitals (HOMO—LUMO) of the substrates; this method is known as the frontier molecular orbitals (FMO) method. Alternatively, the PMO theory based on the Woodward—Hoffmann rule and Hückel-Möbius method is also used to explain the results of pericyclic reactions.

1.5.1 Orbital Symmetry Correlation Diagram Method

The orbital symmetry correlation diagram method was developed by Woodward and Hoffmann and extended by Longuet-Higgins and Abrahamson.

The most important observation in the study of pericyclic reactions is the existence of conservation of molecular orbital symmetry throughout the transformation, meaning thereby that the symmetric orbitals are converted into symmetric orbitals whereas antisymmetric orbitals are converted into antisymmetric orbitals. In this approach, symmetry properties of various molecular orbitals of the bonds that are involved in the bond breaking and formation process during the reaction are considered and identified with respect to C_2 and m elements of symmetry. These properties remain preserved throughout the course of reaction. Then a correlation diagram is drawn in which the molecular orbital levels of like symmetry of the reactant are related to that of the product by drawing lines.

In the ground state, if the symmetry of MOs of the reactant matches that of the products that are nearest in energies, then reaction is *thermally allowed.* However, if the symmetry of MOs of the reactant matches that of the product in the first excited state but not in the ground state, then the reaction is *photochemically allowed* (Figure 1.12). When symmetries of the reactant and product molecular orbitals differ, the reaction does not occur in a *concerted manner.* It must be noted that a symmetry element becomes irrelevant when orbitals involved in the reaction are all symmetric or antisymmetric. In conclusion, we can say that in pericyclic transformations, symmetry properties of the reactants and products remain conserved.

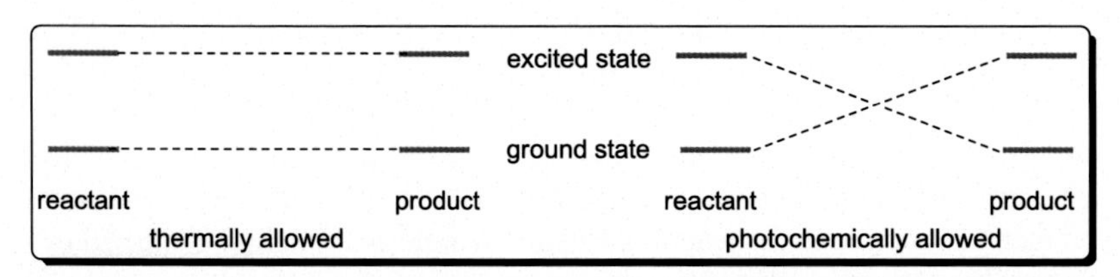

FIGURE 1.12 Correlation between reactant and product MOs under thermal and photochemical conditions.

While drawing the orbital correlation diagram for any system, the following points must be considered:

1. Each reactant molecule must be converted into simpler analogue by removing the substituents attached, if any, because substituent affects only the energy levels of MOs and not the symmetry properties of the π-system. Let us consider the Diels–Alder reaction, a [4 + 2] π-system (Scheme 1.4).

SCHEME 1.4 Conversion of the reactant molecules into simpler analogue.

2. Different processes must be treated separately even if they occur within the same molecule because simultaneous consideration may lead to erroneous outcome. For example, hypothetical two [2 + 2] cycloaddition reactions in cyclooctatetraene have to be considered separately. Similarly, in hexa-2,4-diene, conrotatory and disrotatory electrocyclization processes have to be treated separately while making the orbital diagram (Scheme 1.5).

SCHEME 1.5 Independent processes occurring in the same molecule.

3. Draw and identify the orbitals undergoing change.
4. Arrange the orbitals in order of their increasing energies, and draw them for reactant on left and for product on the right side.
5. Symmetry properties of the various molecular orbitals of the bonds being involved in breaking and formation process during the reaction are considered and identified with respect to elements of symmetry (C_2 and σ) that are preserved throughout the reaction.
6. Orbitals of same symmetry do not cross in the correlation diagram as per *non-cross rule*.
7. After assigning the symmetry element to each orbital, construct an orbital correlation diagram by connecting the orbitals of starting materials to those of the product nearest in energy and having same symmetry.
8. If heteroatoms are present in an alkene component, they have to be replaced by carbon analogues. Interactions in such systems should be considered carefully as they may generate the possibilities of new reaction either by nonbonding electrons or by availability of low energy LUMO.

1.5.2 Frontier Molecular Orbital Method

Although it is more fruitful to construct a correlation diagram for the detailed analysis of a pericyclic reaction, there is, nevertheless, an alternative method that also enables us to reach similar conclusions. It is an easy and extremely simple approach that is based on the interaction of the frontier orbitals, i.e., *the highest occupied molecular orbital* (HOMO) and *the lowest unoccupied molecular orbital* (LUMO) of the components that are involved in a pericyclic reaction.

As shown in Figure 1.13, irradiation of an alkene or conjugated polyene system promotes an electron from its ground state HOMO to the ground state LUMO, which then becomes the highest occupied molecular orbital in the excited state, for example, Ψ_3 of butadiene becomes HOMO upon excitation of an electron from Ψ_2 to Ψ_3 on irradiation.

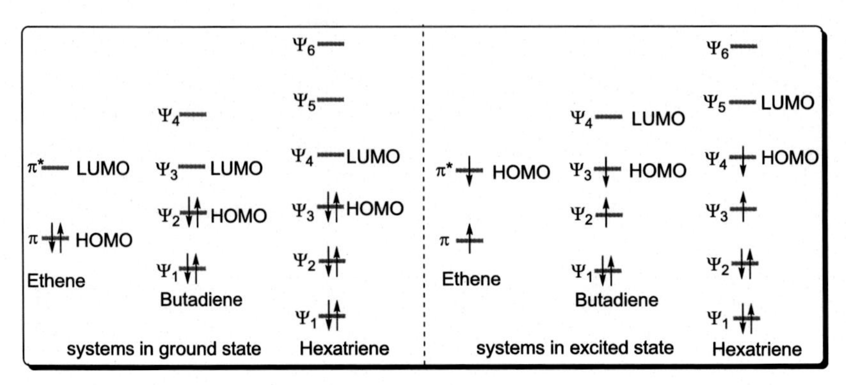

FIGURE 1.13 HOMO and LUMO of alkene systems.

The explanation for this alternative approach is based on the fact that overlapping of wave functions of the same sign is essential for the bond formation. When two systems come close to each other, then their unperturbed molecular orbitals start to interact and those that are close in energy interact more strongly than other orbitals. It is well known that interaction of two filled MOs does not lead to the net energy stabilization of the system but it is the interaction between one filled and other vacant MO that leads to net energy stabilization. This explains why interaction between HOMO and LUMO is considered in this approach (Figure 1.14). If interaction between these two MOs is of bonding type (overlapping of same signed wave functions) in the ground state, then reaction is *thermally allowed*. However, if it is of antibonding type (overlapping of opposite signed wave functions) then it is a *thermally forbidden* reaction. On the other hand, if interaction between HOMO–LUMO is of bonding type in the excited state, then reaction is photochemically allowed. However, it is a photochemically forbidden reaction when it is of antibonding type.

In order to apply the FMO approach in unimolecular pericyclic reactions like electrocyclic reactions and sigmatropic rearrangements, we have to treat a single molecule as having separate components. In such a case, only HOMO of the component has to be considered to predict the feasibility of the reaction under given conditions. Furthermore, this theory does not tell why the energy barrier to forbidden reactions is so high.

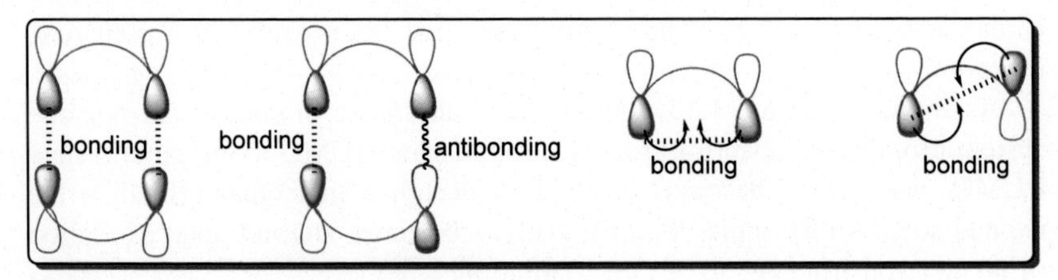

FIGURE 1.14 Interactions in FMOs of alkenes.

1.5.3 Perturbation Molecular Orbital Method

There is yet another qualitative molecular orbital approach, developed by M.J.S. Dewar, that yields simple mnemonics to predict the same stereochemical variations in pericyclic reactions as do the other methods. In the PMO method, aromatic or antiaromatic character of the cyclic transition state is explained by considering the Hückel-Möbius concept of aromaticity. In a Hückel-type system, a cyclic array of all the interacting *p*-orbitals shares a common nodal plane. A Hückel system is aromatic (stabilized by cyclic delocalization) when (4n + 2) π-electrons are present, and antiaromatic (destabilized by cyclic delocalization) when 4n π-electrons are present. However, in a Möbius-type system an extra node is present, introduced by twisting the set of orbitals so that each one forms an angle, theta, with its neighbors. In a Möbius-type system, the molecules and transition states require 4n π-electrons for aromaticity and are antiaromatic with the usual (4n + 2) π-electrons. *It can be generalized and shown that a cyclic array of orbitals with zero or an even number of sign inversions belongs to the Hückel system, and those with an odd number of sign inversions belong to the Möbius system.*

Application of this method to pericyclic reactions led to the generalization that thermal reactions take place via *aromatic or stable transition states* whereas photochemical reactions proceed via *antiaromatic or unstable transition states*. This is the case because a controlling factor in photochemical processes is conversion of excited state reactants into ground state products. Thus, the photochemical reactions convert the reactants into the antiaromatic transition states that correspond to forbidden thermal pericyclic reactions and so lead to corresponding products.

In this approach, we have only to consider a cyclic array of interacting atomic orbitals, representing those orbitals that undergo change in the transition state without considering the symmetry properties and assign signs to the wave functions in the best manner for overlap. Finally, the number of nodes in the array and the number of electrons involved are counted. It should be noted that while counting the number of nodes we ignore sign inversions within any of the basis orbitals (for example, as within a *p*-orbital). The following examples illustrate the construction of orbital interaction diagrams for the [2 + 2] and [4 + 2] cycloadditions by supra–supra and supra–antara modes. (For a detailed description of these terms, refer to Chapter 4). Whether, the reactions are allowed or not are predicted as follows. In the case of [π^2s + π^2s] cycloaddition (4n π-electron system), a supra–supra mode of addition leads to a Hückel array, which is antiaromatic with 4n π-electrons (Figure 1.15). Therefore, the supra–supra mode of reaction is thermally forbidden. However, a supra–antara mode of addition uses a Möbius array, which is aromatic with 4n π-electrons. Therefore, the reaction is thermally allowed in this mode. Similarly, we can analyze the [π^4s + π^2s] cycloaddition having (4n + 2) π-electrons (Figure 1.15). In this case, a supra–supra mode of addition leads to a Hückel array, which is aromatic with (4n + 2) π-electrons. Therefore, [π^4s + π^2s] cycloaddition reaction now becomes thermally allowed. However, a [π^4s + π^2a] cycloaddition uses a Möbius array, which is antiaromatic with (4n + 2) π-electrons. Therefore, the reaction is thermally forbidden in this mode.

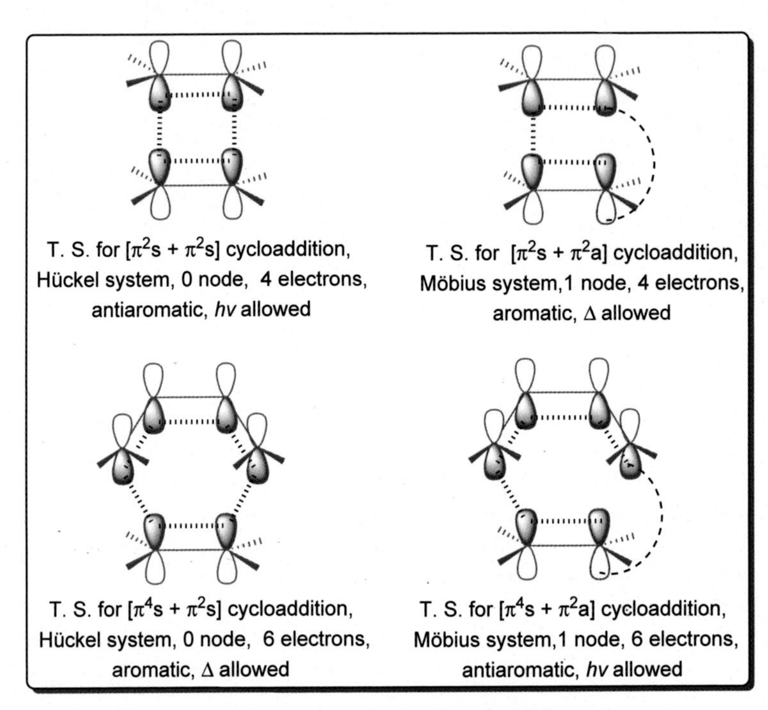

FIGURE 1.15 PMO approach for [2 + 2] and [4 + 2] cycloadditions.

Woodward–Hoffmann rules based on the perturbation molecular orbital method are summarized in Table 1.3.

TABLE 1.3 Woodward–Hoffmann rules based on the perturbation molecular orbital method.

No. of electrons	No. of nodes	T. State type	Aromaticity	Feasibility
4n + 2	0 or Even	Hückel	Aromatic	Δ allowed, *hv* forbidden
4n	0 or Even	Hückel	Antiaromatic	Δ forbidden, *hv* allowed
4n + 2	Odd	Möbius	Antiaromatic	Δ forbidden, *hv* allowed
4n	Odd	Möbius	Aromatic	Δ allowed, *hv* forbidden

Therefore, the prediction of reaction feasibility under thermal or photochemical condition depends upon the extent of stabilization of a cyclic transition state as compared to an open-chain system. The stabilization or destabilization depends upon the aromatic or antiaromatic character of a cyclic transition state in the ground state.

FURTHER READING

[1] R. Hoffmann, R.B. Woodward, The conservation of orbital symmetry, Acc. Chem. Res. 1 (1) (1968) 17–22.

[2] R. Huisgen, Cycloadditions—definition, classification, and characterization, Angew. Chem. Int. Ed. Engl. 7 (1968) 321–328.

[3] R.B. Woodward, R. Hoffmann, The conservation of orbital symmetry, Angew. Chem. Int. Ed. Engl. 8 (11) (1969) 781–853.

[4] R. Hoffmann, R.B. Woodward, Orbital symmetry control of chemical reactions, Science 167 (1970) 825–831.

[5] M.J.S. Dewar, Aromaticity and pericyclic reactions, Angew. Chem. Int. Ed. Engl. 10 (11) (1971) 761–786.

[6] W.C. Herndon, Theory of cycloaddition reactions, Chem. Rev. 72 (2) (1972) 157–179.

[7] J.B. Hendrickson, The variety of thermal pericyclic reactions, Angew. Chem. Int. Ed. Engl. 13 (1) (1974) 47–76.

[8] C.W. Spangler, Thermal [1,j] sigmatropic rearrangements, Chem. Rev. 76 (2) (1976) 187–217.

[9] G. Brieger, J.N. Bennett, The intramolecular Diels–Alder reaction, Chem. Rev. 80 (1) (1980) 63–97.

[10] R. Huisgen, 1,5-Electrocyclizations—an important principle of heterocyclic chemistry, Angew. Chem. Int. Ed. 19 (1980) 947–973.

[11] K. Fukui, The role of frontier orbitals in chemical reactions, Angew. Chem. Int. Ed. Engl. 21 (11) (1982) 801–809.

[12] K.N. Houk, J.D. Evanseck, Transition structures of hydrocarbon pericyclic reactions, Angew. Chem. Int. Ed. Engl. 31 (6) (1991) 682–708.

[13] K. Mikami, M. Shimizu, Asymmetric ene reactions in organic synthesis, Chem. Rev. 92 (5) (1992) 1021–1050.

[14] K.N. Houk, J. Gonzalez, Y. Li, Pericyclic reaction transition states: passions and punctilios 1935–1995, Acc. Chem. Res. 28 (1995) 81–90.

[15] D.M. Birney, S. Ham, G.R. Unruh, Pericyclic and pseudopericyclic thermal cheletropic decarbonylations: when can a pericyclic reaction have a planar, pseudopericyclic transition state? J. Am. Chem. Soc. 119 (1997) 4509–4517.

[16] K.V. Gothelf, K.A. Jorgensen, Asymmetric 1,3-dipolar cycloaddition reactions, Chem. Rev. 98 (1998) 863–909.

[17] H. Ito, T. Taguchi, Asymmetric Claisen rearrangement, Chem. Soc. Rev. 28 (1999) 43–50.

[18] P. Buonora, J.C. Olsen, T. Oh, Recent developments in imino Diels–Alder reactions, Tetrahedron 57 (29) (2001) 6099–6138.

[19] S.M. Allin, R.D. Baird, Development and synthetic applications of asymmetric (3,3) sigmatropic rearrangements, Curr. Org. Chem. 5 (4) (2001) 395–415.

[20] D.M. Hodgson, F.Y.T.M. Pierard, P.A. Stupple, Catalytic enantioselective rearrangements and cycloadditions involving ylides from diazo compounds, Chem. Soc. Rev. 30 (2001) 50–61.

[21] B.R. Bear, S.M. Sparks, K.J. Shea, The type 2 intramolecular Diels–Alder reaction: synthesis and chemistry of bridgehead alkenes, Angew. Chem. Int. Ed. 40 (2001) 820–849.
[22] S. Kanemasa, Metal-assisted stereocontrol of 1,3-dipolar cycloaddition reactions, Synlett (2002) 371–387.
[23] K.C. Nicolaou, S.A. Snyder, T. Montagnon, G. Vassilikogiannakis, The Diels–Alder reaction in total synthesis, Angew. Chem. Int. Ed. 41 (2002) 1668–1698.
[24] E.J. Corey, Catalytic enantioselective Diels–Alder reactions: methods, mechanistic fundamentals, pathways, and applications, Angew. Chem. Int. Ed. 41 (2002) 1650–1667.
[25] E.M. Stocking, R.M. Williams, Chemistry and biology of biosynthetic Diels–Alder reactions, Angew. Chem. Int. Ed. 42 (2003) 3078–3115.
[26] U. Nubbemeyer, Recent advances in asymmetric [3,3]-sigmatropic rearrangements, Synthesis 7 (2003) 961–1008.
[27] C. Najera, J.M. Sansano, Azomethine ylides in organic synthesis, Curr. Org. Chem. 7 (2003) 1105–1150.
[28] I. Coldham, R. Hufton, Intramolecular dipolar cycloaddition reactions of azomethine ylides, Chem. Rev. 105 (7) (2005) 2765–2810.
[29] K.C. Nicolaou, D.J. Edmonds, P.G. Bulger, Cascade reactions in total synthesis, Angew. Chem. Int. Ed. 45 (2006) 7134–7186.
[30] D.H. Ess, G.O. Jones, K.N. Houk, Conceptual, qualitative, and quantitative theories of 1,3-dipolar and Diels–Alder cycloadditions used in synthesis, Adv. Synth. Catal. 348 (2006) 2337–2361.
[31] H. Pellissier, Asymmetric 1,3-dipolar cycloaddition, Tetrahedron 63 (2007) 3235–3285.
[32] H.S. Rzepa, The aromaticity of pericyclic reaction transition states, J. Chem. Educ. 84 (9) (2007) 1535–1540.
[33] J.F. Lutz, 1,3-Dipolar cycloadditions of azides and alkynes: a universal ligation tool in polymer and materials science, Angew. Chem. Int. Ed. Engl. 46 (7) (2007) 1018–1025.
[34] R.K. Bansal, S.K. Kumawat, Diels–Alder reactions involving C=P functionality, Tetrahedron 64 (2008) 10945–10976.
[35] S. Reymond, J. Cossy, Copper-catalyzed Diels–Alder reactions, Chem. Rev. 108 (12) (2008) 5359–5406.
[36] J. Poulin, C.M. Grisé-Bard, L. Barriault, Pericyclic domino reactions: concise approaches to natural carbocyclic frameworks, Chem. Soc. Rev. 38 (2009) 3092–3101.
[37] V.V. Kouznetsov, Recent synthetic developments in a powerful imino Diels–Alder reaction (Povarov reaction): application to the synthesis of *N*-polyheterocycles and related alkaloids, Tetrahedron 65 (14) (2009) 2721–2750.
[38] H. Pellissier, Asymmetric hetero-Diels–Alder reactions of carbonyl compounds, Tetrahedron 65 (2009) 2839–2877.
[39] M. Juhl, D. Tanner, Recent applications of intramolecular Diels–Alder reactions to natural product synthesis, Chem. Soc. Rev. 38 (11) (2009) 2983–2992.
[40] D.J. Tantillo, J.K. Lee, Reaction mechanisms part (ii) pericyclic reactions, Ann. Rep. Prog. Chem. Sect. B 105 (2009) 285–309.
[41] E.A. Ilardi, C.E. Stivala, A. Zakarian, [3,3]-Sigmatropic rearrangements: recent applications in the total synthesis of natural products, Chem. Soc. Rev. 38 (2009) 3133–3148.
[42] M. Kissanea, A.R. Maguire, Asymmetric 1,3-dipolar cycloadditions of acrylamides, Chem. Soc. Rev. 39 (2010) 845–883.
[43] P. Appukkuttan, V.P. Mehta, E.V. Van der Eycken, Microwave-assisted cycloaddition reactions, Chem. Soc. Rev. 39 (2010) 1467–1477.

[44] B. Alcaide, P. Almendros, C. Aragoncillo, Exploiting [2 + 2] cycloaddition chemistry: achievements with allenes, Chem. Soc. Rev. 39 (2010) 783–816.

[45] S. Thompson, A.G. Coyne, P.C. Knipe, M.D. Smith, Asymmetric electrocyclic reactions, Chem. Soc. Rev. 40 (2011) 4217–4231.

[46] A. Ajaz, A.Z. Bradley, R.C. Burrell, W.H.H. Li, K.J. Daoust, L.B. Bovee, K.J. DiRico, R.P. Johnson, Concerted vs stepwise mechanisms in dehydro-Diels–Alder reactions, J. Org. Chem. 76 (22) (2011) 9320–9328.

[47] A. Arrieta, A. de Cozar, F.P. Cossıo, Cyclic electron delocalization in pericyclic reactions, Curr. Org. Chem. 15 (2011) 3594–3608.

[48] H. Pellissier, Asymmetric organocatalytic cycloadditions, Tetrahedron 68 (10) (2012) 2197–2232.

[49] E.M. Greer, C.V. Cosgriff, Reaction mechanisms: pericyclic reactions, Annu. Rep. Prog. Chem. Sect. B: Org. Chem. 108 (2012) 251–271.

[50] V. Nair, R.S. Menon, A.T. Biju, K.G. Abhilash, 1,2-Benzoquinones in Diels–Alder reactions, dipolar cycloadditions, nucleophilic additions, multicomponent reactions and more, Chem. Soc. Rev. 41 (2012) 1050–1059.

[51] J.-A. Funel, S. Abele, Industrial applications of the Diels–Alder reaction, Angew. Chem. Int. Ed. 52 (2013) 3822–3863.

[52] M.G. Memeo, P. Quadrelli, A new life for nitrosocarbonyls in pericyclic reactions, ARKIVOC (i) (2013) 418–423.

[53] A.C. Knall, C. Slugovc, Inverse electron demand Diels–Alder (iEDDA)-initiated conjugation: a (high) potential click chemistry scheme, Chem. Soc. Rev. 42 (2013) 5131–5142.

[54] G. Masson, C. Lalli, M. Benohoud, G. Dagousset, Catalytic enantioselective [4 + 2]-cycloaddition: a strategy to access aza-hexacycles, Chem. Soc. Rev. 42 (2013) 902–923.

[55] X. Jiang, R. Wang, Recent developments in catalytic asymmetric inverse-electron-demand Diels–Alder reaction, Chem. Rev. 113 (7) (2013) 5515–5546.

[56] R.A.A. Foster, M.C. Willis, Tandem *inverse*-electron-demand *hetero-/retro*-Diels–Alder reactions for aromatic nitrogen heterocycle synthesis, Chem. Soc. Rev. 42 (2013) 63–76.

[57] V. Eschenbrenner-Lux, K. Kumar, H. Waldmann, The asymmetric hetero-Diels–Alder reaction in the syntheses of biologically relevant compounds, Angew. Chem. Int. Ed. 53 (2014) 11146–11157.

[58] C.C. Nawrat, C.J. Moody, Quinones as dienophiles in the Diels–Alder reaction: history and applications in total synthesis, Angew. Chem. Int. Ed. 53 (2014) 2056–2077.

[59] A.C. Jones, J.A. May, R. Sarpong, B.M. Stoltz, Toward a symphony of reactivity: cascades involving catalysis and sigmatropic rearrangements, Angew. Chem. Int. Ed. 53 (2014) 2556–2591.

Chapter 2

Electrocyclic Reactions

Chapter Outline

An electrocyclic reaction is a molecular rearrangement that involves the formation of a σ-bond between the termini of a fully conjugated linear π-electron system and a decrease by one in the number of π-bonds, or the reverse of that process. Thus if the open chain partner contains n π-electrons, the cyclic partner has (n – 2) π-electrons and two electrons in a new σ-bond. For example, let us consider electrocyclization of butadiene and hexatriene systems as shown in Scheme 2.1.

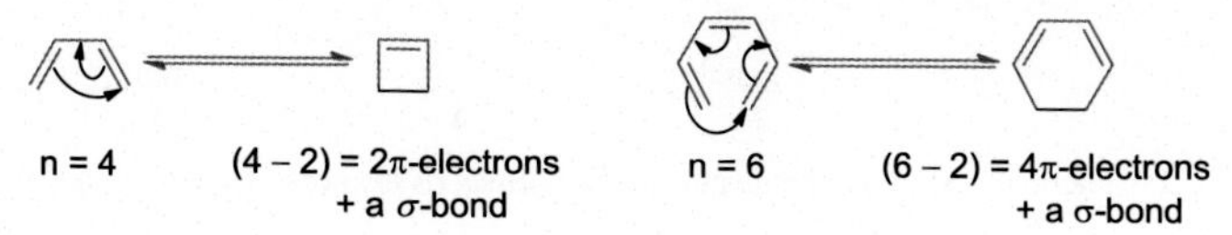

SCHEME 2.1 Electrocyclization of butadiene and hexatriene systems.

2.1 CONROTATORY AND DISROTATORY MODES

A σ-bond of cycloalkene must break to yield the open-chain polyene; this bond may break in two ways. In **conrotatory mode**, the two atomic orbital

Pericyclic Reactions. http://dx.doi.org/10.1016/B978-0-12-803640-2.00002-6

Copyright © 2016 Elsevier Inc. All rights reserved.

components of the σ-bond may both rotate in the same direction, clockwise or anticlockwise (Figure 2.1).

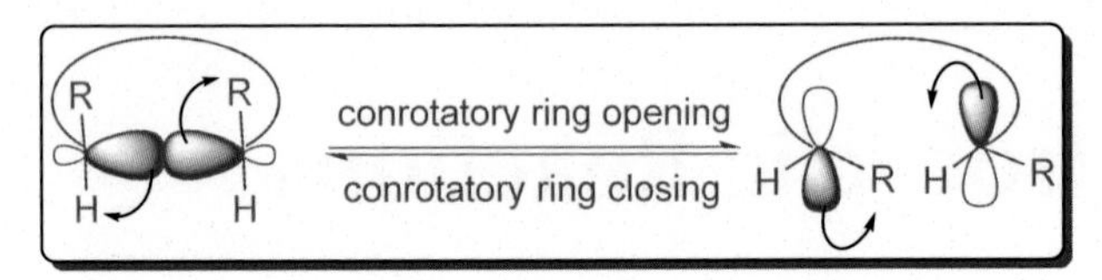

FIGURE 2.1 Conrotatory mode of ring opening and ring closing process.

In **disrotatory mode**, the atomic orbitals may rotate in opposite directions, one clockwise and the other anticlockwise (Figure 2.2).

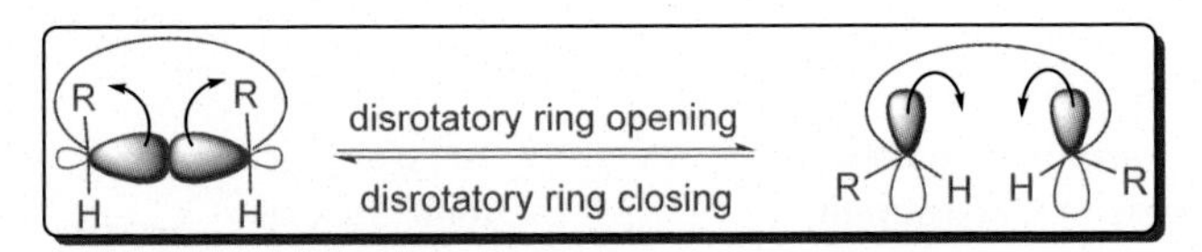

FIGURE 2.2 Disrotatory mode of ring opening and ring closing process.

2.2 STEREOCHEMISTRY OF ELECTROCYCLIC REACTIONS

The stereochemical significance of these two modes of ring opening (or ring closing) becomes apparent when we consider substituted reactants. Thus depending upon these modes, the substituents may rotate in the same direction (conrotatory) or in opposite directions (disrotatory).

For example, during the thermal electrocyclic ring opening of 3,4-dimethylcyclobutene, the *trans*-isomer (**1**) yields only (*2E,4E*)-hexa-2,4-diene (**2**) and the *cis*-isomer (**3**) yields only (*2E,4Z*)-hexa-2,4-diene (**4**). On irradiation, however, the results are opposite. Cyclization of the **2** under photochemical conditions yields the *cis*-product (**3**) (Scheme 2.2).

SCHEME 2.2 Thermal and photochemical transformations of isomeric 3,4-dimethylcyclobutenes.

A similar result is obtained for the octatriene-cyclohexadiene system. For example, during the thermal electrocyclic ring opening of 5,6-dimethylcyclohexa-1,3-diene, the *cis*-isomer (**1**) yields only (*2E,4Z,6E*)-octa-2,4,6-triene (**2**), and the *trans*-isomer (**3**) yields only (*2E,4Z,6Z*)-octa-2,4,6-triene (**4**). On irradiation, however, the results are opposite. Cyclization of the **2** under photochemical conditions yields the *trans*-product (**3**) (Scheme 2.3).

SCHEME 2.3 Thermal and photochemical transformations of isomeric 5,6-dimethylcyclohexa-1,3-dienes.

2.3 SELECTION RULES FOR ELECTROCYCLIC REACTIONS

Empirical Observations: It was noted that under **thermal** conditions, butadiene systems undergo **conrotatory** ring closure, while hexatriene systems undergo **disrotatory** ring closure. The microscopic reverse reactions also occur with the same rotational sense (i.e., on heating, cyclobutene systems open in a conrotatory sense, and cyclohexadiene systems open in a disrotatory sense). It was also noted that changing the conditions from heat to light reversed this reactivity pattern. Under **photochemical** conditions, conjugated polyene systems containing 4π-electrons undergo **disrotatory**, while systems having 6π-electrons undergo **conrotatory** process (Table 2.1).

TABLE 2.1 Selection rules for electrocyclic reactions.

Number of π-electrons	Thermal (Δ)	Photochemical (*hv*)
4n	con	dis
4n + 2	dis	con

2.4 ANALYSIS OF ELECTROCYCLIC REACTIONS

Electrocyclic reactions can be analyzed by correlation-diagram, perturbation molecular orbital (PMO) and frontier molecular orbital (FMO) methods.

2.4.1 Correlation-Diagram Method

An electrocyclic reaction is a concerted and cyclic process in which reactant orbitals transform into product orbitals. The transition state of such reactions should be intermediate between the electronic ground states of starting material and product. Obviously, the most stable transition state will be the one that conserves the symmetry of the reactant orbitals in passing to product orbitals. In other words, a symmetric (S) orbital in the reactant must transform into a symmetric orbital in the product and an antisymmetric (A) orbital must transform into an antisymmetric orbital. If the symmetries of the reactant and product orbitals are not the same, the reaction will not take place in a concerted manner.

Let us exemplify the above principle by analyzing the cyclobutene-butadiene transformation. The symmetry properties of molecular orbitals of cyclobutene and butadiene are expressed in Figure 2.3. The ring opening may be a disrotatory process in which the groups on the saturated carbons rotate in opposite directions or, alternatively, it may proceed via conrotation, involving rotation of the groups in the same direction. In the case of the disrotatory ring opening, cyclobutene preserves a plane of symmetry (*m*) throughout the transformation while a two-fold (C_2) symmetry axis is maintained at all times in the conrotatory mode of ring opening.

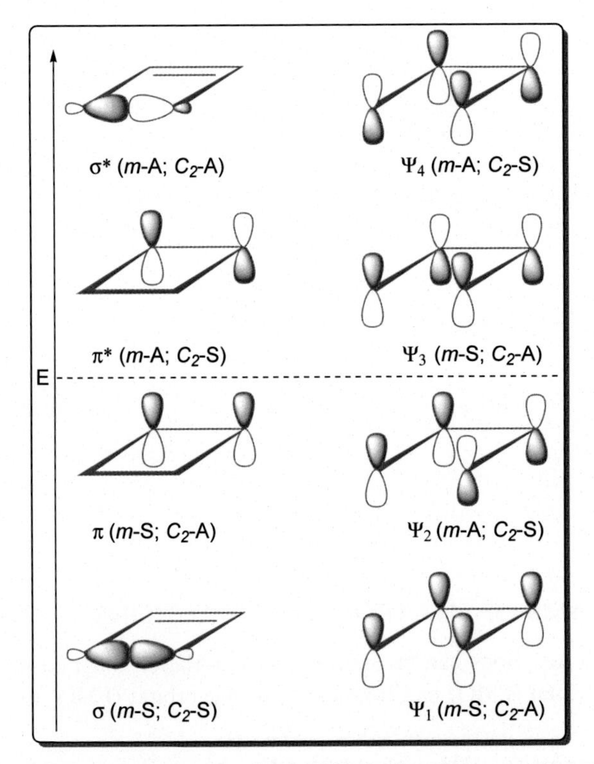

FIGURE 2.3 Symmetry properties of molecular orbitals of cyclobutene and butadiene.

We are now set to analyze the above transformation in terms of the fundamental rule of the conservation of orbital symmetry as proposed by Woodward and Hoffmann. The orbitals of cyclobutene that are directly involved are σ and π, and the related antibonding orbitals are σ* and π*; these orbitals pass on to the four π-molecular orbitals of butadiene, viz., Ψ_1, Ψ_2, Ψ_3, and Ψ_4. All these orbitals are listed in Figure 2.3 in order of increasing energy along with their mirror plane and C_2 symmetry properties. In the ground state of cyclobutene and butadiene, only σ, π and Ψ_1, Ψ_2 orbitals are filled with electron pairs.

It is easy to analyze an electrocyclic reaction by constructing a correlation diagram, which is simply a diagram showing the possible transformation of reactant orbitals to product orbitals. Let us first analyze a disrotatory opening of cyclobutene in which a mirror plane symmetry (*m*) is maintained (Figure 2.4).

In constructing this correlation diagram we have simply connected, by lines, the various orbitals of cyclobutene and butadiene keeping in mind that there is correlation between orbitals of the same symmetry having minimal energy differences. Upon close inspection, the following two conclusions can immediately be drawn:

1. We expect a thermal transformation to take place only when the ground state orbitals of the reactants correlate with the ground state orbitals of the products. Although in Figure 2.4 the cyclobutene ground state σ-orbital correlates with the butadiene ground state orbital Ψ_1, the π-orbital of the former does not correlate with Ψ_2 of the latter. Instead, it correlates with Ψ_3, which is an excited state and an antibonding orbital. Thermal transformation of cyclobutene-butadiene system by disrotatory process is thus *symmetry-forbidden* (Eqn 2.1).

$$\sigma^2\pi^2 \xrightarrow[\times]{\Delta,\ \text{dis}} \Psi_1^2\Psi_3^2 \quad \text{or} \quad \Psi_1^2\Psi_2^2 \xrightarrow[\times]{\Delta,\ \text{dis}} \sigma^2\pi^{*2} \tag{2.1}$$

2. Irradiation of cyclobutene produces the first excited state in which an electron is promoted from π to π* orbital, and in this case σ, π, and π* orbitals of cyclobutene correlate with Ψ_1, Ψ_2, and Ψ_3 orbitals of butadiene. In other words, the first excited state of cyclobutene correlates with the first excited state of butadiene, and hence disrotatory ring opening (ring closing) is photochemically a *symmetry-allowed process* (Eqn 2.2).

$$\underset{\text{ground state}}{\sigma^2\pi^2} \xrightarrow{h\nu} \underset{\text{first excited state}}{\sigma^2\pi^1\pi^{*1}} \overset{\text{dis}}{\rightleftharpoons} \underset{\text{first excited state}}{\Psi_1^2\Psi_2^1\Psi_3^1} \xleftarrow{h\nu} \underset{\text{ground state}}{\Psi_1^2\Psi_2^2} \tag{2.2}$$

Working on similar lines, we can construct another correlation diagram (Figure 2.5) for the conrotatory opening in which a C_2 axis of symmetry is

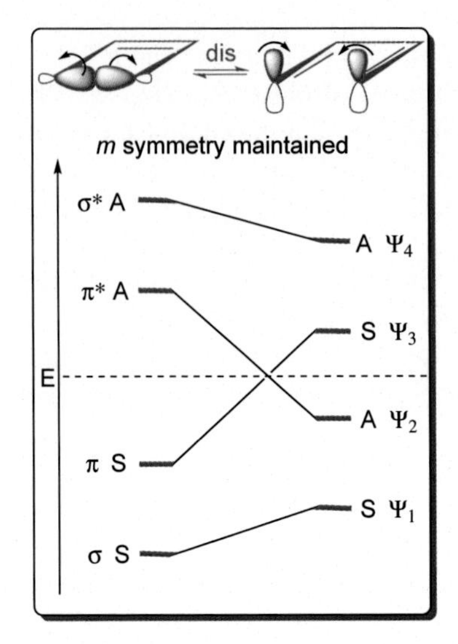

FIGURE 2.4 Correlation diagram for disrotatory interconversion of cyclobutene-butadiene system.

FIGURE 2.5 Correlation diagram for conrotatory interconversion of cyclobutene-butadiene system.

maintained throughout the reaction. Two conclusions may again be drawn from the correlation diagram:

1. Since there is correlation between the ground state orbitals of cyclobutene and butadiene, a thermal conrotatory process in either direction is a *symmetry-allowed* process (Eqn 2.3).

$$\sigma^2\pi^2 \xrightleftharpoons{\Delta,\ \text{con}} \Psi_1^2\,\Psi_2^2 \tag{2.3}$$

2. The first excited state of cyclobutene ($\sigma^2\,\pi^1\,\pi^{*1}$) correlates with the upper excited state ($\Psi_1^2\,\Psi_2^1\,\Psi_4^1$) of butadiene thus making it a high-energy *symmetry-forbidden* process (Eqn 2.4). Similarly, the first excited state of butadiene ($\Psi_1^2\,\Psi_2^1\,\Psi_3^1$) correlates with a high-energy upper excited state ($\sigma^2\,\pi^1\,\sigma^{*1}$) of cyclobutene (Eqn 2.5). In other words, a photochemical conrotatory process in either direction is *symmetry-forbidden*.

$$\underset{\text{ground state}}{\sigma^2\pi^2} \xrightarrow{h\nu} \underset{\text{first excited state}}{\sigma^2\pi^1\pi^{*1}} \xrightarrow[\times]{\text{con}} \underset{\text{upper excited state}}{\Psi_1^2\,\Psi_2^1\,\Psi_4^1} \tag{2.4}$$

$$\underset{\text{ground state}}{\Psi_1^2\,\Psi_2^2} \xrightarrow{h\nu} \underset{\text{first excited state}}{\Psi_1^2\,\Psi_2^1\,\Psi_3^1} \xrightarrow[\times]{\text{con}} \underset{\text{upper excited state}}{\sigma^2\pi^1\sigma^{*1}} \tag{2.5}$$

Thus it becomes clear from the above considerations that thermal opening of the cyclobutene proceeds in a conrotatory process while photochemical interconversion involves a disrotatory mode. These generalizations are true for all systems containing 4n π-electrons where n = 0, 1, 2, etc.

However, for systems containing (4n + 2) π-electrons, theoretical predictions are entirely different and are in conformity with the actual observations. A typical system of this type is the interconversion of cyclohexadiene and hexatriene. In this transformation, six molecular orbitals (Ψ_1 to Ψ_6) of hexatriene and six molecular orbitals (four π and two σ) of cyclohexadiene are actually involved and, therefore, need to be considered. Symmetry properties of the six molecular orbitals of hexatriene and cyclohexadiene are shown in Figure 2.6.

The correlation diagrams for the disrotatory and conrotatory pathway are constructed in the same way as in the case of cyclobutene-butadiene transformation. These are shown in Figures 2.7 and 2.8, respectively.

The following conclusions may be drawn from these correlation diagrams:

1. In the disrotatory mode, ground state bonding orbitals of cyclohexadiene correlate with the ground state bonding orbitals of hexatriene, so it is a *thermally allowed* process (Eqn 2.6).

$$\sigma^2 \pi_1^2 \pi_2^2 \xrightleftharpoons{\Delta,\ \text{dis}} \Psi_1^2 \Psi_2^2 \Psi_3^2 \quad (2.6)$$

2. But in the conrotatory mode (C_2 symmetry), ground state bonding orbitals of cyclohexadiene do not correlate with the ground state bonding orbitals of hexatriene. Since the presence of two electrons in Ψ_4 is a very high-energy process, a conrotatory mode is prohibited under thermal conditions (Eqn 2.7).

$$\sigma^2 \pi_1^2 \pi_2^2 \xrightarrow[\times]{\Delta,\ \text{con}} \Psi_1^2 \Psi_2^2 \Psi_4^2 \quad \text{or} \quad \Psi_1^2 \Psi_2^2 \Psi_3^2 \xrightarrow[\times]{\Delta,\ \text{con}} \sigma^2 \pi_1^2 \pi_3^{*2} \quad (2.7)$$

3. However, if we promote an electron to π_3* in cyclohexadiene (obviously by irradiation), then the orbitals of the reactant with C_2 symmetry correlate with the first excited state of the product (Eqn 2.8).

$$\underset{\text{ground state}}{\sigma^2 \pi_1^2 \pi_2^2} \xrightarrow{hv} \underset{\text{first excited state}}{\sigma^2 \pi_1^2 \pi_2^1 \pi_3^{*1}} \xrightleftharpoons{\text{con}} \underset{\text{first excited state}}{\Psi_1^2 \Psi_2^2 \Psi_3^1 \Psi_4^1} \xleftarrow{hv} \underset{\text{ground state}}{\Psi_1^2 \Psi_2^2 \Psi_3^2} \quad (2.8)$$

Therefore, photochemical interconversion is allowed in the conrotatory pathway. These generalizations are true for all the systems containing (4n + 2) π-electrons, where n = 0, 1, 2, etc. Thus, Woodward–Hoffmann rules for electrocyclic reactions may be summed up as given in Table 2.1.

Woodward and Hoffmann have further explained that under severe thermal conditions, symmetry-forbidden reactions may also take place but then they follow

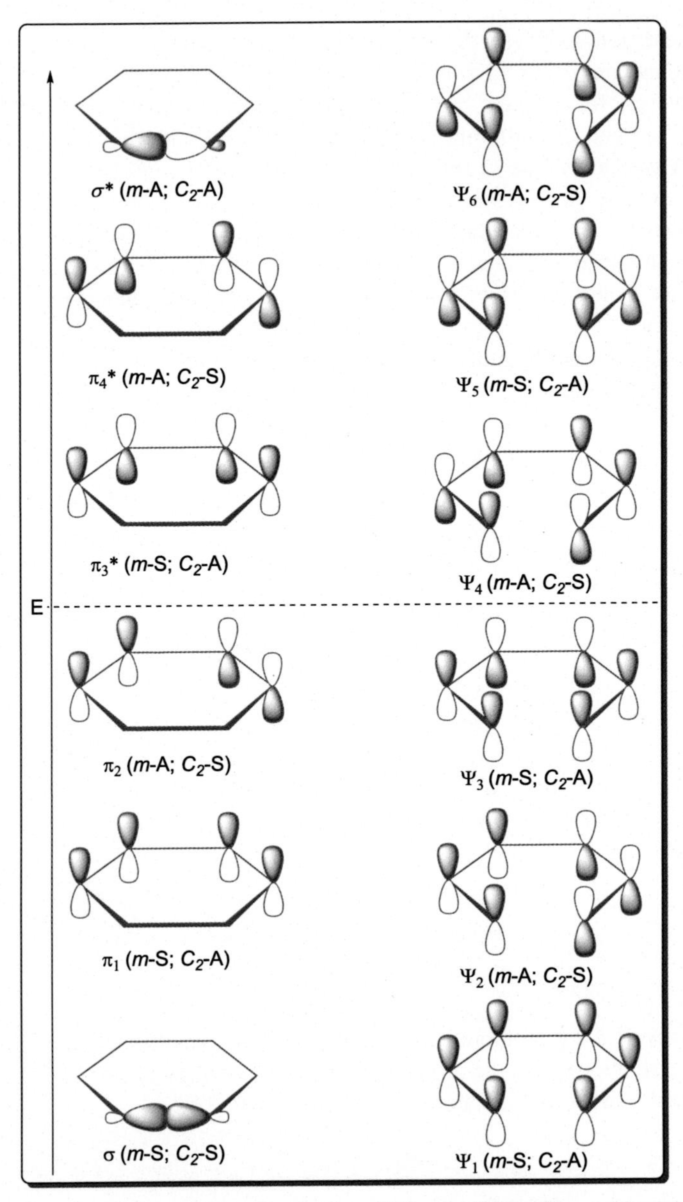

FIGURE 2.6 Symmetry properties of molecular orbitals of cyclohexadiene and hexatriene.

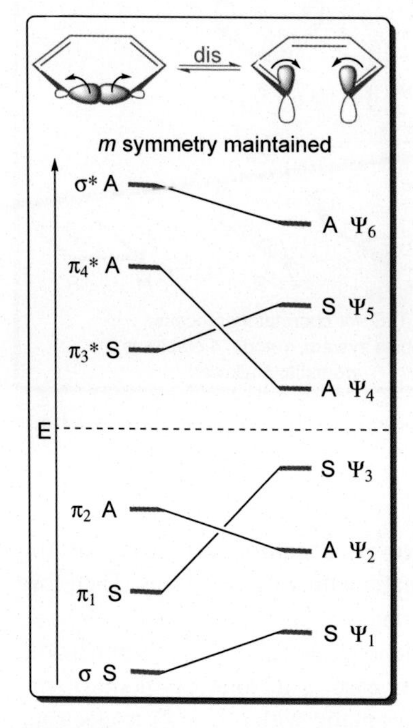

FIGURE 2.7 Correlation diagram for disrotatory interconversion of cyclohexadiene-hexatriene system.

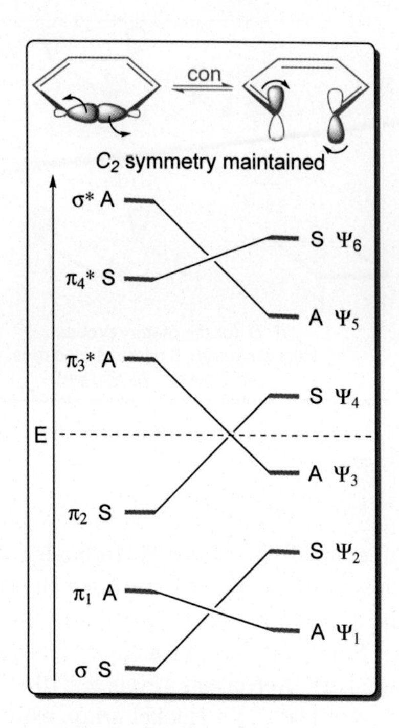

FIGURE 2.8 Correlation diagram for conrotatory interconversion of cyclohexadiene-hexatriene system.

a nonconcerted path and their energy of activation is 10–15 kcal/mol higher than those for symmetry-allowed reactions. *It is because of this energy difference that most of the electrocyclic reactions follow Woodward–Hoffmann rules.*

2.4.2 Perturbation Molecular Orbital Method

In the PMO method, we analyze an electrocyclic reaction through the following steps: (1) Define a basis set of 2*p*-atomic orbitals for all atoms involved (1*s* for hydrogen atoms). (2) Then connect the orbital lobes that interact in the starting materials. (3) Now let the reaction start and then we identify the new interactions that are occurring at the transition state. (4) Depending upon the number of electrons in the cyclic array of orbitals and whether the orbital interaction topology corresponds to a Hückel-type system or Möbius-type system, we conclude about the feasibility of the reaction under thermal and photochemical conditions.

In the case of butadiene to cyclobutene interconversion (4n π-electron system), a disrotatory mode of ring closure leads to a Hückel array, which is antiaromatic with 4n π-electrons (Figure 2.9). Therefore, the disrotatory mode

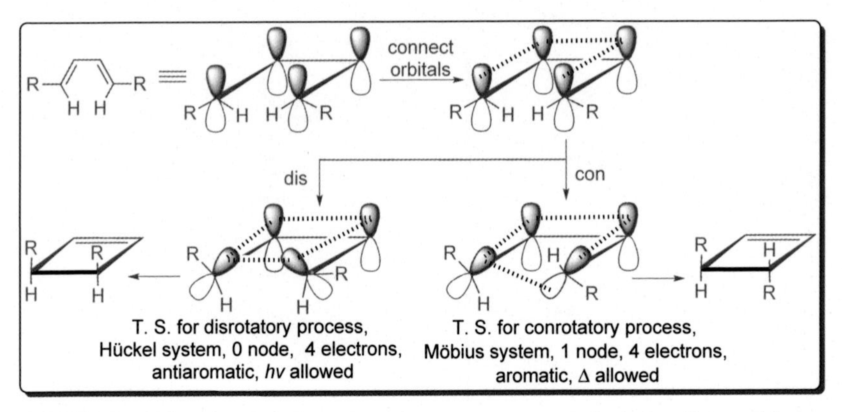

FIGURE 2.9 PMO approach to disrotatory and conrotatory processes for the butadiene-cyclobutene system.

of reaction is thermally forbidden. However, a conrotatory mode of ring closure uses a Möbius array, which is aromatic with 4n π-electrons. Therefore, the reaction is thermally allowed in this mode.

Similarly, we can analyze the hexatriene-cyclohexadiene system having (4n + 2) π-electrons (Figure 2.10). In this case, a disrotatory mode of ring closure leads to a Hückel array, which is aromatic with (4n + 2) π-electrons. Therefore, the disrotatory mode of reaction now becomes thermally allowed. However, a conrotatory mode of ring closure uses a Möbius array, which is antiaromatic with (4n + 2) π-electrons. Therefore, the reaction is thermally forbidden in this mode.

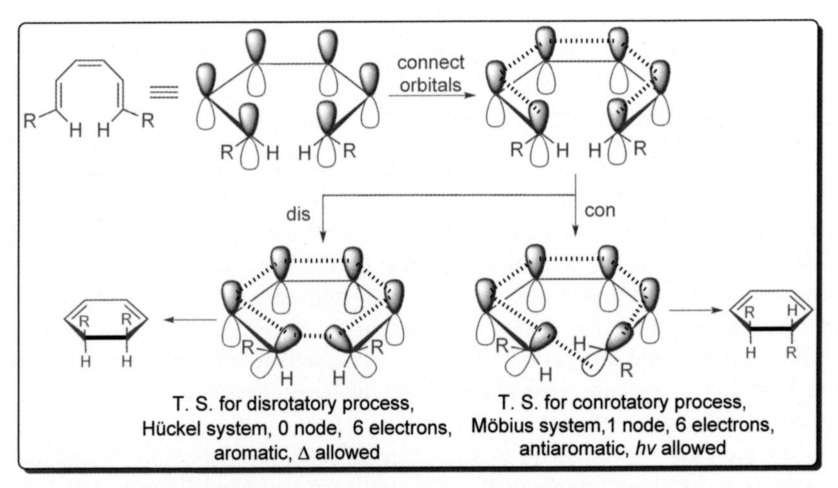

FIGURE 2.10 PMO approach to disrotatory and conrotatory processes for the hexatriene-cyclohexadiene system.

Thus, we reach the same conclusions as described earlier by using the orbital correlation diagram method. For convenience, the selection rules by this approach to electrocyclic reactions are tabulated in Table 2.2.

TABLE 2.2 Selection rules by PMO method.

Array of π-electrons involved	Number of nodes	Aromaticity	Δ allowed	*hν* allowed
4n	0 or Even	Antiaromatic	–	dis
4n	Odd	Aromatic	con	–
4n + 2	0 or Even	Aromatic	dis	–
4n + 2	Odd	Antiaromatic	–	con

2.4.3 Frontier Molecular Orbital Method

Although it is more fruitful to construct a correlation diagram for the detailed analysis of an electrocyclic reaction, there is, nevertheless, an alternative method that also enables us to reach similar conclusions. In this approach, which is extremely simple, *our only guide is the symmetry of the highest occupied molecular orbital (HOMO) of the open-chain partner in an electrocyclic reaction. If this orbital has a C_2 symmetry, then the reaction follows a conrotatory path, and if it has a mirror plane symmetry, a disrotatory mode is observed.* The explanation for this alternative approach is based on the fact that overlapping of wave functions of the same sign is essential for bond formation.

We have already seen that the symmetry of an orbital depends upon the number of nodes, which is equal to n − 1 (Ψ_n = wave function of the MO). *If the number of node(s) is zero or an even integer, the orbital will be symmetric with respect to m and antisymmetric with respect to C_2. However, the symmetry properties are reversed if the number of nodes is an odd integer.* For example, in the ground state of butadiene, which is the open-chain partner in the butadiene-cyclobutene interconversion, Ψ_2 is the highest occupied molecular orbital, and since it has one node and displays C_2 symmetry, thermal ring closure is a conrotatory process. Irradiation of butadiene promotes an electron from Ψ_2 to Ψ_3, which then becomes the highest occupied molecular orbital, and since Ψ_3 has mirror symmetry (two nodes), disrotation is required for photochemical ring closure (Figure 2.11).

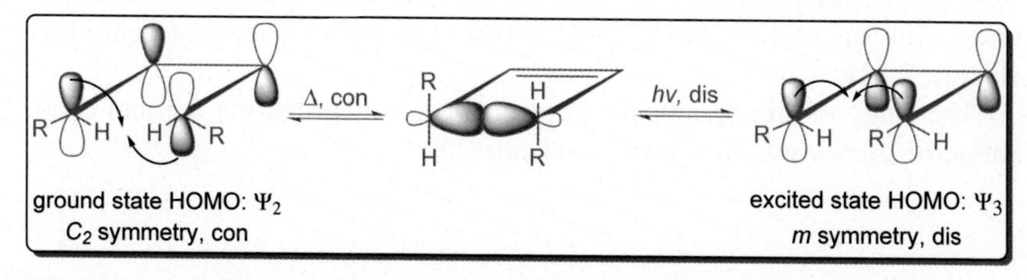

FIGURE 2.11 Butadiene-cyclobutene interconversion on the basis of FMO approach.

For example, in the ground state of (2*Z*,4*E*)-hexa-2,4-diene, HOMO is Ψ_2 and obviously cyclization is possible only through conrotation (Scheme 2.4).

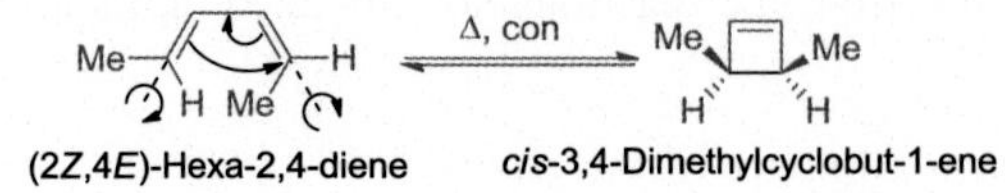

Δ, con

(2*Z*,4*E*)-Hexa-2,4-diene *cis*-3,4-Dimethylcyclobut-1-ene

SCHEME 2.4 Electrocyclization of (2*Z*,4*E*)-hexa-2,4-diene under thermal conditions.

Similarly, in the case of (2*E*,4*E*)-hexa-2,4-diene, we get *trans*-3,4-dimethylcyclobut-1-ene (Scheme 2.5).

Δ, con

(2*E*,4*E*)-Hexa-2,4-diene *trans*-3,4-Dimethylcyclobut-1-ene

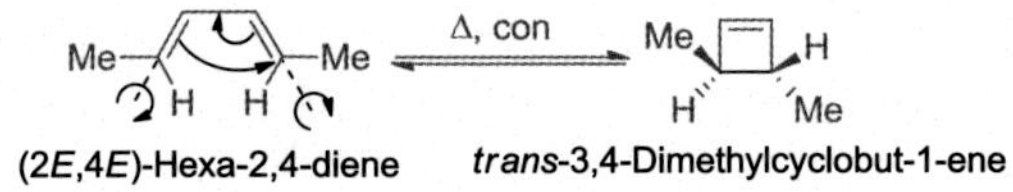

SCHEME 2.5 Electrocyclization of (2*E*,4*E*)-hexa-2,4-diene under thermal conditions.

On the other hand, irradiation of (2*Z*,4*E*)-hexa-2,4-diene (butadiene system) promotes an electron to Ψ_3, which then becomes HOMO, and bond formation is possible only through disrotation (Scheme 2.6).

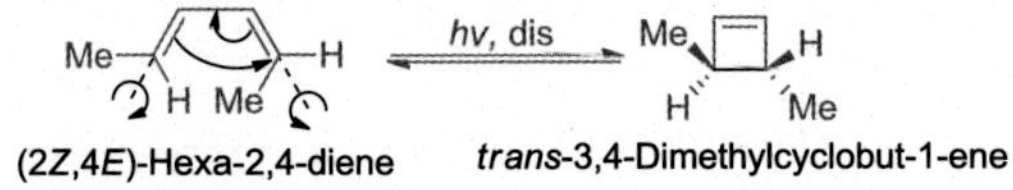

hv, dis

(2*Z*,4*E*)-Hexa-2,4-diene *trans*-3,4-Dimethylcyclobut-1-ene

SCHEME 2.6 Photochemical electrocyclization of (2*Z*,4*E*)-hexa-2,4-diene.

In a similar way, in the case of (2*E*,4*E*)-hexa-2,4-diene, we get *cis*-3,4-dimethylcyclobut-1-ene under photochemical conditions (Scheme 2.7).

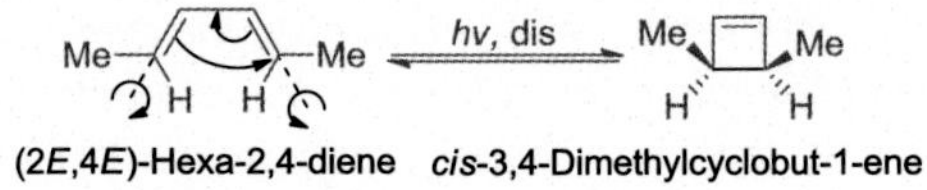

hv, dis

(2*E*,4*E*)-Hexa-2,4-diene *cis*-3,4-Dimethylcyclobut-1-ene

SCHEME 2.7 Photochemical electrocyclization of (2*E*,4*E*)-hexa-2,4-diene.

Similarly, in the hexatriene-cyclohexadiene transformation, the HOMOs of the open-chain partner under thermal and photochemical conditions are Ψ_3 and Ψ_4, respectively. As may be expected, the reaction proceeds by disrotation on heating and by conrotation under photochemical conditions (Figure 2.12).

The study of simple model compounds confirmed that the thermal cyclization of trienes was disrotatory (Scheme 2.8).

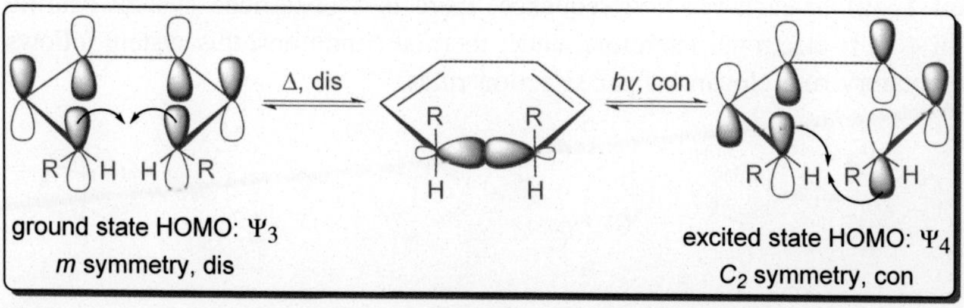

Δ, dis
hv, con
R H H R
R R
H H
R H R H
ground state HOMO: Ψ_3
m symmetry, dis
excited state HOMO: Ψ_4
C_2 symmetry, con

FIGURE 2.12 Hexatriene-cyclohexadiene interconversion on the basis of FMO approach.

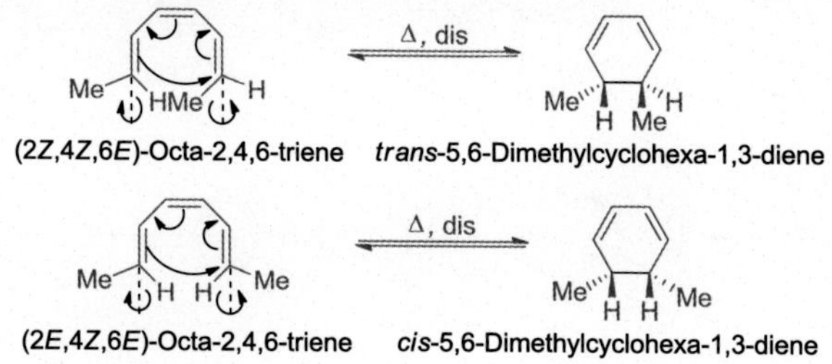

Δ, dis
Me HMe H
Me H H Me
(2*Z*,4*Z*,6*E*)-Octa-2,4,6-triene
trans-5,6-Dimethylcyclohexa-1,3-diene

Δ, dis
Me H H Me
Me H H Me
(2*E*,4*Z*,6*E*)-Octa-2,4,6-triene
cis-5,6-Dimethylcyclohexa-1,3-diene

SCHEME 2.8 Electrocyclization of (2*Z*,4*Z*,6*E*) or (2*E*,4*Z*,6*E*)-octa-2,4,6-triene under thermal conditions.

On the other hand, irradiation of (2*Z*,4*Z*,6*E*)-octa-2,4,6-triene (hexatriene system) promotes an electron to Ψ_4, which then becomes HOMO, and bond formation is possible only through conrotation (Scheme 2.9).

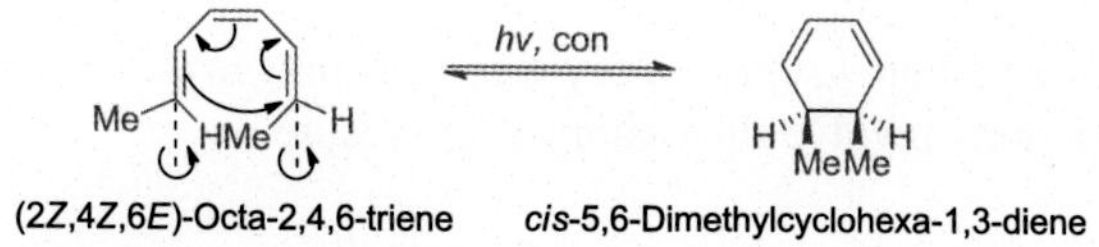

hv, con
Me HMe H
H H MeMe
(2*Z*,4*Z*,6*E*)-Octa-2,4,6-triene
cis-5,6-Dimethylcyclohexa-1,3-diene

SCHEME 2.9 Photochemical electrocyclization of (2*Z*,4*Z*,6*E*)-octa-2,4,6-triene.

2.4.4 Solved Problems (Multiple Choice Questions)

Q 1. The direction of rotation of the following thermal electrocyclic ring closures, respectively, is:

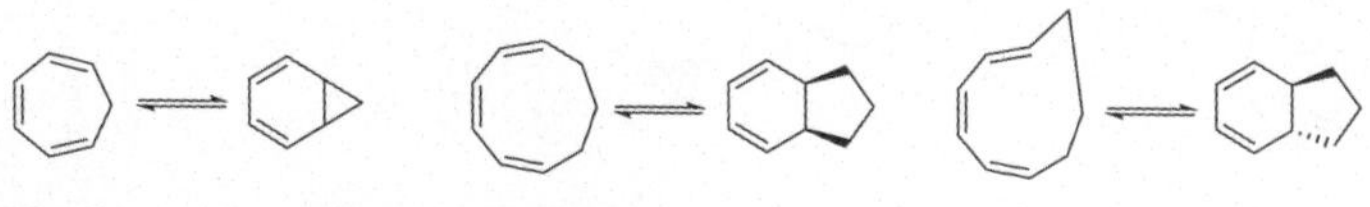

(a) Disrotatory, disrotatory, disrotatory
(b) Conrotatory, conrotatory, conrotatory
(c) Disrotatory, disrotatory, conrotatory
(d) Disrotatory, conrotatory, disrotatory

Sol 1. (a) In each reaction sequence, there is a hexatriene system bearing (4n + 2) π-electrons. Therefore, under thermal conditions, this system follows disrotatory ring closure as per selection rules.

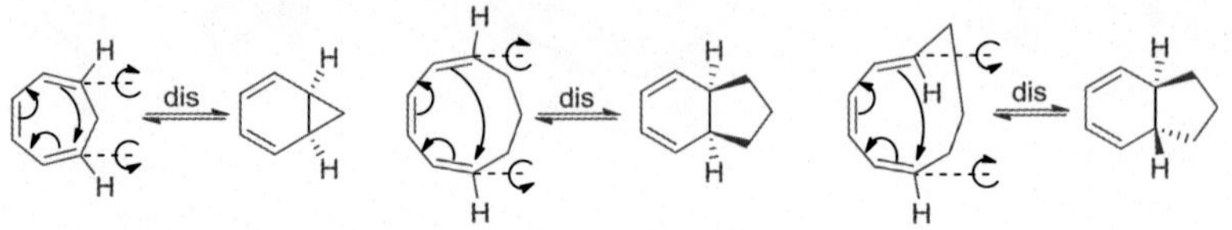

Q 2. Consider the following electrocyclic reactions:

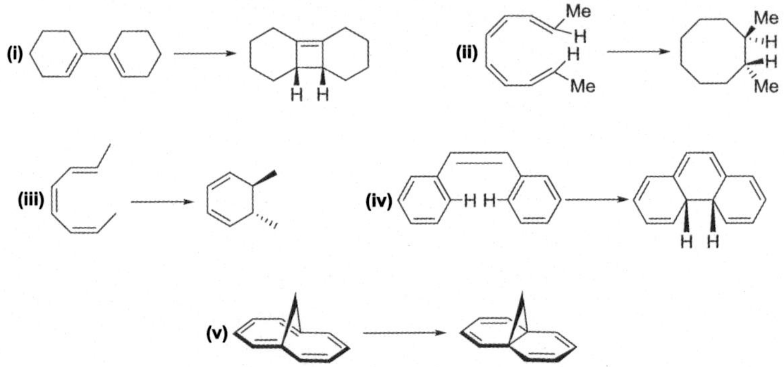

(i) (ii) (iii) (iv) (v)

Conrotatory ring closure is involved in:

(a) i (b) ii (c) iii (d) iv (e) v

Sol 2. (b) As shown below, conrotatory ring closure is involved only in (ii); the rest of the reactions involve disrotatory ring closure.

(i) dis (ii) con (iii) dis (iv) dis (v) dis

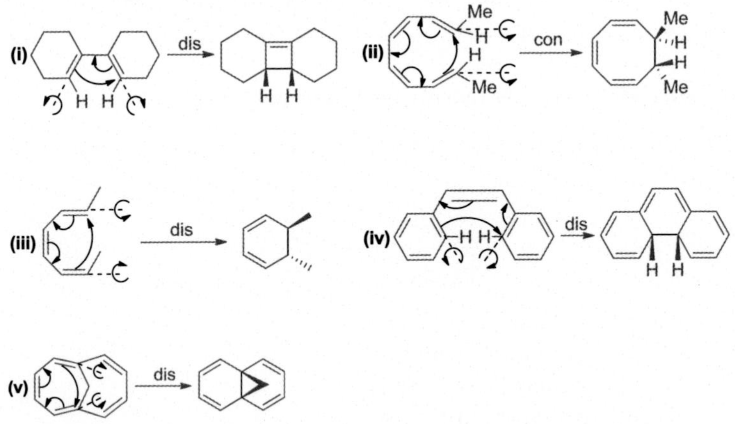

Q 3. The transformation below is feasible by a:

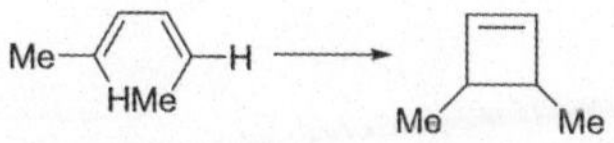

(a) Thermal disrotatory process
(b) Photochemical disrotatory process
(c) Thermal conrotatory process
(d) Photochemical conrotatory process

Sol 3. (c) The given transformation involves conrotatory cyclization of (*2Z,4E*)-hexa-2,4-diene (**I**) to give *cis*-3,4-dimethylcyclobut-1-ene (**II**). For a 4n π-electron system, conrotatory process is feasible only under thermal conditions.

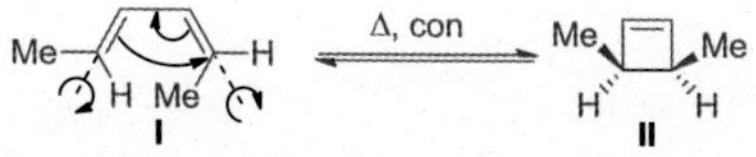

Q 4. Look at the reaction and identify the processes involved:

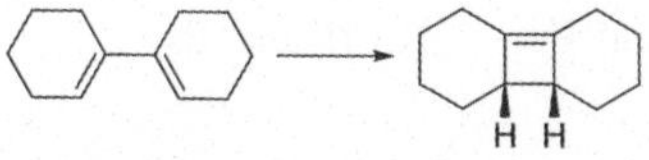

(a) 4π-Electron thermal conrotatory electrocyclic reaction
(b) 4π-Electron photochemical disrotatory electrocyclic reaction
(c) (2π + 2π) Photochemical cycloaddition reaction
(d) (2π + 2π) Thermal cycloaddition reaction

Sol 4. (b) The reaction involves 4π-electron photochemical disrotatory electrocyclic reaction.

excited state HOMO: Ψ_3
m symmetry, dis

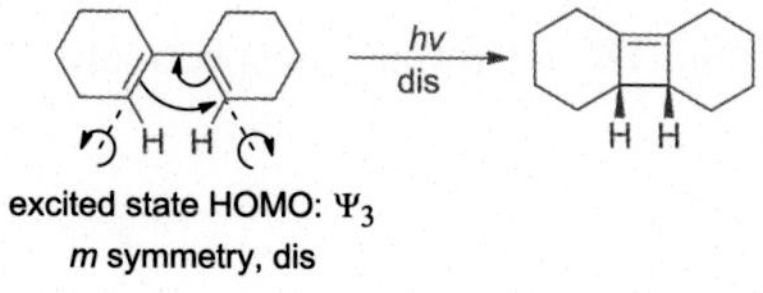

Q 5. Identify the photoproduct obtained by the irradiation of *trans*-stilbine in presence of I_2 or O_2.

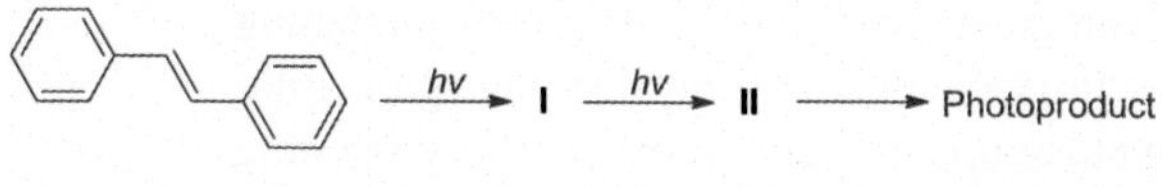

(a) Phenanthrene (b) Naphthalene (c) Anthracene (d) Phenylnaphthalene

Sol 5. (a) *Trans*-stilbine undergoes photochemical *cis-trans* isomerization to give *cis*-stilbine (**I**). Irradiation of *cis*-stilbine gives dihydrophenanthrene (**II**) by

6π-electron conrotatory cyclization. **II** is further oxidized by I_2 or O_2 to give phenanthrene (**III**).

H H *hv* H H *hv* con H H
I II
excited state HOMO: Ψ_4
C_2 symmetry, con
I_2 or O_2
III

Q 6. Select the correct classification for the following reaction from options I to IV given below:

CHO CH_3 Δ CHO CH_3

(I) Conrotatory electrocyclic reaction (II) Disrotatory electrocyclic reaction
(III) Valence isomerization (IV) $[\pi^4s + \pi^2a]$ Cycloaddition reaction
(a) I and III (b) II and IV (c) II and III (d) I and IV

Sol 6. (c) Disrotatory electrocyclic reaction and valence isomerization.

H CHO CH_3 H Δ dis H CHO CH_3 H

ground state HOMO: Ψ_3
m symmetry, dis

Q 7. The products **II** and **III** are formed, respectively, from **I** via:

D H D H D D D H H D
II I III

(a) *hv*, Conrotatory opening and Δ, disrotatory opening
(b) *hv*, Disrotatory opening and Δ, conrotatory opening
(c) Δ, Conrotatory opening and *hv*, disrotatory opening
(d) Δ, Disrotatory opening and *hv*, conrotatory opening

Sol 7. (a) A (4n + 2) π-electron system, under photochemical conditions undergoes conrotatory opening and under thermal conditions undergoes disrotatory opening.

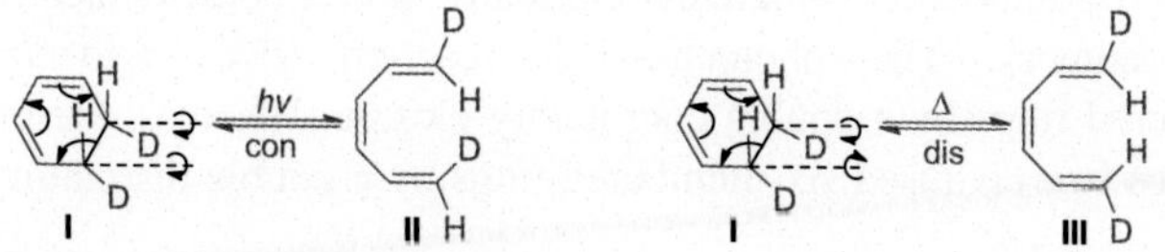

H D I hv con D H D II H H D I Δ dis D H H III D

Q 8. Electrocyclization of (*2E,4Z,6E*)-octa-2,4,6-triene under photochemical conditions, gives:

(a) *trans*-5,6-Dimethylcyclohexa-1,3-diene
(b) *cis*-5,6-Dimethylcyclohexa-1,3-diene
(c) A mixture of *trans*- and *cis*-5,6-dimethylcyclohexa-1,3-diene
(d) 1,2-Dimethylcyclohexa-1,3-diene

Sol 8. (a) (*2E,4Z,6E*)-Octa-2,4,6-triene (**I**) is a (4n + 2) π-electron system, so under photochemical conditions it undergoes conrotatory electrocyclization to give *trans*-5,6-dimethylcyclohexa-1,3-diene (**II**).

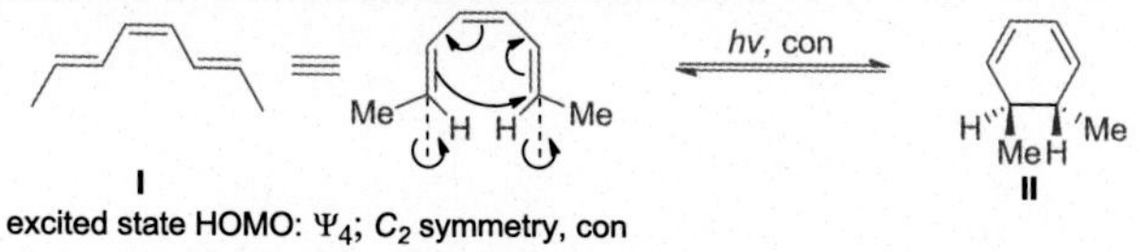

≡ Me H H Me hv, con H Me Me H II

I

excited state HOMO: Ψ_4; C_2 symmetry, con

Q 9. The products **I** and **II** obtained during the following reactions are:

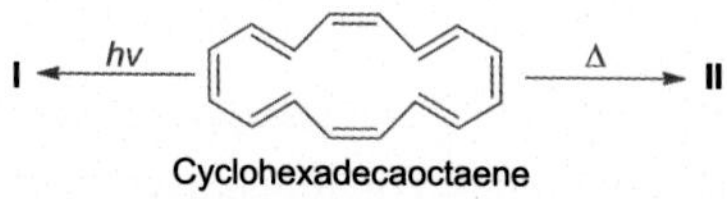

I ← hv — Δ → II

Cyclohexadecaoctaene

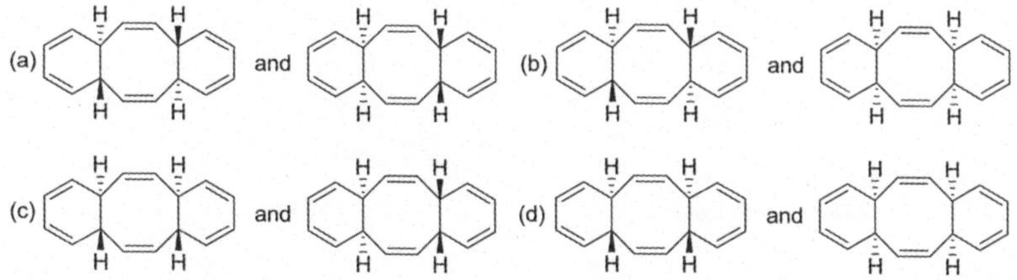

(a) H H H H and H H H H (b) H H H H and H H H H

(c) H H H H and H H H H (d) H H H H and H H H H

Sol 9. (a) The arrow pushing mechanism reveals that the reaction involves the ring closure of two hexa-1,3,5-triene systems. Thus, two electrocyclic reactions involving three electron pairs each, take place in this isomerization. The two pairs of electrons in the eight-membered ring do not take part in electrocyclization.

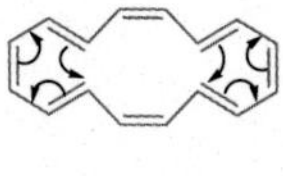

Thus [16]-annulene isomerizes thermally and photochemically to two different isomers. The photochemical reaction gives two *trans*-fused six-membered rings by a double conrotatory closure. Whereas, the thermal reaction gives two *cis*-fused six-membered rings by a double disrotatory closure.

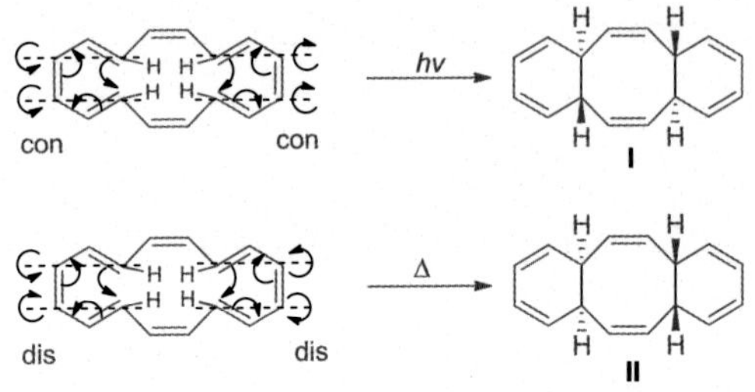

It should be noted that under both photochemical and thermal conditions, we can also get one more isomer. However, the two possible isomers that could be obtained photochemically or thermally are less stable due to steric hindrance.

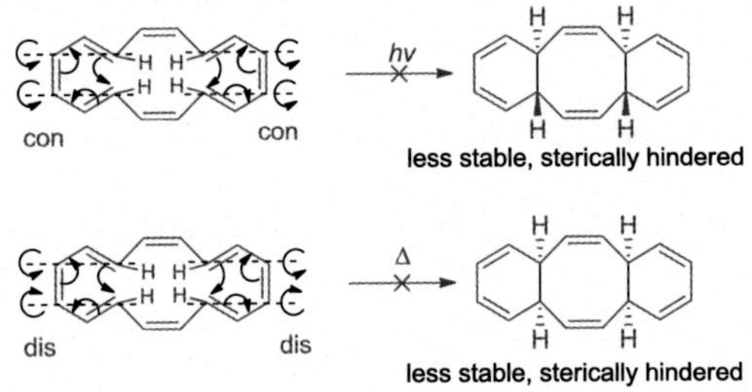

Q 10. The two-step conversion of 7-dehydrocholesterol to vitamin D_3 proceeds through:

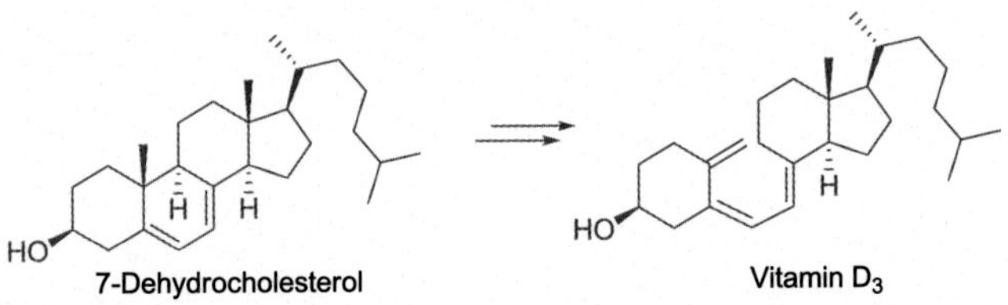

(a) Photochemical electrocyclic disrotatory ring opening and thermal antarafacial [1,7] H shift
(b) Photochemical electrocyclic conrotatory ring opening and thermal antarafacial [1,7] H shift
(c) Thermal electrocyclic conrotatory ring opening and photochemical suprafacial [1,7] H shift
(d) Thermal electrocyclic disrotatory ring opening and thermal suprafacial [1,7] H shift

Sol 10. **(b)** Vitamin D_3 (cholecalciferol) is produced through the action of ultraviolet irradiation (UV) on its precursor 7-dehydrocholesterol. The transformation occurs in two steps. In the first step, 7-dehydrocholesterol is photolyzed by ultraviolet light in a 6π-electron conrotatory electrocyclic reaction to give previtamin D_3. In the second step, previtamin D_3 spontaneously isomerizes to vitamin D_3 in a thermal antarafacial sigmatropic [1,7] hydrogen shift.

excited state HOMO: Ψ_4
C_2 symmetry, con

$h\nu$ con

Previtamin D_3

Δ [1,7] H shift

Vitamin D_3

Q 11. The following tetraene system upon photolysis gives:

$h\nu$

(a) (b) (c) (d)

Sol 11. **(b)** **Periselectivity** is a special kind of site selectivity involving a conjugated system. When such a system undergoes a pericyclic reaction, in some cases more than one pathway (involving the whole of the conjugated array of electrons, or a smaller part of it) may be symmetry-allowed according to the Woodward–Hoffmann rules. However, the system preferentially selects the least energy pathway over all the other available pathways. In general, pericyclic reactions use the longest part of a conjugated system that is compatible with the orbital-symmetry rules. This is because the ends of conjugated systems carry the largest coefficients in the frontier orbitals, which make these reactions go faster. For example, in the given problem the octatetraene undergoes disrotatory 8π-electron cyclization to give a cyclooctatriene and not a vinylcyclohexadiene or a divinylcyclobutene.

$h\nu$ dis 8π-electron cyclization

excited state HOMO: Ψ_5
m symmetry, dis

$h\nu$ con 6π-electron cyclization

not formed

excited state HOMO: Ψ_4
C_2 symmetry, con

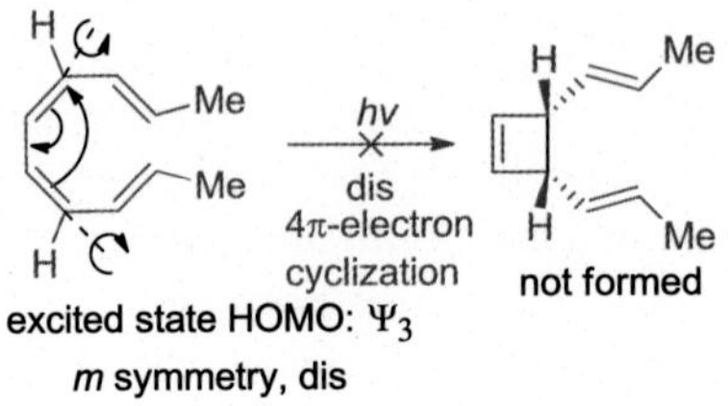

Q 12. The conditions **X** and **Y** required for the following pericyclic reactions are:

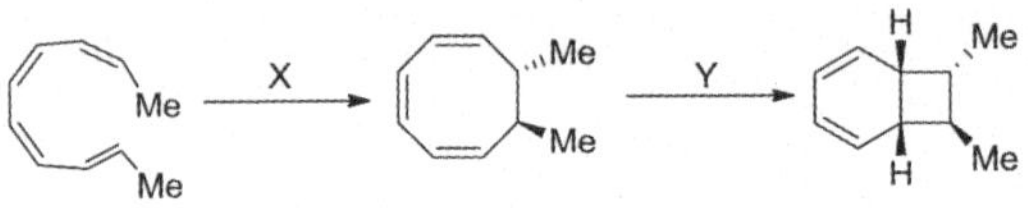

(a) X-Δ; Y-Δ (b) X-*hv*; Y-Δ (c) X-*hv*; Y-*hv* (d) X-Δ; Y-*hv*

Sol 12. (b) The product, 7,8-dimethyl cycloocta-1,3,5-triene (**I**) with *trans* stereochemistry is obtained upon disrotatory ring closure of octatetraene system, which occurs only under photochemical conditions. Similarly, a bicyclic product (**II**) with *cis* stereochemistry at fused carbons is obtained when a triene system undergoes disrotatory electrocyclization under thermal conditions.

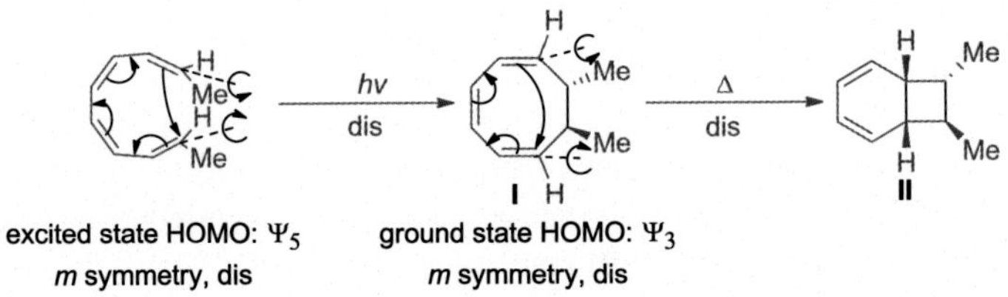

Q 13. In the following sequence of pericyclic reactions, **X** and **Y** are:

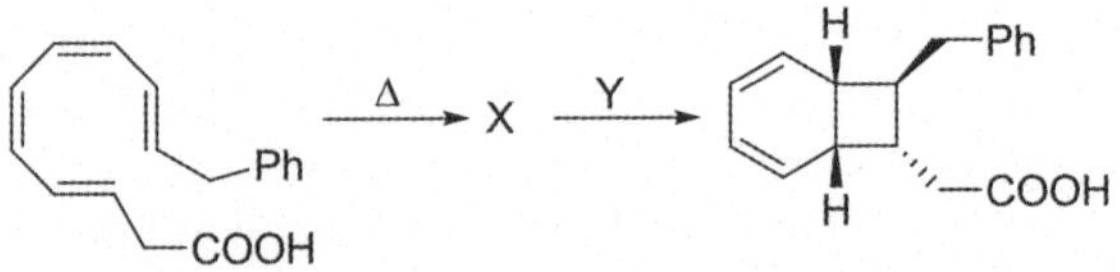

(a) X = 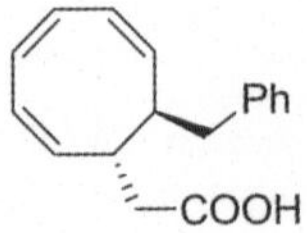 Y = *hv* / dis

(b) X = 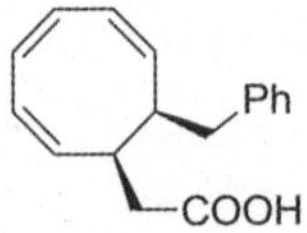 Y = *hv* / con

(c) X = [structure: cyclooctatriene with Ph and COOH] Y = Δ/ dis

(d) X = [structure: cyclooctatriene with Ph and COOH] Y = Δ / con

Sol 13. (c)

Ph, H, H, HOOC —Δ, con→ H, Ph, COOH, H —Δ, dis→ H, Ph, H, COOH

ground state HOMO: Ψ_4
C_2 symmetry, con

ground state HOMO: Ψ_3
m symmetry, dis

Q 14. In the following concerted reaction, the product is formed by a:

—Δ→ H, H

(a) 6π-Electron disrotatory electrocyclization
(b) 4π-Electron disrotatory electrocyclization
(c) 6π-Electron conrotatory electrocyclization
(d) 4π-Electron conrotatory electrocyclization

Sol 14. (a) The *cis*-orientation of hydrogens in the product indicates that cyclization involves disrotatory mode. Under thermal conditions, cyclooctatetraene preferentially undergoes a 6π-electron disrotatory electrocyclization forming only the *cis*-isomer. It should be noted that cyclooctatetraene does not undergo the thermally allowed 4π-electron conrotatory electrocyclization as the *trans*-fused bicyclic structure is highly strained.

H, H —Δ, dis→ H, H

ground state HOMO: Ψ_3
m symmetry, dis

H, H —Δ, con ✕→ H, H

ground state HOMO: Ψ_2
C_2 symmetry, con

However, under photochemical conditions, cyclooctatetraene preferentially undergoes a 4π-electron disrotatory electrocyclization forming only the *cis*-isomer. Cyclooctatetraene does not undergo the photochemically allowed 6π-electron conrotatory electrocyclization, which generates the less stable *trans*-fused bicyclic structure.

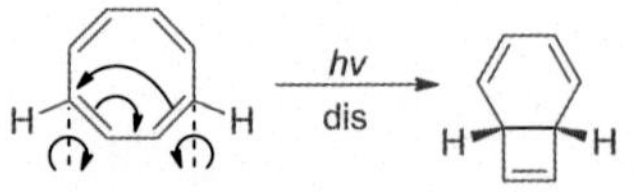

excited state HOMO: Ψ_3
m symmetry, dis

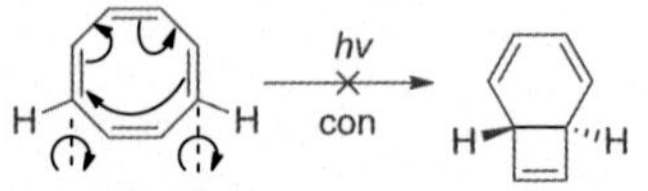

excited state HOMO: Ψ_4
C_2 symmetry, con

2.4.5 Solved Problems (Subjective)

Q 1. What are the major products obtained by the ring opening of cyclobutenes (**I–IV**) shown below in thermal conditions as concerted process? Explain while writing structures of the products.

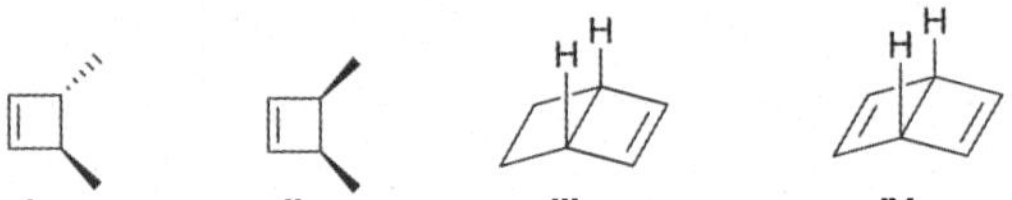

I **II** **III** **IV**

Sol 1. For a 4n π-electron system, under thermal conditions, ring opening occurs by conrotatory process. In the case of *trans*-3,4-dimethylcyclobut-1-ene (**I**), the ring can open by two conrotatory paths. Clockwise opening yields the *E*,*E*-isomer, whereas anticlockwise opening yields the *Z*,*Z*-isomer. However, in actual practice, only *E*,*E*-isomer is obtained. This is due to the fact that in the case of *Z*,*Z*-isomer, the transition state leading to the open product is less stable because of the steric strain between the methyl groups.

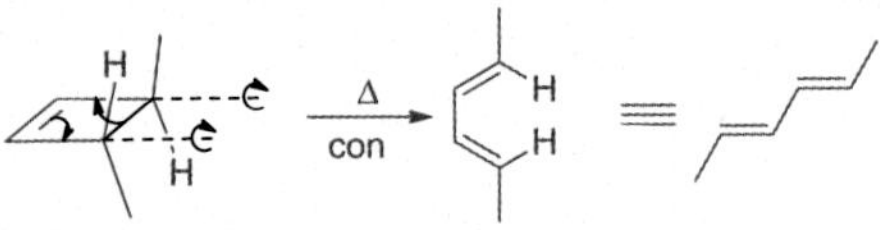

(clockwise ring opening) (2*E*,4*E*)-Hexa-2,4-diene

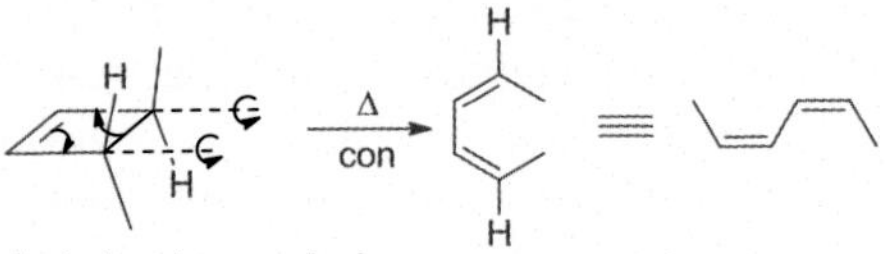

(anticlockwise ring opening) (2*Z*,4*Z*)-Hexa-2,4-diene (not obtained)

In the case of *cis*-3,4-dimethylcyclobut-1-ene (**II**), ring opening by both paths yields the *Z*,*E*-isomer.

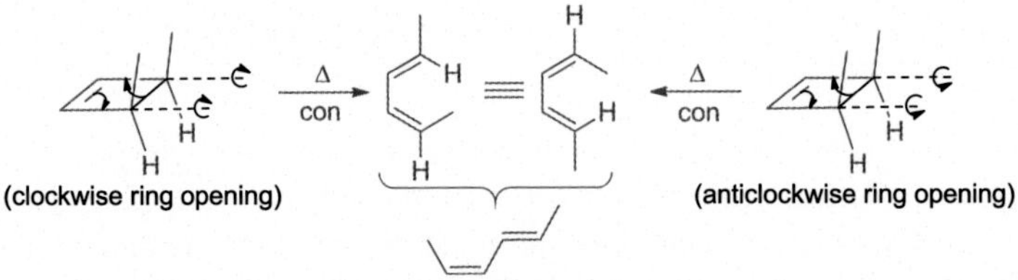

(clockwise ring opening) (anticlockwise ring opening)

(2*Z*,4*E*)-Hexa-2,4-diene

In the case of bicyclo[2.2.0]hex-2-ene (**III**) and bicyclo[2.2.0]hexa-2,5-diene or **Dewar benzene** (**IV**), ring opening should occur by a conrotatory process. However, in both cases, conrotatory ring opening places a *trans*-double bond in the six-membered ring, which is geometrically unstable. Therefore, ring opening occurs by a higher-energy nonconcerted pathway.

III —Δ, con→ geometrically unstable; hydrogen placed in a six-membered ring —π-bond rotation→ Cyclohexa-1,3-diene

IV —Δ, con→ geometrically unstable; hydrogen placed in a six-membered ring —π-bond rotation→ Benzene

The relative kinetic stability of Dewar benzene is unusually high as its conversion to benzene is exothermic by 71 kcal/mol. In addition, the central bond is not only strained but also *bis*-allylic. The unusual stability of Dewar benzene is related to the orbital symmetry requirements for concerted electrocyclic transformations. A concerted thermal pathway would be conrotatory and, therefore, leads to the formation of a highly strained *Z*,*Z*,*E*-cyclohexatriene. A disrotatory process, which would lead directly to benzene, is forbidden.

Q 2. Explain the difference in the pyrolytic stabilities of the following laterally fused *cis*- and *trans*-cyclobutene systems.

Laterally fused *cis*- and *trans*-cyclobutenes	Ea (kcal/mol)	K = 10^{-4} at °C
H H	29	87
H H	42	261
H H	27	109
H H	45	273

Sol 2. For a 4n π-electron system, under thermal conditions, ring opening occurs by conrotatory process. Therefore, laterally fused *trans*-cyclobutenes readily undergo electrocyclic cleavage to give stable compounds having *cis*-double bonds in the six- or seven-membered rings. However, in the case of laterally fused *cis*-cyclobutenes, the symmetry-allowed conrotatory ring opening leads to the formation of unstable products having a *trans*-double bond in a six- or a seven-membered ring. Therefore, such transformations require a high temperature and may follow a nonconcerted pathway.

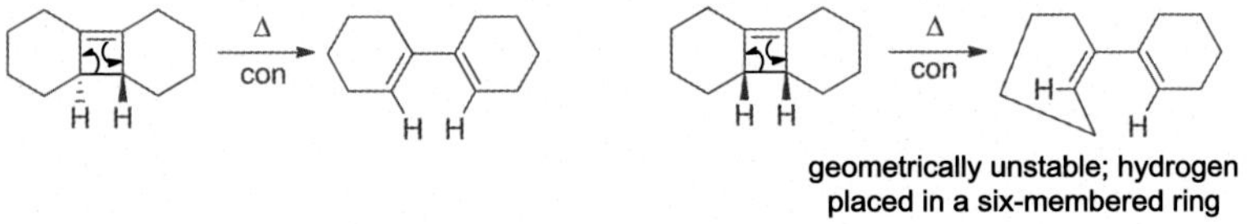

geometrically unstable; hydrogen placed in a six-membered ring

Q 3. Explain selective formation of products in the following electrocyclic reactions:

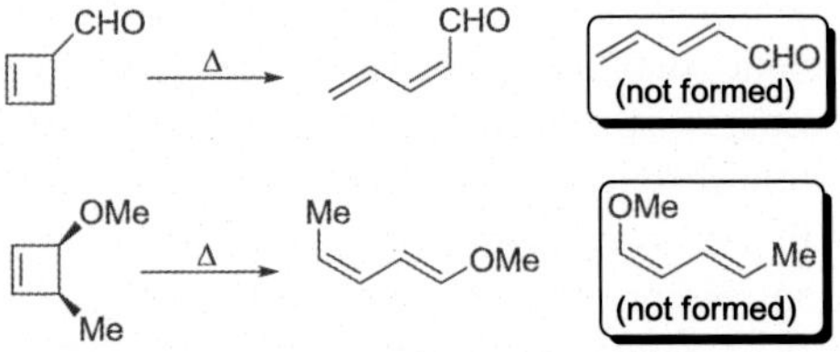

(not formed)

(not formed)

Sol 3. Two distinct stereochemical consequences exist for an electrocyclic ring opening of cyclobutenes following a conrotatory process. The substituent located at C−3 may move *away* or *toward* the bond undergoing the fission. *Torquoselectivity is defined as the predisposition of a given R substituent for a given conrotatory motion.* Steric factors should cause a preference for the larger group to move outward. It is observed, however, that π-donor substituents prefer con-out mode whereas π-acceptor substituents prefer con-in mode. For example, the inward rotation of the formyl group (π-acceptor) in 3-formylcyclobutene generates (Z)-penta-2,4-dienal exclusively. Similarly, the outward rotation of methoxy group (π-donor) in *cis*-3-methoxy-4-methylcyclobut-1-ene generates (1*E*,3*Z*)-1-methoxypenta-1,3-diene preferentially.

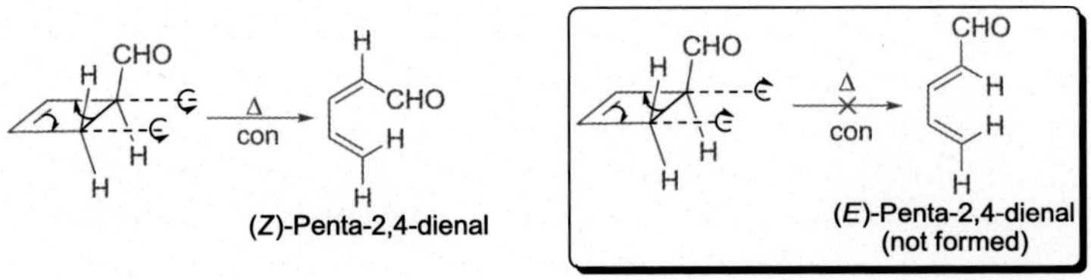

(*Z*)-Penta-2,4-dienal

(*E*)-Penta-2,4-dienal (not formed)

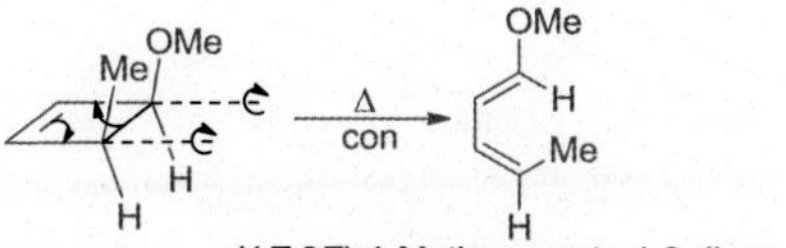

(1*E*,3*Z*)-1-Methoxypenta-1,3-diene

Q 4. Explain the following transformations:

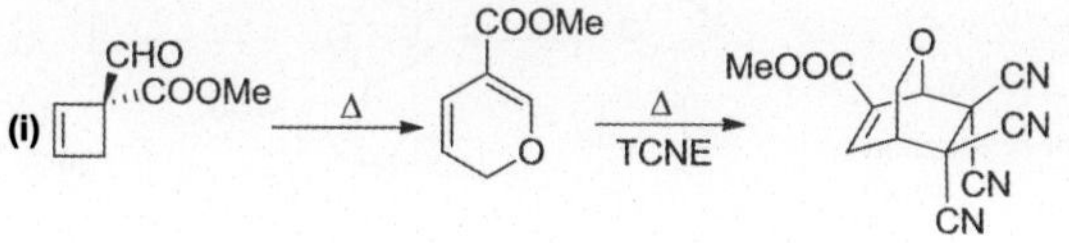

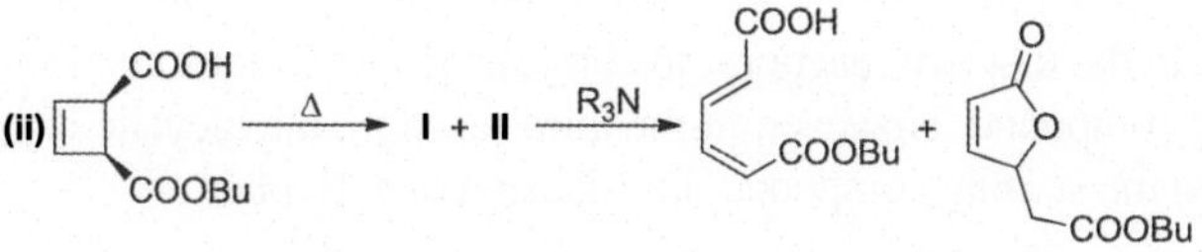

(i)

(ii)

Sol 4. (i) For a 4n π-electron system, under thermal conditions, ring opening occurs by conrotatory process. In the case of methyl 1-formylcyclobut-2-enecarboxylate (**I**), the ring can open by two conrotatory paths. However, in actual practice, the formyl group (π-acceptor) undergoes inward rotation (torquoselectivity) to give the dienal intermediate (**II**), which then undergoes electrocyclic ring closure to give methyl 2*H*-pyran-5-carboxylate (**III**). On heating with tetracyanoethylene (TCNE), **III** undergoes Diels−Alder reaction to give **IV**.

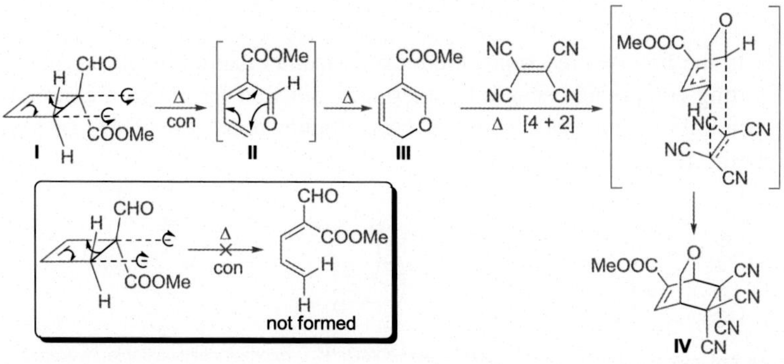

not formed

(ii) For a 4n π-electron system, under thermal conditions, ring opening occurs by conrotatory process. As the two groups are similar but not the same, there is no selection, and hence the ring opens up by both the conrotatory paths leading to the formation of two diastereomeric dienes (**I** and **II**). On treatment with tertiary amine, both **I** and **II** form anions. However, only one of the anions undergoes a 1,4-addition to the unsaturated ester to form a lactone (**III**), but the other is too far away and cannot react.

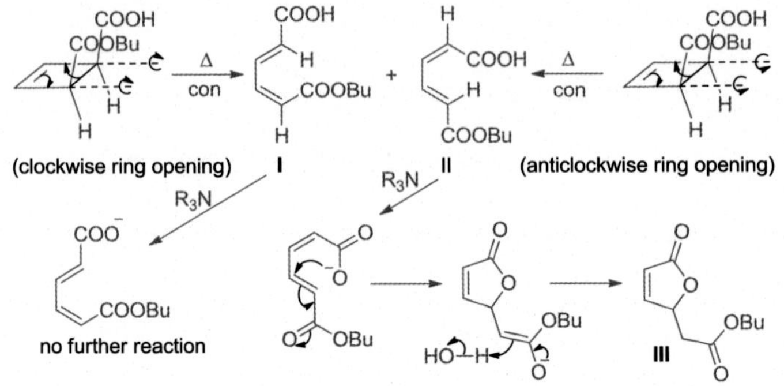

(clockwise ring opening) I II (anticlockwise ring opening)

no further reaction III

Q 5. In the three reactions shown below, the compounds readily undergo symmetry-allowed disrotatory butadiene—cyclobutene interconversion under photochemical conditions. In the first two reactions, the bicyclic isomer cannot revert thermally to the starting compound. However, in the third reaction, the bicyclic isomer reverts thermally to the starting compound, i.e., diazepinone. Explain.

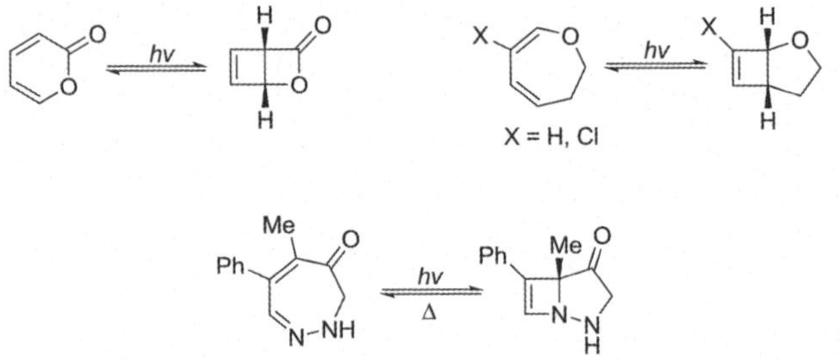

X = H, Cl

Sol 5. In the first two reactions, the bicyclic isomer cannot revert thermally to the starting compound because the thermally allowed 4π-electron conrotatory ring openings would have to put a *trans*-double bond into a six- or seven-membered ring.

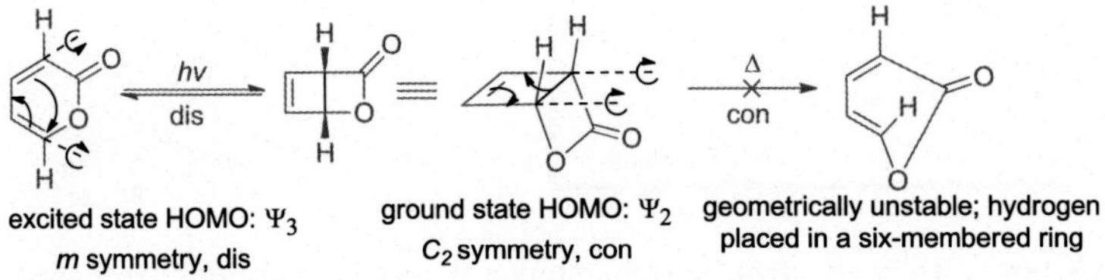

excited state HOMO: Ψ_3 *m* symmetry, dis

ground state HOMO: Ψ_2 C_2 symmetry, con

geometrically unstable; hydrogen placed in a six-membered ring

As direct *cis-trans* interconversion is not possible in fused carbocyclic systems, thermal reversibility in the conrotatory fashion has been explained by an inversion at the nitrogen atom in the diazepinone.

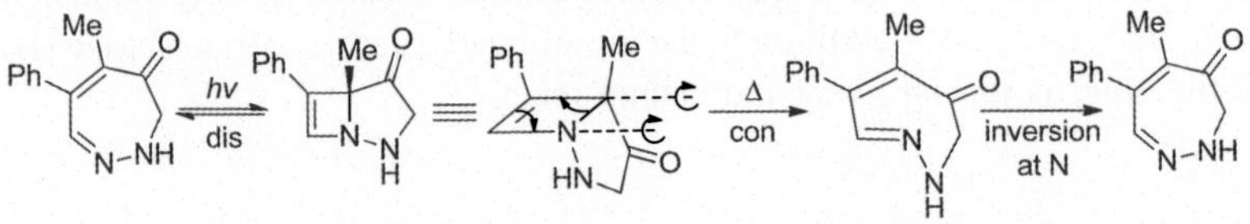

inversion at N

Q 6. How do you explain that on heating (2*E*,4*E*)-butadiene derivative equilibrates with (2*Z*,4*Z*)-butadiene isomer without forming (2*Z*,4*E*)-butadiene derivative?

Ph Ph Ph Ph (2*E*,4*E*) Δ Ph Ph Ph Ph (2*Z*,4*Z*) Ph Ph Ph Ph (2*Z*,4*E*) not formed

Sol 6. Thermal isomerization of the substituted dienes takes place through the formation of the cyclobutene intermediate by a thermally allowed conrotatory electrocyclization. Under thermal conditions, for a 4n π-electron system, ring opening also occurs by conrotatory process. As the two groups are the same there is no selection, and hence the ring opens up by both the conrotatory paths, leading to the formation of two diastereomeric dienes i.e., (2*E*,4*E*)- and (2*Z*,4*Z*)-butadiene derivatives. (2*Z*,4*E*)-Butadiene derivative can be formed only by disrotatory process, which is disallowed under thermal conditions.

(2*E*,4*E*) (2*Z*,4*Z*)

Δ con (clockwise ring closure) (anticlockwise ring opening) or (clockwise ring opening) Δ con (anticlockwise ring closure)

Q 7. Compound **I** is photochromic. The process is reversible, giving back the starting material. Propose a structure for the isomer **II**.

F F F F F F S MeMe S **I** Here, $h\nu_2 < h\nu_1$ $h\nu_1$ $h\nu_2$ **II**

Sol 7. Photochromism is the reversible transformation of a chemical species between the two forms by the absorption of light, where the two forms have different absorption spectra. It can be described as a reversible change of color upon exposure to light. Such molecules find use as photoswitches in optoelectronic devices. For example, the given diarylethene derivative is capable of undergoing reversible chemical changes. They operate by means of a 6π-electrocyclic reaction, the thermal analogue of which is impossible due to steric hindrance.

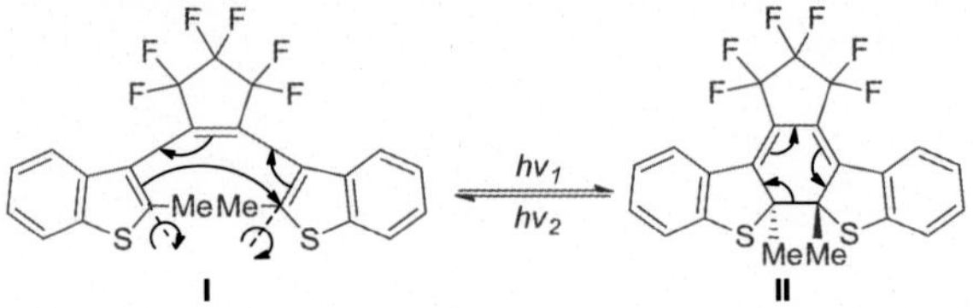

Q 8. When compound **I** is heated at 150 °C, it is transformed to compounds **III** and **IV**, via a cyclic tetraene (**II**). Propose a structure for **II**, and describe the processes involved in the conversion of **I** to **II**, **II** to **III**, and **III** to **IV**.

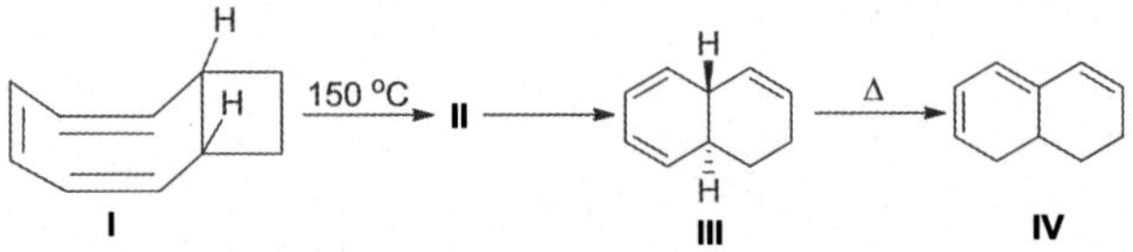

Sol 8. Under thermal conditions, compound **II** is formed by the ring opening of **I** via conrotatory mode, which is converted thermally into compound **III** via disrotatory electrocyclization. Compound **III** undergoes [1,5] hydrogen shift to yield **IV**.

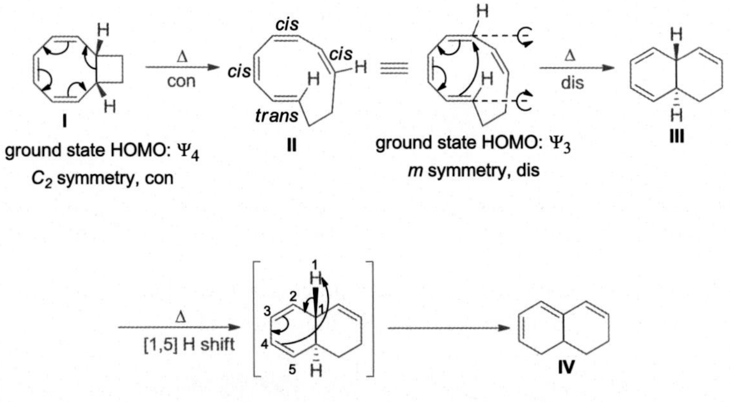

ground state HOMO: Ψ_4
C_2 symmetry, con

ground state HOMO: Ψ_3
m symmetry, dis

Q 9. A hydrocarbon **I** (C_9H_{12}) absorbs 2 mol of hydrogen. On photolysis in methanol at −20 °C, **I** isomerizes to **II**. **II** absorbs 3 mol of hydrogen to form cyclononane. In its UV spectrum, **II** exhibits λ_{max} at 290 nm. On warming to 30 °C, **II** isomerizes to **III** whose structure is given below. Write structures of **I** and **II**, and explain each step.

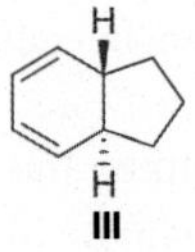

Sol 9. Compound **II** absorbs 3 mol of hydrogen to form cyclononane; therefore, **II** must contain three double bonds. Moreover, higher value of λ_{max} (i.e., 290 nm) indicates that three double bonds are conjugated, which means **II** must be cyclonona-1,3,5-triene. Ground state HOMO of **II** is Ψ_3, having mirror symmetry, therefore, it must undergo disrotatory ring closure under thermal conditions. As two hydrogens in **III** are *trans*, **II** must be *(1E,3Z,5Z)*-cyclonona-1,3,5-triene. First excited state HOMO of **II** is Ψ_4, having C_2 symmetry, therefore, it must be formed by photochemical conrotatory ring opening of **I**, which must have two *cis* hydrogens. The complete sequence of reaction is shown below:

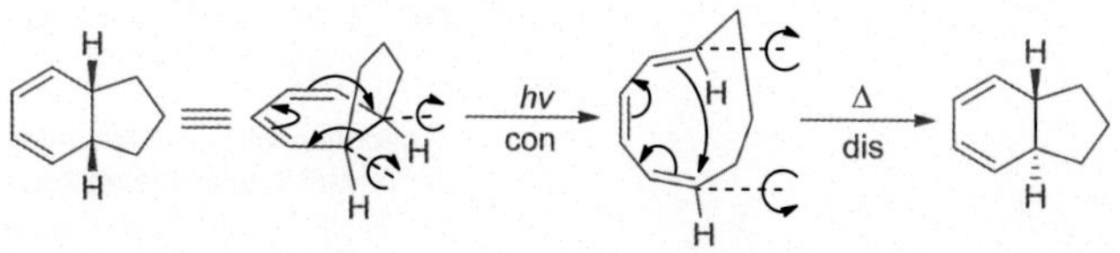

H H ≡ H H hv con H H Δ dis H H
I II III

Q 10. Explain the sequence of symmetry-allowed changes, which take place in the following reactions:

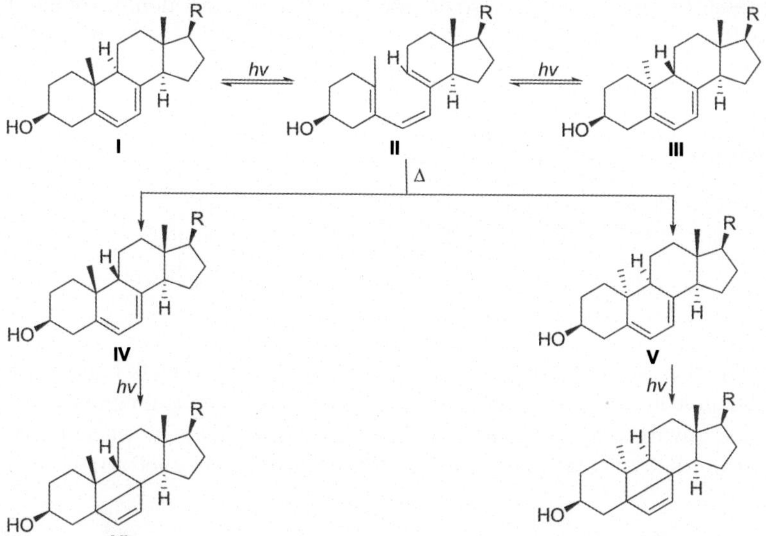

R H H HO I hv R H H HO II hv R H H HO III
Δ
R H H HO IV hv R H H HO VI
R H H HO V hv R H H HO VII

Sol 10. The given reaction involves the following sequence of symmetry-allowed changes: Product **II** arises by a photochemical 6π-electron conrotatory opening of **I**. It is in photochemical equilibrium with **I** (its starting material) and **III** (product of alternative conrotatory ring closure). On heating, **II** gives a mixture of two other cyclohexadiene derivatives, i.e., **IV**

and **V**, by a 6π-electron disrotatory cyclization. When compounds **IV** and **V** are irradiated, they do not undergo symmetry-allowed 6π-electron conrotatory opening because this would put a *trans*-double bond either into ring A or ring C.

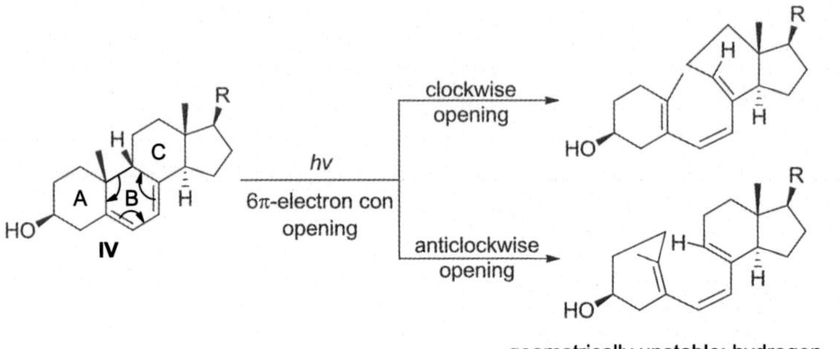

geometrically unstable; hydrogen placed in a six-membered ring

Therefore, compounds **IV** and **V** undergo a different symmetry-allowed photochemical reaction, namely 4π-electron disrotatory cyclization to yield cyclobutene derivatives **VI** and **VII**. These compounds in turn are comparatively thermally stable because the thermally allowed 4π-electron conrotatory ring openings would have to put a *trans*-double bond into ring B.

Q 11. Ethyl 1*H*-azonine-1-carboxylate gives different products under thermal and photochemical conditions. Explain the mechanism for their formation.

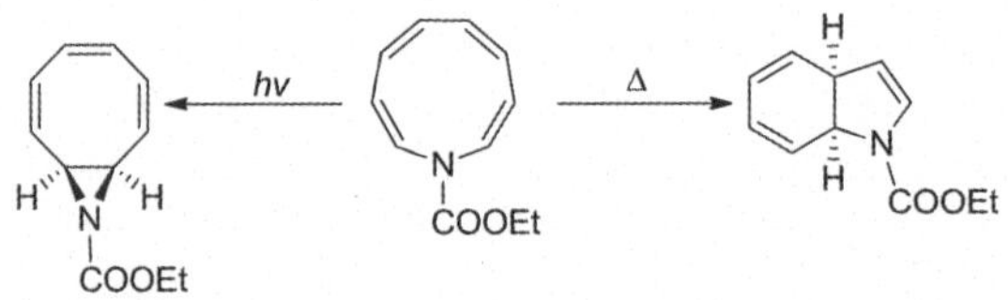

Sol 11. The *cis*-orientation of hydrogens in both the products indicates that cyclization involves disrotatory mode. Under photochemical conditions, the process involves 8π-electron disrotatory electrocyclization, whereas under thermal conditions it is a 6π-electron disrotatory electrocyclization.

excited state HOMO: Ψ_5
m symmetry, dis

ground state HOMO: Ψ_3
m symmetry, dis

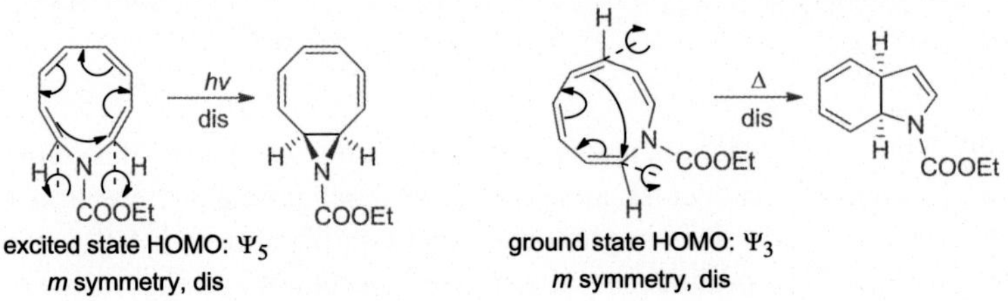

Q 12. The key reaction that results in the formation of the highly substituted phenolic ring of an antifungal agent mycophenolic acid is provided below:

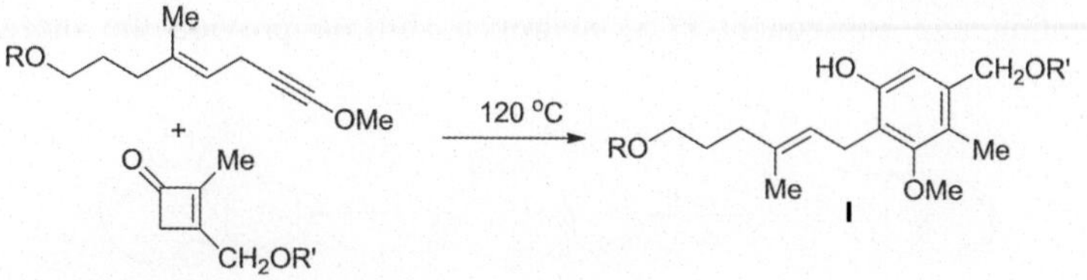

Provide a mechanism for this reaction.

Sol 12. The vinylketene-based benzannulation strategy proceeds via a "cascade" process involving four successive pericyclic reactions. Thermolysis or irradiation of cyclobutenone (**II**) serves as the driving force for the cascade, effecting a reversible electrocyclic ring opening to generate the transient vinylketene intermediate (**III**). Vinylketene **III** is immediately intercepted by the alkyne annulation partner in a regioselective [2 + 2] cycloaddition to afford a new cyclobutenone (**IV**), which under the reaction conditions undergoes reversible electrocyclic cleavage to generate dienylketene (**V**). Rapid 6π-electrocyclic closure and then tautomerization furnishes the desired aromatic product (**I**).

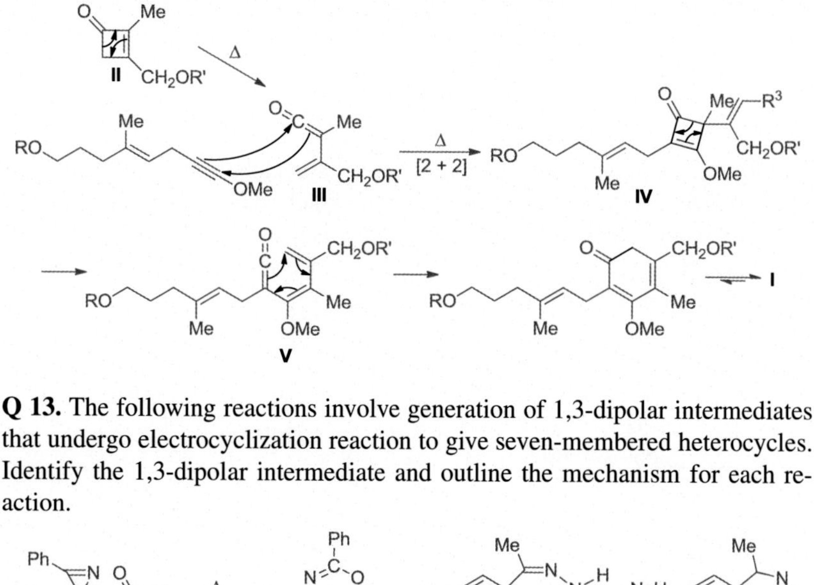

Q 13. The following reactions involve generation of 1,3-dipolar intermediates that undergo electrocyclization reaction to give seven-membered heterocycles. Identify the 1,3-dipolar intermediate and outline the mechanism for each reaction.

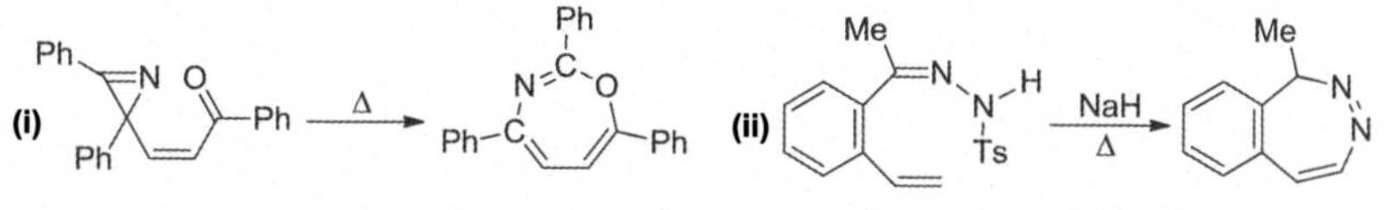

Sol 13. (i) Thermal ring opening of an azirine gives a nitrile ylide that undergoes a 8π-electron cyclization to provide the 1,3-oxazepine derivative.

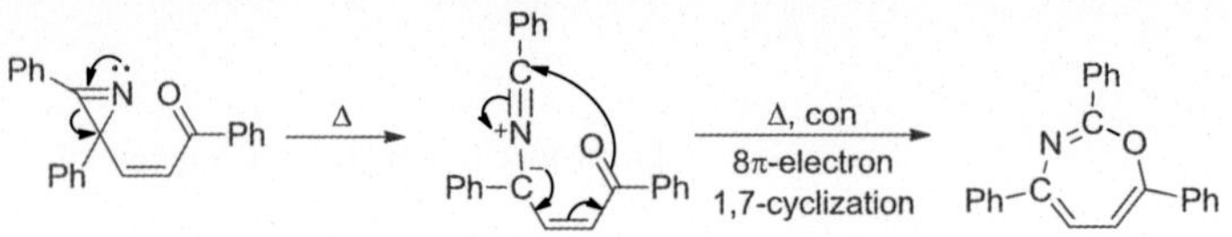

(ii) The given tosyl hydrazone derivative on heating with a strong base gives *ortho* vinyl diazoalkene intermediate, which undergoes an 8 π-electron cyclization and subsequent [1,5] hydrogen shift to provide the diazepine derivative.

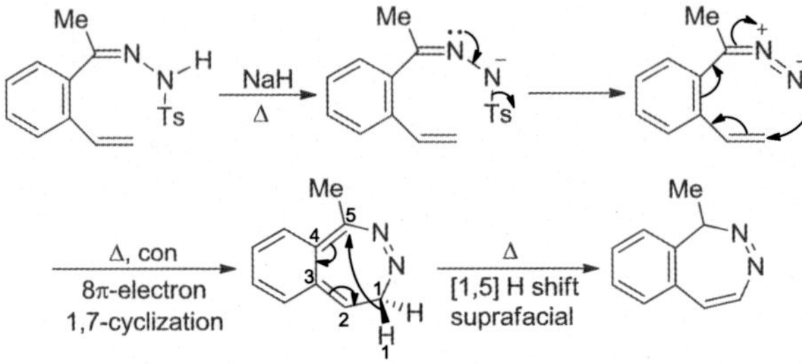

Q 14. While benzo analogues **II–IV** undergo cycloreversion of the central cyclohexane ring by all disrotatory opening, the triscyclobutenocyclohexane **I** does so by stepwise conrotatory cyclobutene rupture. How do you explain these observations?

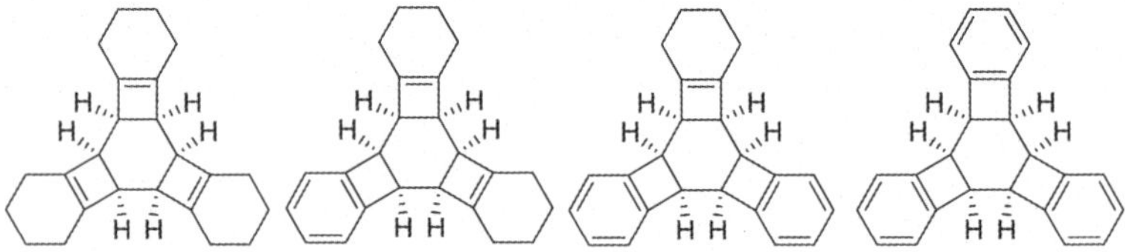

Sol 14. The compound **I** undergoes sequential cyclobutene ring opening to give cyclohexenofused [12]annulene (**V**), which then undergoes thermal Möbius π-bond shifts to substituted di-*trans* isomer **VI**. The compound **VI** then forms the cage compound **VIII** in two steps: thermal electrocyclization followed by [4 + 2] cycloaddition.

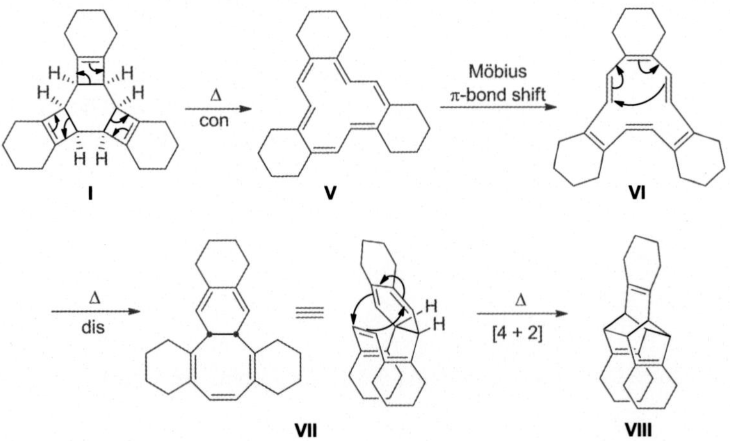

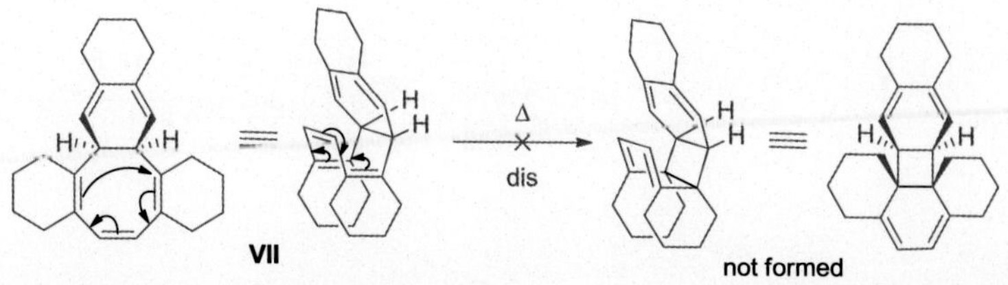

VII

not formed

In contrast, benzo analogues **II–IV** undergo cycloreversion of the central cyclohexane ring by all-disrotatory opening. This is due to the fact that in these cases π-bond shift is energetically prohibited by loss of benzene ring aromaticity. For example, the benzofused analogue of **I** (in which the three peripheral cyclohexene rings are replaced by benzene) is unraveled by concerted $[\sigma^2s + \sigma^2s + \sigma^2s]$ retro-cyclization to the corresponding all-*cis*-tribenzo-[12]-annulene.

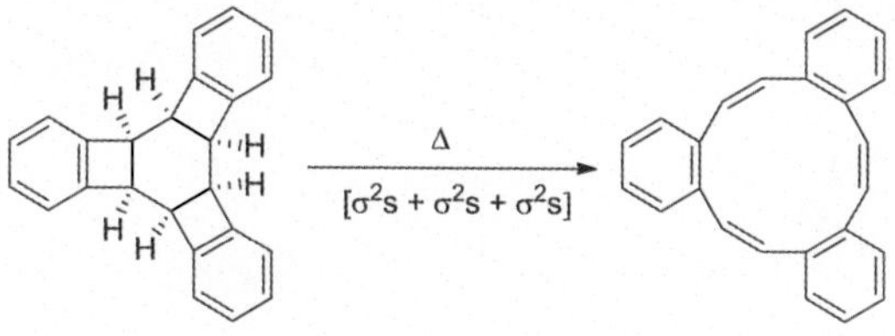

$[\sigma^2s + \sigma^2s + \sigma^2s]$

Q 15. Write the mechanism of the following reactions:

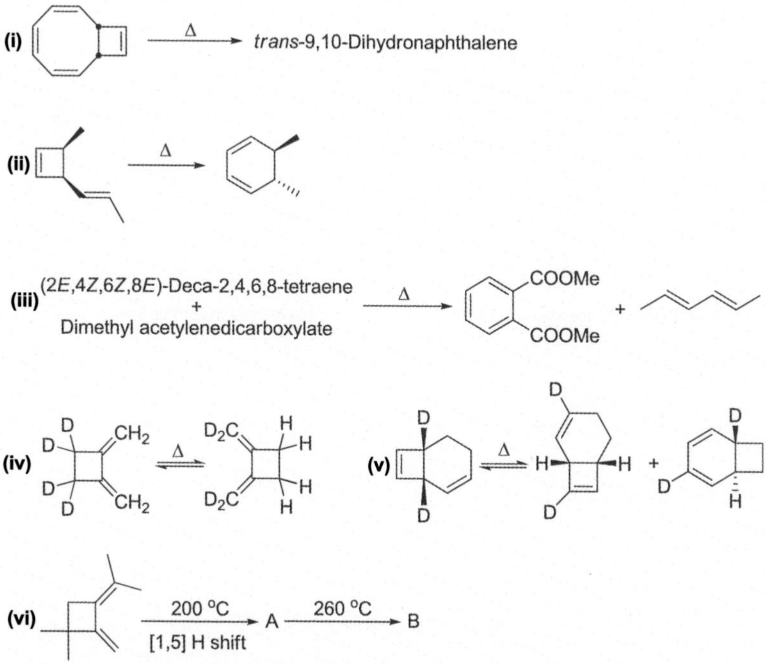

(i) Δ → *trans*-9,10-Dihydronaphthalene

(ii) Δ →

(iii) (2*E*,4*Z*,6*Z*,8*E*)-Deca-2,4,6,8-tetraene + Dimethyl acetylenedicarboxylate Δ → COOMe, COOMe +

(iv) Δ ⇌

(v) Δ ⇌ +

(vi) 200 °C, [1,5] H shift → A; 260 °C → B

(vii) H H $h\nu$ Δ H H

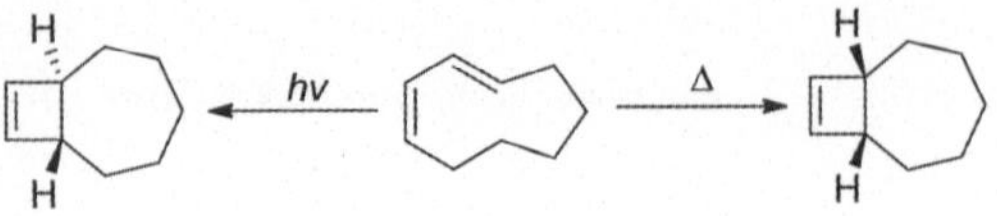

(viii) H H Δ $h\nu$ 254 nm 60:40 H H $h\nu$ 300 nm H H H H

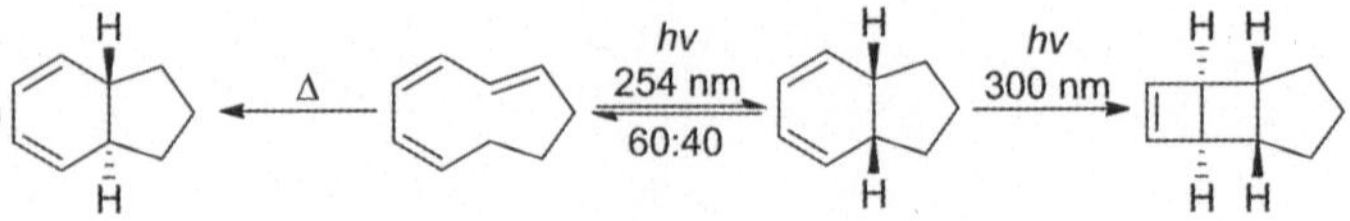

(ix) H MePhN Δ

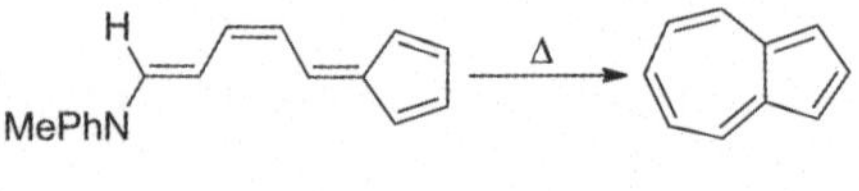

(x) NC CN Ph $h\nu$ Δ NC CN Ph Ph CN CN

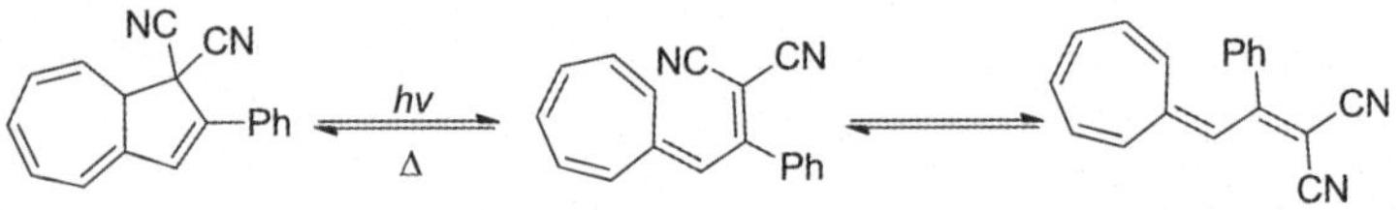

(xi) MeOOC Ph (i) H_2, Pd/$BaSO_4$, quinoline (ii) toluene, Δ

Ph H H H H H H H COOMe + MeOOC Ph

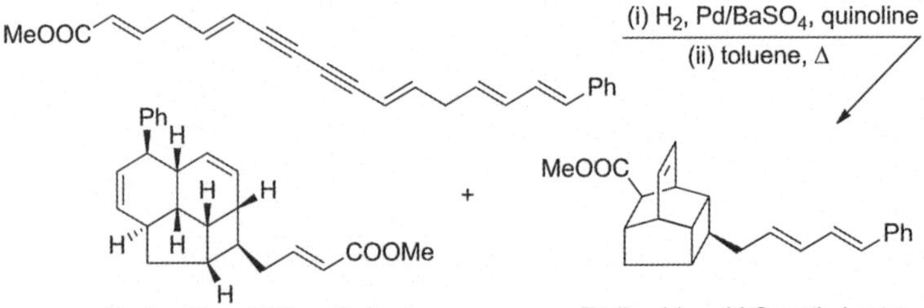

Endiandric acid B methyl ester

Endiandric acid C methyl ester

(xii) $h\nu$

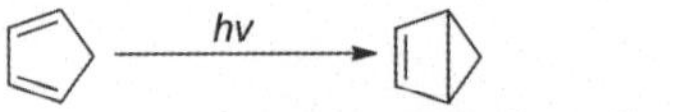

relatively stable thermally

(xiii) Δ

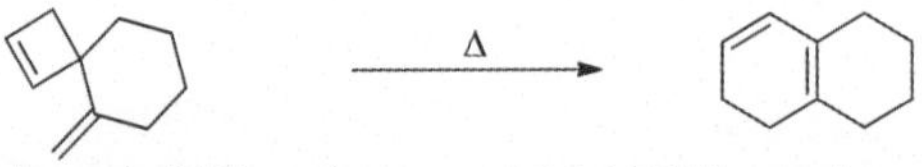

5-Methylenespiro[3.5]non-1-ene

1,2,3,4,5,6-Hexahydronaphthalene

(xiv) Me * O O Δ Me O O *

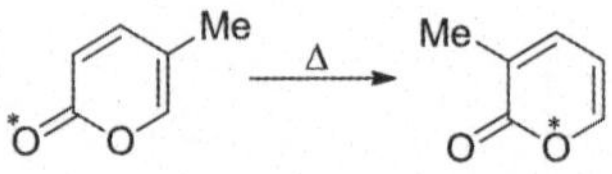

(xv) * 100 °C *

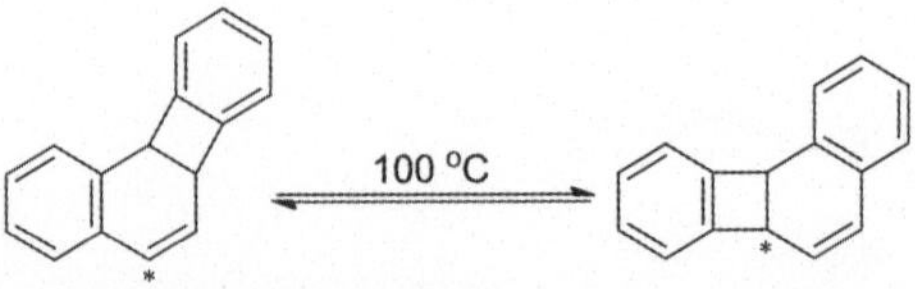

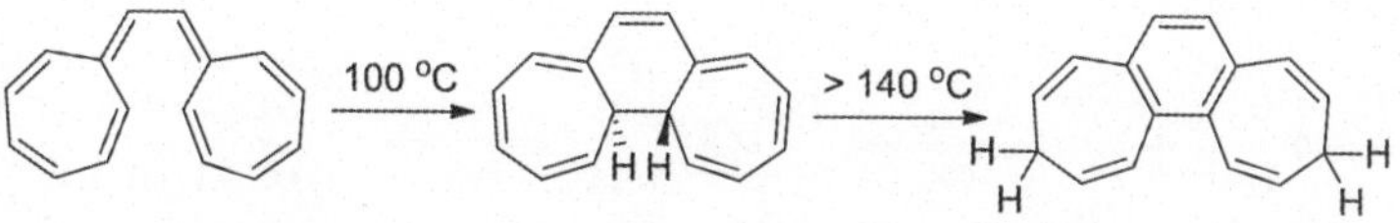

Sol 15. (i) Pyrolysis of bicyclo[6.2.0]deca-2,4,6,9-tetraene to *trans*-9,10-dihydronaphthalene takes place in two steps. In the first step, a 4n π-electron system (eight or four electrons) undergoes thermal conrotatory electrocyclic ring opening. In the second step, (4n + 2) π-electron system undergoes thermal disrotatory electrocyclization to give *trans*-9,10-dihydronaphthalene.

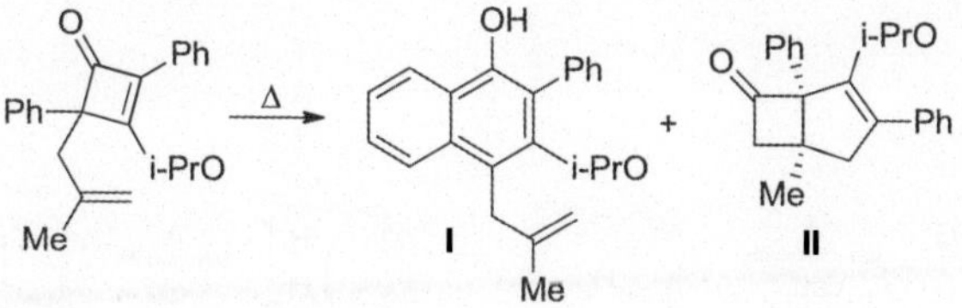

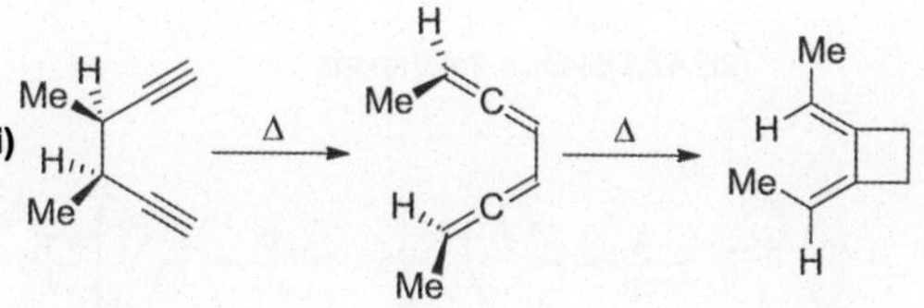

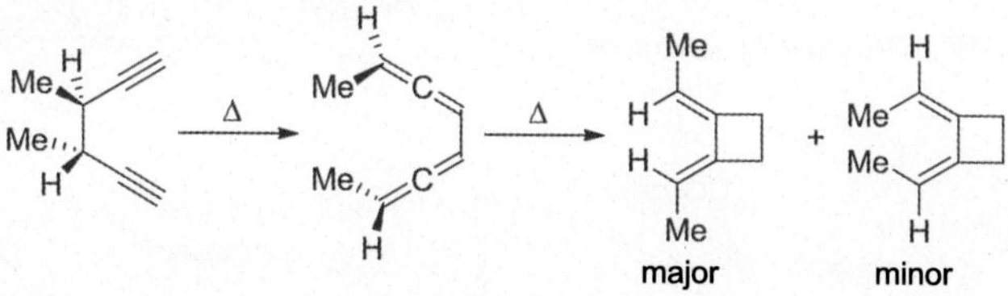

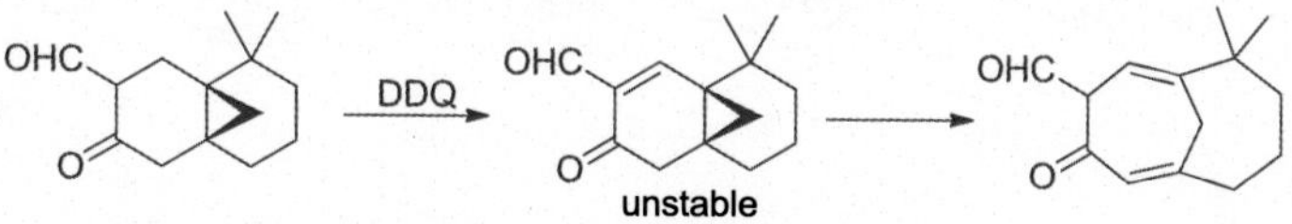

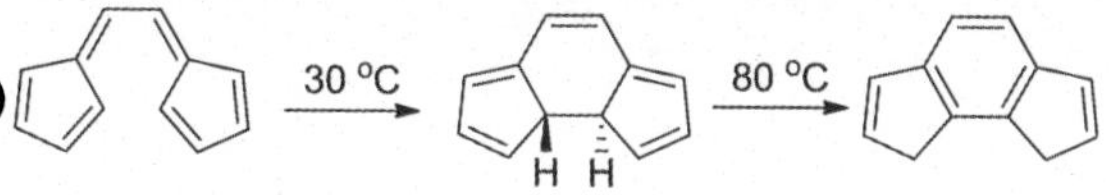

identical intermediate

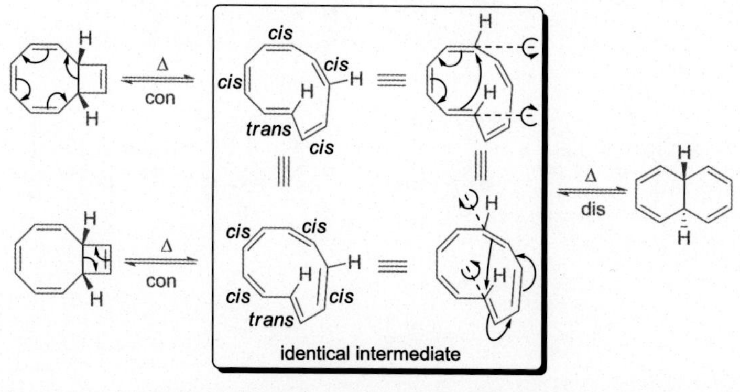

(ii) H H Δ con H H Δ dis H H

(2*E*,4*Z*,6*E*)-Octa-2,4,6-triene

(iii) Δ con H H Δ dis H H

(2*E*,4*Z*,6*Z*,8*E*)-Deca-2,4,6,8-tetraene

MeOOC COOMe

COOMe MeOOC

Δ Diels-Alder

Δ retro Diels-Alder

COOMe COOMe +

H H Δ con H H H H

(2*E*,4*E*)-Hexa-2,4-diene

(iv) D D D D CH_2 CH_2 Δ con D D D D H H H H Δ con D_2C D_2C H H H H

(v) D D Δ con D cis trans cis D D cis trans D H cis H Δ con D D H H D H H D

D D Δ con trans D cis cis D D D H Δ dis D D H

(vi) 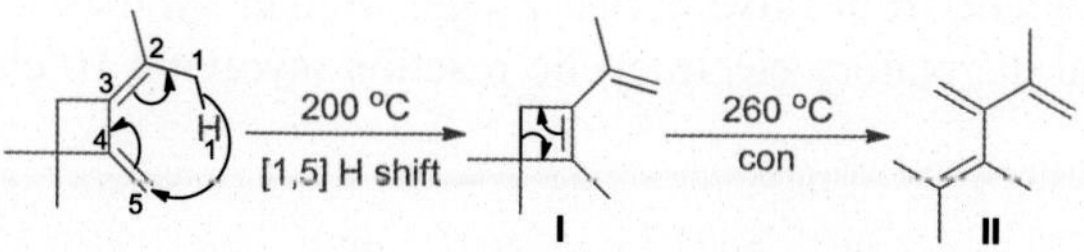 200 °C [1,5] H shift → I → 260 °C con → II

(vii) 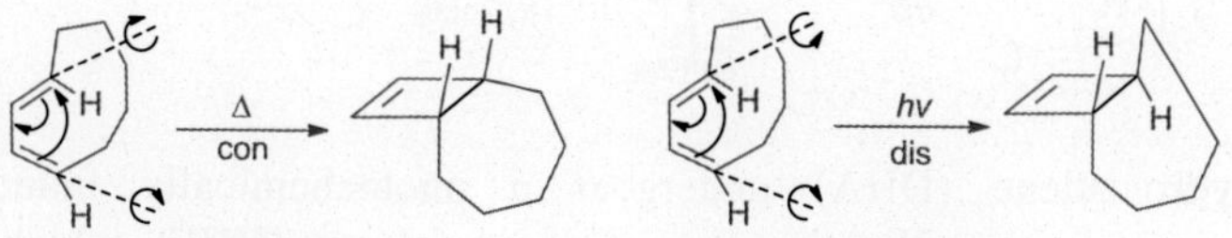 Δ con; *hv* dis

ground state HOMO: Ψ_2
C_2 symmetry, con

excited state HOMO: Ψ_3
m symmetry, dis

(viii) Irradiation of (*1Z,3Z,5E*)-cyclonona-1,3,5-triene by 254 nm light establishes a 60:40 photoequilibrium with the *cis*-fused bicyclononadiene. This equilibrium is shifted to the right by longer-wavelength 300 nm light, which also induces further electrocyclization to tricyclo[4.3.0.0]non-3-ene. As a rule, cyclohexadiene systems absorb light at longer wavelength than similar noncyclic hexatriene systems.

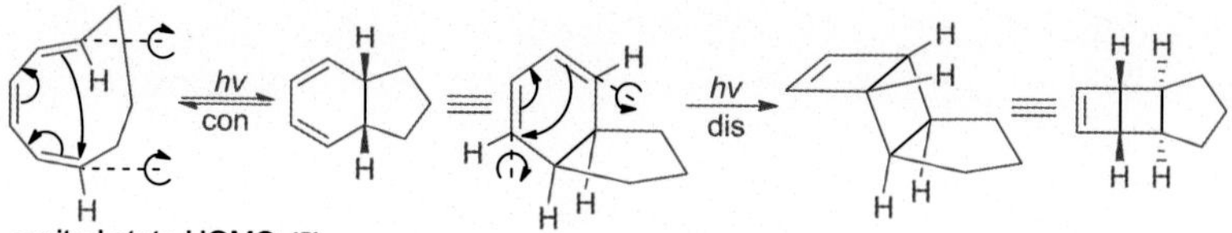

excited state HOMO: Ψ_4
C_2 symmetry, con

excited state HOMO: Ψ_3
m symmetry, dis

Alternative mode of disrotatory ring closure gives less stable product.

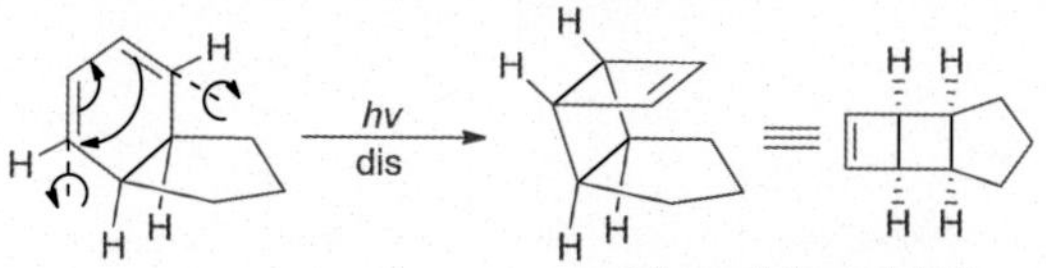

not formed (less stable)

On heating, (*1Z,3Z,5E*)-cyclonona-1,3,5-triene, undergoes disrotatory electrocyclization to give *trans*-fused bicyclononadiene.

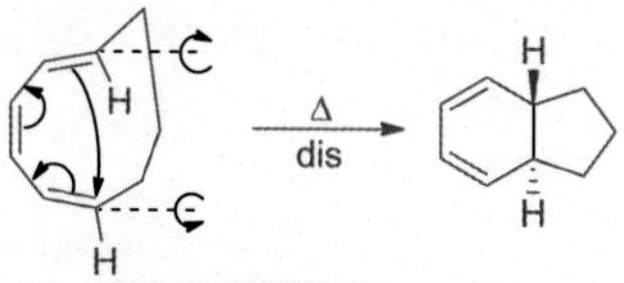

ground state HOMO: Ψ_3
m symmetry, dis

(ix) Formation of azulene from fulvene (the Ziegler-Hafner synthesis), is most probably a thermal disrotatory electrocyclic reaction involving 10 electrons.

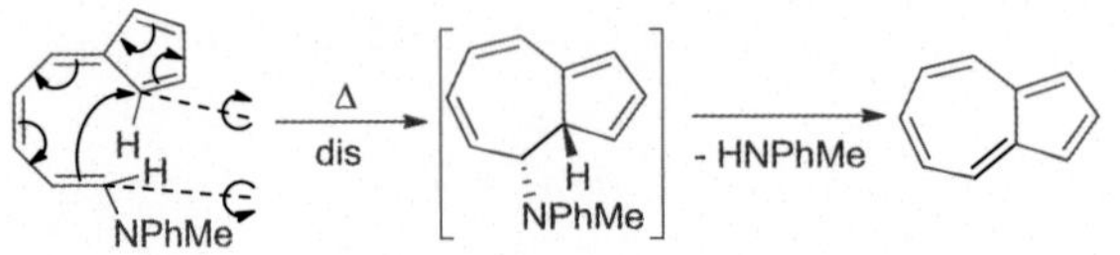

H
H
NPhMe
Δ
dis
H
NPhMe
- HNPhMe

(x) Dihydroazulene (DHA) undergoes a photochemically induced 10-electron retro-electrocyclization to vinyl heptafulvene (VHF), which in turn undergoes a thermally assisted ring closure back to DHA. The initially formed *s-cis*-VHF is in equilibrium with the generally more stable *s-trans*-VHF.

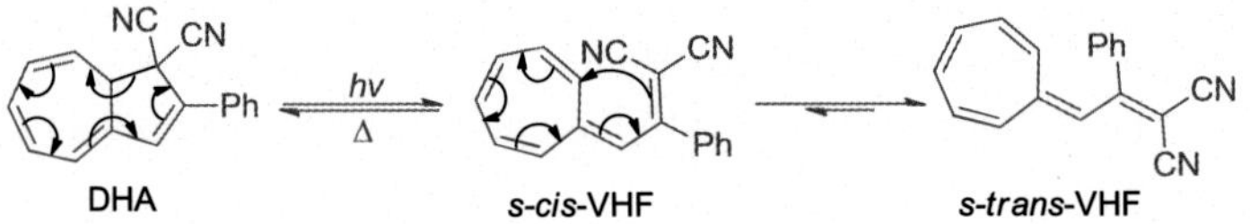

NC CN
Ph
DHA
hv
Δ
NC CN
Ph
s-cis-VHF
Ph
CN
CN
s-trans-VHF

(xi)

Ph
COOMe
H_2, Pd/$BaSO_4$, quinoline
Lindlar's catalyst
COOMe
H
H
Ph
ground state HOMO: Ψ_4
C_2 symmetry, con
Δ
con

H
COOMe
Ph
H
ground state HOMO: Ψ_3
m symmetry, dis
Δ
dis
H
COOMe
H
Ph

Ph
H
H
H
H
H
H
COOMe
Endiandric acid B methyl ester
Δ
[4 + 2]
Ph
H
H
H
COOMe
Endiandric acid F methyl ester

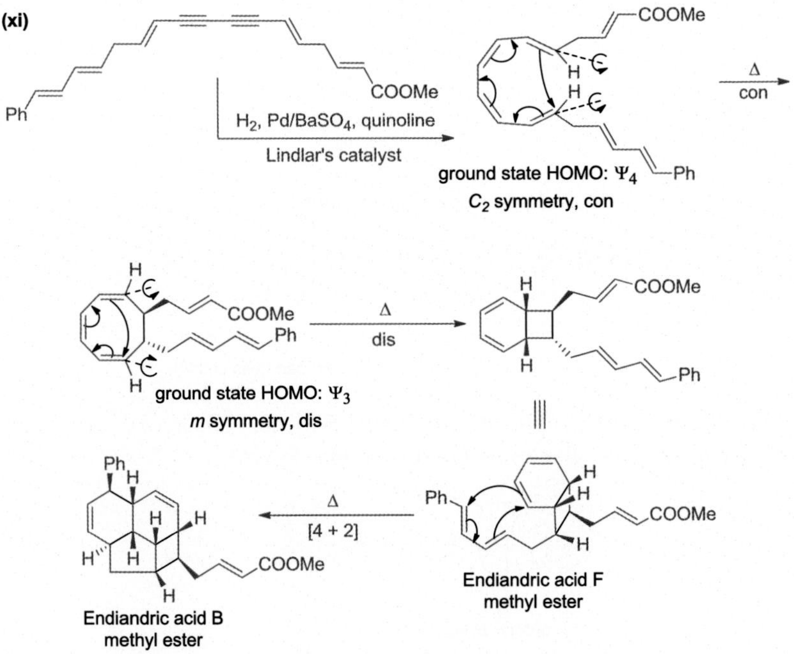

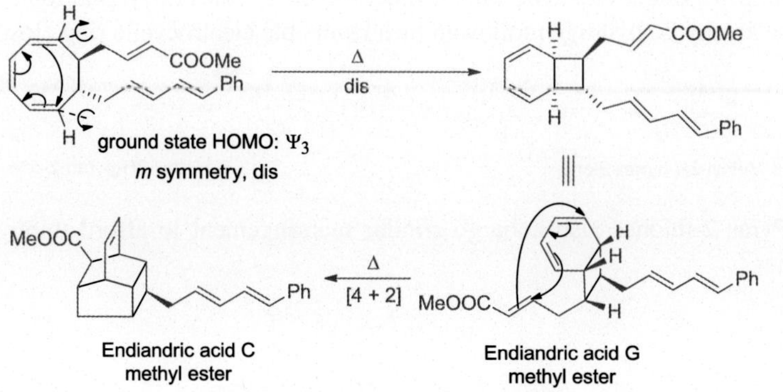

(xii) Under photochemical conditions, cyclopenta-1,3-diene undergoes disrotatory ring closure to give bicyclo[2.1.0]pent-2-ene, which at first sight might seem incapable of isolation because of the possibility of immediate rearrangement to cyclopenta-1,3-diene. This rearrangement does occur, but not so fast as to preclude isolation of the substance. This is due to the fact that under thermal conditions, ring opening should occur by a conrotatory process, which places a *trans*-double bond in the five-membered ring, which is geometrically impossible. Therefore, ring opening occurs by a higher energy nonconcerted pathway.

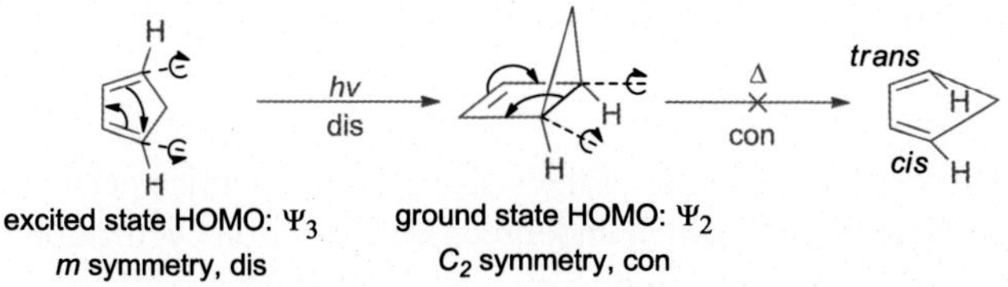

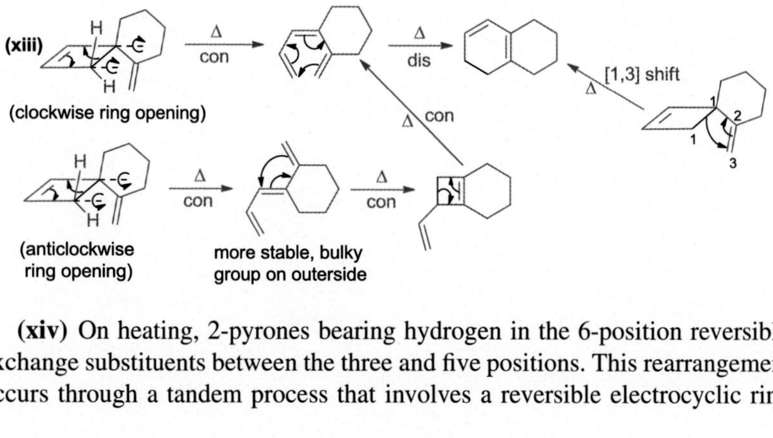

(xiv) On heating, 2-pyrones bearing hydrogen in the 6-position reversibly exchange substituents between the three and five positions. This rearrangement occurs through a tandem process that involves a reversible electrocyclic ring opening to a ketene aldehyde, which undergoes reversible [1,5] sigmatropic shift of the aldehydic hydrogen followed by a reversible electrocyclic ring closure.

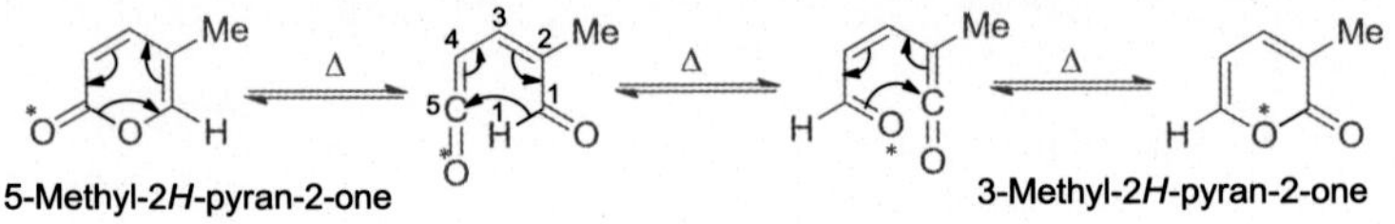

5-Methyl-2H-pyran-2-one

3-Methyl-2H-pyran-2-one

Pyran-2-thiones also undergo similar rearrangement to afford thiopyran-2-ones.

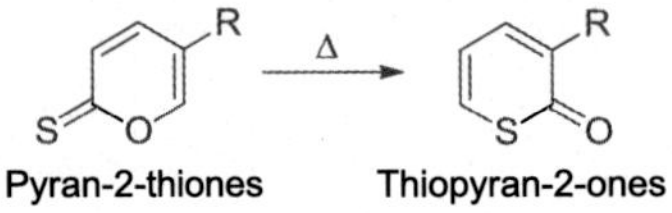

Pyran-2-thiones Thiopyran-2-ones

(xv) As proved by deuterium-labeling experiments, 2,3;7,8-dibenzobicyclo [4.2.0]octa-2,4,7-triene undergoes a degenerate rearrangement at 100 °C. The reaction involves 6π-electron electrocyclic ring opening to dibenzocyclooctatetraene, which again undergoes ring closure. Here apparently the two fully aromatic benzene rings in the tetracyclic compound make it more stable than the "less" aromatic benzene rings of dibenzocyclooctatetraene.

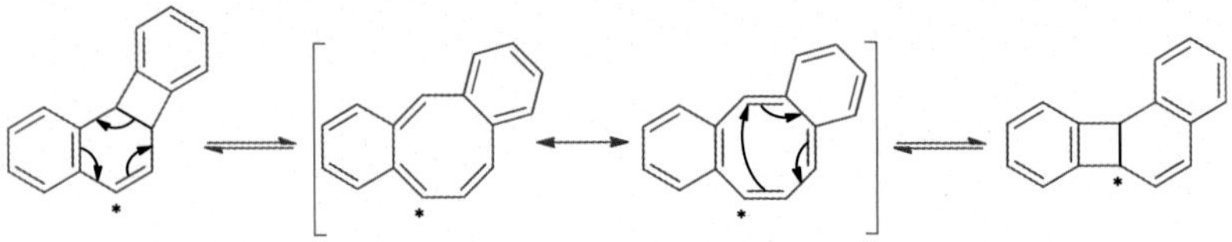

(xvi) Thermal ring opening of the starting cyclobutenone generates a vinyl ketene intermediate (**III**), which undergoes a 6π-electron cyclization involving a phenyl group at the 4-position to give the naphthalene derivative (**I**). However, as an allyl group is also present at the 4-position, **III** also undergoes a [2 + 2] cycloaddition to give **II**.

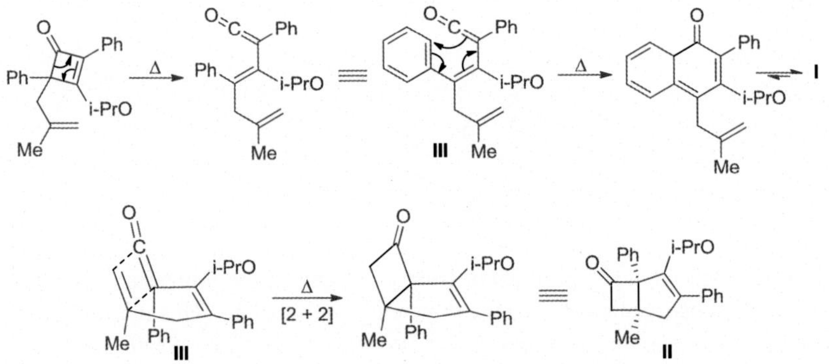

(xvii) Me H 1 2 3 H Me 1 2 3 — Δ [3,3] shift → H Me C H C Me — Δ con → Me H Me H

(xviii) Me H 1 2 3 Me H 1 2 3 — Δ [3,3] shift → H Me C Me C H — Δ con → Me H H Me major

H Me C Me C H — Δ con → H Me Me H minor

(xix) The compound obtained after oxidation with DDQ can tautomerize to give the cyclohexadiene system containing $(4n+2)$ π-electrons, which undergoes disrotatory electrocyclic ring opening under thermal conditions to give a seven-membered triene system. The triene system is finally converted into its more stable tautomeric form.

OHC O — DDQ → OHC O ⇌ OHC HO

OHC O ⇌ OHC HO ← Δ dis — OHC HO

(xx) On heating, pentafulvalene (**I**) gives 8a,8b-dihydro-as-indacene (**II**). The reaction is a 12π-electron cyclization and should proceed in a conrotatory manner and, therefore, results in the *trans* fusion. At 80 °C, **II** undergoes two [1,5] hydrogen shifts to give 1,8-dihydro-as-indacene (**III**), which is thermodynamically more stable due to the benzenoid system.

H H **I** — 30 °C con 12π-electron cyclization → 3 2 1 4 5 H H 1 1 2 3 1 5 4 **II** — 80 °C [1,5] H shift → **III**

(xxi) Under thermal conditions, heptafulvalene undergoes a 16π-electron cyclization in a conrotatory manner followed by two [1,5] hydrogen shifts to give a thermodynamically stable benzenoid system.

30 °C, con, 16π-electron cyclization

80 °C, [1,5] H shift

2.5 ELECTROCYCLIC REACTIONS OF IONIC SPECIES

2.5.1 Three-Atom Electrocyclizations

Woodward–Hoffmann orbital symmetry rules can be applied to the charged systems as well. The conversion of a **cyclopropyl cation** to an allylic cation is the simplest one, which involves only 2π-electrons (Figure 2.13). This is an electrocyclic reaction of $(4n + 2)$ type $(n = 0)$ and should, therefore, be a disrotatory process.

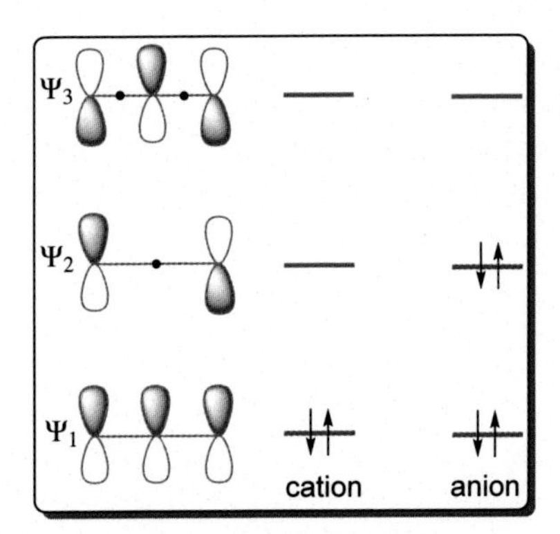

FIGURE 2.13 Molecular orbitals and electron occupancy in allylic cation and anion.

Due to the strain imposed by the three-membered ring, the cyclopropyl cation is not a stable intermediate and electrocyclic ring opening occurs readily. Therefore, in the solvolysis of cyclopropyl tosylate in acetic acid, allyl acetate is obtained rather than cyclopropyl acetate. Solvolysis reactions of other cyclopropyl halides, sulphonates, and diazotization of cyclopropylamine in aqueous solution also give the allylic products.

Solvolysis can proceed via cyclopropyl cation followed by rearrangement to allyl cation (case-I), or it can proceed directly to allyl cation (case-II). However, experimental observations indicate that ring opening is a concerted process in conjunction with the rupture of the bond to the leaving group, i.e., case-II (Scheme 2.10).

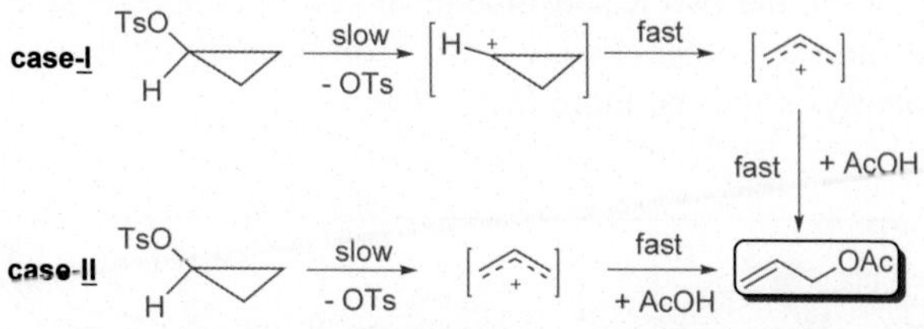

SCHEME 2.10 Solvolysis of cyclopropyl tosylate.

Let us consider the solvolysis of some cyclopropyl derivatives as shown in Scheme 2.11:

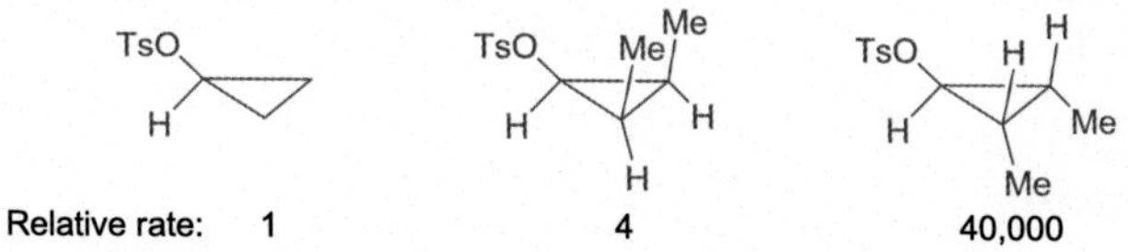

SCHEME 2.11 Relative rates of solvolysis of some cyclopropyl tosylates.

Breaking of the C−X bond and cyclopropane ring opening are concerted process and involves the backside attack on the C−X bond by the electrons of the opening σ-bond of the cyclopropane ring (just like the backside attack by a nucleophile in an S_N2 reaction). As the cationic transition state has two electrons, therefore, under thermal conditions the disrotatory modes of ring opening are symmetry-allowed. There are two ways of disrotatory ring opening, however, due to the backside attack, ring opening takes place only by one mode (*torquoselectivity*). Hence, when the two R substituents are on the same side of X, they rotate toward each other (Figure 2.14). Consequently, if these substituents are bulky, solvolysis will take place slowly due to steric reasons.

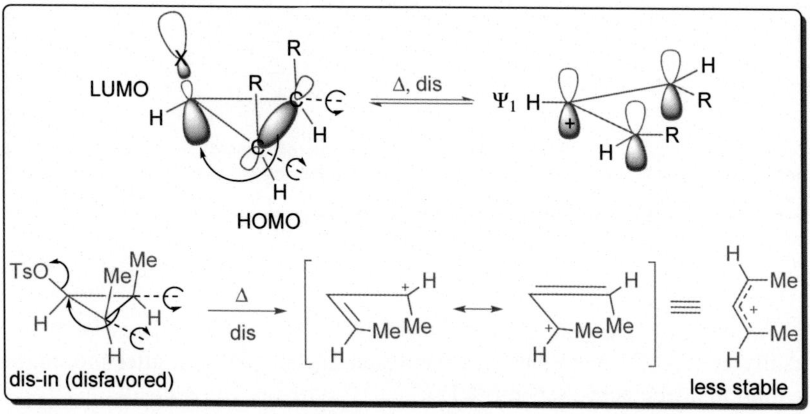

FIGURE 2.14 Solvolysis of disubstituted cyclopropyl tosylate via dis-in mode.

Similarly, when the two R substituents are on the opposite side of X, they rotate away from each other (Figure 2.15). Consequently, if these substituents are bulky, solvolysis will be more facile due to steric reasons.

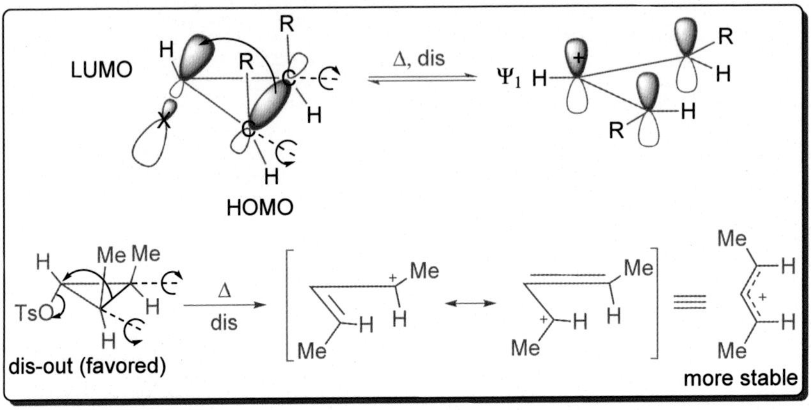

FIGURE 2.15 Solvolysis of disubstituted cyclopropyl tosylate via dis-out mode.

However, when the two R substituents are the part of a small ring and are on the same side of X (i.e., the leaving group is in the *endo* conformation), opening of the cyclopropane ring takes readily as it leads to the formation of a more stable ring with *cis*-double bond. On the other hand, when the two R substituents are on the opposite side of X (i.e., the leaving group is in the *exo* conformation), solvolysis will not take place as it leads to the formation of a small ring with *trans*-double bond, the formation of which is inhibited in the transition state (Figure 2.16).

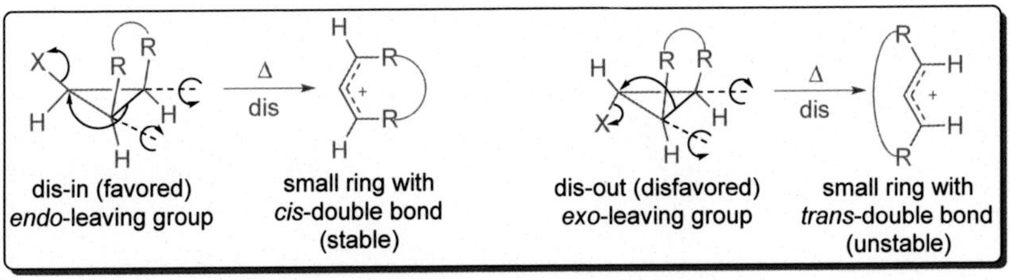

For example, the *endo*-7-chlorobicyclo[4.1.0]heptane undergoes solvolysis readily at 125 °C; its epimer was recovered unchanged even after prolonged heating in acetic acid (Scheme 2.12).

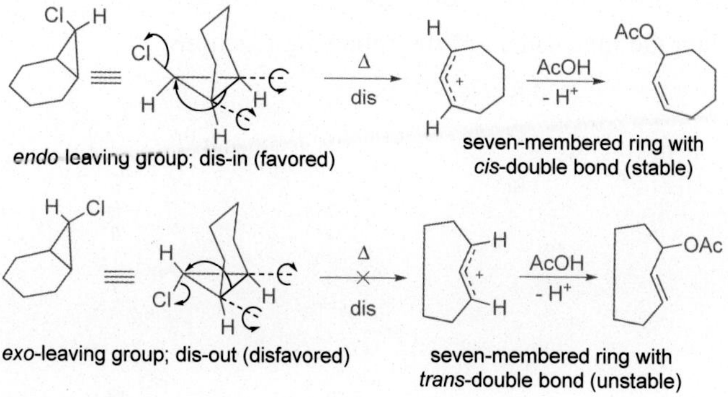

SCHEME 2.12 Solvolysis of *endo*-7-chlorobicyclo[4.1.0]heptane and its epimer.

Cyclopropyl anion: The HOMO of the reactant in ground state is Ψ_2 having C_2 symmetry, therefore, under thermal conditions cyclization will take place in a conrotatory manner (Figure 2.17).

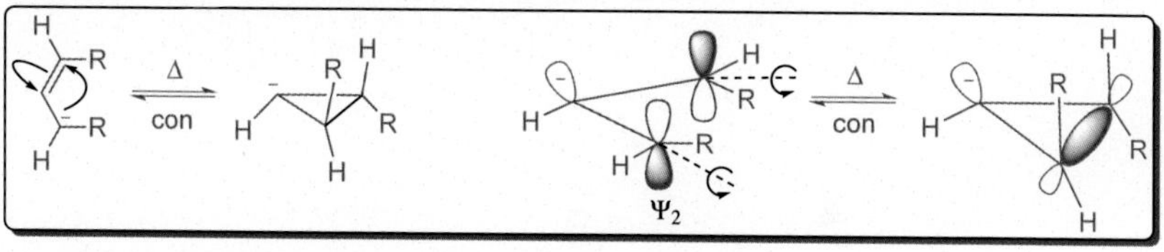

FIGURE 2.17 Thermal cyclization of cyclopropyl anion in a conrotatory manner.

Thermolysis or photolysis of suitably substituted aziridines: Aziridines undergo isomerization to azomethine ylides under thermal or photochemical conditions. It is observed that thermal isomerization involves a conrotatory ring opening, whereas a disrotatory ring opening takes place under photochemical conditions. Such a behavior is expected as aziridines are isoelectronic with the cyclopropyl anion and follow orbital symmetry rules. In a typical example, the *cis*-dicarboxylic acid ester undergoes conrotatory ring opening on heating to give the *trans*-azomethine ylide and disrotatory ring opening on irradiation to yield the *cis*-azomethine ylide. The *trans*-dicarboxylic acid ester behaves in a similar manner (Scheme 2.13).

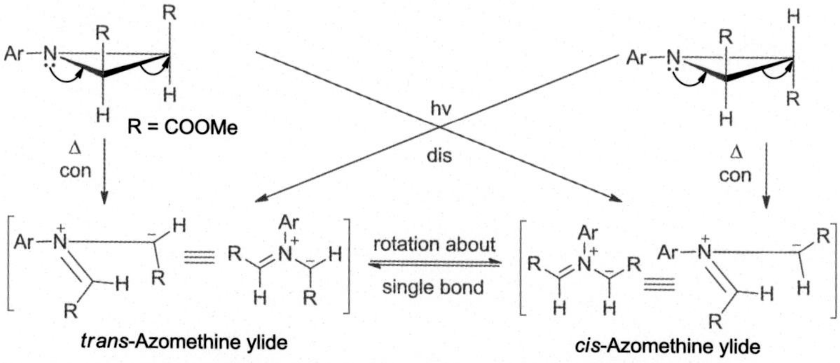

SCHEME 2.13 Thermolysis or photolysis of suitably substituted aziridines.

2.5.1.1 Solved Problems

Q 1. Explain the mechanism of the following reactions.

(i) [reaction scheme: cyclopentene + :CClBr → I + II; I —AgOH, Δ→ III; II —AgOH, Δ→ IV]

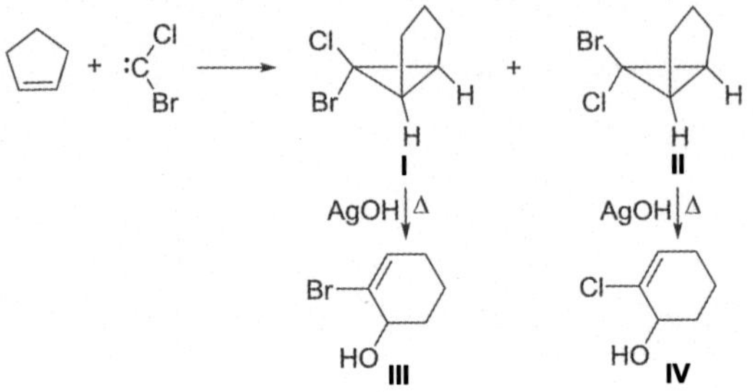

(ii) [reaction scheme: norbornene + :CClF → two cyclopropane adducts; each —Δ→ ring-expanded product]

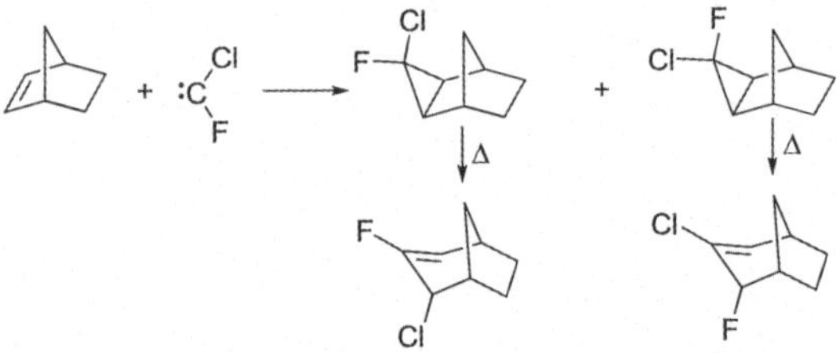

Sol 1. (i) First step of the reaction involves cheletropic addition of chlorobromocarbene to cyclopentene to give two isomers **I** and **II**. The two isomers can be distinguished on the basis of difference in *endo*- and *exo*-leaving groups. These results can be explained on the basis that the *endo* group is preferentially expelled by the dis-in mode of opening of cyclopropane σ-bond to give the more stable *cis*-cyclohexene. The alternative dis-out mode of opening of cyclopropane σ-bond will generate the less stable *trans*-cyclohexene.

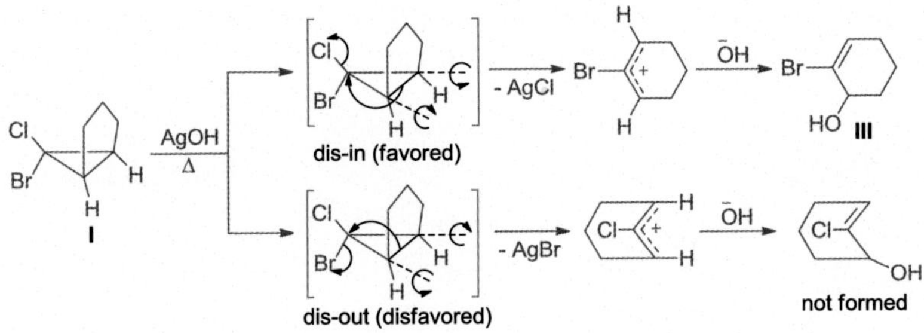

[reaction scheme: I —AgOH, Δ→ dis-in (favored) —(- AgCl)→ allyl cation —$\bar{O}H$→ III; dis-out (disfavored) —(- AgBr)→ cation —$\bar{O}H$→ not formed]

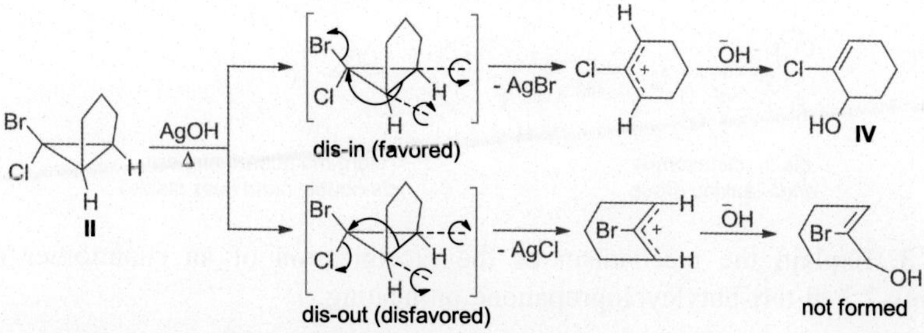

(ii) In both the cases, the *endo* group is preferentially expelled by the *dis-in* mode of opening of cyclopropane σ-bond and then recombines at the adjacent position to give the more stable *cis*-cycloheptene. The alternative *dis-out* mode of opening of cyclopropane σ-bond will generate the less stable *trans*-cycloheptene.

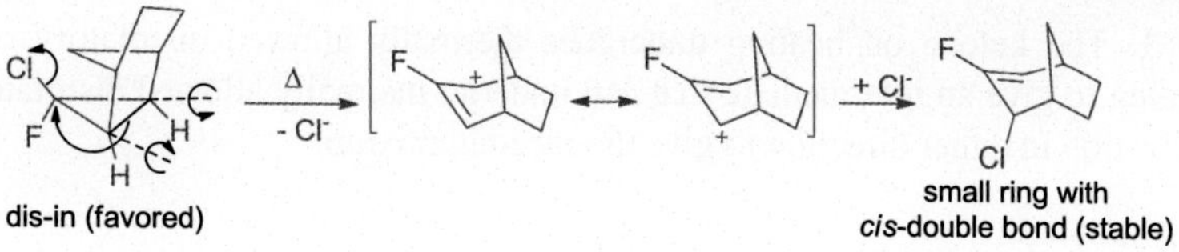

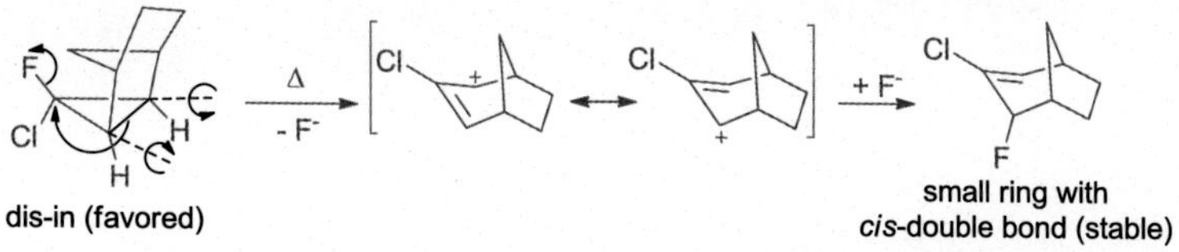

Q 2. Give the products of the following reactions and explain which one will form faster.

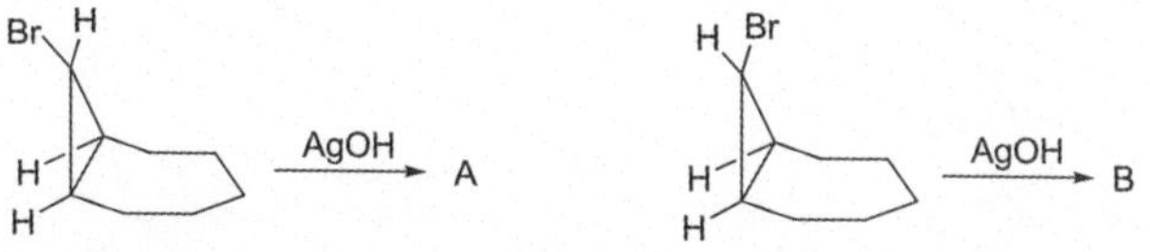

Sol 2. *Trans*-cyclooctene system is more stable than the *cis*-isomer; therefore, in this case the *exo*-isomer undergoes solvolysis at a much faster rate than the corresponding *endo*-isomer.

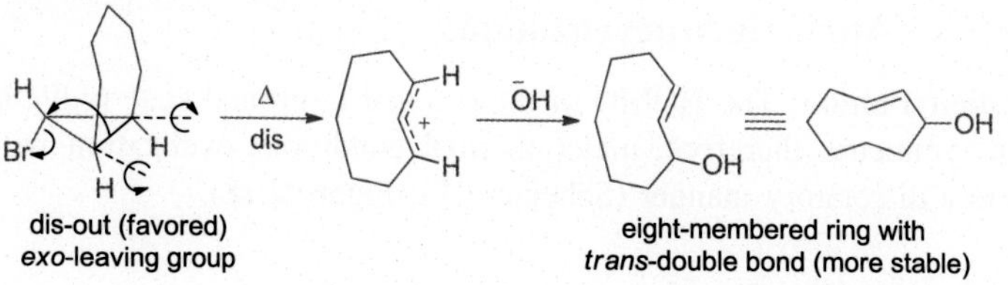

dis-in (disfavored)
endo-leaving group

eight-membered ring with
cis-double bond (less stable)

Q 3. Explain the mechanism of the racemization of an enantiomer of *trans*-2,3-di-tert-butylcyclopropanone on heating.

80 °C

(2*S*,3*S*)-2,3-di-*tert*-Butylcyclopropanone (2*R*,3*R*)-2,3-di-*tert*-Butylcyclopropanone

Sol 3. The ketone on heating undergoes thermally allowed disrotatory ring opening to give an intermediate that can undergo thermally allowed disrotatory cyclization in either direction to give the racemic mixture

2.5.2 Five-Atom Electrocyclizations

Pentadienyl anion: The HOMO of the reactant in ground state is Ψ_3 having mirror symmetry, therefore, under thermal conditions cyclization will take place in a disrotatory manner (Scheme 2.14; Figure 2.18).

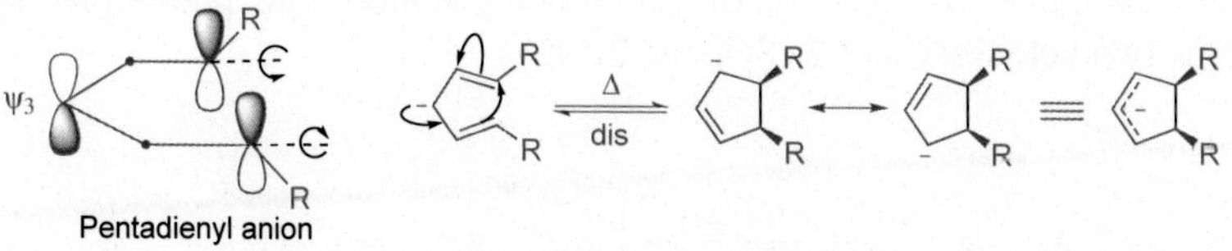

SCHEME 2.14 Thermal cyclization of pentadienyl anion in a disrotatory manner.

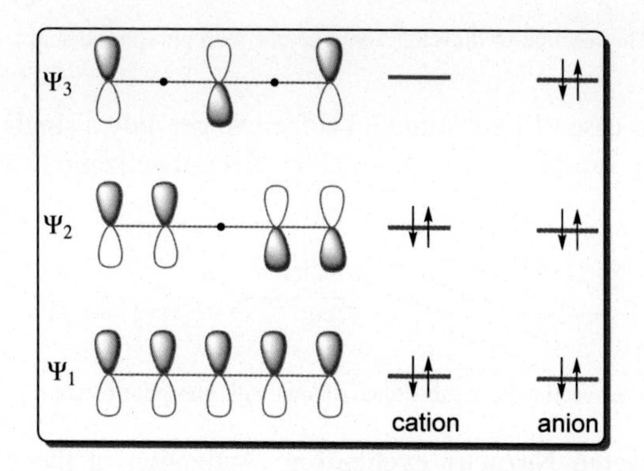

FIGURE 2.18 Molecular orbitals and electron occupancy in pentadienyl cation and anion.

Pentadienyl cation: In this case, the HOMO of the reactant in ground state is Ψ_2 having C_2 symmetry, therefore, under thermal conditions cyclization will take place in a conrotatory manner (Scheme 2.15).

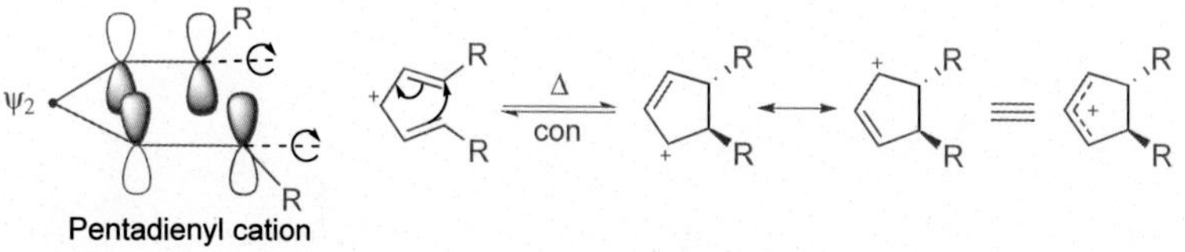

SCHEME 2.15 Thermal cyclization of pentadienyl cation in a conrotatory manner.

The **Nazarov cyclization reaction** may be defined as an acid (protic or Lewis) induced cationic 4π-electrons electrocyclic ring closure reaction of α,α′-divinyl ketones to form cyclopentenones.

Mechanism: Activation of the ketone by an acid catalyst generates a pentadienyl cation, which undergoes a thermally allowed 4π-electron conrotatory electrocyclization (Scheme 2.16). This generates an oxyallyl cation, which undergoes an elimination reaction to lose a β-hydrogen. The subsequent tautomerization of the enolate produces the cyclopentenone product.

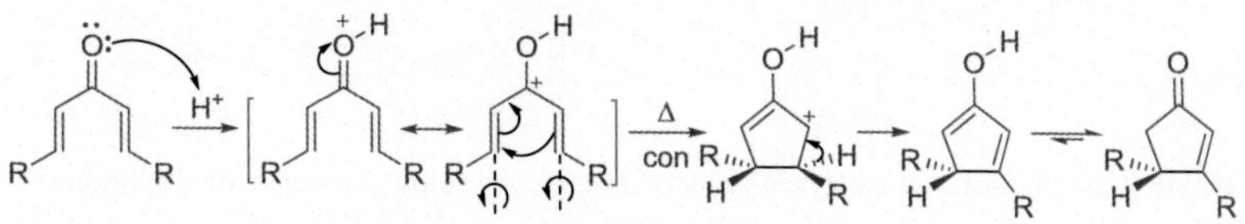

SCHEME 2.16 Mechanism of Nazarov cyclization reaction.

For example, treatment of dicyclohexenyl ketone with phosphoric acid affords two ketones **1** and **2** (Scheme 2.17).

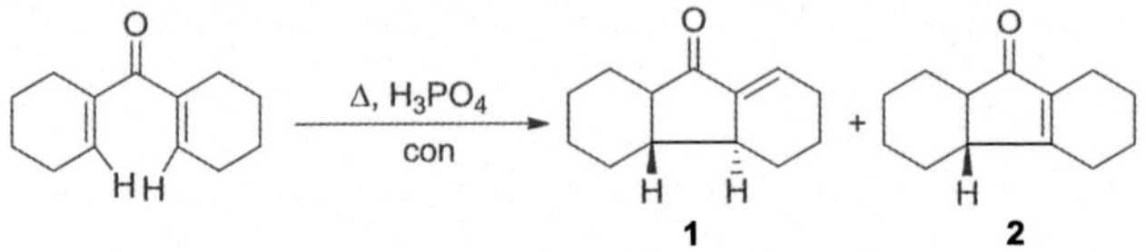

SCHEME 2.17 The reaction of dicyclohexenyl ketone with phosphoric acid.

However, in case of a substituted ketone, we get only a single product with stereochemistry resulting from conrotatory electrocyclization (Scheme 2.18).

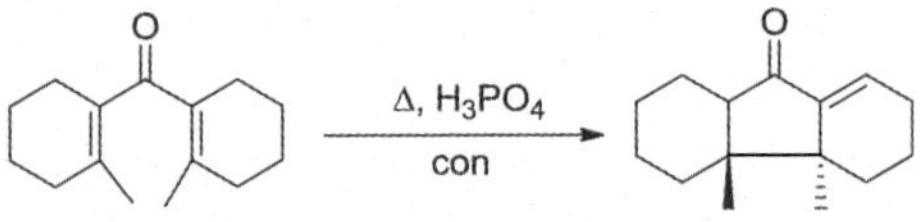

SCHEME 2.18 The reaction of a substituted ketone with phosphoric acid.

Silicon-directed Nazarov cyclization: Activation of the ketone **1** by a Lewis acid catalyst generates a pentadienyl cation, which undergoes a thermally allowed 4π-electron conrotatory electrocyclization (Scheme 2.19). This generates a silicon-stabilized cation, which undergoes an elimination reaction of silyl group to give the cyclopentadienol. Subsequent tautomerization of cyclopentadienol produces the cyclopentenone product **2**.

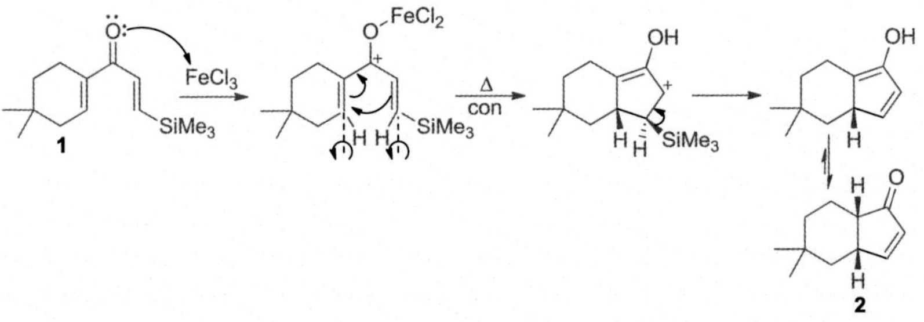

SCHEME 2.19 Silicon-directed Nazarov cyclization reaction in presence of Lewis acid.

However, in the absence of silyl group (ketone **3**) that hydrogen will be lost preferably, which leads to the formation of more substituted alkene **4** (Scheme 2.20).

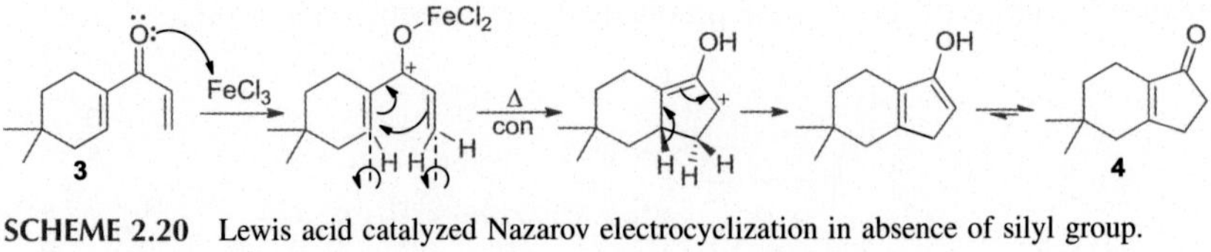

SCHEME 2.20 Lewis acid catalyzed Nazarov electrocyclization in absence of silyl group.

2.5.2.1 Solved Problems

Q 1. Explain the mechanism of the following reactions:

(i) n-BuLi, THF, - 78 °C; 35 °C

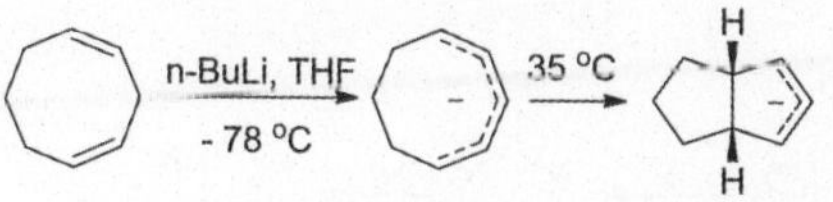

(ii) PhLi, - 130 °C

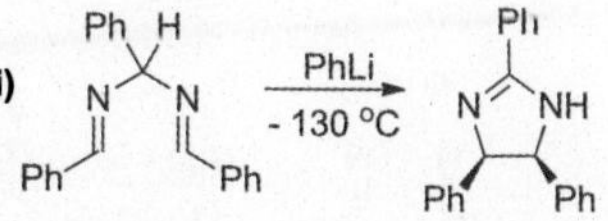

(iii) (i) KNH_2, liq. NH_3 (ii) H_2O

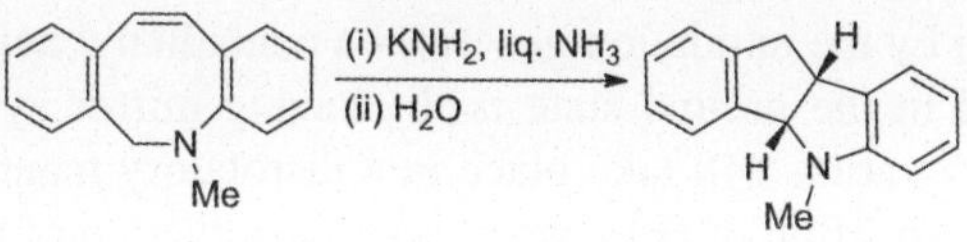

(iv) *hv*; Δ — R = H, Me

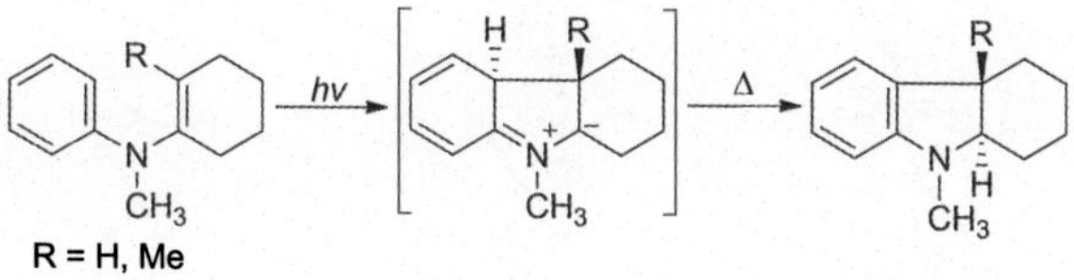

(v) $NaOAm^t$, Δ

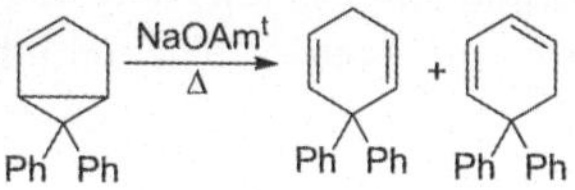

(vi) CH_3OH, H_2O, $HgSO_4$, H_2SO_4, Δ

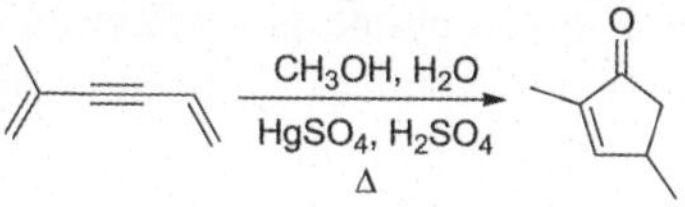

(vii) $TiCl_4$

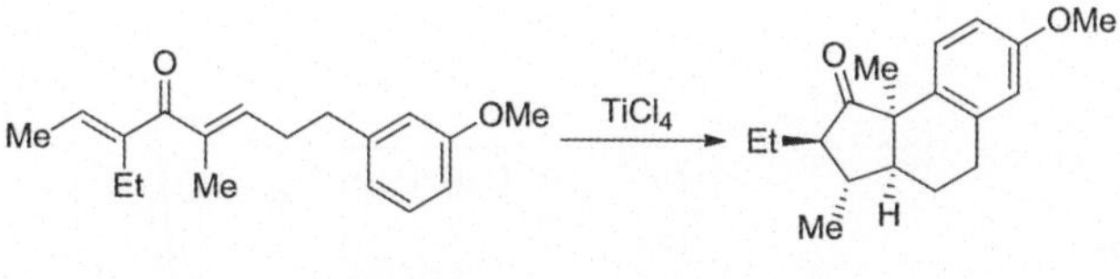

Sol 1. **(i)** In a pentadienyl anion system, the HOMO of the reactant in ground state is Ψ_3 having mirror symmetry, therefore, under thermal conditions, cyclization will take place in a disrotatory manner.

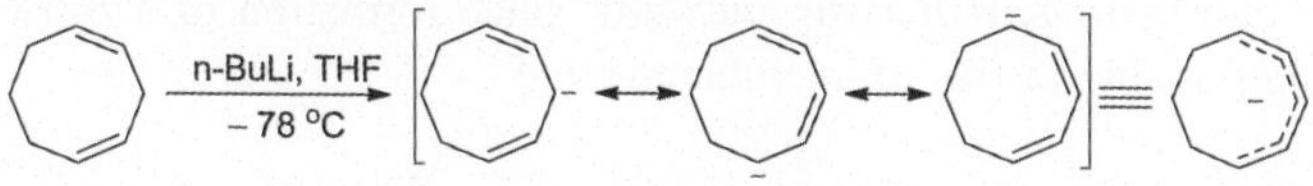

n-BuLi, THF, – 78 °C

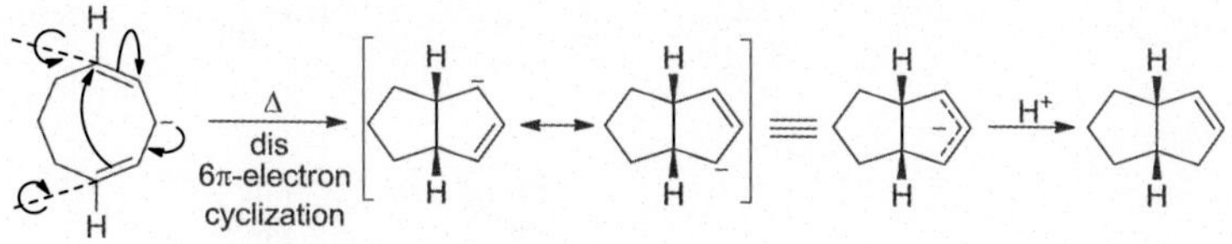

Δ, dis, 6π-electron cyclization; H^+

(ii) Hydrobenzamide on treatment with phenyllithium gives a pentadienyl anion system that closes in a disrotatory manner to give the thermodynamically less favored product having the two large *cis*-substituents.

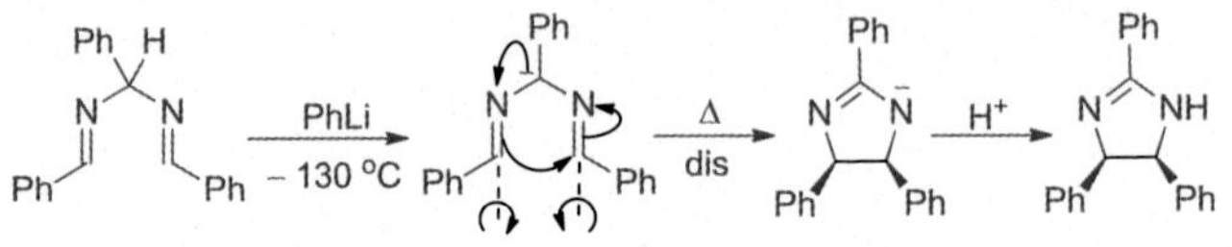

(iii) Abstraction of proton by the amide ion generates a pentadienyl anion system (**I**). The HOMO of **I** in the ground state is Ψ_3 having mirror symmetry, therefore, thermal cyclization will take place in a disrotatory manner to give another anion (**II**). Protonation of **II** by water with preservation of the benzene ring at the left hand results in the formation of the final product.

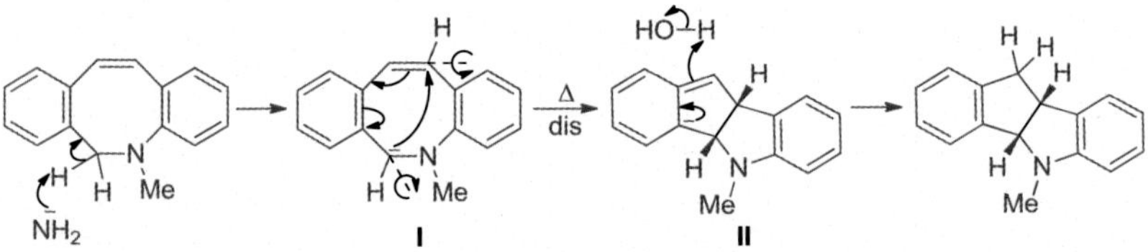

(iv) The given amine is isoelectronic with pentadienyl anion, and hence undergoes photochemical cyclization by a conrotatory course. The initially formed intermediate (**I**) on coming to the ground state follows a symmetry-allowed suprafacial [1,4] shift to generate a stable product.

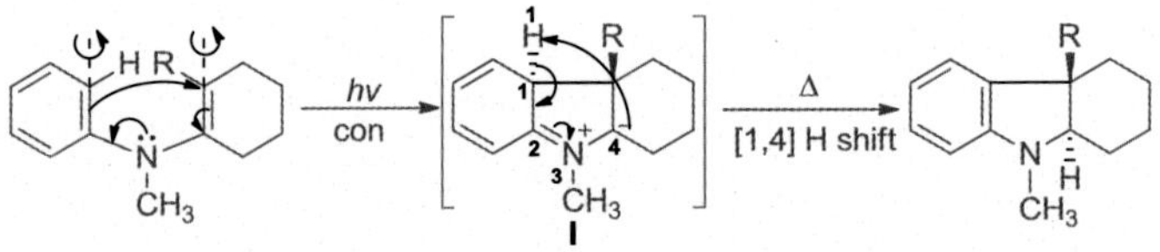

(v) The cyclopentenyl anion opens to provide the hexadienyl anion due to relief of strain in a cyclopropane ring. The thermal ring opening is disrotatory with the two hydrogens moving outward, since formation of a *trans*-double bond is impossible in the six-membered ring.

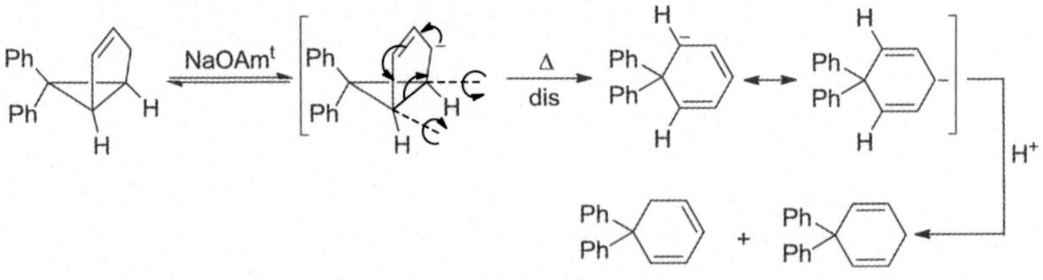

(vi) First step of the reaction involves an alkyne oxymercuration reaction to give the allylic olefin, which isomerizes in situ to form a divinyl ketone before ring closure to the cyclopentenone product (**Nazarov cyclization**).

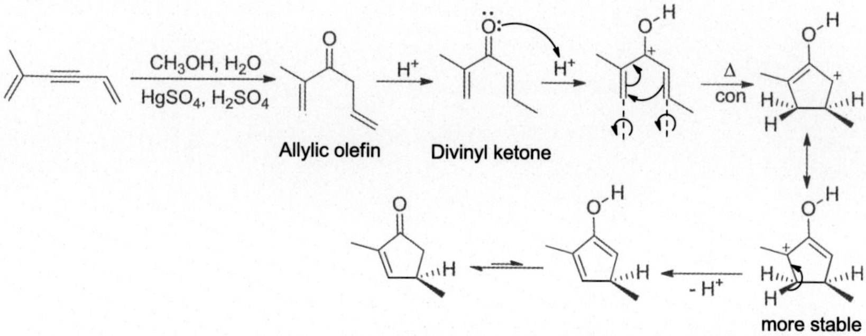

(vii) Activation of the dienone (**I**) by $TiCl_4$ generates a pentadienyl cation (**II**), which undergoes a thermally allowed 4π-electron conrotatory electrocyclization. This generates a cation (**III**), which undergoes an intramolecular electrophilic substitution reaction at the activated *para* position of the benzene ring to give the enolate (**IV**). Protonation followed by subsequent tautomerization of **V** produces the cyclopentenone product **VI**. In the last step, ethyl group moves *anti* to the neighboring methyl group to avoid steric congestion.

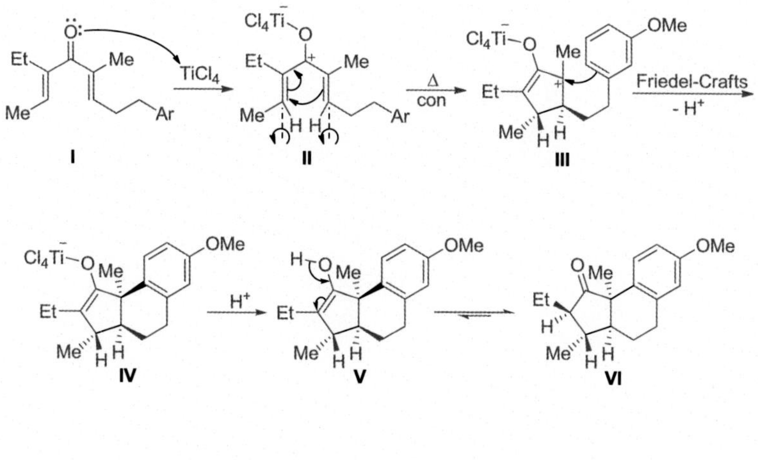

Chapter 3

Sigmatropic Rearrangements

Chapter Outline

Pericyclic Reactions. http://dx.doi.org/10.1016/B978-0-12-803640-2.00003-8
Copyright © 2016 Elsevier Inc. All rights reserved.

Many thermal (or photochemical) rearrangements involve the shifting of a σ-bond, flanked by one or more π-electron systems, to a new position [i,j] within the molecule in an uncatalyzed intramolecular process. Since it is rearrangement of a σ-bond, these reactions are called *sigmatropic rearrangements* of order [i,j]. These reactions are often classified with two numbers, i and j, set in brackets [i,j] and the system is numbered by starting with the atoms forming the migrating σ-bond. These numbers [i and j] indicate the new positions of the σ-bond whose termini are i − 1 and j − 1 atoms removed from the original bonded loci (Scheme 3.1).

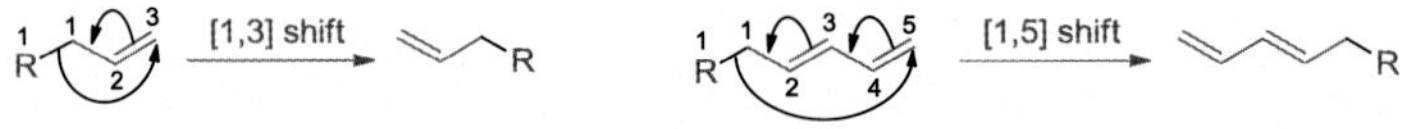

SCHEME 3.1 Sigmatropic rearrangements of order [1,3] and [1,5].

Very often, the migrating σ-bond is situated in between two π-bond systems as in the Cope and the Claisen rearrangements (Scheme 3.2).

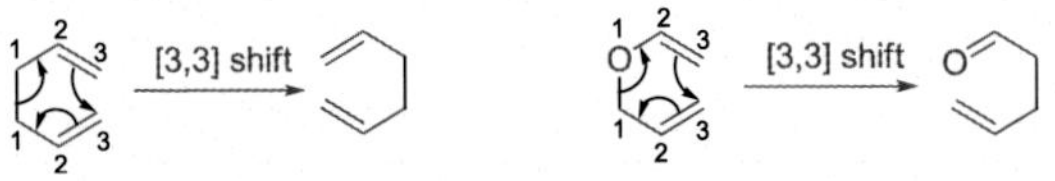

SCHEME 3.2 Sigmatropic rearrangements of order [3,3].

3.1 SUPRAFACIAL AND ANTARAFACIAL PROCESSES

Since a sigmatropic reaction involves the migration of a σ-bond across the π-electron system, there are two different stereochemical courses by which the process may occur. *When the migrating σ-bond moves across the same face of the conjugated system, it is called a* **suprafacial process** *whereas in* **antarafacial process** *the migrating σ-bond is reformed on the opposite π-electron face of the conjugated system.* The following [1,5] sigmatropic shifts illustrate both these processes and their stereochemical consequences (Figure 3.1).

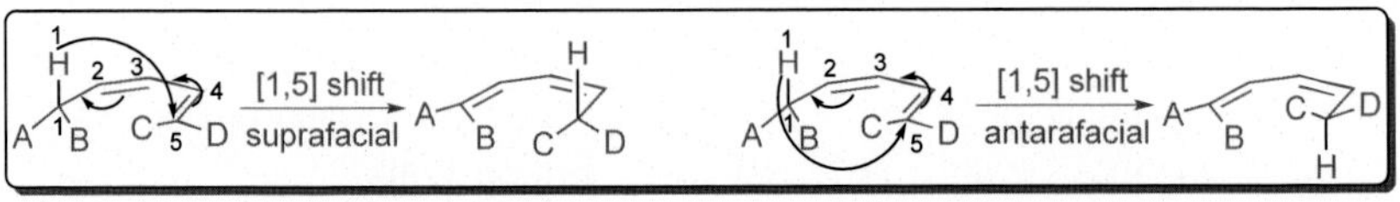

FIGURE 3.1 Different modes of hydrogen atom migration in [1,5] sigmatropic rearrangements.

The shift may occur with retention or inversion at the migrating group R. For example, the four possible stereochemical outcomes of 1,3-sigmatropic shift are illustrated below. Carbon atom C-4 can migrate to the top (suprafacial) or bottom (antarafacial) of C-1 and in the process may undergo retention or inversion. The four possibilities give rise to distinct products (Figure 3.2).

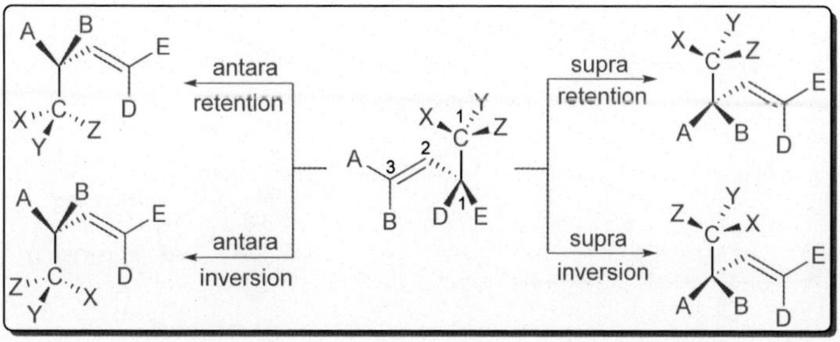

FIGURE 3.2 Different modes of carbon atom migration in [1,3] sigmatropic rearrangements.

3.2 ANALYSIS OF SIGMATROPIC REARRANGEMENTS OF HYDROGEN

For the analysis of sigmatropic rearrangements, the correlation diagrams are not relevant since it is only the transition state and not the reactants or products that may possess molecular symmetry elements. However, these reactions can be analyzed satisfactorily by using frontier molecular orbital (**FMO**) and perturbation molecular orbital (**PMO**) methods.

3.2.1 FMO Analysis of [1,3] Sigmatropic Rearrangements of Hydrogen

One of the ways to analyze sigmatropic rearrangements is to assume that the migrating bond undergoes homolytic cleavage resulting in the formation of a pair of radicals. As bonding characters are to be maintained throughout the course of the rearrangements, the most important bonding interaction will be between the highest occupied molecular orbitals (HOMOs) of the two species produced by this cleavage. This is to be expected, as it is these orbitals that contain the unpaired electrons. We shall illustrate this analysis by examining a [1,3] sigmatropic shift of hydrogen in which the homolytic cleavage results in the production of a hydrogen atom and allyl radical (Figure 3.3).

The ground state electronic configuration of allyl radical is $\Psi_1{}^2\ \Psi_2{}^1$. HOMO (Ψ_2) of this radical has opposite sign on the terminal lobes (C_2 symmetry). Suprafacial [1,3] hydrogen shift under thermal condition is forbidden because there is no question of inversion at this atom, which is bonded to the carbon atom through its spherically symmetrical 1*s*-orbital. Antarafacial [1,3] hydrogen shift is allowed only by the principles of orbital symmetry. The transition state is a highly contorted species and the reaction is forbidden because of the steric inhibition involved in such a process.

In the presence of light, lowest unoccupied molecular orbital (LUMO) of the ground state becomes HOMO of the excited state known as photochemical HOMO. The excited state electronic configuration of allyl radical is $\Psi_1{}^2\ \Psi_3{}^1$.

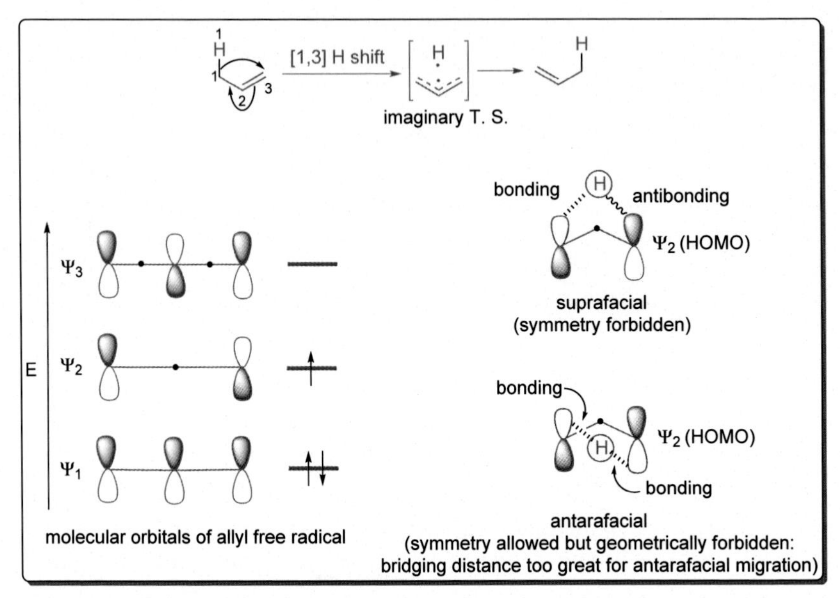

FIGURE 3.3 FMO analysis of [1,3] H shifts.

Photochemical HOMO (Ψ_3) of this radical has same sign on the terminal lobes (*m* symmetry), which permits suprafacial [1,3] hydrogen shift (Figure 3.4).

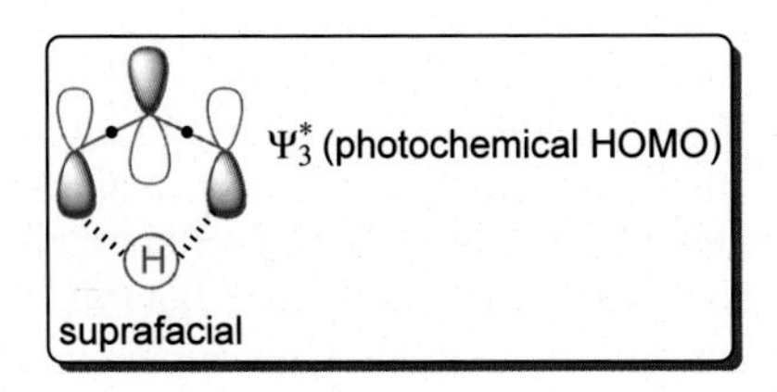

FIGURE 3.4 FMO analysis of [1,3] H shifts continued.

3.2.2 FMO Analysis of [1,5] Sigmatropic Rearrangements of Hydrogen

The [1,5] sigmatropic shift of hydrogen or deuterium atom occurs via a suprafacial pathway under thermal conditions. The reaction can be analyzed by assuming that a homolytic cleavage results in the production of a hydrogen atom and pentadienyl radical. The ground state electronic configuration of pentadienyl radical is $\Psi_1^2\ \Psi_2^2\ \Psi_3^1$. Since HOMO ($\Psi_3$) of this radical has similar sign on the terminal lobes (mirror symmetry), [1,5] suprafacial migration will be a thermally allowed process. The first excited state of the pentadienyl radical has the configuration $\Psi_1^2\ \Psi_2^2\ \Psi_4^1$ and the symmetry characteristics of HOMO (Ψ_4) are thus reversed (C_2 symmetry). Therefore, under photochemical conditions [1,5] suprafacial migration is no longer possible and the shift has to be an antarafacial process (Figure 3.5).

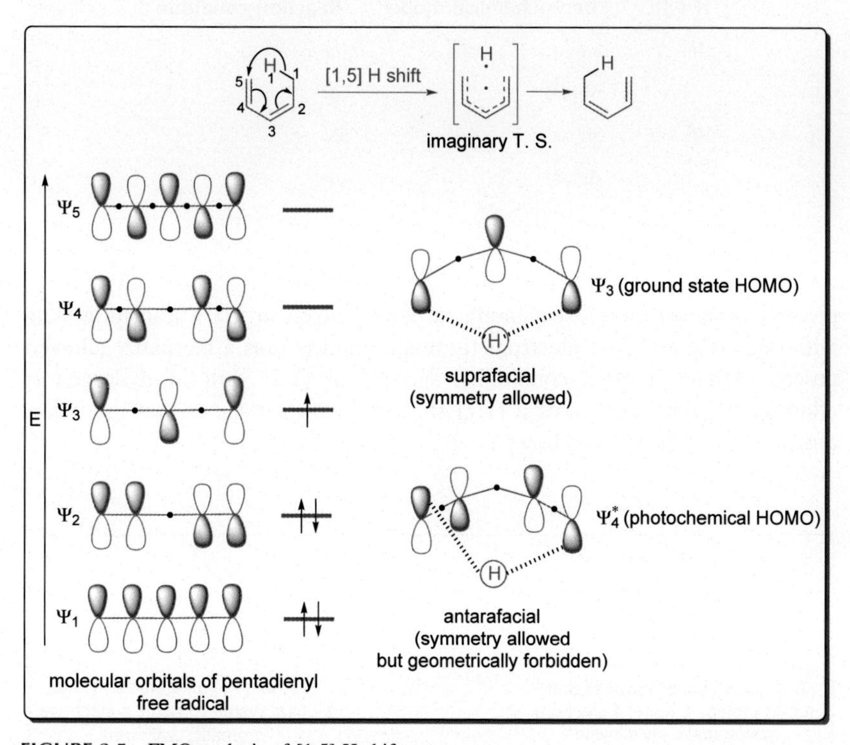

FIGURE 3.5 FMO analysis of [1,5] H shifts.

A similar analysis of such systems has led to the formulation of selection rules that state that if a sigmatropic reaction of the order [i,j] (for hydrogen migration $i = 1$) has $i + j = 4n + 2$, then thermal reaction is *suprafacial* and photochemical reaction will be *antarafacial*. However, for those cases in which $i + j = 4n$, the rules are reversed and the thermal reactions are *antarafacial* while the photochemical reaction will be *suprafacial*. The selection rules for the sigmatropic shift of hydrogen by FMO method are given in Table 3.1.

3.2.3 PMO Analysis of [i,j] Sigmatropic Rearrangements of Hydrogen

Sigmatropic reactions can be treated successfully by PMO method, and similar conclusions are arrived at as by other approaches (Figure 3.6). For instance, [1,3] suprafacial shift of hydrogen occurs via a transition state with zero node and four electrons (antiaromatic) and thus is a photochemically allowed

TABLE 3.1 Selection rules for the sigmatropic shift of hydrogen by FMO method.

[i + j]	Stereochemical mode	Reaction condition
4n	Supra	*hν*
	Antara	Δ
4n + 2	Supra	Δ
	Antara	*hν*

process, whereas the [1,3] antarafacial shift can occur via a transition state with one node and four electrons (aromatic) and is thus a thermally allowed process. Under thermal conditions, suprafacial [1,5] shift of hydrogen is allowed. Similar analysis of a [1,7] shift of hydrogen shows that the antarafacial shift is thermally allowed.

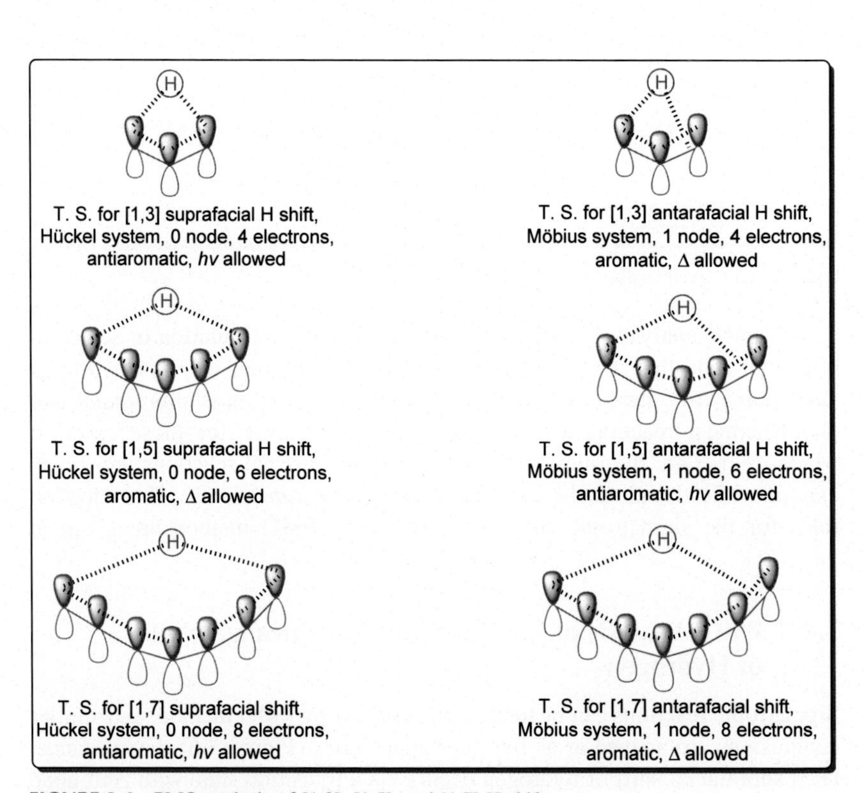

FIGURE 3.6 PMO analysis of [1,3], [1,5], and [1,7] H shifts.

TABLE 3.2 **Selection rules for the sigmatropic shift of hydrogen by PMO method.**

No. of electrons in the T. S.	Stereochemical mode	No. of nodes in the T. S.	Aromaticity of the T. S.	Reaction condition
4n	Supra	0 or Even	Antiaromatic	*hv*
	Antara	Odd	Aromatic	Δ
4n + 2	Supra	0 or Even	Aromatic	Δ
	Antara	Odd	Antiaromatic	*hv*

The selection rules for the sigmatropic shift of hydrogen by PMO method are given in Table 3.2.

There are many examples of 1,5-hydrogen migrations in molecules having a pentadienyl fragment. The example given below shows the transformation of a less stable diene bearing exocyclic double bonds to a more stable diene having endocyclic double bond (Scheme 3.3).

SCHEME 3.3 [1,5] Hydrogen shift in diene bearing exocyclic double bonds.

The equilibration among the various isomeric methylcycloheptadienes (**1–4**) can be explained by [1,5] hydrogen shifts (Scheme 3.4).

SCHEME 3.4 [1,5] Hydrogen shifts in isomeric methylcycloheptadienes.

To study the stereochemistry of the hydrogen shift, optically active 1-deuterio-1-methyl-3-tert-butylindene was heated and it was found to give products with optical and deuterium labeling consistent with a suprafacial deuterium shift (Scheme 3.5).

Due to steric reasons, suprafacial migrations are more common than antarafacial shifts. However, with the lengthening of the conjugated system, sometimes it is possible for a σ-bond to migrate to the opposite π-electron face. Like the thermal [1,3] hydrogen shift, a [1,7] hydrogen shift follows the

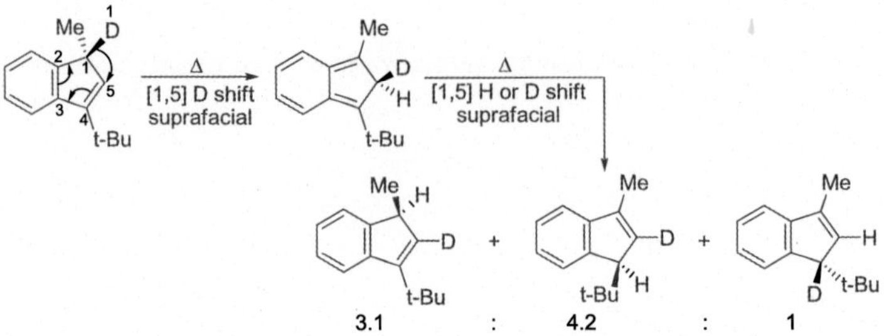

SCHEME 3.5 [1,5] Hydrogen shifts in 1-deuterio-1-methyl-3-tert-butylindene.

same symmetry rules and is allowed when antarafacial. The geometrical restrictions on the antarafacial transition state of [1,7] hydrogen shift are not as severe as in the [1,3] case because a π-system involving seven carbon atoms is more flexible than the one involving only three carbon atoms (Scheme 3.6).

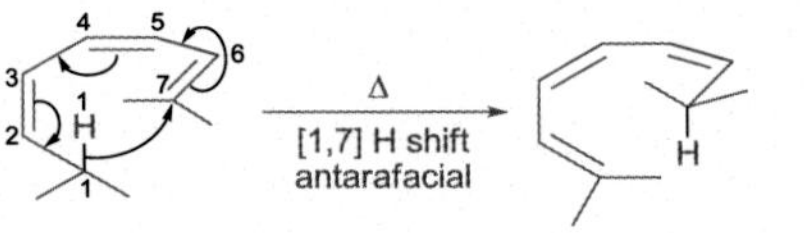

SCHEME 3.6 Allowed [1,7] hydrogen shift in antarafacial manner.

[1,7] Hydrogen shift can be illustrated in the thermal equilibrium between precalciferol (previtamin D_2) and calciferol (vitamin D_2) (Scheme 3.7).

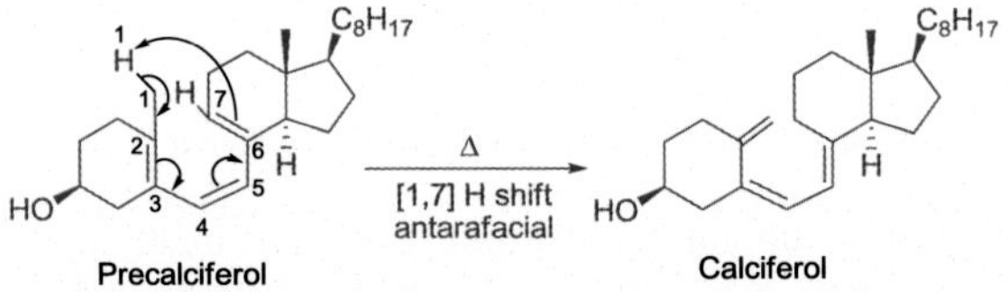

SCHEME 3.7 Conversion of precalciferol into vitamin D_2 by [1,7] hydrogen shift.

In cycloheptatriene, an antarafacial [1,7] hydrogen shift is impossible. Consequently, [1,7] hydrogen shifts within this system must be initiated photochemically. For example, the interconversion of isomers of 1,4-di(cycloheptatrienyl)benzene involves the sequence of thermal [1,5] and photochemical [1,7] sigmatropic hydrogen shifts (Scheme 3.8).

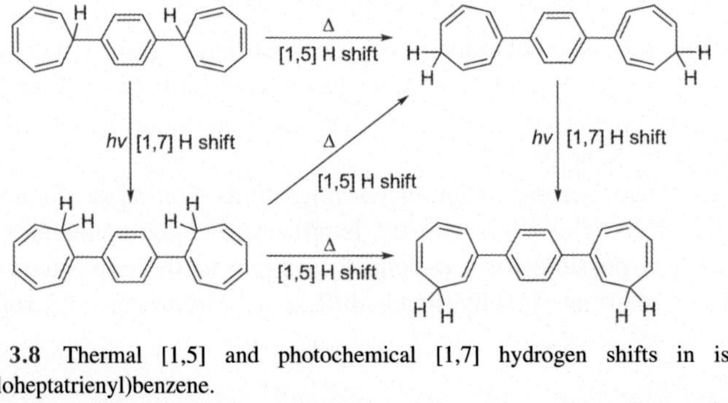

SCHEME 3.8 Thermal [1,5] and photochemical [1,7] hydrogen shifts in isomers of 1,4-di(cycloheptatrienyl)benzene.

3.2.4 Solved Problems

Q 1. How will you classify the reaction given below?

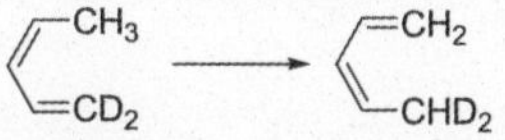

Sol 1. A suprafacial [1,5] hydrogen shift that is symmetry allowed under thermal conditions.

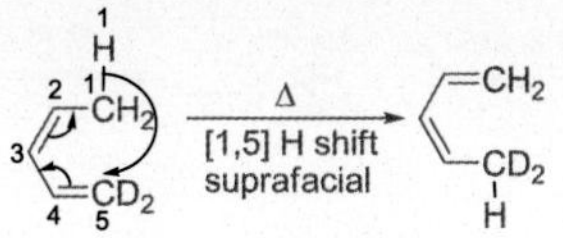

Q 2. The reaction given below is an example of:

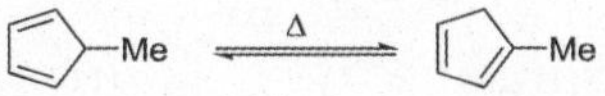

(a) [1,3] sigmatropic hydrogen shift (b) [1,3] sigmatropic methyl shift
(c) [1,5] sigmatropic hydrogen shift (d) [1,5] sigmatropic methyl shift

Sol 2. (c)

Q 3. The product of the following reaction is:

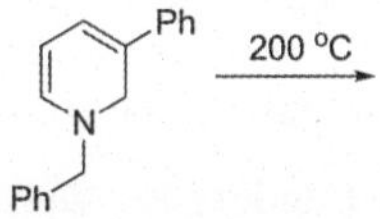

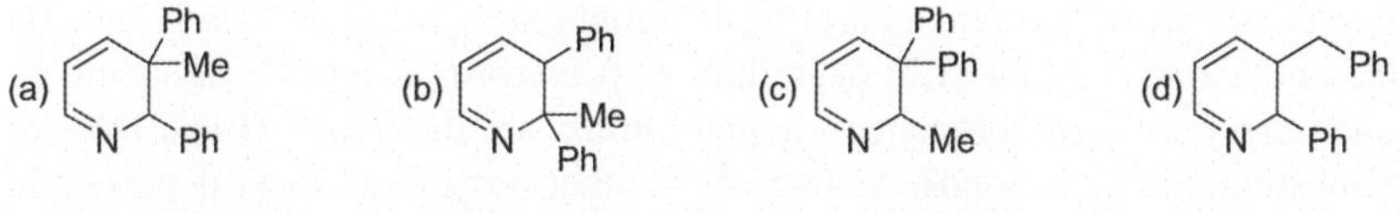

Sol 3. (a) Electrocyclic ring opening of 1,2-dihydropyridine derivative (**I**) gives 1-azatriene intermediate (**II**), which then undergoes [1,7] hydrogen shift to give **III**. Electrocyclization of **III** then affords **IV**.

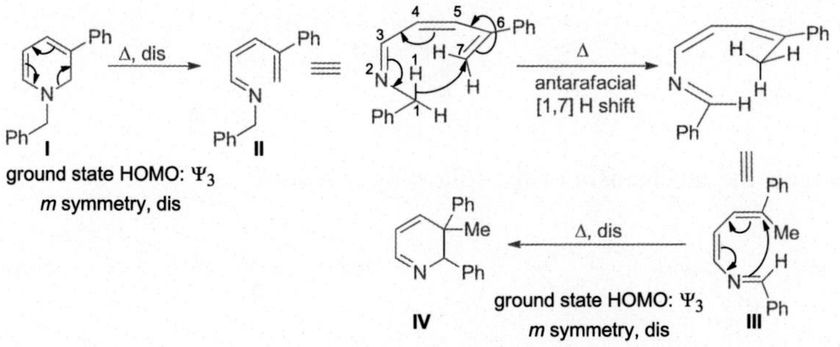

Q 4. Explain the mechanism of following transformation:

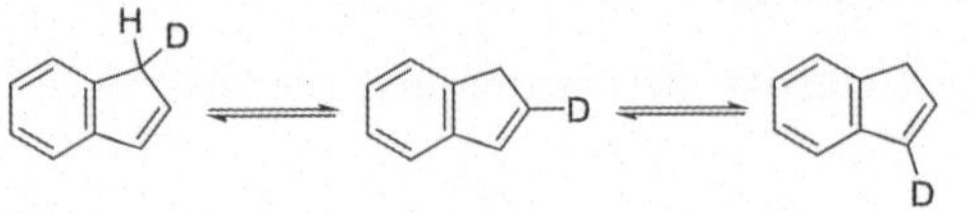

Sol 4. This scrambling of deuterium label of 1-deuteroindene is a result of a series of thermally allowed [1,5] deuterium and hydrogen shifts.

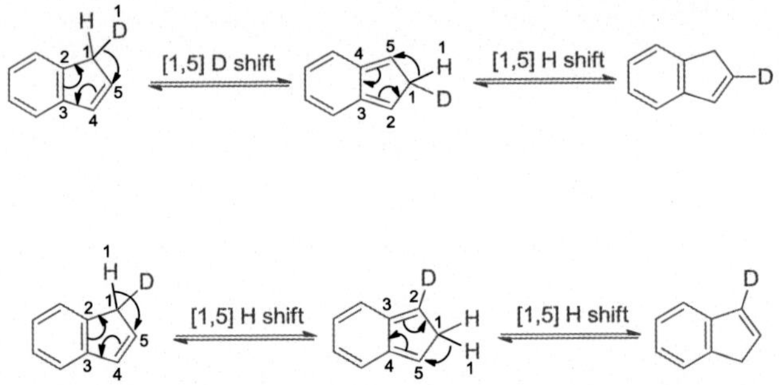

[1,5] D shift

[1,5] H shift

[1,5] H shift

[1,5] H shift

Q 5. Upon heating, the diene shown below undergoes suprafacial [1,5] hydrogen shifts.

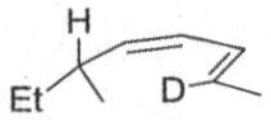

Give the structures of the dienes present in the reaction mixture.

Sol 5. Upon heating, the diene **I** undergoes suprafacial [1,5] hydrogen shift to give the diene **II**, a suprafacial [1,5] deuterium shift then converts **II** into **III** and another suprafacial [1,5] deuterium shift converts it into **IV**. It should be noted that the major components at equilibrium are the dienes **II** and **IV** with trisubstituted double bonds. Neither of the other possible isomers is present in the equilibrium mixture, showing that no [1,5] antarafacial shifts had occurred.

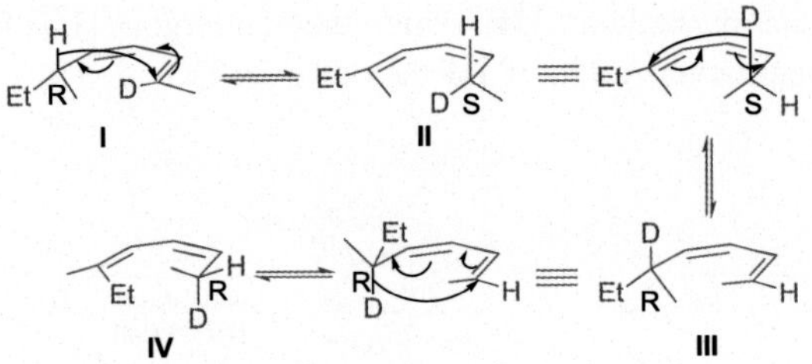

Q 6. Give the mechanism of the following reactions:

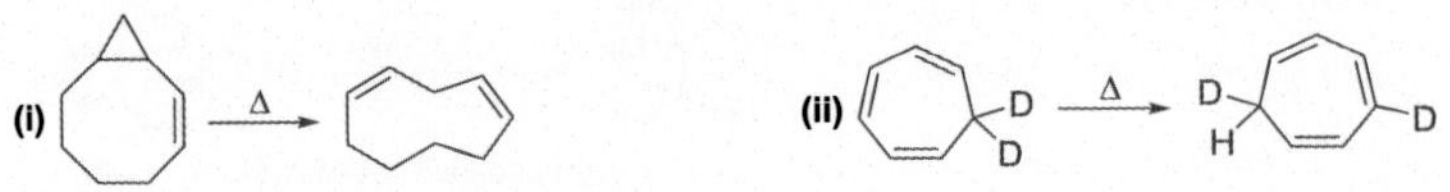

(i)

(ii)

(iii) D D Δ H D D

(iv) OH O H + H_3PP Δ O CH_3

(v) O N Ph O N N H Δ O N Ph O N N

(vi) CH_3 O H_3C CH_3 100 °C D_2O CH_3 O H_3C CD_3

(vii) hv major + minor

(viii) O * COOMe H hv H COOMe * OH

Sol 6. **(i)** The reaction involves a thermally allowed [1,5] hydrogen shift via a folded conformation.

≡ H H 3 4 5 2 1 1H Δ [1,5] H shift H H H ≡

(*Z*)-Bicyclo[6.1.0]non-2-ene (1*Z*,4*Z*)-Cyclonona-1,4-diene

(ii) 5 4 3 2 1 D D 1 Δ [1,5] D shift D H D

(iii) The reaction involves a [1,5] deuterium shift in the first step and a [1,5] hydrogen shift in the second step. It should be noted that in the second step, only a [1,5] hydrogen shift (and not a [1,5] deuterium shift) takes place because of the stereoelectronic demands for such reactions.

1 D D 2 1 3 4 5 Δ [1,5] D shift suprafacial D 5 4 3 2 1 D H 1 Δ [1,5] H shift suprafacial H D D

(iv) The mechanism involves the initial formation of *o-cis*-butadienylphenol (**I**) by way of normal Wittig reaction. The compound **I** on [1,7] sigmatropic shift of hydrogen gives an intermediate (**II**) bearing methylene quinonoid structure, which on electrocyclic ring closure yields benzopyran (**III**) instantaneously.

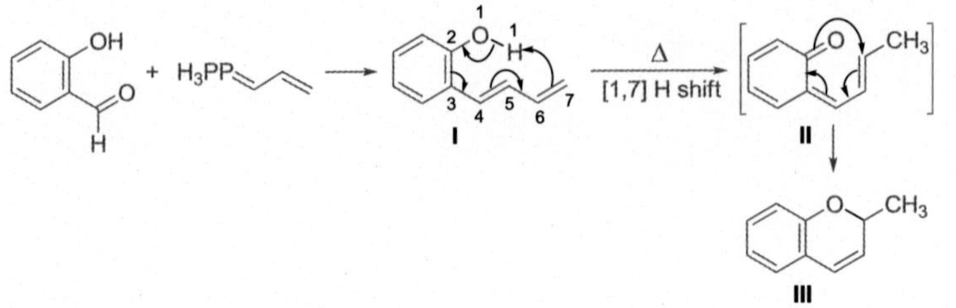

OH
O
H
+ H3PP
1
2 O 1 H
3 4 5 6 7
I
Δ
[1,7] H shift
O CH3
II
O CH3
III

(v) The reaction involves two consecutive thermally allowed [1,5] hydrogen shifts.

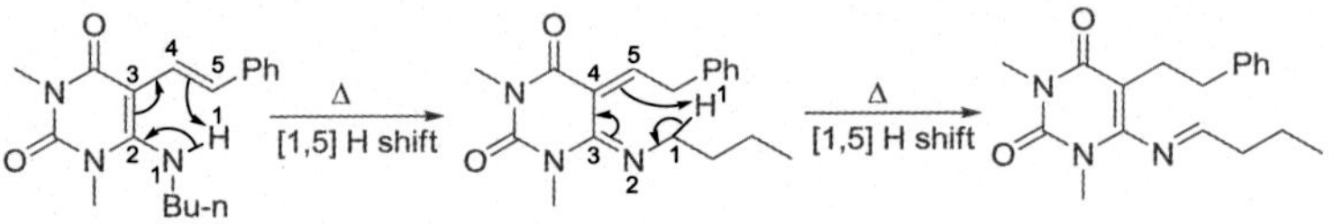

O
N 3 4 5 Ph
O N 2 1 N 1 H
Bu-n
Δ
[1,5] H shift
O
N 4 5 Ph
H 1
O N 3 N 1
2
Δ
[1,5] H shift
O
N Ph
O N N

(vi)

CH3
3 4 O 5
H3C 2 1 H 1
Δ
[1,5] H shift
CH3
O H
H3C CH2
D2O
CH3
3 2 1 D
1 O
4
H3C CH2
5
Δ
[1,5] D shift
CH3
O
H3C CH2 D

CH3
O
H3C CD3

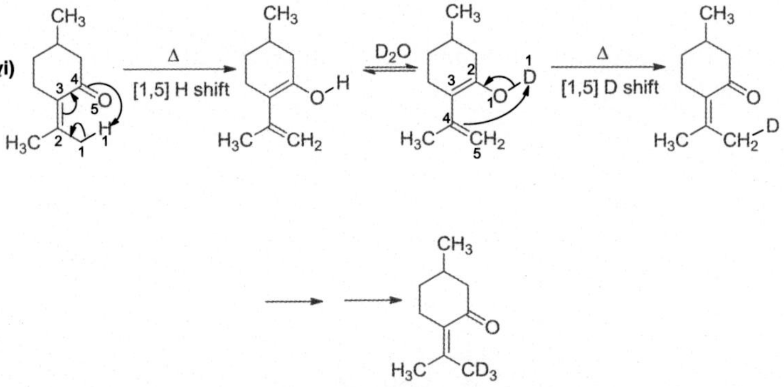

(vii) The reaction involves photochemically allowed suprafacial [1,3] hydrogen shift.

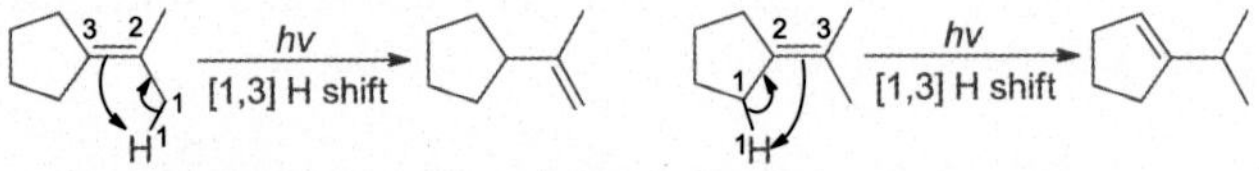

3 2
1
H 1
hv
[1,3] H shift
2 3
1
1 H
hv
[1,3] H shift

(viii) The reaction involves photochemically allowed suprafacial [1,3] alkyl shift with retention at the migrating center.

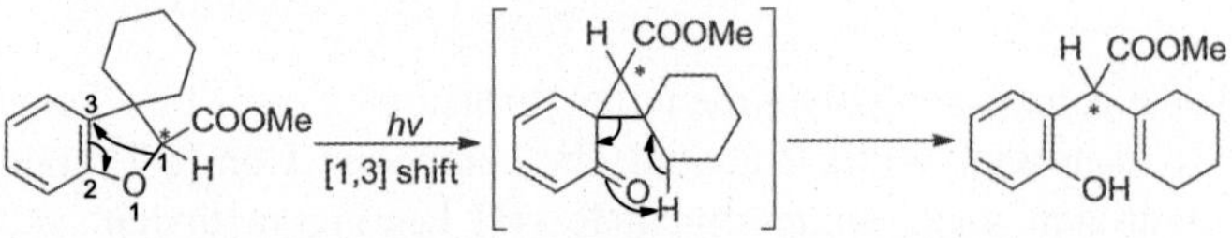

3
COOMe
2 O 1 H
1
hv
[1,3] shift
H COOMe
*
O H
H COOMe
*
OH

Q 7. Rationalize the following transformations:

(i) A $\xrightarrow{\Delta}$ Toluene $\xleftarrow{\Delta}$ B

(ii) $\xrightarrow{\Delta}$; $\xrightarrow{H_3O^+}$

(iii) $\xrightarrow{\Delta}$ + (Not formed)

(iv) $\xrightarrow{\Delta}$ A + B

(v) $\xrightarrow{\Delta}$

(vi) $\xrightarrow{145\ ^{\circ}C}$

Sol 7. (i) Compound **I** can be converted into toluene by thermally allowed suprafacial [1,5] hydrogen shift.

I $\xrightarrow[\text{[1,5] H shift suprafacial}]{\Delta}$ Toluene

However, compound **II** can be converted into toluene by [1,3] or [1,7] hydrogen shift. Under thermal conditions, both of these rearrangements are symmetry forbidden in the suprafacial mode and geometrically forbidden in the antarafacial mode. It should be noted that antarafacial [1,7] hydrogen shift can occur in flexible systems. But in rigid systems it also becomes geometrically forbidden.

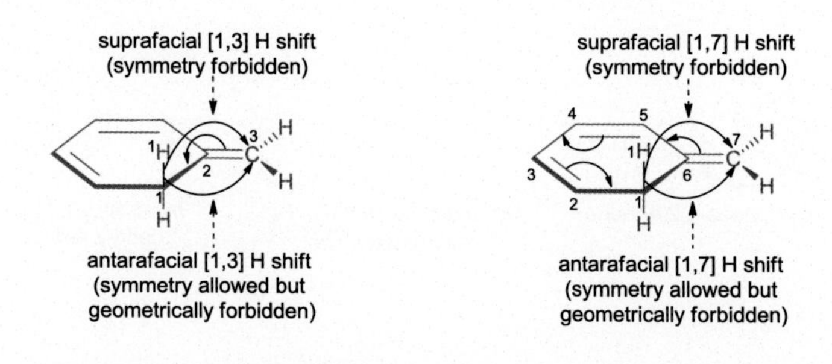

(ii) The given compound does not undergo aromatization on heating because in this case also both [1,3] or [1,7] hydrogen shifts are forbidden.

However, in the acidic medium, aromatization takes place through a nonconcerted mechanism as shown below:

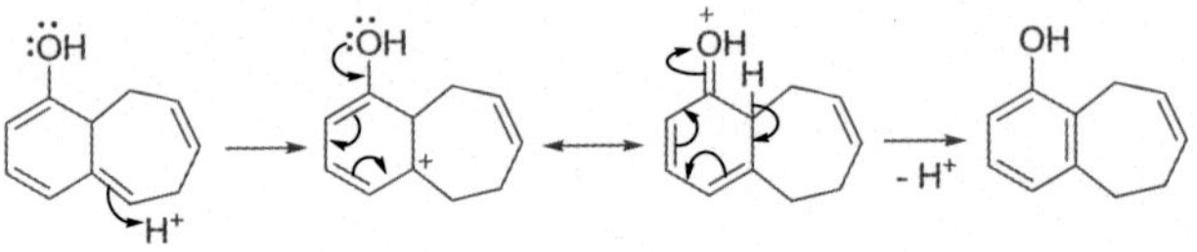

(iii) The reaction involves thermally allowed suprafacial [1,5] hydrogen shift across a conjugated diene wherein a cyclopropane ring replaces one of the double bonds of the diene system. The exclusive formation of the *Z*-isomer is due to the preference for the chair-like transition state during [1,5] hydrogen shift.

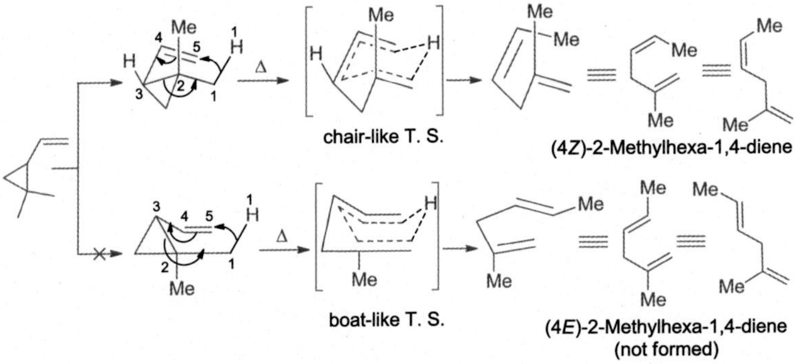

chair-like T. S.

(4*Z*)-2-Methylhexa-1,4-diene

boat-like T. S.

(4*E*)-2-Methylhexa-1,4-diene (not formed)

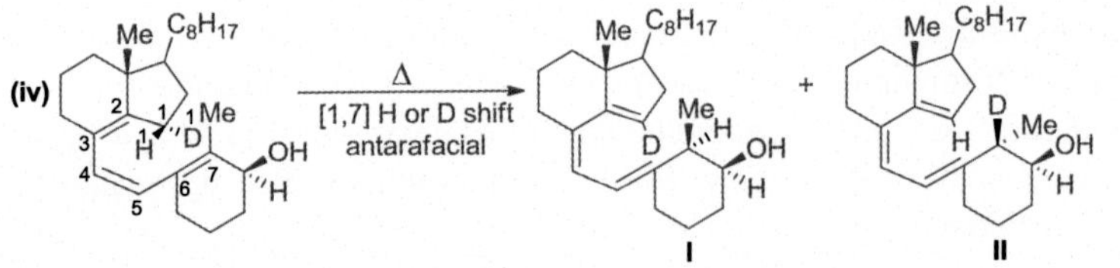

(iv) [1,7] H or D shift antarafacial

I + II

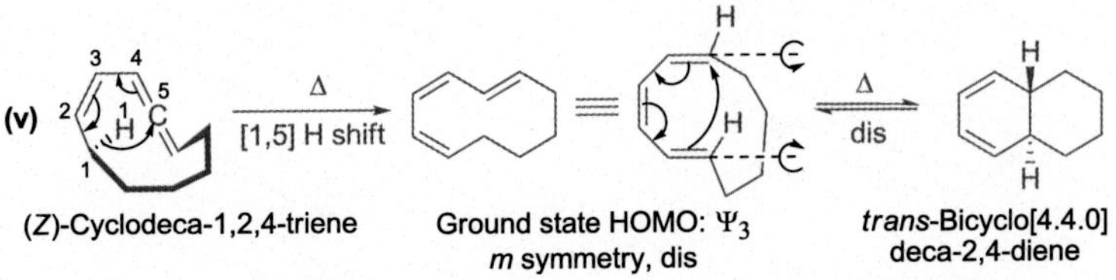

(v) [1,5] H shift

(*Z*)-Cyclodeca-1,2,4-triene

Ground state HOMO: Ψ_3 *m* symmetry, dis

trans-Bicyclo[4.4.0] deca-2,4-diene

(vi) The process involves cyclization of the cycloheptatriene moiety followed by the opening of bicyclo[2.1.0]pent-2-ene and a [1,5] hydrogen shift.

Δ dis — Δ con — Δ [1,5] H shift

Ground state HOMO: Ψ_3 *m* symmetry, dis — Ground state HOMO: Ψ_2 C_2 symmetry, con

3.3 ANALYSIS OF SIGMATROPIC REARRANGEMENTS OF ALKYL GROUP

With elements other than hydrogen, more specifically with a carbon, there is a possibility of the sigmatropic shift with inversion of configuration in the migrating group. This was impossible with a hydrogen atom, which has its electron in a 1*s* orbital that has only one lobe. However, a carbon free radical has its odd electron in a *p*-orbital, which has two lobes of opposite sign.

3.3.1 FMO Analysis of [1,3] Sigmatropic Rearrangements of Alkyl Group

It has been observed that in a [1,3] thermal suprafacial process, symmetry is conserved only if the migrating carbon has opposite lobes. In other words, if the migrating carbon was originally bonded via its negative lobe the new carbon–carbon bond will be formed by using the positive lobe. The whole process results in the inversion of configuration in the migrating group (Figure 3.7).

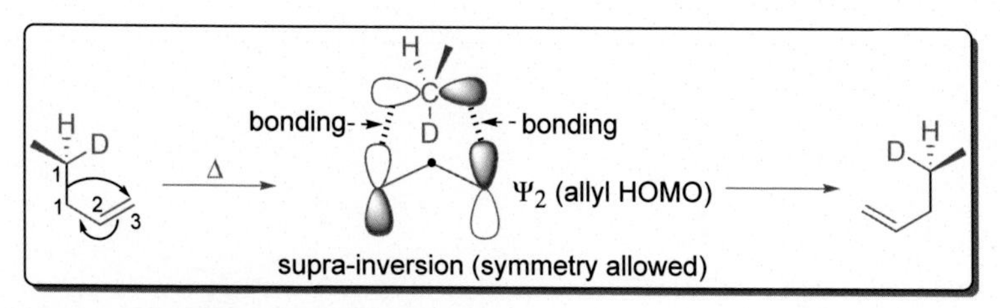

FIGURE 3.7 FMO analysis of [1,3] alkyl shifts.

However, the [1,3] thermal suprafacial process that involves the same lobe gives the product with retention of configuration in the migrating group, but the process is not symmetry allowed. On the other hand, symmetry-allowed [1,3] thermal antarafacial process with retention is geometrically forbidden and antarafacial process with inversion is symmetry forbidden (Figure 3.8).

H, C–D
bonding
antibonding
H D
Δ
1
1 2 3
Ψ_2 (allyl HOMO)
H D
supra-retention (symmetry forbidden)
C–D
bonding
bonding
Ψ_2 (allyl HOMO)
antara-retention (symmetry allowed; geometrically forbidden)
H C D
bonding
antibonding
Ψ_2 (allyl HOMO)
antara-inversion (symmetry forbidden)

FIGURE 3.8 FMO analysis of [1,3] alkyl shifts continued.

3.3.2 FMO Analysis of [1,5] Sigmatropic Rearrangements of Alkyl Group

During [1,5] thermal suprafacial process, symmetry can be conserved only if the migrating carbon involves the same lobe. In other words, if the migrating carbon atom was originally bonded via its positive lobe, it will have to use its positive lobe to form the new C—C bond. The entire process results in the retention of configuration in the migrating group. However, if migrating carbon uses the opposite lobe, it may give the product with inversion of configuration, but the process becomes symmetry forbidden. On the other hand, [1,5] thermal antarafacial process with retention is symmetry forbidden and antarafacial process with inversion is symmetry allowed (Figure 3.9).

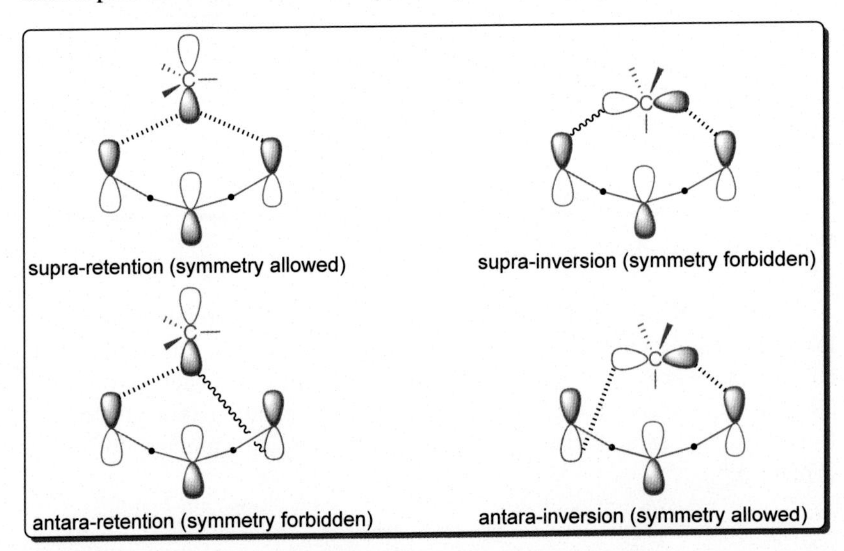

FIGURE 3.9 FMO analysis of [1,5] H shifts.

TABLE 3.3 Selection rules for the sigmatropic alkyl shifts by FMO method.

[i + j]	Δ Allowed ($h\nu$ forbidden)	$h\nu$ Allowed (Δ forbidden)
4n	Supra-inversion (s_i)	Supra-retention (s_r)
	Antara-retention (a_r)	Antara-inversion (a_i)
4n + 2	Supra-retention (s_r)	Supra-inversion(s_i)
	Antara-inversion (a_i)	Antara-retention (a_r)

The selection rules for the sigmatropic shift of carbon by FMO method are given in the Table 3.3, where n is zero or an integer.

3.3.3 PMO Analysis of [i,j] Sigmatropic Rearrangements of Alkyl Group

Analysis of [1,3] and [1,5] alkyl shifts by PMO method is summarized in Figures 3.10 and 3.11, respectively.

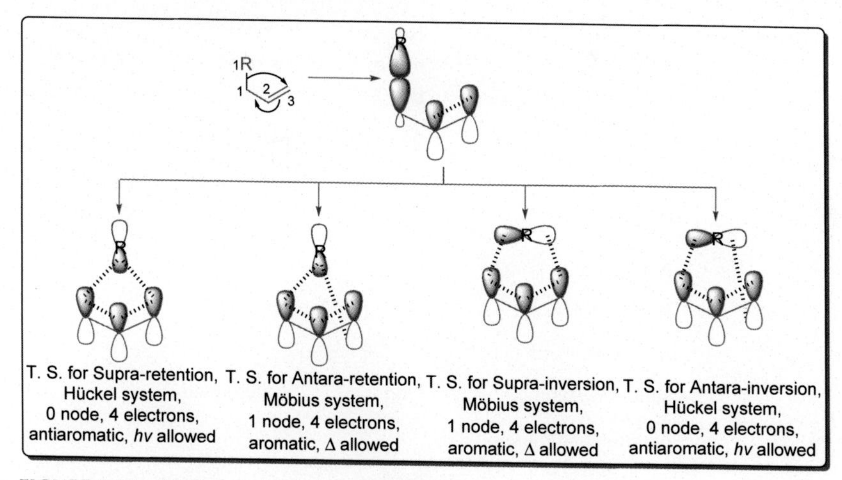

FIGURE 3.10 PMO analysis of [1,3] alkyl shifts.

The selection rules for the sigmatropic shift of carbon by PMO method are given in Table 3.4, where n is zero or an integer.

3.3.4 Solved Problems

Q 1. The conversion of vinylcyclopropane to cyclopentene upon heating is an example of which of the following sigmatropic rearrangements?

(a) [3,3] (b) [1,3] (c) [1,5] (d) [1,2]

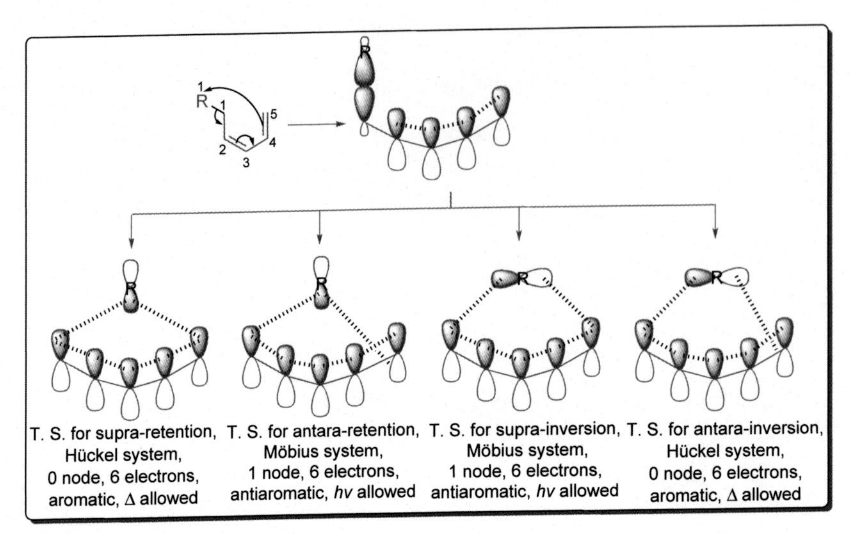

FIGURE 3.11 PMO analysis of [1,5] alkyl shifts.

TABLE 3.4 Selection rules for the sigmatropic alkyl shifts by PMO method.

No. of electrons in the T. S.	Stereochemical mode	No. of nodes in the T. S.	Aromaticity of the T. S.	Reaction condition
4n	Supra-retention (s_r)	0 or Even	Antiaromatic	*hν*
	Antara-inversion (a_i)			
	Supra-inversion (s_i)	Odd	Aromatic	Δ
	Antara-retention (a_r)			
4n + 2	Supra-retention (s_r)	0 or Even	Aromatic	Δ
	Antara-inversion (a_i)			
	Supra-inversion (s_i)	Odd	Antiaromatic	*hν*
	Antara-retention (a_r)			

Sol 1. (b) The thermal expansion of a vinylcyclopropane to a cyclopentene ring involves a [1,3] sigmatropic migration of carbon. The reaction is known as a *vinylcyclopropane rearrangement*. The reaction involves the formation of a strained four-membered transition state and usually occurs at a temperature higher than 300 °C.

Vinylcyclopropane —Δ, [1,3] shift→ [ ] ——→ Cyclopentene

Q 2. Explain the following transformations:

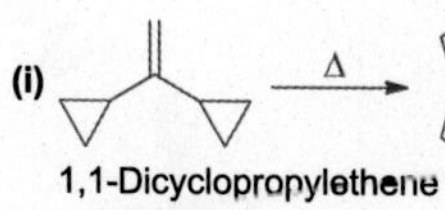

(i) 1,1-Dicyclopropylethene $\xrightarrow{\Delta}$

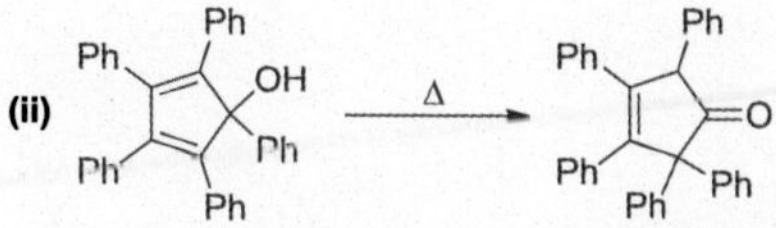

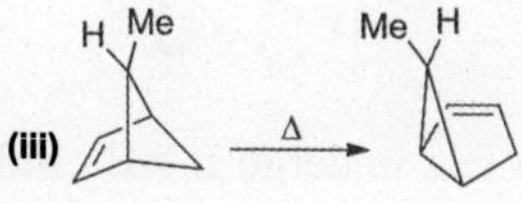

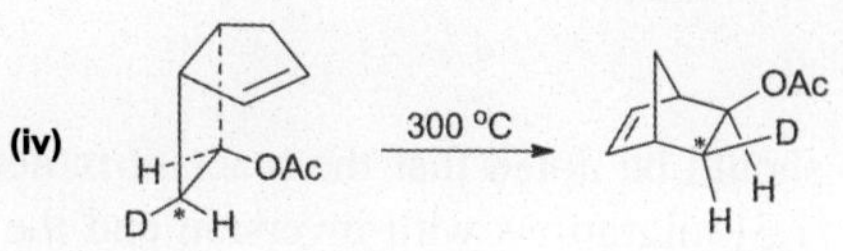

(ii) $\xrightarrow{\Delta}$

(iii) $\xrightarrow{\Delta}$

(iv) Deuterated bicyclo[3.2.0]heptene $\xrightarrow{300\ ^{\circ}C}$ *Exo*-norbornal

(v) $\underset{hv}{\rightleftharpoons}$

(vi) $\xrightarrow{\Delta}$ $\xrightarrow{\Delta}$

(vii) I $\xrightarrow{\Delta}$ III $\xleftarrow{\Delta}$ II

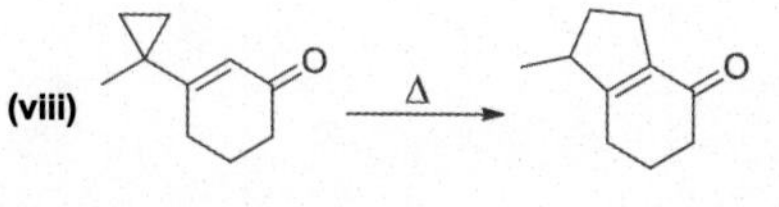

(viii) $\xrightarrow{\Delta}$

Sol 2. (i)

[1,3] C shift, [1,3] C shift

(ii) [1,5] phenyl shift; enol ⇌ keto

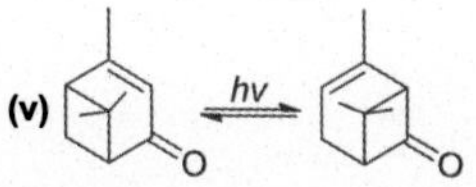

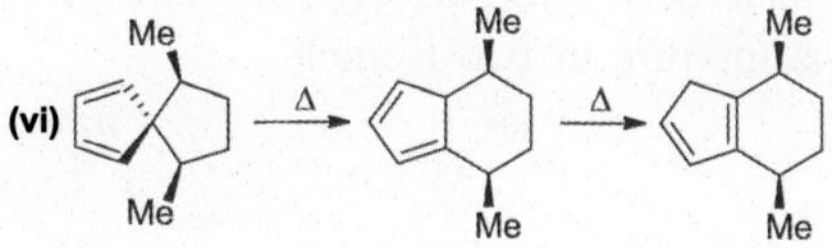

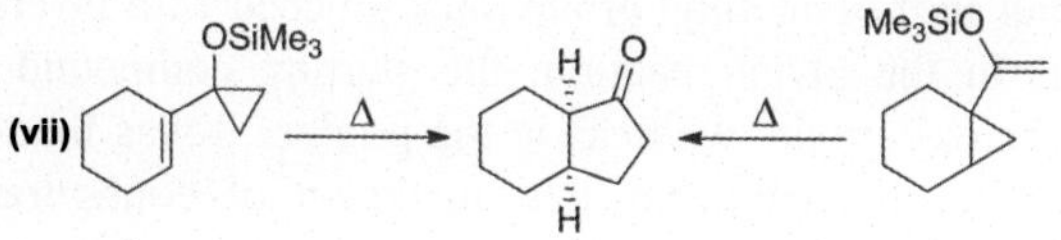

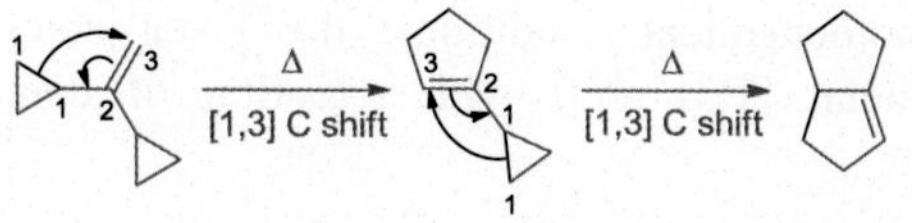

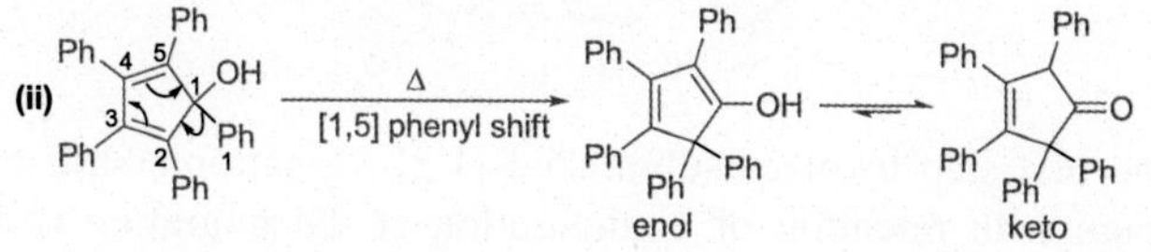

(iii) The reaction involves [1,3] sigmatropic shift, which under thermal conditions should be suprafacial with inversion at the migrating apical carbon.

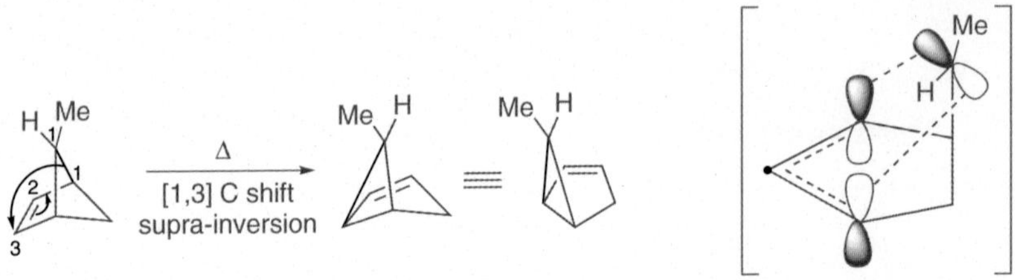

Δ
[1,3] C shift
supra-inversion

It should be noted that the reactant molecule can undergo a full circle of such [1,3] migrations with inversion and the hydrogen and the methyl group will throughout remain *endo* and *exo*, respectively. However, if the [1,3] sigmatropic shift proceeds with retention, the hydrogen and the methyl group would have been alternatively *endo* and *exo* with each migration step, giving rise to a mixture of two isomers.

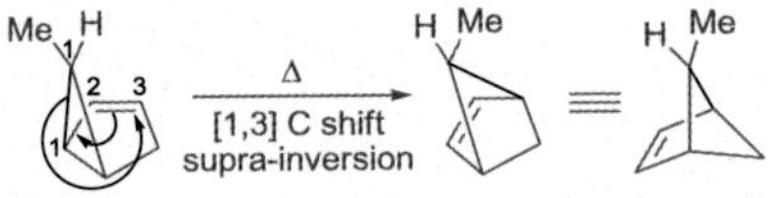

Δ
[1,3] C shift
supra-inversion

(iv) A [1,3] suprafacial shift of an alkyl group must proceed with inversion at the migrating center. In the given reaction, the starting compound has deuterium *trans* to the acetoxy group, whereas in the product it was found to be exclusively *cis*. This result establishes that inversion of configuration occurred at C-7 during the migration, in accord with the stereochemistry required by the Woodward–Hoffmann rules.

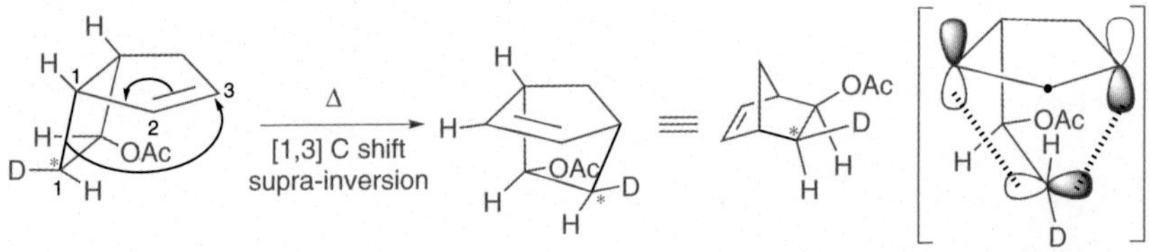

Δ
[1,3] C shift
supra-inversion

(v) Under photochemical conditions, the given compound undergoes reversible suprafacial [1,3] shift with retention of configuration in the migrating group.

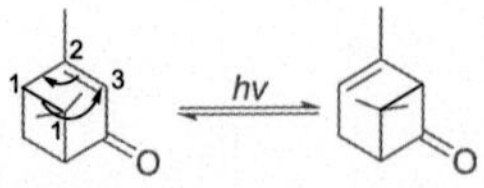

hv

(vi) The first step involves suprafacial [1,5] sigmatropic shift of carbon, which occurs with retention of configuration at the migrating carbon. The second step involves [1,5] sigmatropic shift of hydrogen.

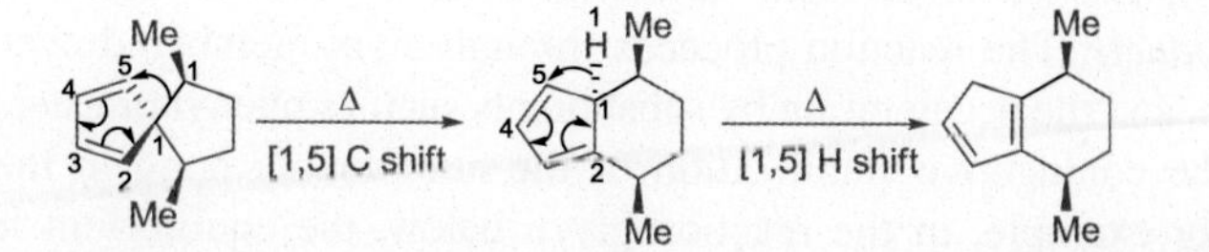

(vii) The compounds **I** and **II** on heating undergo a [1,3] sigmatropic migration of carbon to give silyl enol ether (**IV**). On workup, **IV** gives the ketone **III**, which preferentially adopts the *cis* stereochemistry. It should be noted that the compound **II** can also undergo [1,3] sigmatropic migration of carbon in an alternate way, but in that case we get an unstable bridge-head alkene.

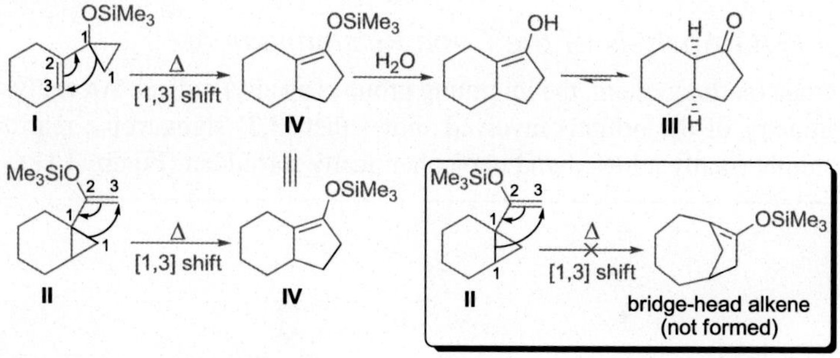

(viii) The compound **I** on heating undergoes a [1,3] sigmatropic migration of carbon to give β,γ-unsaturated ketone (**II**). Formation of the extended enol (**III**) followed by movement of the double bond converts **II** into α,β-unsaturated ketone (**IV**).

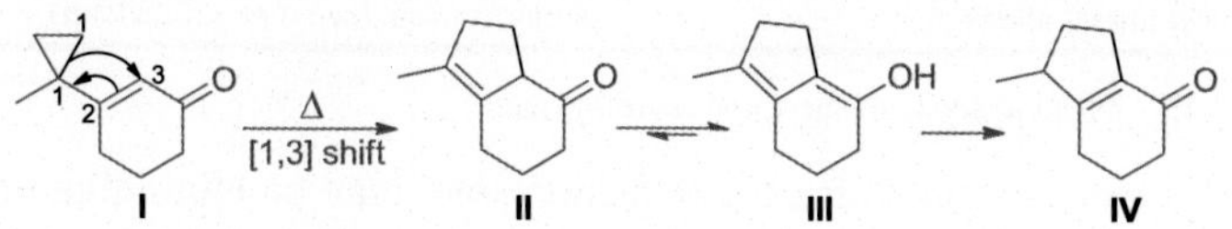

3.4 [3,3] SIGMATROPIC REARRANGEMENTS

3.4.1 The Cope Rearrangement

The thermal [3,3] sigmatropic rearrangement of 1,5-dienes is called the *Cope rearrangement*. The reaction proceeds through a cyclic transition state (Scheme 3.9).

The Cope rearrangement is a reversible process; the equilibrium mixture has a great proportion of the thermodynamically stable isomer. It follows

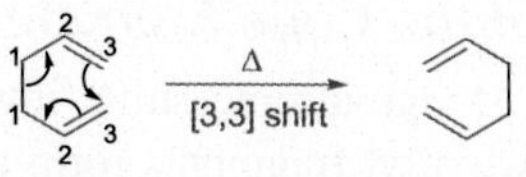

SCHEME 3.9 The Cope rearrangement of hexa-1,5-diene.

first-order kinetics. It is an intramolecular process because rearrangement of a mixture of two different hexa-1,5-dienes does not lead to the formation of *cross products*. The reaction proceeds through a six-membered cyclic transition state, and the acceleration by substituents such as phenyl or ester groups is due to the conjugative stabilization of the new double bond in the product diene. For example, in the reaction given below, the equilibrium favors the product because the double bond is in conjugation with the carbonyl and cyano groups in the product (Scheme 3.10).

SCHEME 3.10 Cope rearrangement of (*E*)-ethyl 2-allyl-2-cyanopent-3-enoate.

3.4.1.1 FMO Analysis of the Cope Rearrangement

In the Cope rearrangement, the migrating group is an allyl radical. An analysis of the symmetry of the orbitals involved shows that [3,3] sigmatropic rearrangements are thermally allowed and photochemically forbidden (Figure 3.12).

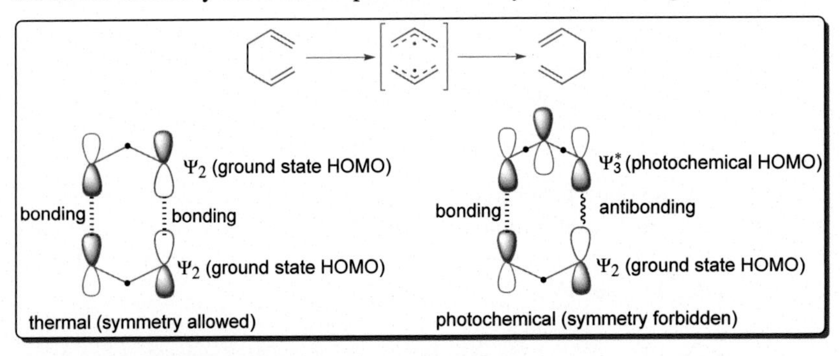

FIGURE 3.12 FMO analysis of the Cope rearrangement.

In [3,3] sigmatropic shifts, the transition state may be **chair-like** or **boat-like**. Both of these transitions states are symmetry allowed (Figure 3.13).

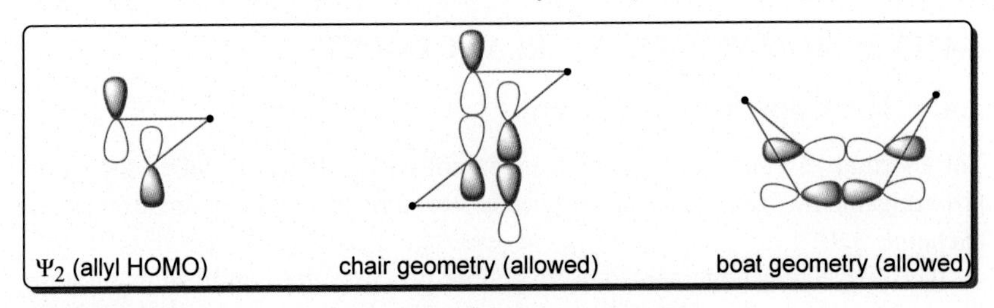

FIGURE 3.13 FMO analysis of the Cope rearrangement continued.

3.4.1.2 PMO Analysis of the Cope Rearrangement

The transition state for [3,3] sigmatropic shifts involves interaction between two allyl fragments. The two allyl fragments form a closed loop of π-orbitals with no sign inversions in the supra–supra case and one inversion in the supra–antara case (Figure 3.14).

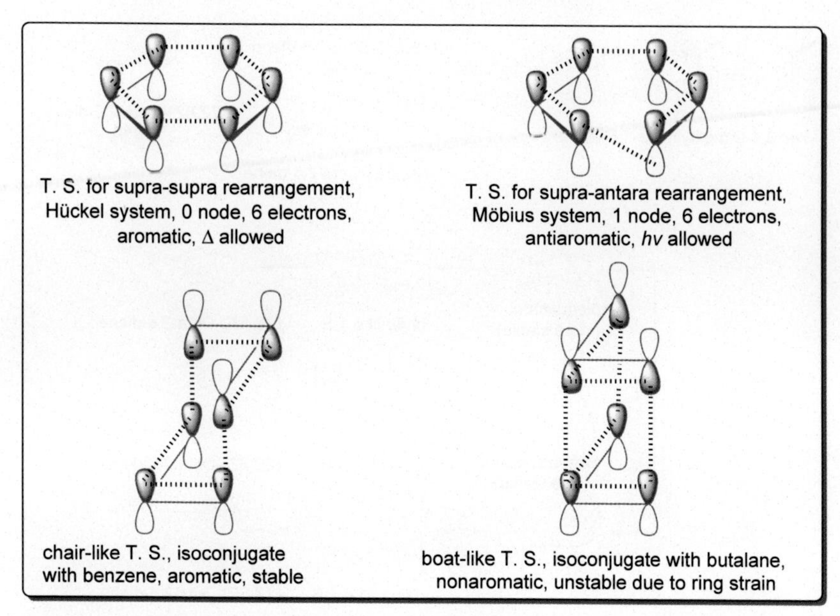

T. S. for supra-supra rearrangement,
Hückel system, 0 node, 6 electrons,
aromatic, Δ allowed
T. S. for supra-antara rearrangement,
Möbius system, 1 node, 6 electrons,
antiaromatic, *hv* allowed
chair-like T. S., isoconjugate
with benzene, aromatic, stable
boat-like T. S., isoconjugate with butalane,
nonaromatic, unstable due to ring strain

FIGURE 3.14 PMO analysis of the Cope rearrangement.

3.4.1.3 Stereochemistry of the Cope Rearrangement: Boat versus Chair Transition State

In [3,3] sigmatropic shifts, a *chair-like transition state* is energetically preferred to a *boat-like structure*. Rearrangement of the *meso* diene through such a transition state then would give the *cis–trans* isomer (Scheme 3.11), while in the case of the rearrangement of the racemic mixture the *trans–trans* isomer is the major product (Scheme 3.12).

meso 3,4-Dimethylhexa-1,5-diene

Δ

(2*Z*,6*E*)-Octa-2,6-diene (99.7%) + (2*E*,6*E*)-Octa-2,6-diene (0.3%)

Δ chair-like T.S. (2*Z*,6*E*)-Octa-2,6-diene

Δ boat-like T.S. (2*E*,6*E*)-Octa-2,6-diene

SCHEME 3.11 Stereoselectivity in the Cope rearrangement of *meso* 3,4-dimethylhexa-1,5-diene.

rac 3,4-Dimethylhexa-1,5-diene

Δ

(2*E*,6*E*)-Octa-2,6-diene (90%)

+

(2*Z*,6*E*)-Octa-2,6-diene (< 1%)

+

(2*Z*,6*Z*)-Octa-2,6-diene (10%)

diequatorial (more stable)

chair-like T.S.

(2*E*,6*E*)-Octa-2,6-diene

axial-equatorial (less stable)

chair-like T.S.

(2*Z*,6*Z*)-Octa-2,6-diene

boat-like T.S.

(2*Z*,6*E*)-Octa-2,6-diene

SCHEME 3.12 Stereoselectivity in the Cope rearrangement of racemic 3,4-dimethylhexa-1,5-diene.

3.4.1.4 Solved Problems

Q 1. The major product formed when (*3R,4S*)-3,4-dimethylhexa-1,5-diene is heated at 240 °C is:

(a) (*2Z,6Z*)-Octa-2,6-diene (b) (*2E,6E*)-Octa-2,6-diene
(c) (*2E,6Z*)-Octa-2,6-diene (d) (*3Z,5E*)-Octa-2,6-diene

Sol 1. (c)

Q 2. Formation of the ketone **II** from the diazoketone **I** involves:

N_2 PhCOOAg, heat; Me **I** → Me **II**

(a) Generation of carbene and a [2,3] sigmatropic rearrangement
(b) Generation of carbene and an electrocyclic ring closing reaction
(c) Generation of ketene and a [2 + 2] cycloaddition
(d) Generation of ketene and a [3,3] sigmatropic rearrangement

Sol 2. (d)

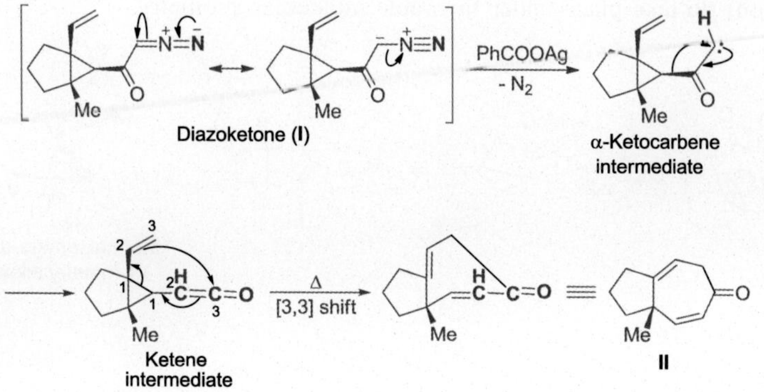

Me, O, =N̄=N (+/−) ↔ Me, O, −N≡N (+)

Diazoketone (I)

PhCOOAg, - N_2

H, Me, O

α-Ketocarbene intermediate

3, 2, 1, 2H, 1, C=C=O, 3, Me

Ketene intermediate

Δ [3,3] shift

H, =C−C=O, Me ≡ Me, =O

II

Q 3. The rearrangement product of the following reaction is:

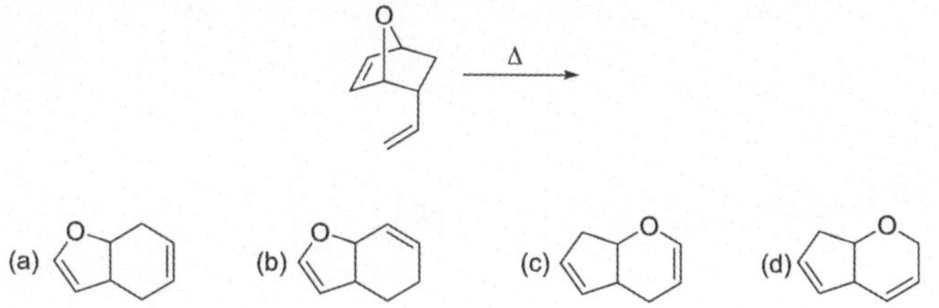

O, Δ

(a) (b) (c) (d)

Sol 3. (a)

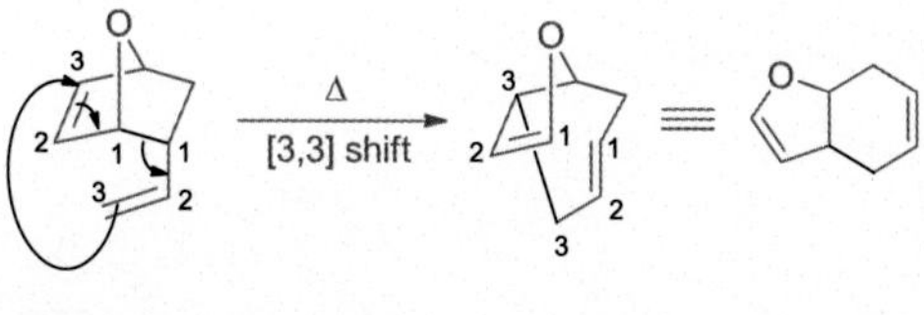

O, 3, 2, 1, 1, 3, 2

Δ [3,3] shift

O, 3, 2, 1, 1, 2, 3 ≡

Q 4. In the reactions given below, the Cope rearrangement takes place in an antara–antara manner. Explain.

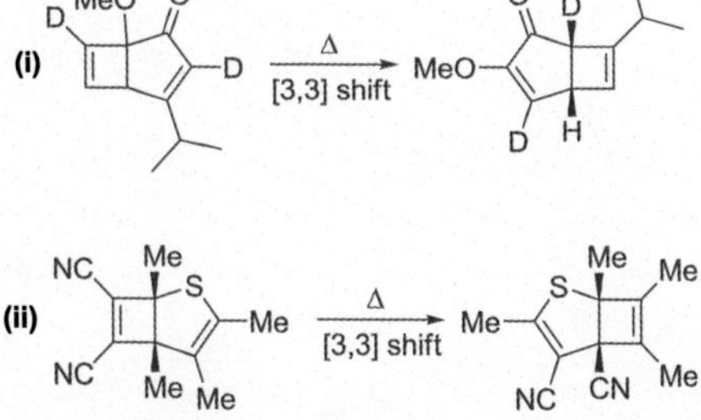

(i) MeO, O, D, D — Δ [3,3] shift → MeO, O, D, D, H

(ii) NC, NC, Me, S, Me, Me, Me — Δ [3,3] shift → Me, S, Me, Me, Me, NC, CN, Me

Sol 4. The Cope rearrangement is allowed both in a supra–supra and an antara–antara manner. One may conclude that the transition state for an antara–antara process is rather difficult to attain due to steric reasons. However, such reactions actually have been observed. Thus, symmetry-allowed reactions do take place under favorable molecular geometry.

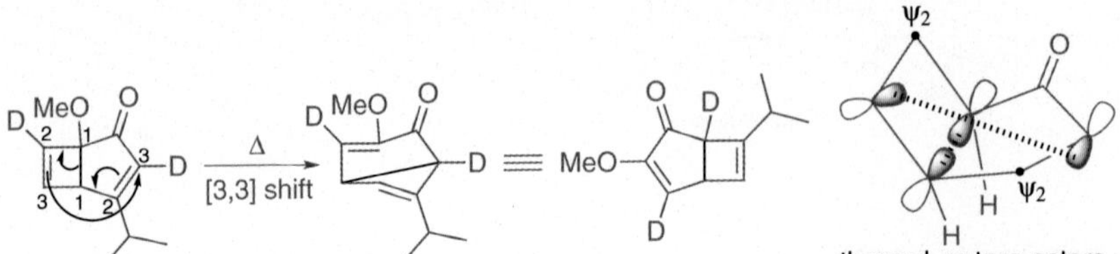

D MeO O 2 1 3 D 3 1 2 —Δ [3,3] shift→ D MeO O D ≡ MeO O D D

ψ2 O H ψ2 H

thermal, antara-antara
(symmetry allowed)

Similarly,

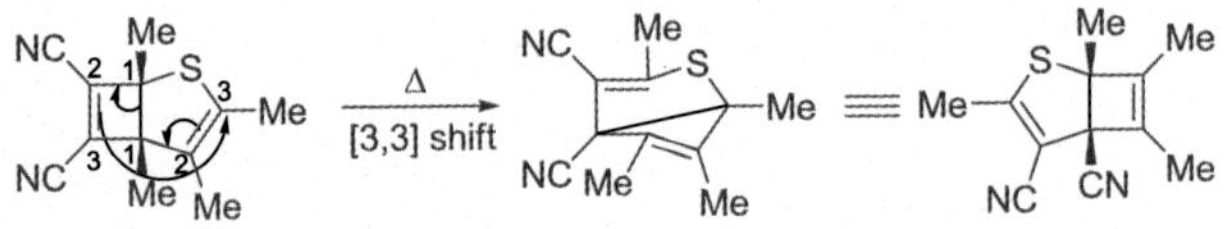

NC Me 2 1 S 3 Me NC 3 1 2 Me Me —Δ [3,3] shift→ NC Me S Me NC Me Me ≡ Me S Me Me Me NC CN Me

Q 5. Rationalize the following transformations:

(i) 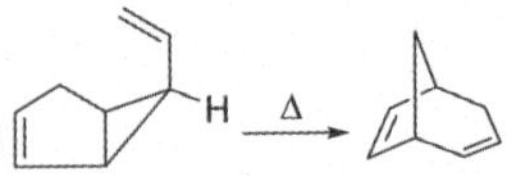 H —Δ→

(ii) 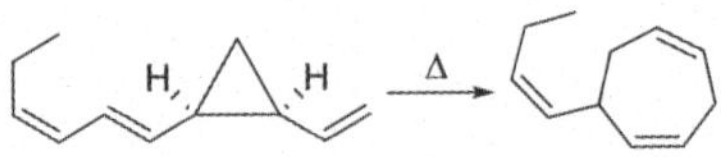 H H —Δ→

(iii) 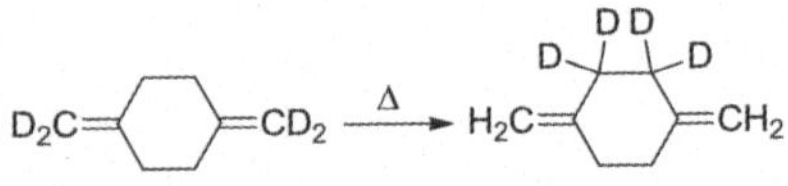 D_2C= =CD_2 —Δ→ H_2C= D D D D =CH_2

(iv) 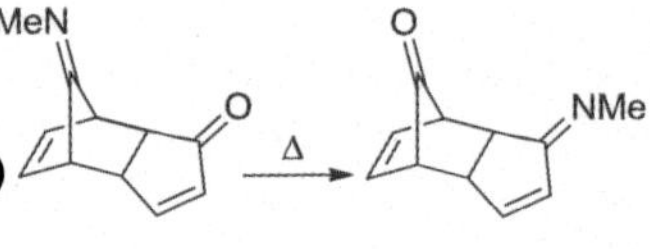 MeN O —Δ→ O NMe

(v) 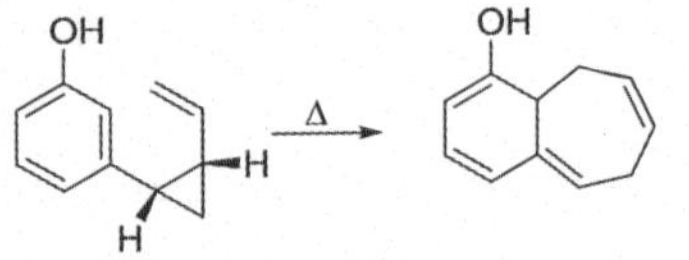 OH H H —Δ→ OH

(vi) 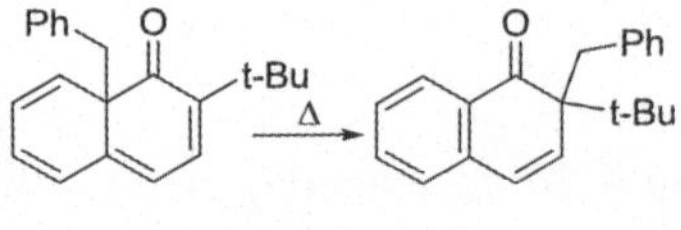 Ph O t-Bu —Δ→ O Ph t-Bu

(vii) 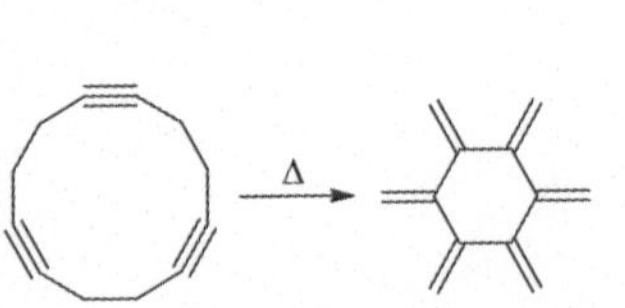 —Δ→

(viii) —Δ→ H H ⇌Δ

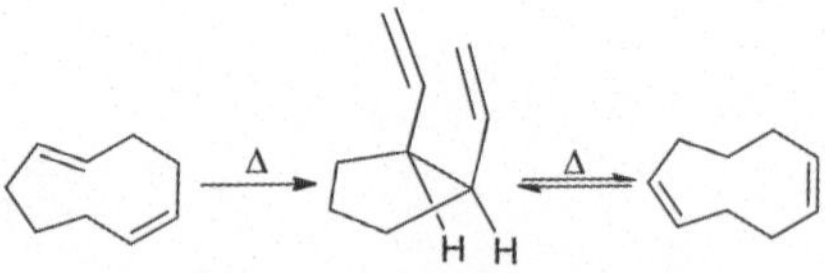

(ix) 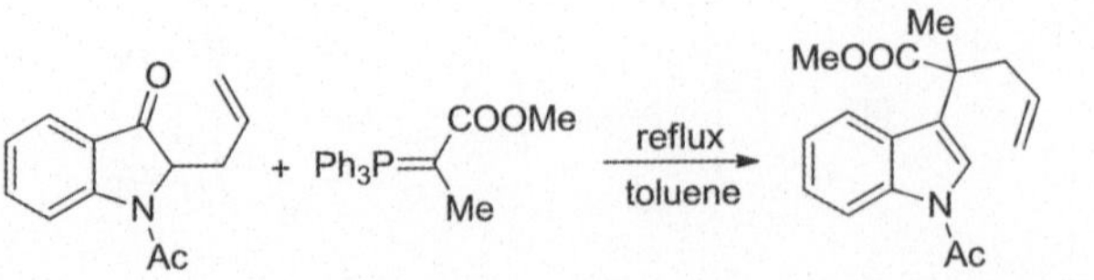 O N Ac + Ph_3P= COOMe Me —reflux, toluene→ Me MeOOC N Ac

(x) I → AgOBz (0.1 eq.), Et_3N (1.0 eq.), THF → II → Δ → III

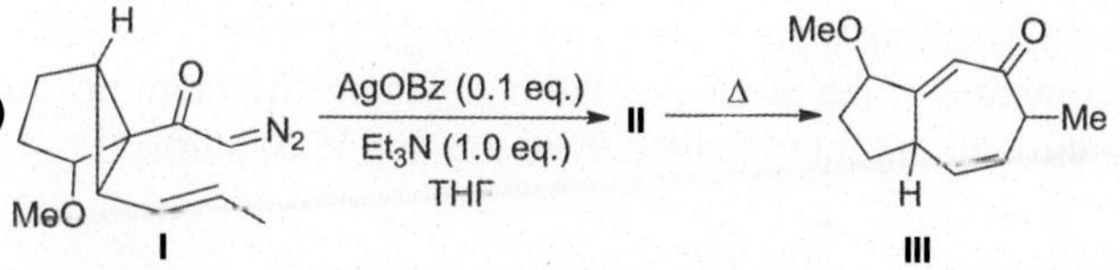

(xi) I ⇌ 70° C, 2.5 h ⇌ II → 135° C, prolonged heating → III

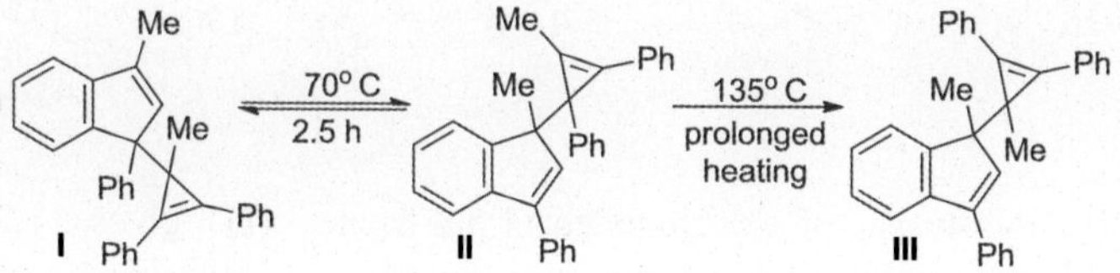

Sol 5.

(i) Δ [3,3] shift

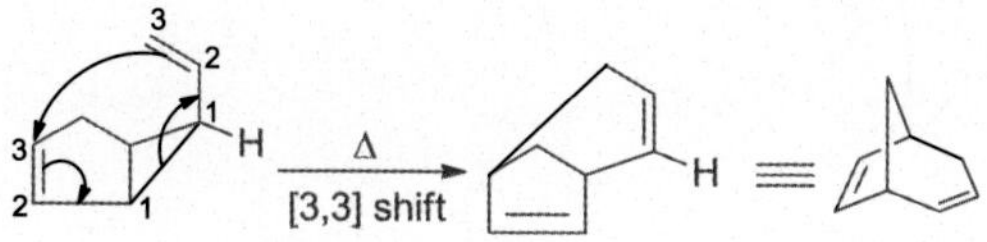

(ii) Δ [3,3] shift

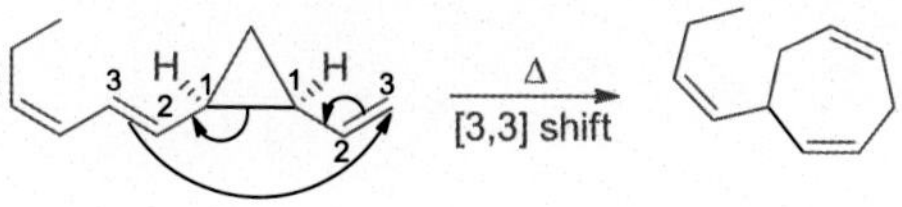

(iii) Δ [3,3] shift, boat-like T. S.

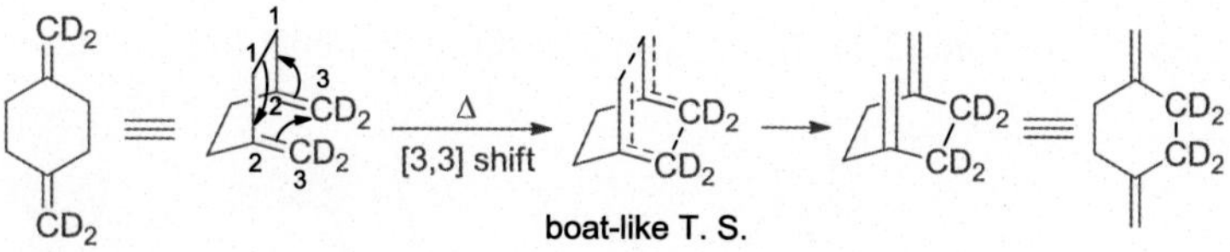

(iv) Δ [3,3] shift

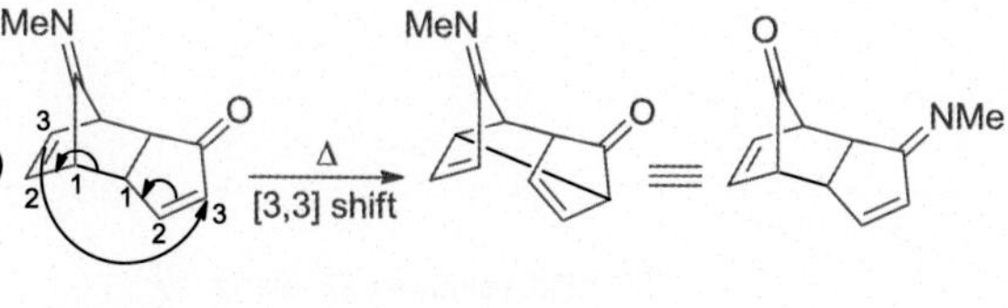

(v) Δ [3,3] shift

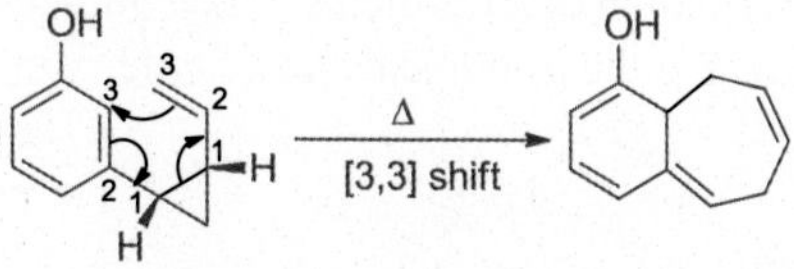

(vi) The product can be formed by a direct [1,5] sigmatropic shift or two consecutive [3,3] sigmatropic shifts. However, [1,5] shift is favored because the direct formation of the product is thermodynamically more favorable than proceeding through the intermediate from the first [3,3] shift.

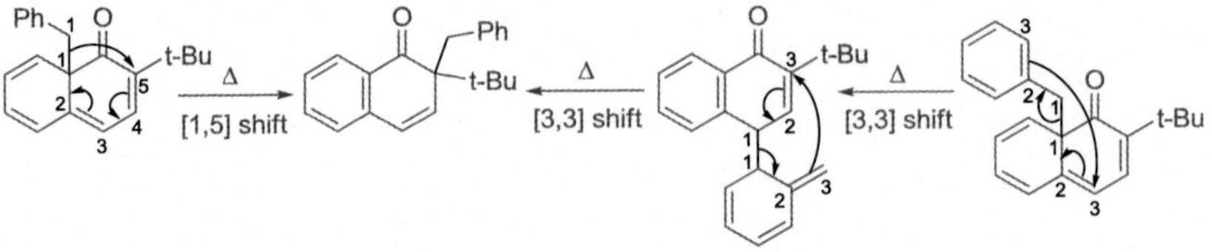

(vii) Pyrolysis of cyclododeca-1,5,9-triyne (**I**) gives 1,2,3,4,5,6-hexamethylenecyclohexane, commonly known as [6]-radialene (**II**). The mechanism involves a series of [3,3] shifts.

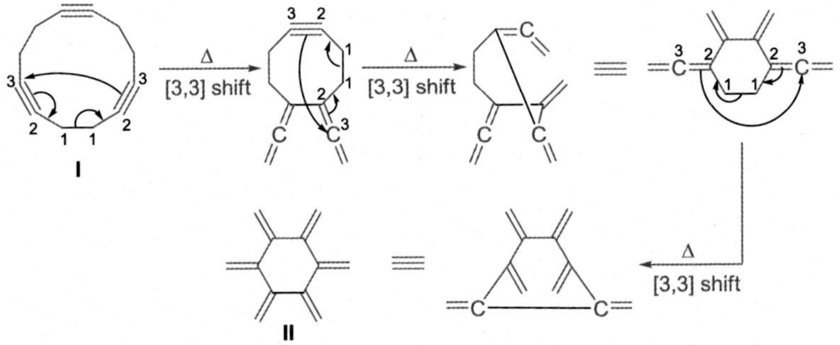

(viii) The first [3,3] shift takes place through a chair-like transition state, whereas the second [3,3] shift takes place through a boat-like transition state.

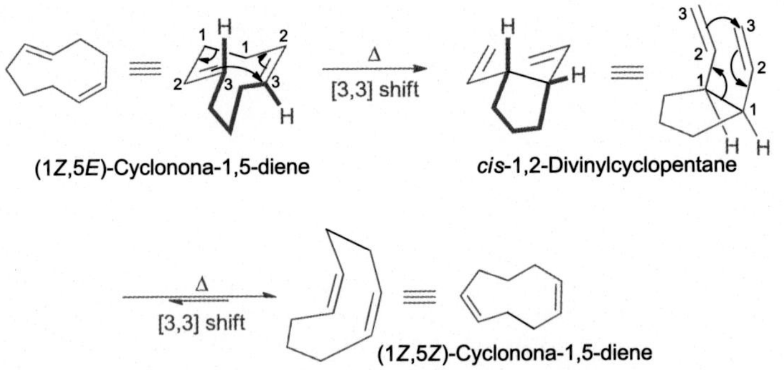

(1*Z*,5*E*)-Cyclonona-1,5-diene

cis-1,2-Divinylcyclopentane

(1*Z*,5*Z*)-Cyclonona-1,5-diene

(ix) Wittig reaction results in the formation of a 1,5-diene, which immediately undergoes the Cope rearrangement to give the 3-indoleacetic acid derivative.

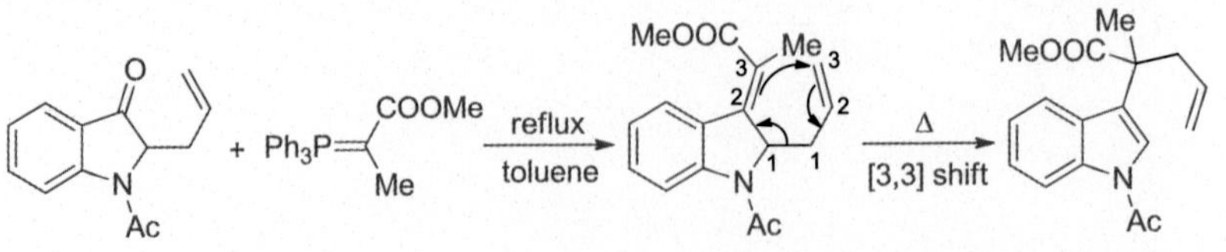

(x) Wolff rearrangement of diazoketone (**I**) gives cyclopropylketene (**II**), which immediately undergoes the Cope rearrangement to give a highly functionalized cycloheptadienone (**III**).

I —AgOBz (0.1 eq.), Et_3N (1.0 eq.), THF→ II —Δ, [3,3] shift→ III

(xi) On heating at 70 °C, [3,3] sigmatropic shifts establish an equilibrium between **I** and **II**. However, upon prolonged heating equilibrium shifts toward the more stable isomer (**III**) having all the phenyl substituents in conjugation with the double bonds.

I ⇌ (Δ, [3,3] shift) II ⇌ (Δ, [3,3] shift) [ ] ⇌ (Δ, [3,3] shift) III

Q 6. The following diene racemizes with a half-life of 24 h at 50 °C. Explain.

Sol 6. The optically active diene changes into its mirror image due to facile [3,3] shift. The rearrangement proceeds through a chair-like symmetrical transition state.

⇌ (Δ, [3,3] shift) ⇌ (Δ, [3,3] shift)

chair-like T. S.

Q 7. At 230 °C, *erythro cis,trans*-1,3,4,6-tetradeuteriohexa-1,5-diene (**I**) undergoes degenerate Cope rearrangement, but at temperatures above 260 °C, it gives *threo trans,trans*-1,3,4,6-tetradeuteriohexa-1,5-diene (**II**). Explain.

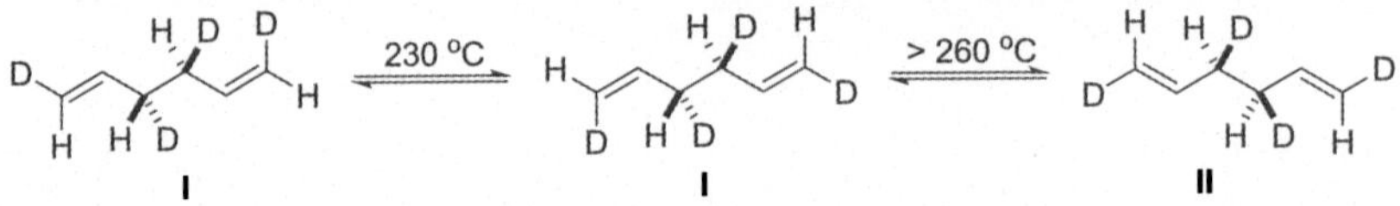

Sol 7. At 230 °C, **I** undergoes Cope rearrangement through the more favorable chair-like transition state, thereby maintaining the stereochemical degeneracy (erythro from erythro). But at temperatures above 260 °C, the rearrangement also starts taking place through the less favorable boat-like transition state to give the other diastereomer of the starting material (threo from erythro). Thus, the Cope rearrangement cannot be stereochemical degenerate when it proceeds through the boat-like transition state.

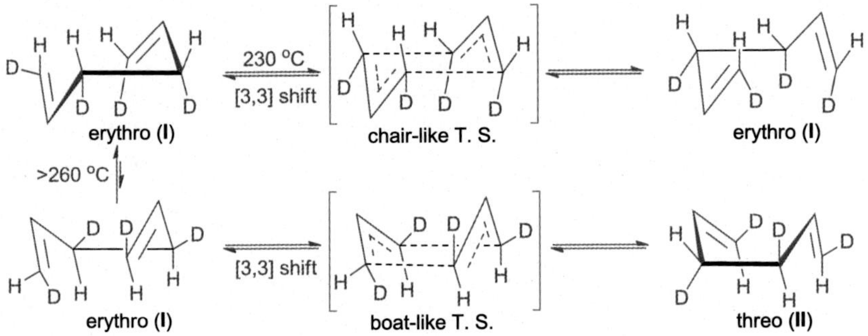

3.4.2 Degenerate Cope Rearrangement

cis-1,2-Divinylcyclopropane (**1**) rapidly undergoes Cope rearrangement to (*1Z,4Z*)-cyclohepta-1,4-diene (**2**). A favorable geometry for the interaction of the diene termini is provided by the *cis*-orientation of the molecule. In such an arrangement, transition state can be easily reached as the loss in entropy is smaller than for an acyclic diene. The entropy of activation is further reduced as the bond being broken is strained. Indeed, *cis*-1,2-divinylcyclopropanes give this rearrangement so rapidly that they generally cannot be isolated at room temperature. However, in case of *trans*-1,2-divinylcyclopropane (**3**), rearrangement to **2** cannot be concerted since the ends of the molecule where bonding is needed are too far apart (Scheme 3.13).

It should be noted that in the most stable conformation of *cis*-1,2-divinylcyclopropane, the two vinyl groups are directed away from the cyclopropane ring and *syn* ring hydrogens, whereas in the less stable conformation, the two vinyl groups point back toward the cyclopropane ring and are closer to the *syn* ring hydrogens. However, *cis*-1,2-divinylcyclopropane undergoes Cope rearrangement with less favored conformation.

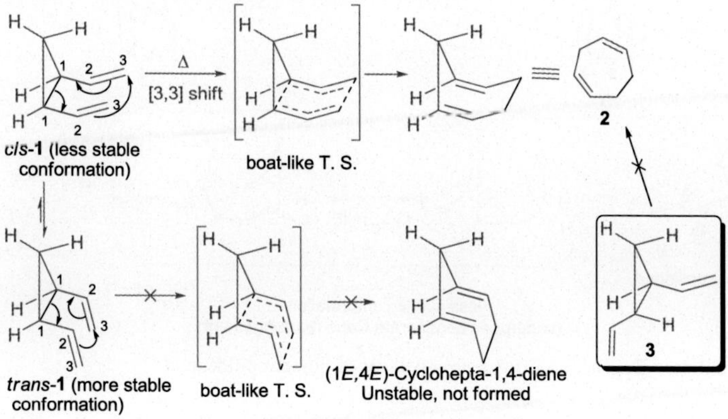

SCHEME 3.13 The Cope rearrangement of *cis*-1,2-divinylcyclopropane.

cis-1,2-Divinylcyclopropane (**1**) does not undergo Cope rearrangement through the chair-like transition state because it gives a less stable seven-membered ring with a *trans*-double bond (Scheme 3.14). Hence, the reaction proceeds through a boat-like transition state as shown in Scheme 3.13.

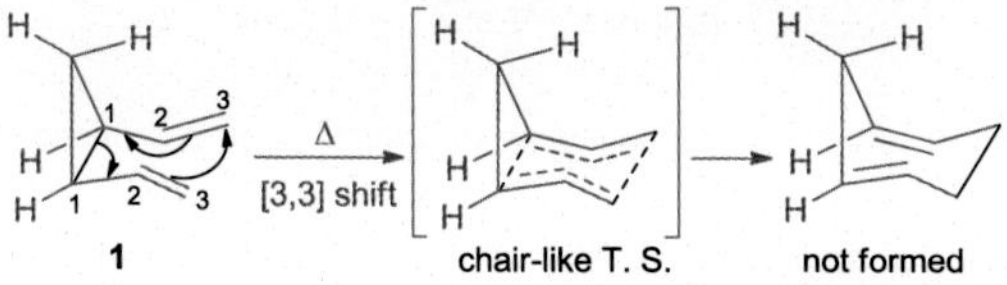

SCHEME 3.14 The Cope rearrangement of *cis*-1,2-divinylcyclopropane continued.

In case both the vinyl groups are incorporated in the ring, divinylcyclopropane rearrangement proceeds more rapidly due to the further decrease in the entropy of activation. The system then becomes homotropilidene, which undergoes a degenerate Cope rearrangement. *A* **degenerate rearrangement** *leads to the formation of a product that is indistinguishable from the reactant.* The degenerate Cope rearrangements should not be confused with resonance. Resonance structures are conceptual and are not structures in equilibrium. Like *cis*-1,2-divinylcyclopropane, homotropilidene or bicyclo[5.1.0]octa-2,5-diene also undergoes Cope rearrangement with less favored conformation (Scheme 3.15).

Although homotropilidene undergoes Cope rearrangement very rapidly, the rate can be further enhanced by eliminating the possibility of the formation of the more stable but unproductive conformation in the above equilibria. This has been done by bridging the two methylene groups in homotropilidene directly or by a methylene or an ethylene bridge to get semibullvalene, barbaralane and bullvalene, respectively (Scheme 3.16).

A **fluxional** molecule is the one that undergoes a dynamic molecular process that interchanges two or more chemically and/or magnetically

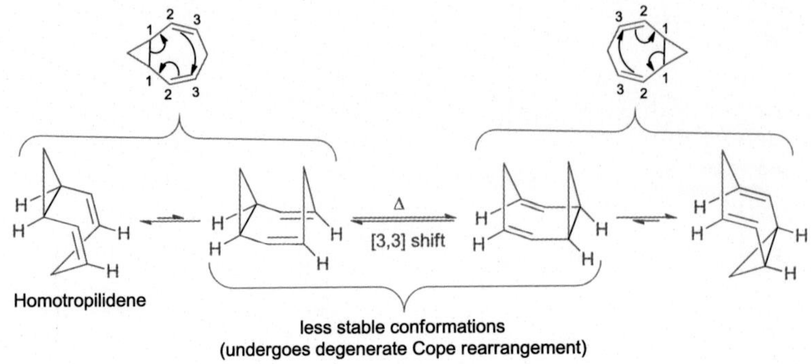

Homotropilidene

less stable conformations
(undergoes degenerate Cope rearrangement)

SCHEME 3.15 Degenerate Cope rearrangement of homotropilidene.

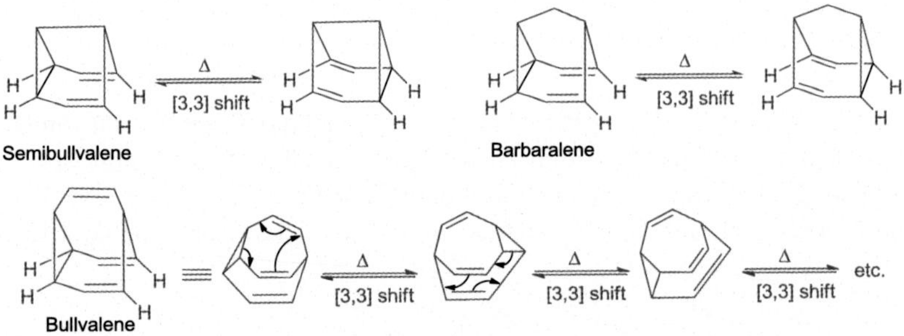

Semibullvalene

Barbaralene

Bullvalene

SCHEME 3.16 Degenerate Cope rearrangement in bridged homotropilidene systems.

different groups in a molecule. If the rate of this exchange is faster than the time scale of spectroscopic observations, these two different groups can appear to be identical. Bullvalene is the best example of fluxional molecules. The bullvalene molecule is a cyclopropane platform with three vinyl arms conjoined at a methine group. As a fluxional molecule, bullvalene is subject to degenerate Cope rearrangement. Bullvalene has been synthesized and its ^{1}H NMR spectrum was determined. At 100 °C, the rate of rearrangements is very fast so that all hydrogen atoms appear equivalent and only one signal is observed at 4.2 ppm. But at −25 °C, the rate of rearrangements becomes slow and hence two signals in the ratio 4:6 are observed. This is in accord with a single nontautomeric structure. The four are the allylic protons and the six are the vinylic ones. The ^{13}C NMR spectrum of bullvalene also shows only one peak at 100 °C.

3.4.2.1 Solved Problem

Q 1. Among the following dienes, the one that undergoes a degenerate Cope rearrangement is:

(a) 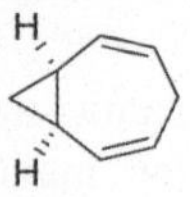 (b)  (c) 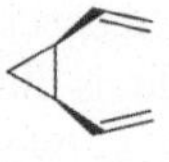 (d)

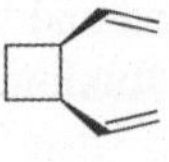

Sol 1. (a) Only the diene (a) leads to a product that is indistinguishable from the reactant, i.e., undergoes degenerate Cope rearrangement.

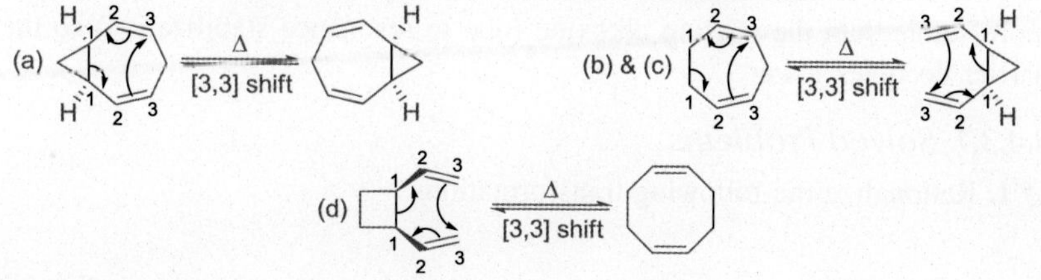

(a) H 1 2 3 H 1 2 3 Δ [3,3] shift H H
(b) & (c) 1 2 3 1 2 3 Δ [3,3] shift 3 2 1 H 3 2 1 H
(d) 1 2 3 1 2 3 Δ [3,3] shift

3.4.3 Oxy-Cope and Anionic Oxy-Cope Rearrangements

In the 1,5-diene system, a hydroxyl substituent at C-3 gives the enol product. Subsequent tautomerization of enol providing the carbonyl compound is the driving force for the reaction. The process is termed as the *Oxy-Cope* rearrangement (Scheme 3.17).

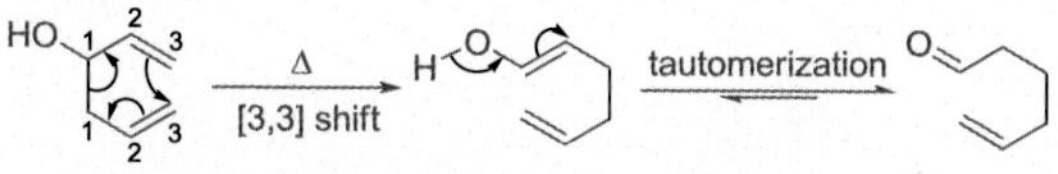

HO 1 2 3 1 2 3 Δ [3,3] shift H O tautomerization O

SCHEME 3.17 Oxy-Cope rearrangement of hexa-1,5-dien-3-ol.

If the C-3 hydroxyl group is deprotonated first, the subsequent Oxy-Cope is much faster and is known as the *anionic Oxy-Cope reaction*. The reaction is greatly accelerated when the C-3 hydroxyl group is converted to the alkoxide (Scheme 3.18).

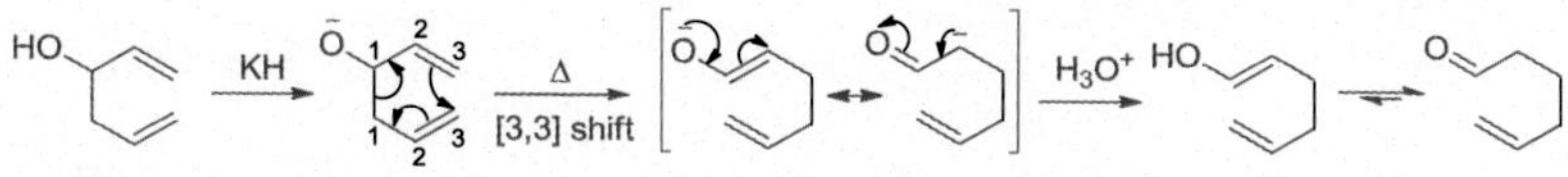

HO KH O 1 2 3 1 2 3 Δ [3,3] shift O O H_3O^+ HO O

SCHEME 3.18 Anionic Oxy-Cope rearrangement of hexa-1,5-dien-3-ol.

As shown in Figure 3.15, enhancement in the rate of anionic Oxy-Cope rearrangements is due to the following reasons: (1) In the ground state, the anion is localized on oxygen; in the transition state it is spread out over a highly delocalized system. This stabilizes the transition state relative to the

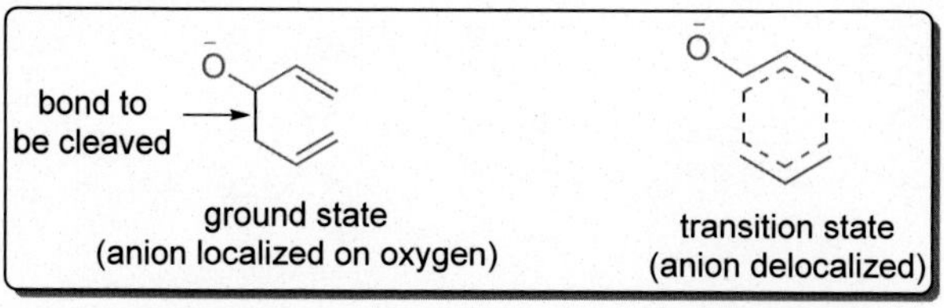

bond to be cleaved
ground state (anion localized on oxygen)
transition state (anion delocalized)

FIGURE 3.15 Origin of anion acceleration in the Oxy-Cope rearrangement.

ground state (a smaller intrinsic barrier). (2) Donation of the lone pairs of electrons by oxygen into the adjacent C–C σ* antibonds weakens the bonds to be cleaved (a smaller intrinsic barrier). (3) The product alkoxide (enolate) is more stable than the starting alkoxide (due to resonance stabilization) so the barrier becomes lower.

3.4.3.1 Solved Problems

Q 1. Rationalize the following transformations:

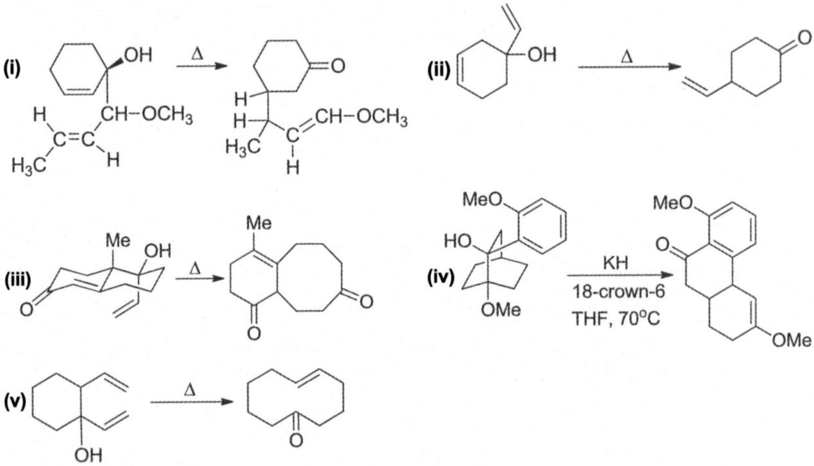

(i) OH, H, CH–OCH_3, C=C, H_3C, H — Δ → O, H, H, H_3C, C, CH–OCH_3, H

(ii) OH — Δ → O

(iii) Me, OH, O — Δ → Me, O, O

(iv) MeO, HO, OMe — KH, 18-crown-6, THF, 70°C → MeO, O, OMe

(v) OH — Δ → O

Why the new double bond in the product is *trans*?

Sol 1.

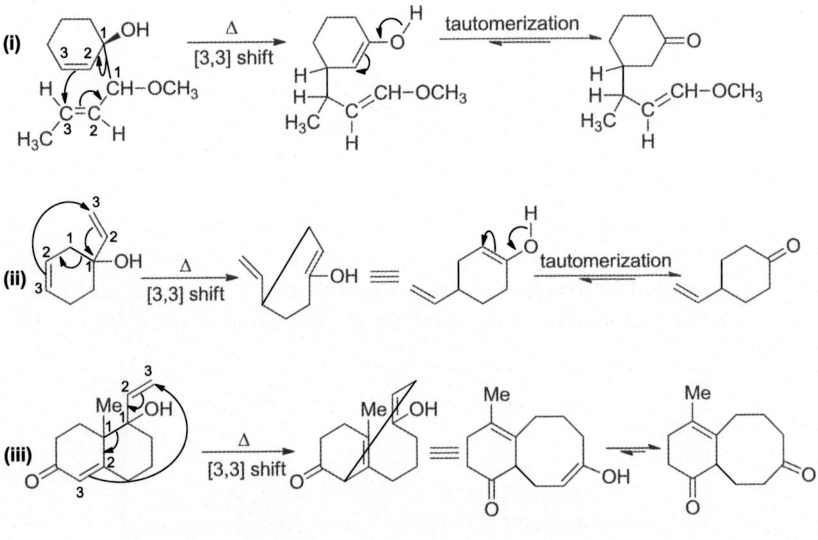

(i) 1, OH, 3, 2, 1, H, CH–OCH_3, C=C, H_3C, 3, 2, H — Δ [3,3] shift → H, O, H, H, H_3C, C, CH–OCH_3, H — tautomerization ⇌ O, H, H, H_3C, C, CH–OCH_3, H

(ii) 3, 2, 1, 2, 1, 3, OH — Δ [3,3] shift → OH ≡ H, O — tautomerization ⇌ O

(iii) 3, 2, Me, OH, 1, 1, 2, O, 3 — Δ [3,3] shift → Me, OH, O ≡ Me, O, OH ⇌ Me, O, O

(iv) The given compound on treatment with KH and 18-crown-6 undergoes an anionic oxy-aromatic-Cope rearrangement. Use of KH in the presence of crown ether helps in achieving maximum charge separation and acceleration of the reaction.

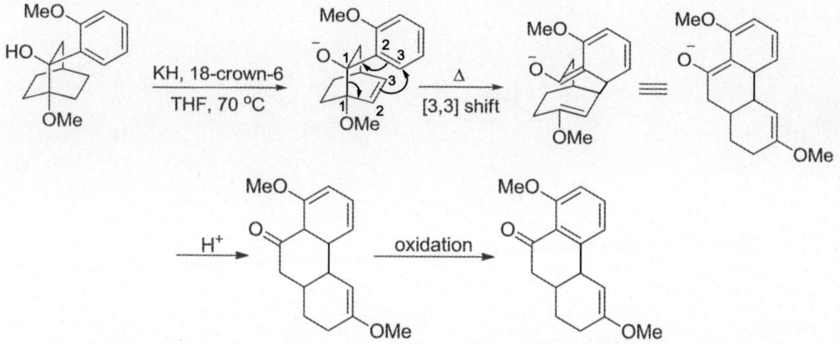

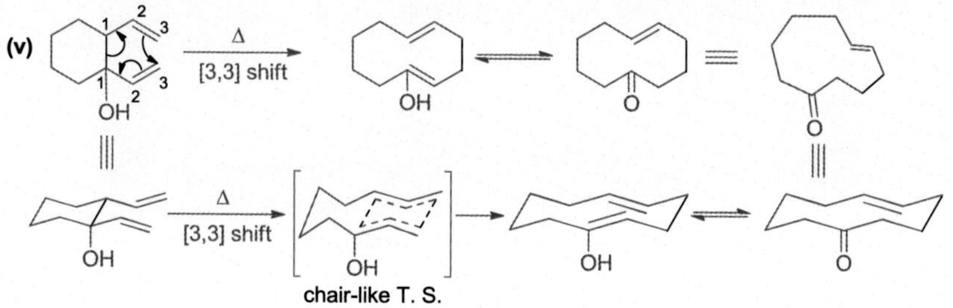

MeO HO OMe
KH, 18-crown-6
THF, 70 °C
MeO $^{-}$O 1 2 3 3 1 2 OMe
Δ
[3,3] shift
MeO $^{-}$O OMe ≡ MeO $^{-}$O OMe

H^+
MeO O OMe
oxidation
MeO O OMe

(v) 1 2 3 1 2 3 OH
Δ
[3,3] shift
OH ⇌ O ≡ O

|||

OH
Δ
[3,3] shift
OH
chair-like T. S.
OH ⇌ O

Q 2. The major product in the following reaction is:

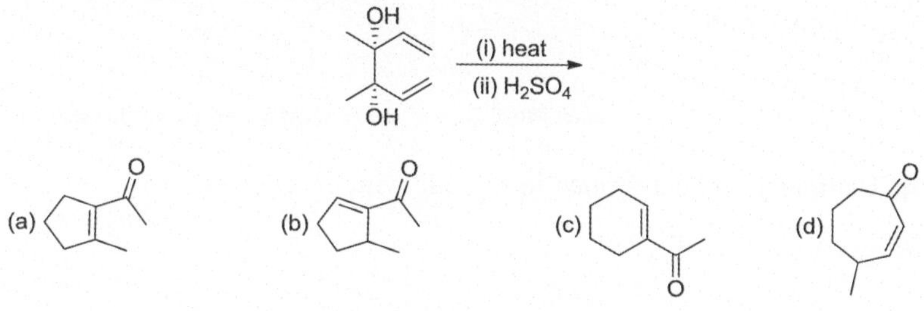

OH OH
(i) heat
(ii) H_2SO_4

(a) (b) (c) (d)

Sol 2. (a)

OH_2 3 1 1 OH 2 3
Δ
[3,3] shift
OH OH
tautomerization
O O
H^+
Aldol
HO^+ H O
HO
H^+
H_2O^+ H O
$- H_2O$
$- H^+$
O

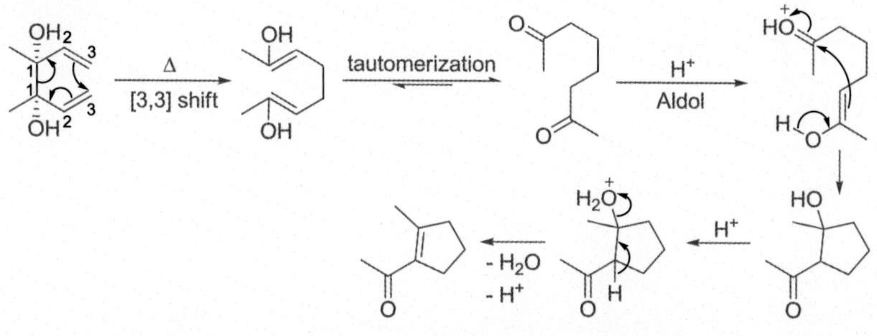

Q 3. The major product in the following reaction is:

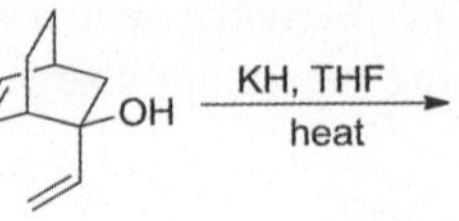

OH KH, THF heat

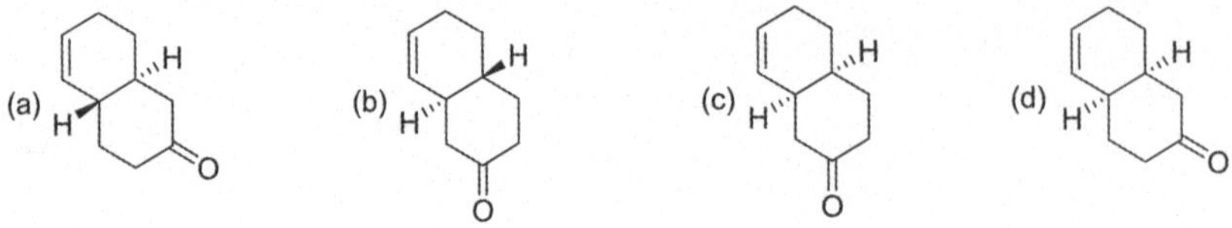

(a) H H O (b) H H O (c) H H O (d) H H O

Sol 3. (d)

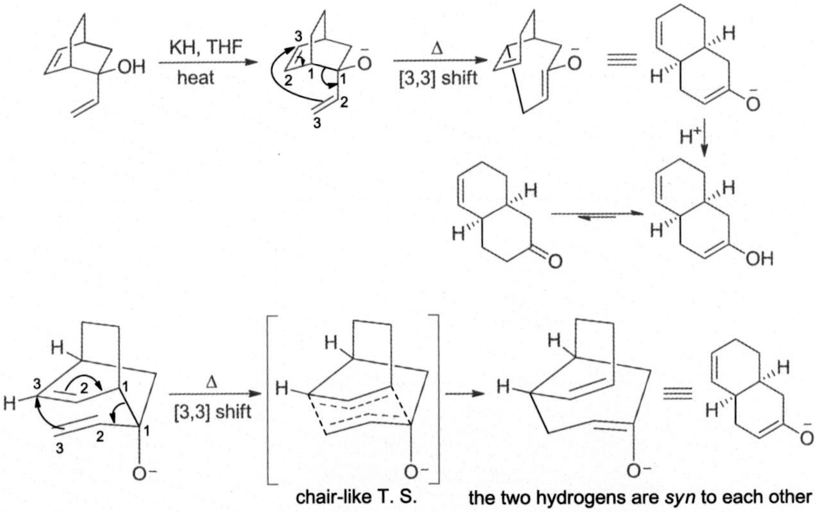

OH KH, THF heat 3 2 1 1 2 3 O⁻ Δ [3,3] shift O⁻ ≡ H H O⁻

H^+

H H O ⇌ H H OH

H 3 2 1 1 2 3 H O⁻ Δ [3,3] shift H H O⁻ → H H O⁻ ≡ H H O⁻

chair-like T. S.

the two hydrogens are *syn* to each other

Q 4. The major product formed in the following reaction is:

OH (i) KH/THF, Δ (ii) H_3O^+

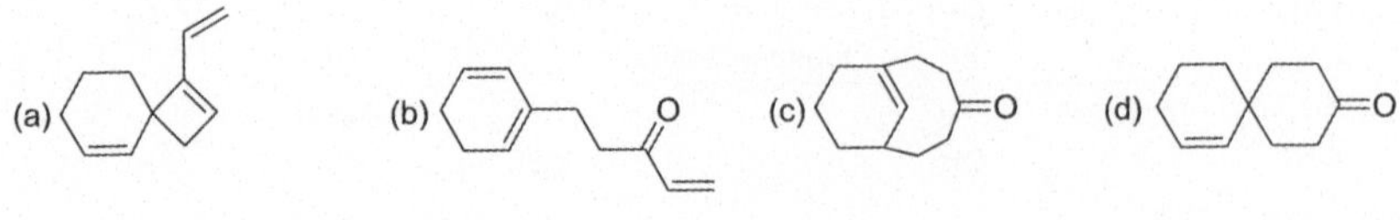

(a) (b) O (c) O (d) O

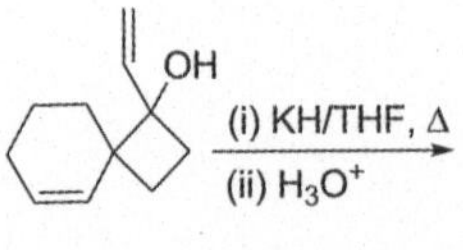

Sol 4. (c)

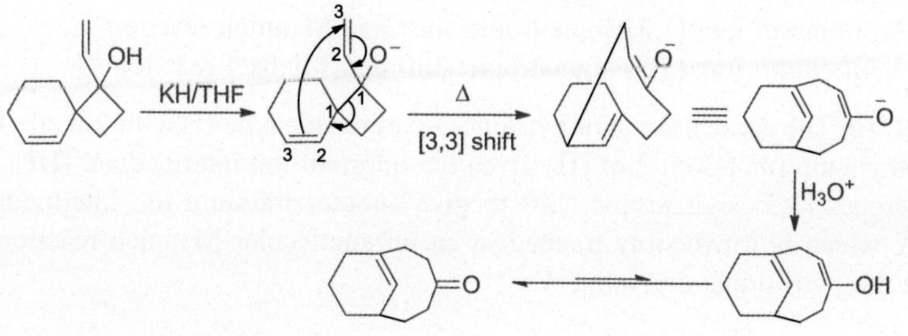

OH KH/THF 3 2 O⁻ 1 1 3 2 Δ [3,3] shift O⁻ ≡ O⁻ H_3O^+ O OH

3.4.4 Aza-Cope Rearrangement

It is just like the Oxy-Cope but now we have a nitrogen located strategically in this molecule. The Aza-Cope rearrangement is induced by the nitrogen normally present as an iminium ion in the diene. The unsaturated iminium ions rearrange significantly more rapidly than their corresponding neutral all-carbon analogues (Scheme 3.19).

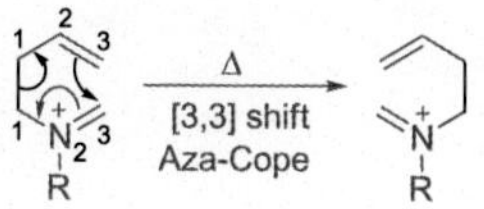

1 2 3 1 N 2 3 R Δ [3,3] shift Aza-Cope N R

SCHEME 3.19 Generalized Aza-Cope rearrangement.

If the Aza-Cope rearrangement leads to proximal enol and iminium groups, then after rearrangement, an irreversible Mannich cyclization may occur to afford acyl-substituted pyrrolidines (Scheme 3.20).

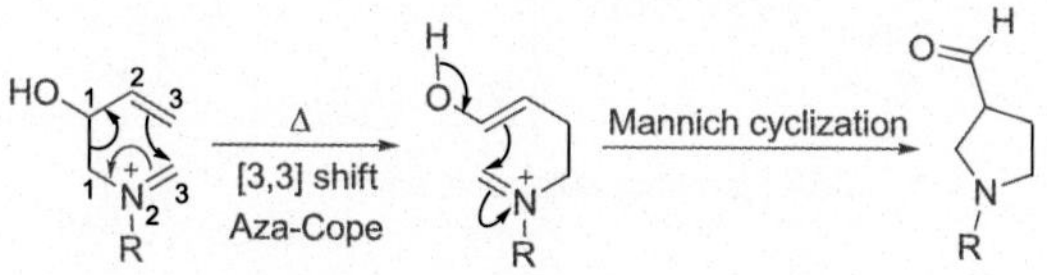

HO 1 2 3 1 N 2 3 R Δ [3,3] shift Aza-Cope H O N R Mannich cyclization O H N R

SCHEME 3.20 The Aza-Cope rearrangement followed by Mannich cyclization.

This sequence, known as the **Aza-Cope/Mannich reaction**, has become a synthetically useful method for pyrrolidine synthesis.

3.4.4.1 Solved Problems

Q 1. The following transformation involves:

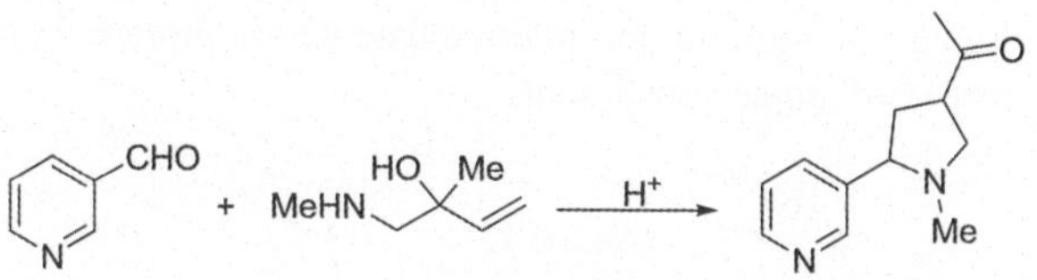

CHO N + MeHN HO Me H^+ O N Me N

(a) An iminium ion, [3,3] sigmatropic shift and Mannich reaction
(b) A nitrenium ion, [3,3] sigmatropic shift and Michael reaction
(c) An iminium ion, [1,3] sigmatropic shift and Mannich reaction
(d) A nitrenium ion, [1,3] sigmatropic shift and Michael reaction

Sol 1. (a) The condensation of pyridine-3-carboxaldehyde (**I**) with 2-methyl-1-(methylamino)but-3-en-2-ol (**II**) gives the iminium ion intermediate (**III**). **III** undergoes [3,3] sigmatropic shift to give another iminium ion intermediate (**IV**), which is irreversibly trapped in an intramolecular Mannich reaction to give the pyrrolidine derivative **V**.

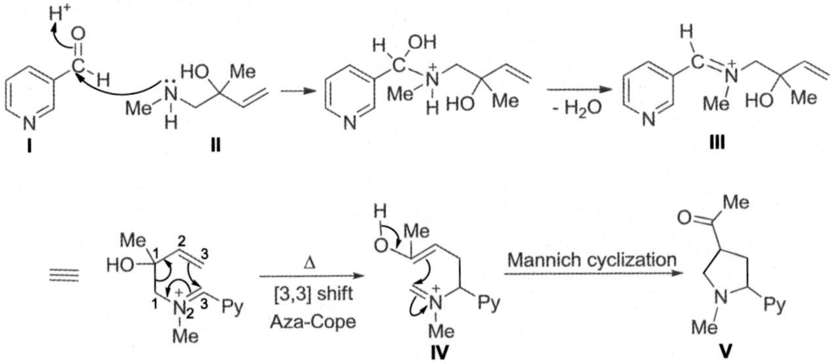

H^+ I II $-H_2O$ III

Δ [3,3] shift Aza-Cope IV Mannich cyclization V

Q 2. Show the mechanism of the following transformations:

(i) MeCHO, CSA EtOH, 78° C; R = Me; R = $CHPh_2$

CSA (camphor sulphonic acid)

(ii) Δ

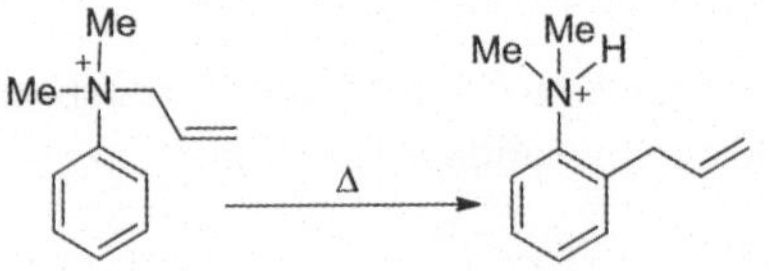

Sol 2. (i) The condensation of acetaldehyde with 2-(alkylamino)-1-(1-phenylvinyl) cyclopentanol gives the iminium ion intermediate whose preferred configuration depends on the size of amine substituents.

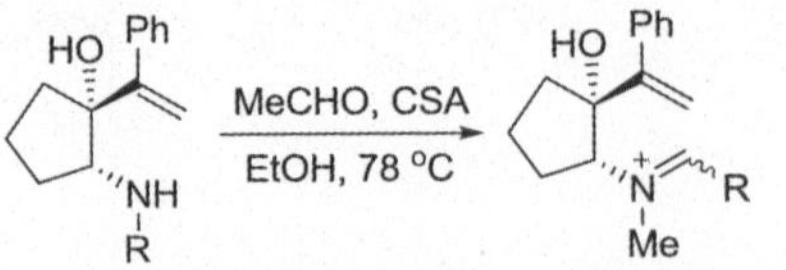

MeCHO, CSA EtOH, 78 °C

As shown below, small methyl substituent gives the (*Z*)-iminium ion (**I**), which reacts to give the *anti* diastereomer (**II**) preferentially, whereas, bulky −$CHPh_2$ substituent gives the (*E*)-iminium ion (**III**), which reacts to give the *anti* diastereomer (**IV**) preferentially.

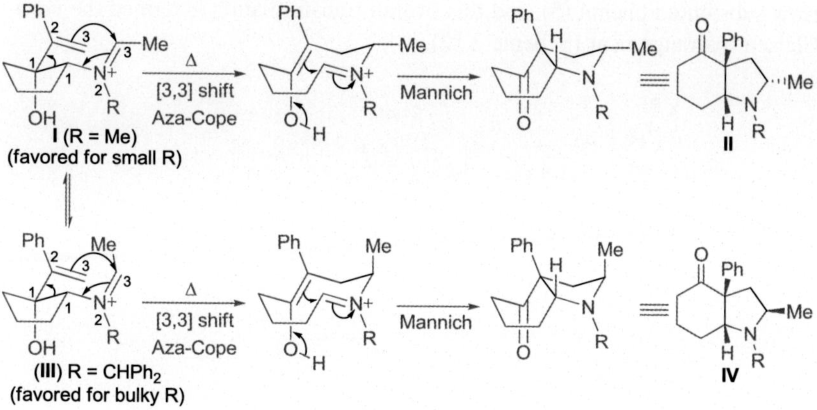

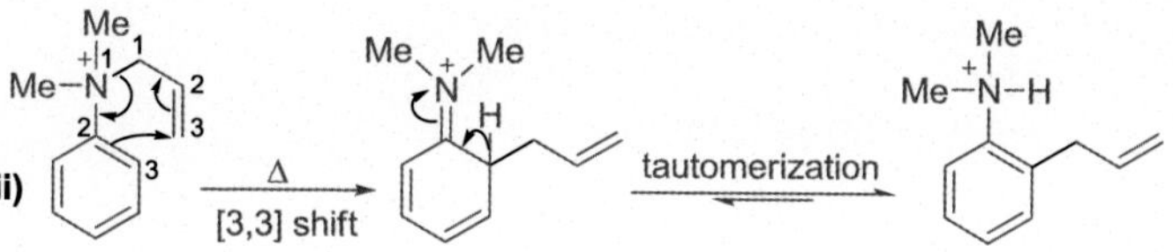

3.4.5 The Claisen Rearrangement

The **Claisen rearrangement** is another example of a thermal reaction in which the fragments do not separate during the course of the rearrangement. It is a [3,3] sigmatropic rearrangement in which an allyl vinyl ether is converted thermally to an unsaturated carbonyl compound. A typical example of the Claisen rearrangement is the conversion of allyl vinyl ether to pent-4-enal on heating (Scheme 3.21).

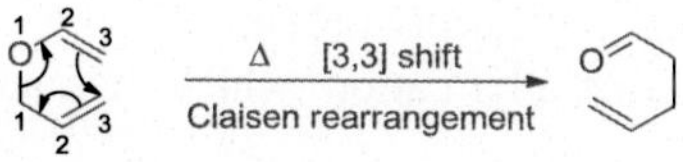

SCHEME 3.21 The Claisen rearrangement of allyl vinyl ether.

The aromatic variation of the Claisen rearrangement is the [3,3] sigmatropic rearrangement of an allyl phenyl ether to an intermediate, which quickly tautomerizes to an *ortho*-substituted phenol. In this rearrangement, an allyl group migrates from a phenolic oxygen to a carbon atom *ortho* to it.

The **aromatic Claisen rearrangement** involves the thermal transformation of a prop-2-enyl phenyl ether (**1**) into a 2-(prop-2-enyl)-cyclohexa-3,5-dienone

(**2**). When $R^3 = H$, the dienone (**2**), usually known as the *ortho*-dienone, rapidly tautomerizes to the 2-(prop-2-enyl)-phenol (**3**). This overall process is known as the *ortho*-Claisen rearrangement. If $R^3 \neq H$, the dienone (**2**) undergoes a Cope rearrangement to give a 4-(prop-2-enyl)-cyclohexa-2,5-dienone (**4**), usually termed the *para*-dienone. The *para*-dienone (**4**) rapidly tautomerizes to the *para*-substituted phenol (**5**), and this overall transformation is termed the *para*-Claisen rearrangement (Scheme 3.22).

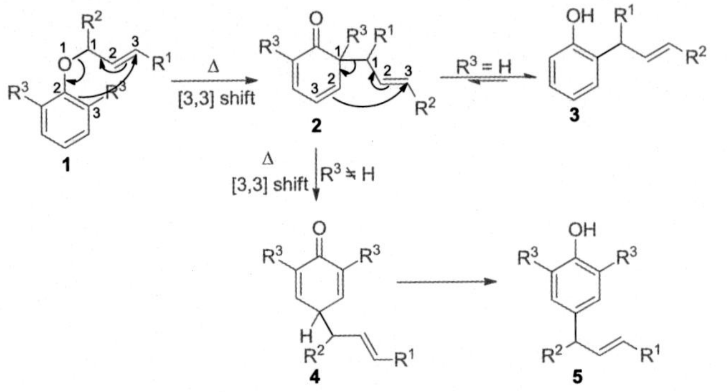

SCHEME 3.22 The aromatic Claisen rearrangement.

3.4.5.1 Solved Problems

Q 1. Thermolysis of allyl phenyl ether generates.
(a) *o*-Allylphenol only
(b) *o*- and *p*-Allylphenols
(c) *o*-, *m*- and *p*-Allylphenols
(d) *m*-Allylphenol only

Sol 1. (b)

Q 2. Thermal reaction of allyl phenyl ether generates a mixture of *ortho*- and *para*-allyl phenols. The *para*-allyl phenol is formed via:
(a) A [3,5] sigmatropic shift
(b) First *ortho*-allyl phenol is formed, which then undergoes a [3,3] sigmatropic shift
(c) Two consecutive [3,3] sigmatropic shifts
(d) Dissociation to generate allyl cation, which than adds at *para*-position

Sol 2. (c)

Q 3. The major product formed in the thermal reaction given below is:

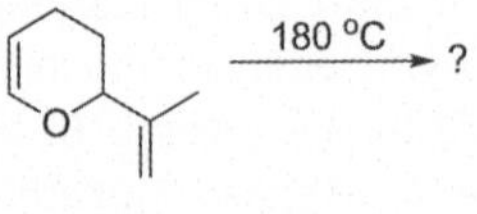

180 °C → ?

(a) 4*H*-Pyran (b) 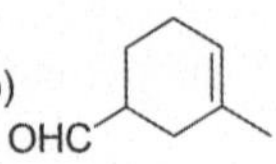 OHC (c) 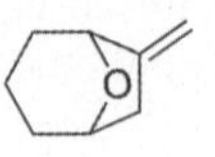 (d)

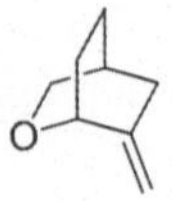

Sol 3. (b)

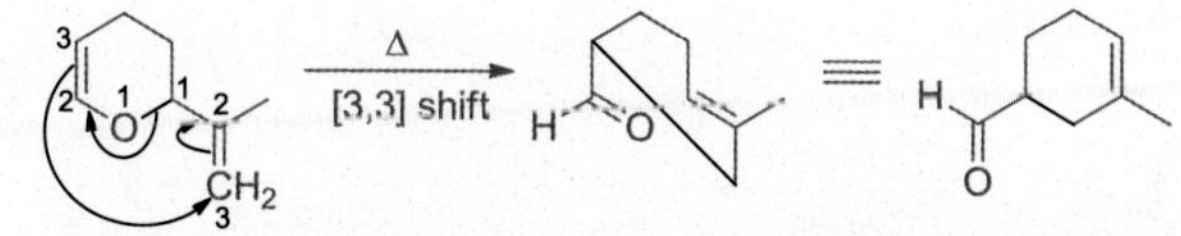

Q 4. The product obtained by heating allyl ether of 2-naphthol is:

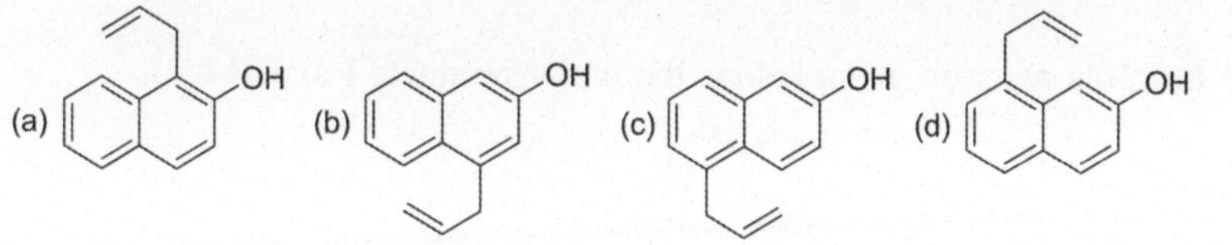

(a) (b) (c) (d)

Sol 4. (a) Allyl ether of 2-naphthol (**I**) on heating undergoes Claisen rearrangement to give 1-allylnaphthalen-2-ol (**II**) and 3-allylnaphthalen-2-ol (**III**). Reaction favors the formation of **II** as the aromaticity in the intermediate enone is lost only in one ring. However, in the formation of **III** aromaticity in the intermediate enone is lost in both the rings.

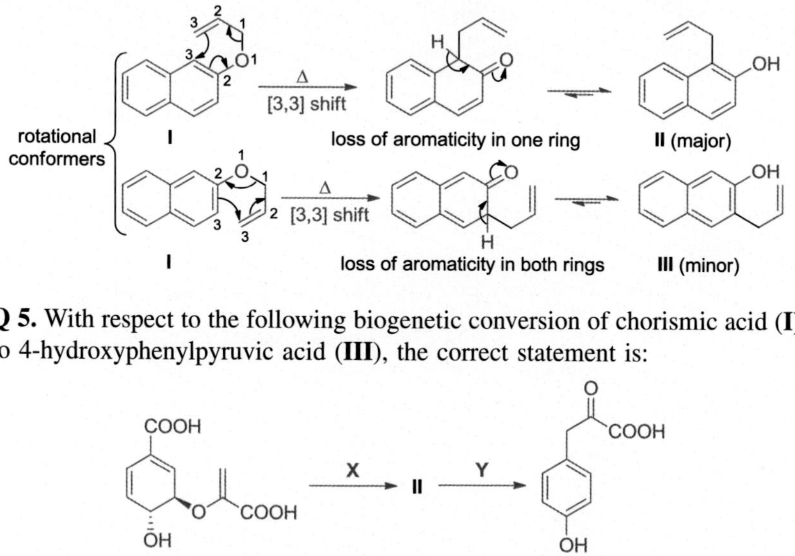

rotational conformers

I → Δ [3,3] shift → loss of aromaticity in one ring → II (major)

I → Δ [3,3] shift → loss of aromaticity in both rings → III (minor)

Q 5. With respect to the following biogenetic conversion of chorismic acid (**I**) to 4-hydroxyphenylpyruvic acid (**III**), the correct statement is:

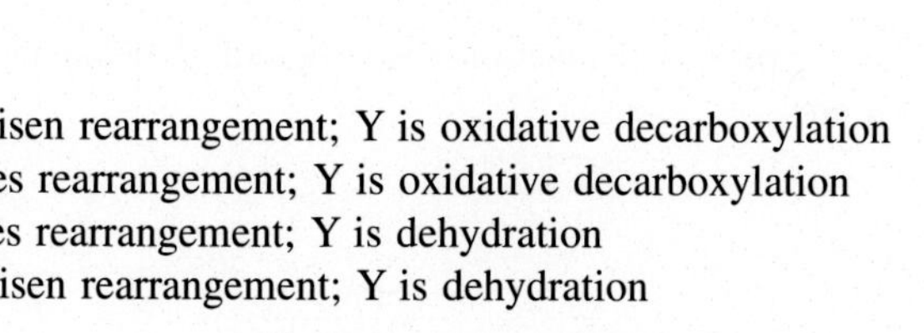

I → X → II → Y → III

(a) X is Claisen rearrangement; Y is oxidative decarboxylation
(b) X is Fries rearrangement; Y is oxidative decarboxylation
(c) X is Fries rearrangement; Y is dehydration
(d) X is Claisen rearrangement; Y is dehydration

Sol 5. (a)

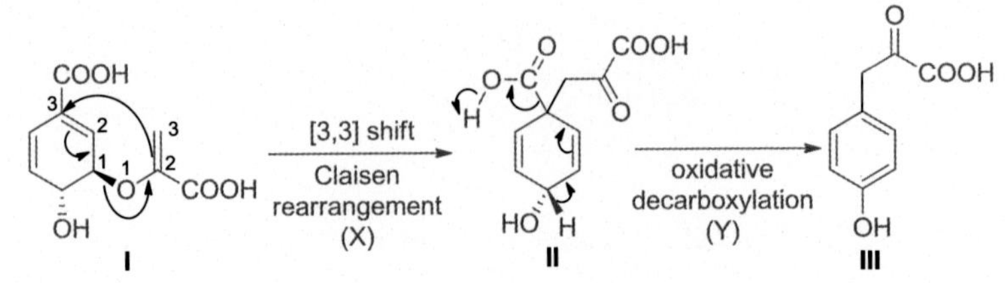

COOH, OH, COOH, I → [3,3] shift, Claisen rearrangement (X) → II (HO, H, COOH) → oxidative decarboxylation (Y) → III (OH, COOH)

Q 6. In the reaction given below, the major products **I** and **II** are:

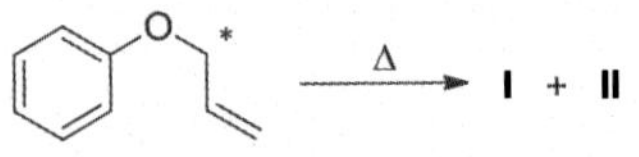

Δ → I + II

(* = ^{13}C labelled carbon)

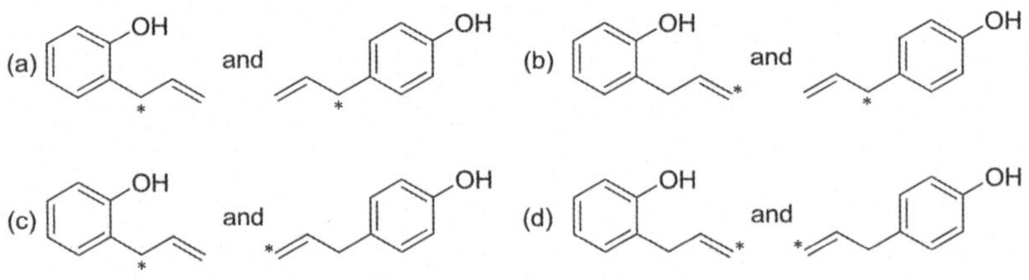

(a) ... and ... (b) ... and ...

(c) ... and ... (d) ... and ...

Sol 6. (b) Use of ^{14}C-labeled allyl phenyl ether gives an important clue regarding the mechanism of the Claisen rearrangement. The reaction was found to be consistent with a cyclic mechanism. ^{14}C-labelling has demonstrated that C-3 of the allyl group becomes bonded to the ring carbon *ortho* to the phenolic oxygen.

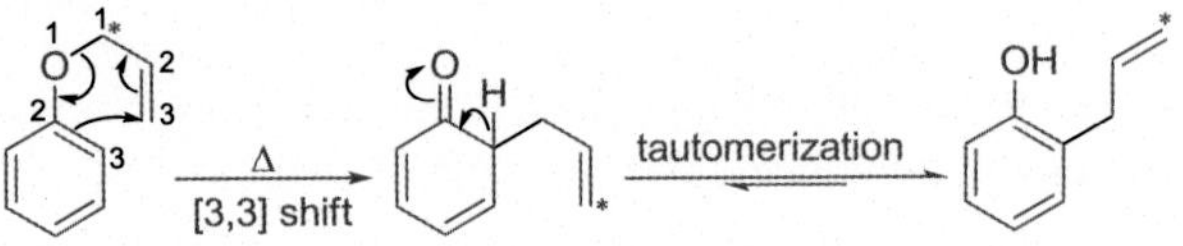

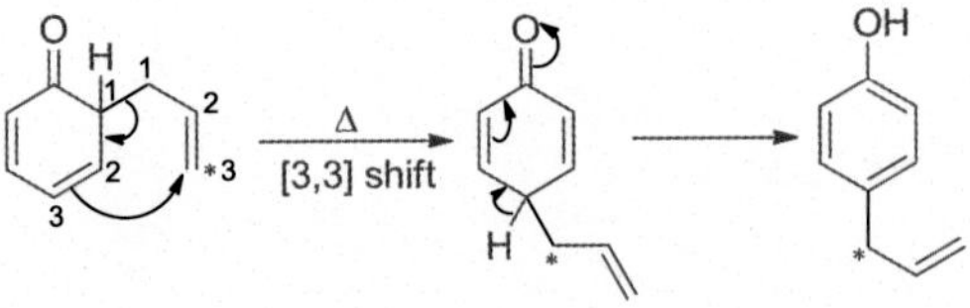

Δ [3,3] shift → tautomerization

Δ [3,3] shift →

Q 7. In the following pericyclic reaction, the structure of the allene formed and its configuration are:

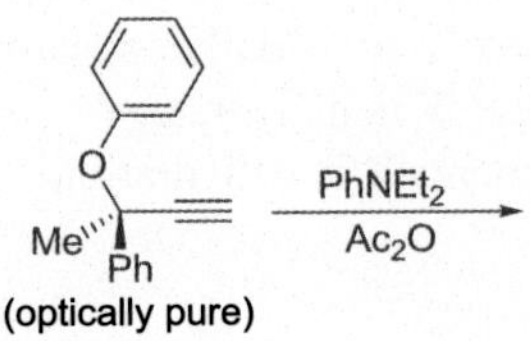

Me, Ph (optically pure) → $PhNEt_2$, Ac_2O

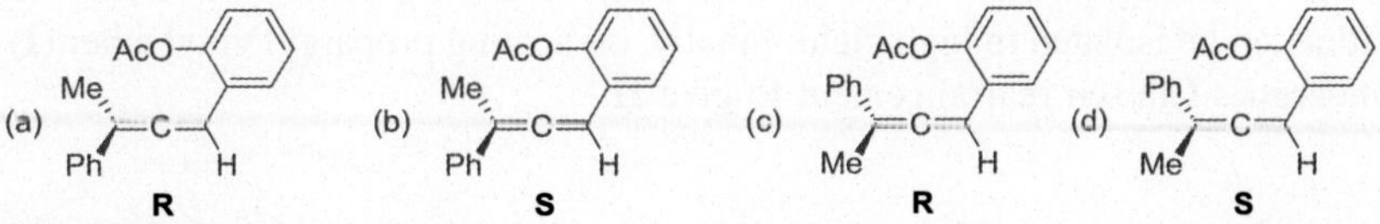

(a) R (b) S (c) R (d) S

Sol 7. (a)

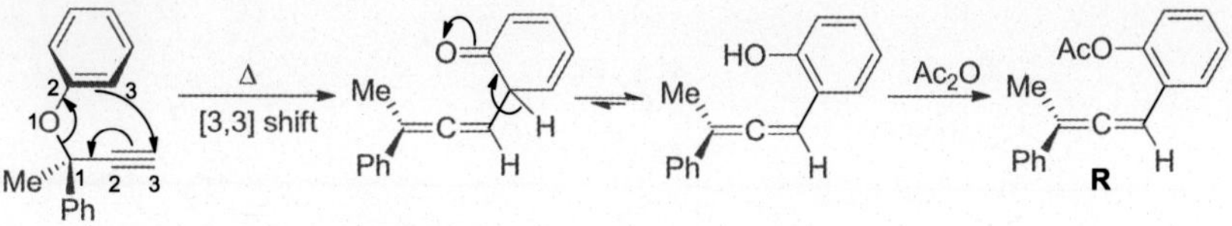

Q 8. In the following reaction sequence, the correct structures of the major products **I** and **II** are:

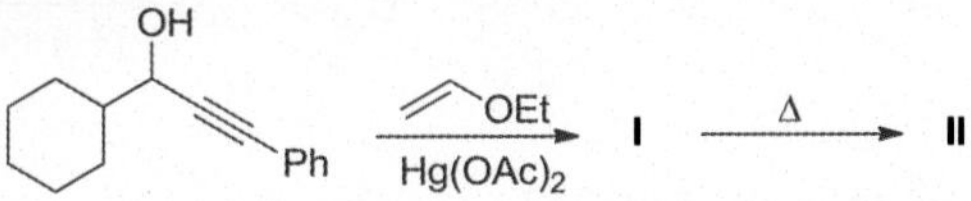

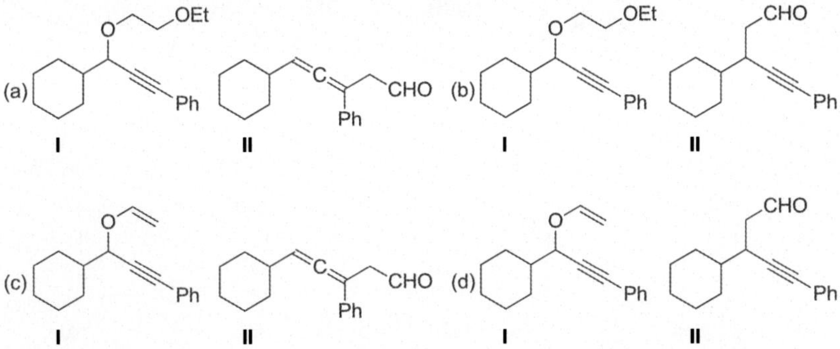

(a) I II (b) I II

(c) I II (d) I II

Sol 8. (c) Propargyl vinyl ether Claisen substrate (**I**) can be prepared by the interchange of the propargylic alcohols with alkyl vinyl ethers under the conditions of oxymercuration reaction. The propargyl alcohol is treated with a large excess of ethyl vinyl ether in the presence of catalytic amount of $Hg(OAc)_2$. The preparation involves (kind of) an oxymercuration of the C=C double bond of the ethyl vinyl ether. The attacking electrophile, i.e., $Hg(OAc)^+$ ion, forms an open-chain cation **III** as an intermediate rather than a cyclic mercurinium ion. The open-chain cation **III** is more stable than the mercurinium ion because it can be stabilized by way of carboxonium resonance. Next, cation **III** takes up the propargyl alcohol, and a protonated mixed acetal **IV** is formed. Compound **IV** eliminates EtOH and $Hg(OAc)^+$ in an E1 process, and the product **I** results. The propargyl vinyl ether **I** is in equilibrium with the substrate alcohol and ethyl vinyl ether. However, the use of a large excess of the ethyl vinyl ether shifts the equilibrium to the side of **I** so that the latter can be isolated in high yield. Finally, on heating propargyl vinyl ether (**I**) undergoes Claisen rearrangement to give **II**.

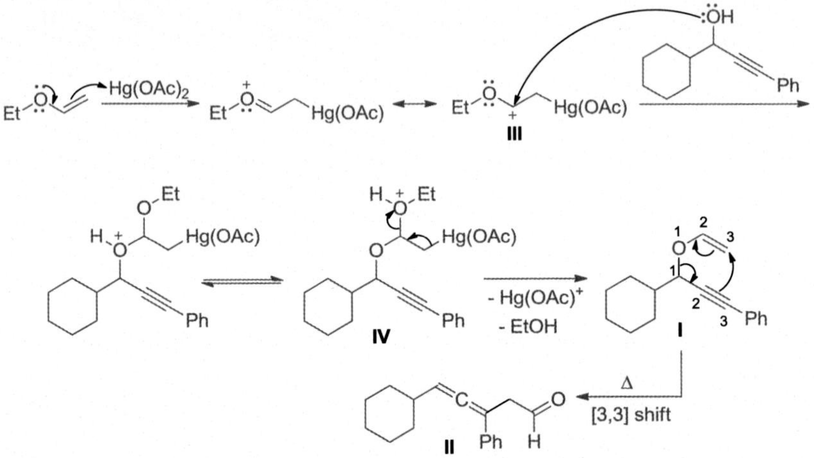

Et–O–CH=CH$_2$ + Hg(OAc)$_2$ → Et–O$^+$=CH–CH$_2$Hg(OAc) ↔ Et–O–C$^+$H–CH$_2$Hg(OAc) (**III**) → (cyclohexyl propargyl alcohol, Ph) → oxonium adduct ⇌ **IV** → (– Hg(OAc)$^+$, – EtOH) → **I** → Δ [3,3] shift → **II**

Q 9. Identify **I** and **II** in the following reaction sequence:

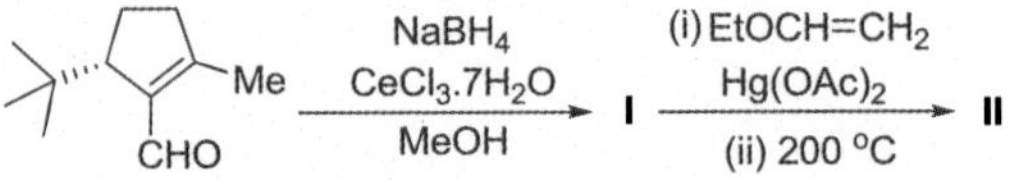

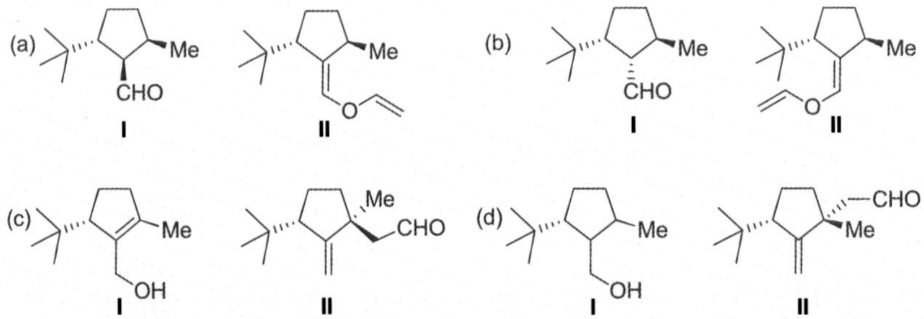

Substrate (CHO, Me) → NaBH$_4$, CeCl$_3$.7H$_2$O, MeOH → **I** → (i) EtOCH=CH$_2$, Hg(OAc)$_2$; (ii) 200 °C → **II**

(a) **I**, **II** (b) **I**, **II**

(c) **I**, **II** (d) **I**, **II**

Sol 9. (c) Sodium borohydride in the presence of $CeCl_3$ selectively reduces α,β-unsaturated carbonyl compounds into allylic alcohols. It is called **Luche's reduction**. The selective 1,2-addition is facilitated by the strongly oxophilic Ce^{3+} ion, which coordinates with the carbonyl oxygen. The allylic alcohol **I** then undergoes mercury catalyzed *trans*-etherification to give allyl vinyl ether (Claisen substrate), which readily undergoes [3,3] shift on heating to give **II**.

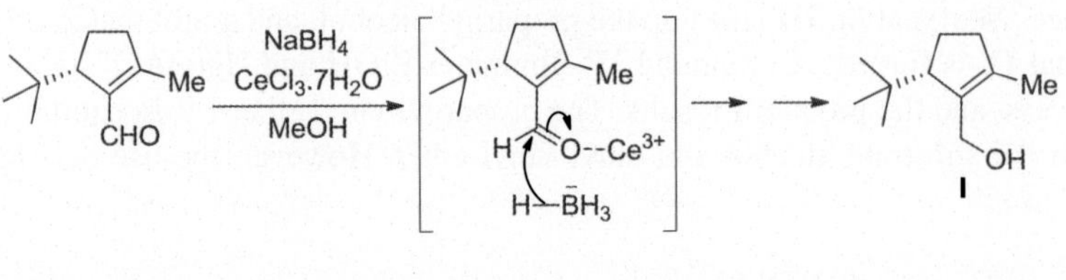

Substrate (CHO, Me) → NaBH$_4$, CeCl$_3$.7H$_2$O, MeOH → [H–C=O--Ce^{3+}, H–$\bar{B}$H$_3$] → **I** (OH, Me)

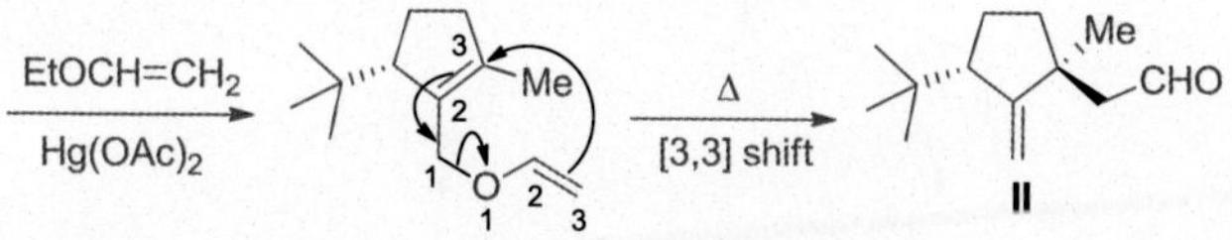

Q 10. Select the products obtained when a mixture of **I** and **II** is heated.

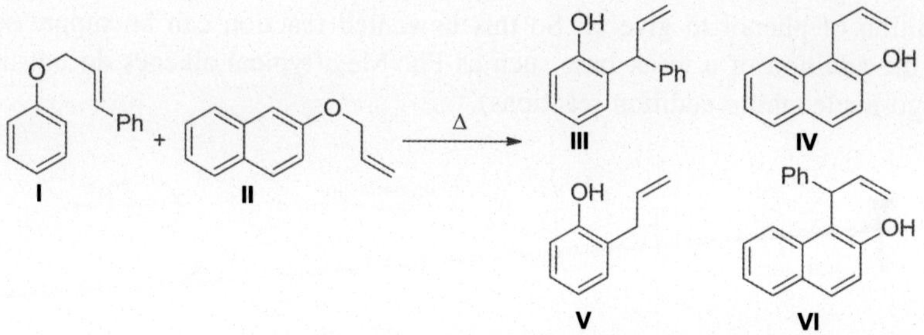

(a) **III** and **IV** (b) **V** and **VI** (c) **III**, **IV** and **V** (d) **III**, **IV**, **V** and **VI**

Sol 10. (a) The reaction was shown to be intramolecular as there was absence of the crossover products **V** and **VI** on heating a mixture of **I** and **II**. As expected, there was formation of only products **III** and **IV** in this reaction.

Q 11. Explain the formation of citral through the following reaction sequence via sigmatropic shifts.

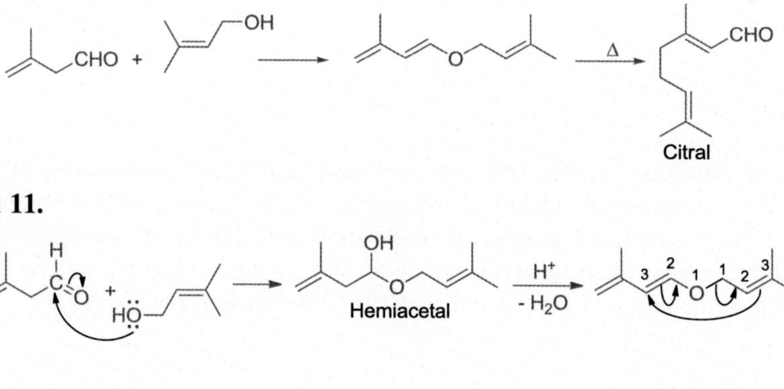

Sol 11.

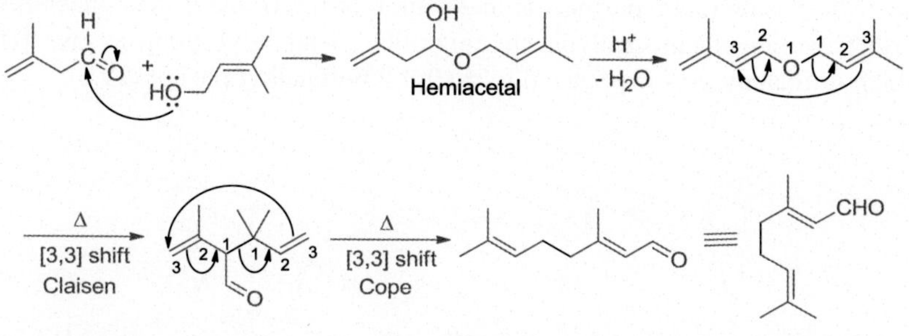

Q 12. Claisen rearrangement of allyl ether **I** often gives a product **II** (instead of the expected product). How is this unwanted product formed? Addition of a small amount of a weak base such as $PhNMe_2$ helps to prevent the unwanted reaction. What could be the role of the base? Explain.

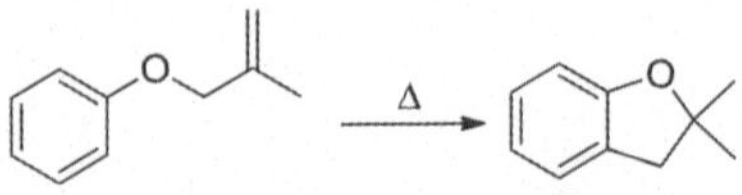

Sol 12. In the first step, expected Claisen rearrangement of allyl ether **I** takes place to give **III**. However, **III** being an alkene can undergo acid-catalyzed addition of phenol to give **II**. So this unwanted reaction can be suppressed by the addition of a weak base such as $PhNMe_2$ (typical alkenes do not undergo nucleophilic addition reactions).

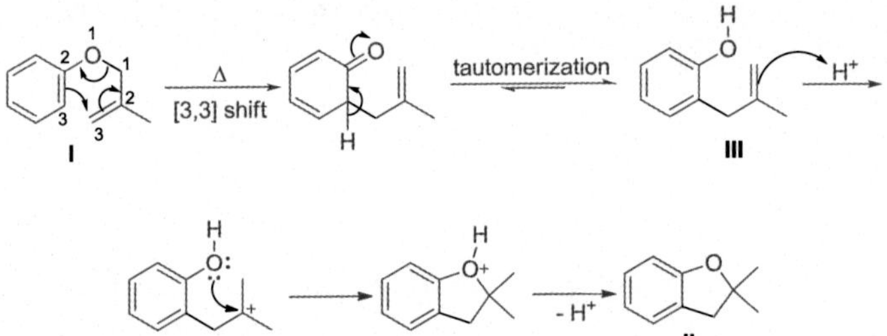

Q 13. Suggest the possible mechanism for the conversion of 2,6-dimethyl phenyl allyl ether to 4-allyl-2,6-dimethyl phenol. How will you prove that it passes through a dienone intermediate?

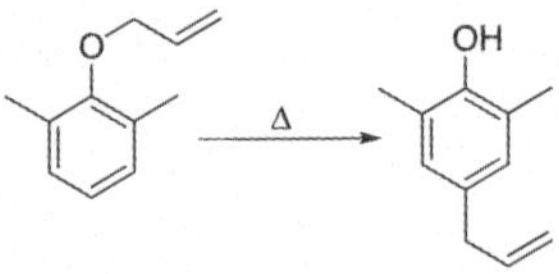

Sol 13. Mechanism: The first step involves *ortho* migration resulting in the formation of *ortho* substituted cyclohexadienone (**II**). As hydrogen is absent at the *ortho* position, this prevents aromatization of the **II** and it then undergoes further rearrangement involving the migration of the allyl group to give **III**, which ultimately gets aromatized to 4-allyl-2,6-dimethyl phenol (**IV**).

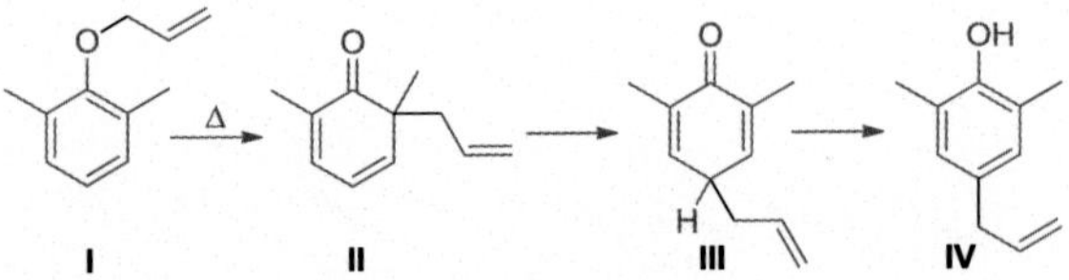

Evidence for the mechanism: (i) When the above reaction was carried out in the presence of maleic anhydride, Diels–Alder adducts are formed by the trapping of the intermediate **II**. Their formation proves that dienone is an intermediate in the thermal rearrangement of **I** to **IV**.

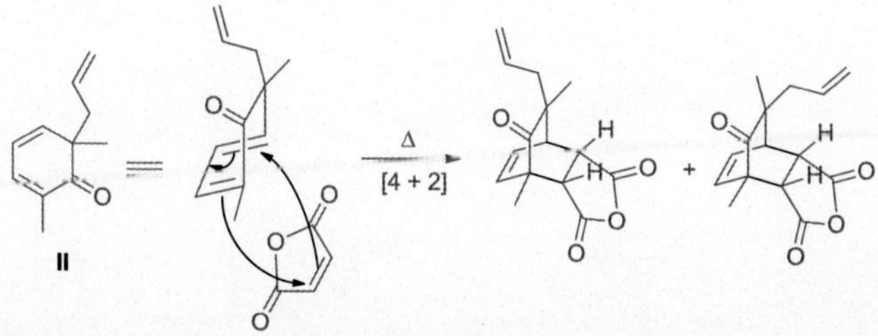

(ii) Reaction of sodium 2,6-dimethylphenoxide and prop-2-enyl bromide in benzene at 15 °C gave the *ortho*-substituted cyclohexadienone (**II**), which rearranged at 75 °C in cyclohexane to give **I** and **IV**. This showed that the **II** must be an intermediate in the aromatic Claisen rearrangement and that the reaction must occur reversibly.

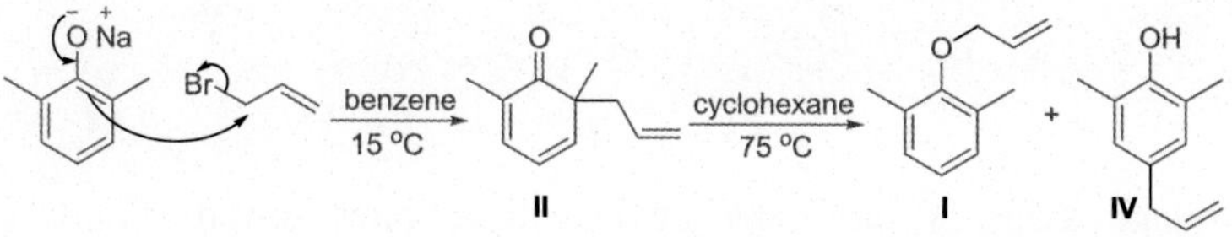

(iii) The stable 2-methyl-2-(prop-2-enyl)-3,6-di(1,1-dimethylethyl)-cyclohexa-3,5-dienone (**V**) rearranged to the *para*-dienone (**VI**) between 70 and 100 °C. The latter tautomerized to the *para*-substituted phenol (**VII**) on heating above 200 °C. The stability of the *para*-dienone (**VI**) is due to the large steric strain that develops in the aromatized product as a result of having an adjacent prop-2-enyl and 1,1-dimethylethyl group in the same plane. This reaction showed that *para*-dienones are intermediates in the *para*-Claisen rearrangement.

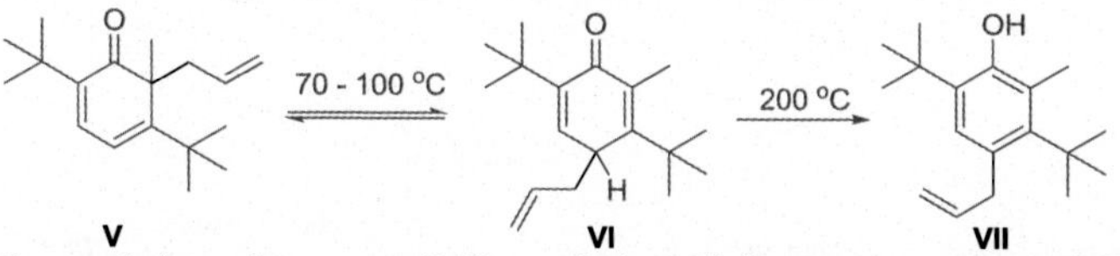

Q 14. In studying the reversibility of the Claisen rearrangement, it was observed that on heating ^{14}C-labeled 2,4,6-trimethylphenyl prop-2-enyl ether, the radioactivity got distributed between the C-1 and C-3 of the prop-2-enyl group. Explain the results.

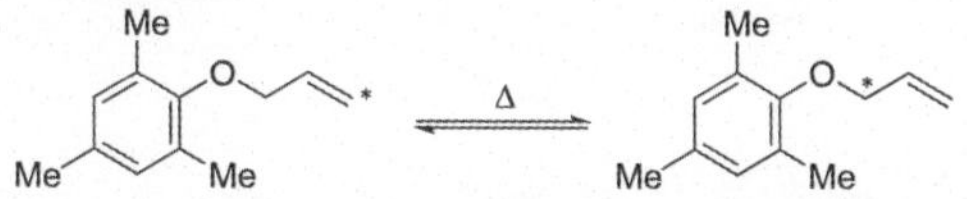

Sol 14. This phenomenon is best accounted for by invoking the intermediacy of an intramolecular Diels–Alder adduct. Cleavage of the four-membered ring in Diels–Alder adduct results in the formation of the rearranged isomer.

Diels-Alder adduct

Q 15. The four stereoisomers of 1-(prop-1-en-1-yloxy)but-2-ene are given below:

E,E *Z,E* *E,Z* *Z,Z*

Arrange them in the order of increasing rates toward the Claisen rearrangement.

Sol 15. The Claisen rearrangement shows preference for a chair-like transition state. The most favorable geometry is that in which the two larger methyl substituents are placed in the pseudoequatorial position in the transition state. This is possible only in the case of *E,E*-isomer. Therefore, the order of rates of reaction of four stereoisomers is: *E,E* > *Z,E* = *E,Z* > *Z,Z*.

Δ [3,3] shift

Claisen rearrangement

Δ [3,3] shift chair-like T. S. *E,E* *threo*

Δ [3,3] shift chair-like T. S. *Z,E* *erythro*

Δ [3,3] shift chair-like T. S. *E,Z* *erythro*

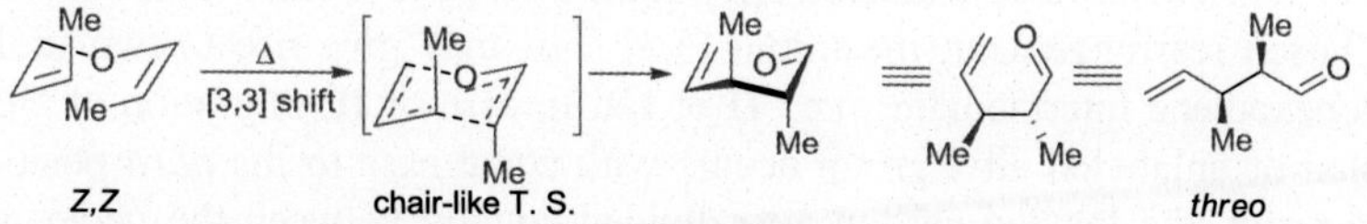

Z,Z — Δ [3,3] shift → chair-like T. S. → threo

Q 16. Predict the products in the following reactions and explain their formation.

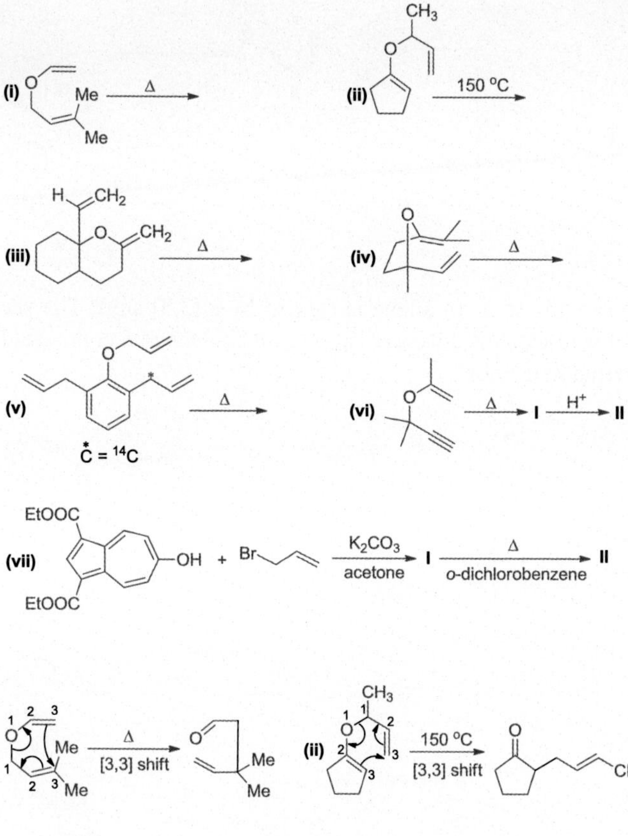

(i) Δ →

(ii) 150 °C →

(iii) Δ →

(iv) Δ →

(v) Δ → $\overset{*}{C} = {}^{14}C$

(vi) Δ → I $\xrightarrow{H^+}$ II

(vii) + Br–CH₂CH=CH₂ $\xrightarrow[\text{acetone}]{K_2CO_3}$ I $\xrightarrow[\text{o-dichlorobenzene}]{\Delta}$ II

Sol 16.

(i) Δ [3,3] shift →

(ii) 150 °C [3,3] shift →

(iii) Δ [3,3] shift →

(iv) Δ [3,3] shift →

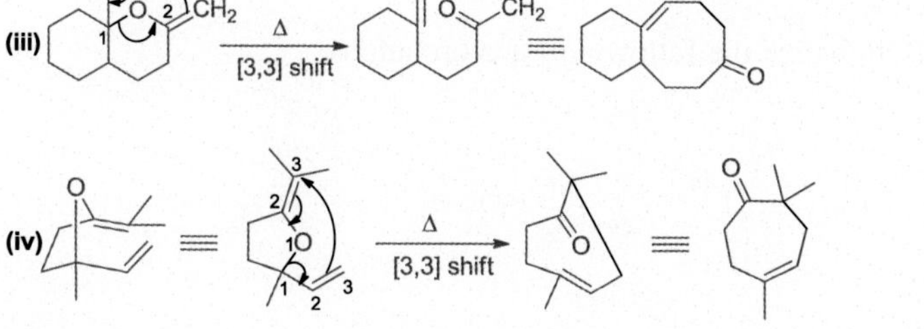

(v) When *ortho*-α-C^{14}-labeled allyl-2,6-diallyl phenyl ether (**I**) is subjected to Claisen rearrangement, the initial [3,3] shift may give rise to two triallyl cyclohexadiene intermediates, i.e., **II** or **III**. In case of **II**, migration of either labeled or unlabeled allyl group occurs with equal ease to the *para* position. As a result, the labeled carbon gets divided equally between the *ortho*- and *para*-positions in the rearranged products **IV** and **V**. This experiment proves the intermediacy of cyclohexadiene intermediates.

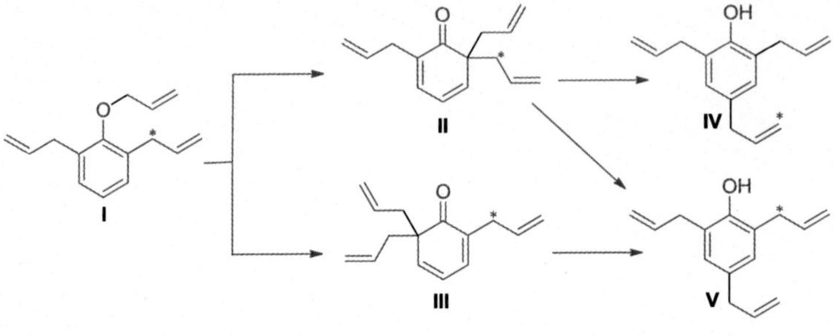

(vi) In the first step, an allene is formed by a [3,3] shift. The second step involves an acid-catalyzed isomerization of the allene to a conjugated dienone via an intermediate enol.

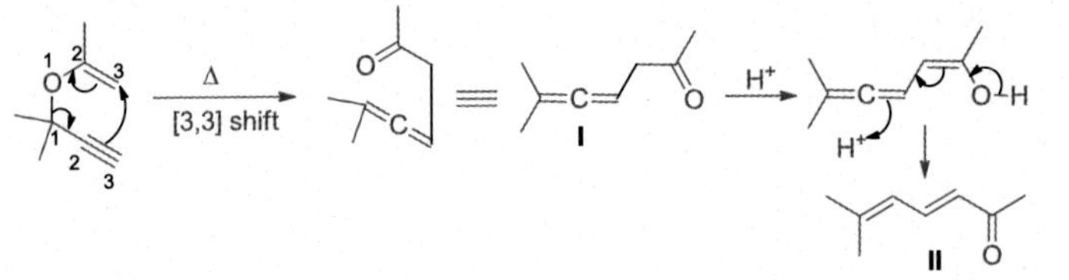

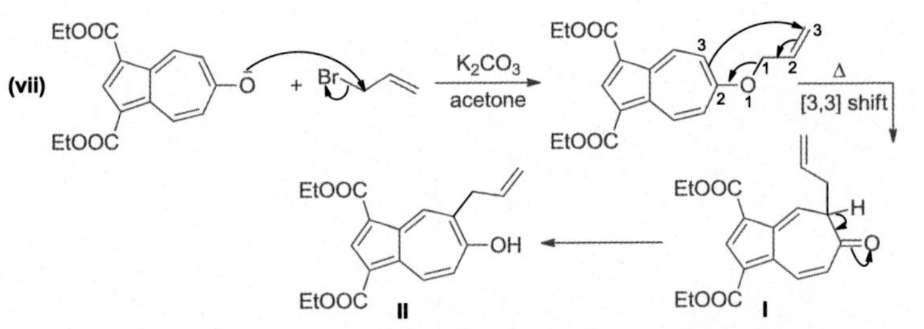

Δ [3,3] shift

H^+

(vii) K_2CO_3 acetone

Δ [3,3] shift

Q 17. Rationalize the following transformations:

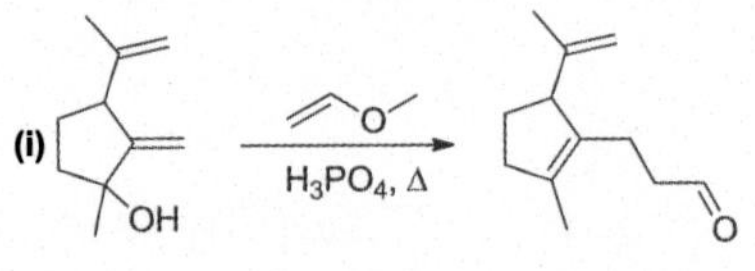

(i) H_3PO_4, Δ

(ii) 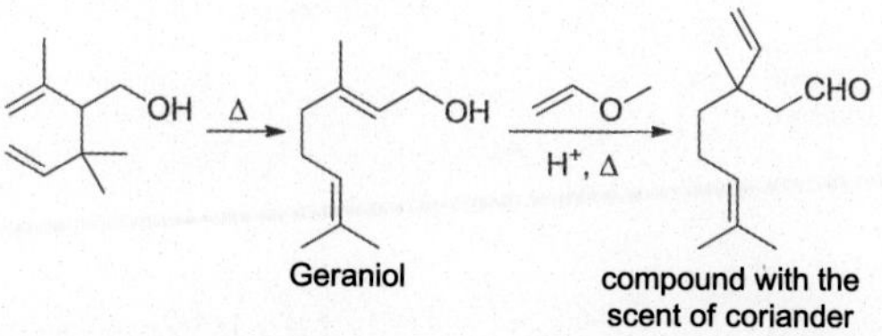 OH Δ → OH; O, H^+, Δ → CHO

Geraniol

compound with the scent of coriander

(iii) 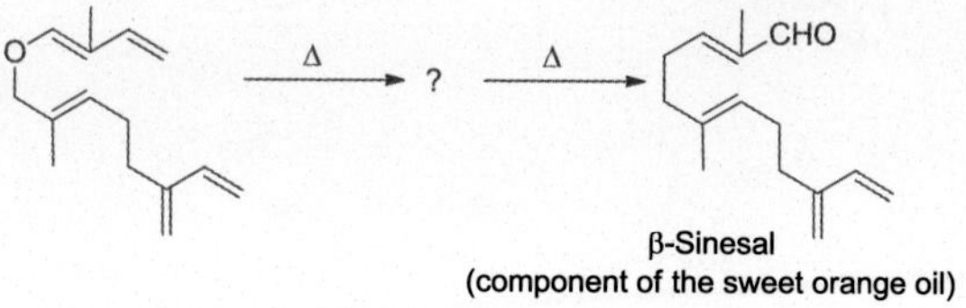 O Δ → ? Δ → CHO

β-Sinesal (component of the sweet orange oil)

(iv) 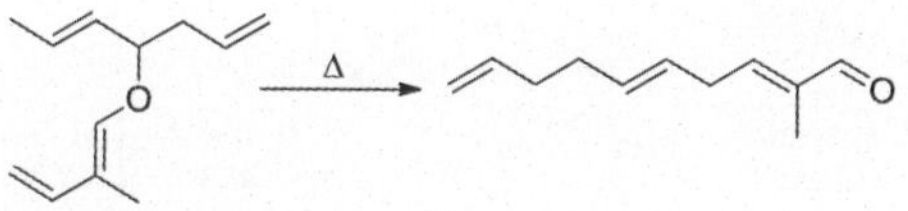 O Δ → O

(v) 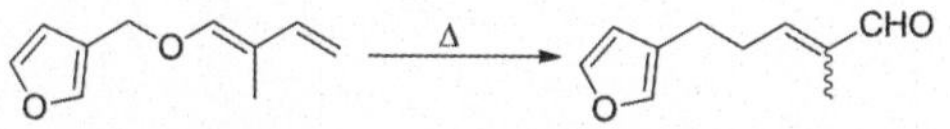 O O Δ → O CHO

(vi) 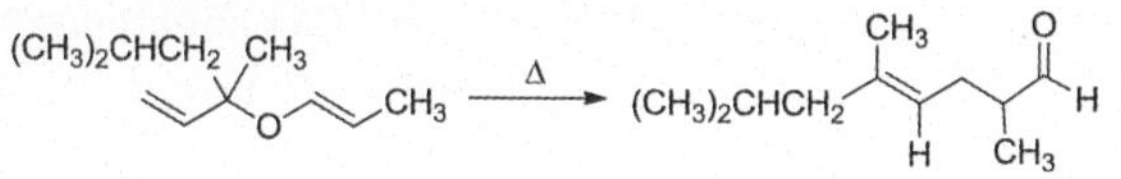 $(CH_3)_2CHCH_2$ CH_3 O CH_3 Δ → $(CH_3)_2CHCH_2$ CH_3 H O H CH_3

(vii) 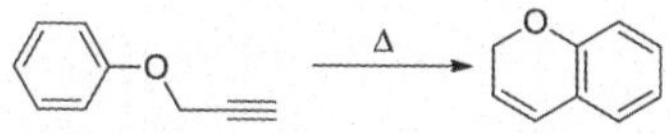 O Δ → O

Sol 17.

(i) 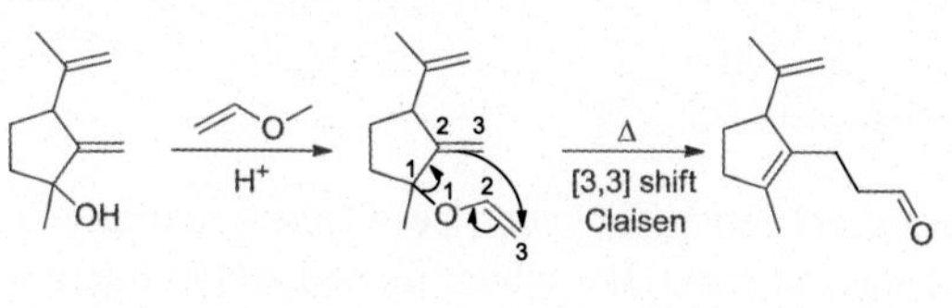 OH; O, H^+ → 2 3 1 1 2 O 3; Δ [3,3] shift Claisen → O

(ii) 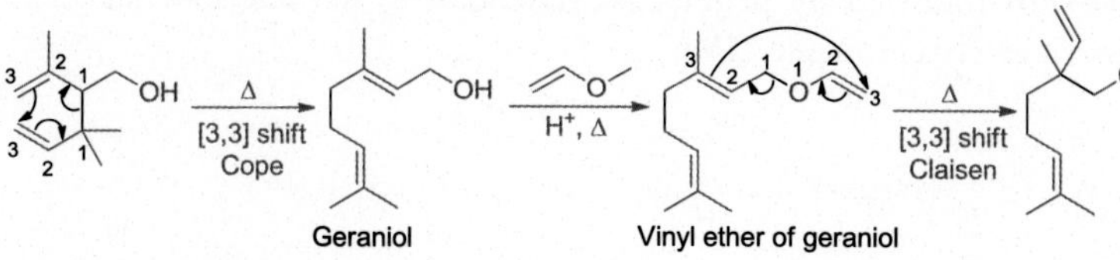 3 2 1 OH 3 2 1; Δ [3,3] shift Cope → OH; O, H^+, Δ → 3 2 1 1 2 O 3; Δ [3,3] shift Claisen → CHO

Geraniol

Vinyl ether of geraniol

(iii) [3,3] shift Claisen; [3,3] shift Cope

(iv) [3,3] shift Claisen; [3,3] shift Cope; [3, 3] shift Cope

(v) [3,3] shift Claisen; [3,3] shift Cope

(vi) The Claisen rearrangement shows preference for a chair-like transition state. Therefore, the major product has the *E*-configuration at the newly formed double bond because of the preference for placing the larger substituent in the pseudoequatorial position in the transition state.

[3,3] shift; chair-like T.S.

(*E*)-2,5,7-Trimethyloct-4-enal

(vii) On heating, propargyl aryl ether (**I**) undergoes Claisen rearrangement to give *ortho*-allenylphenol intermediate (**II**), which immediately undergoes a [1,5] H shift followed by 6π-electrocyclization of the intermediate dienone (**III**) to yield 2H-chromene (**IV**).

I → [3,3] shift → II → [1,5] H shift → III → 6π-electrocyclization → IV

3.4.6 Some Clever Variants of the Claisen Rearrangement

(a) The **Carroll–Claisen rearrangement** involves the transformation of a β-ketoallylester into α-allyl-β-ketocarboxylic acid accompanied by decarboxylation. The final product is an allyl ketone (Scheme 3.23).

β-Ketoallylester ⇌ [Δ, [3,3] shift] α-Allyl-β-ketocarboxylic acid [Δ, $-CO_2$] ⇌ Allyl ketone

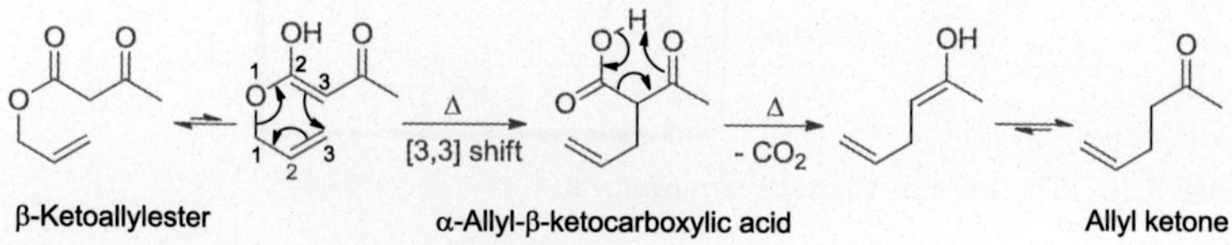

SCHEME 3.23 The Carroll–Claisen rearrangement.

(b) The **Eschenmoser–Claisen rearrangement** is the reaction of an allylic alcohol with dimethyl acetamide dimethylacetal (DMA–DMA) to generate an enamine intermediate that undergoes rearrangement resulting in the formation of γ,δ-unsaturated amide product (Scheme 3.24).

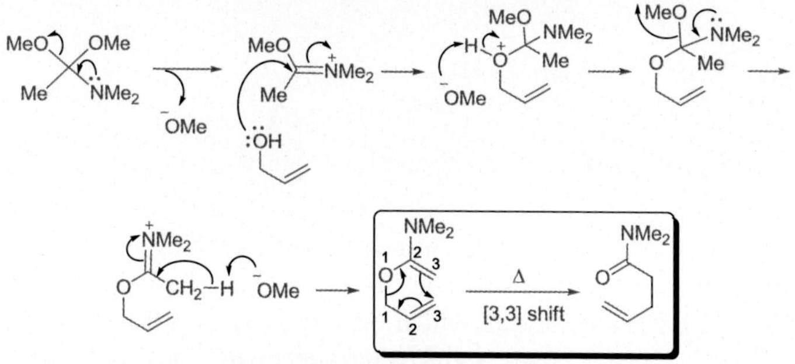

SCHEME 3.24 The Eschenmoser–Claisen rearrangement.

(c) The **Johnson–Claisen rearrangement** is the reaction of an allylic alcohol with trimethylorthoacetate to give an γ,δ-unsaturated ester (Scheme 3.25).

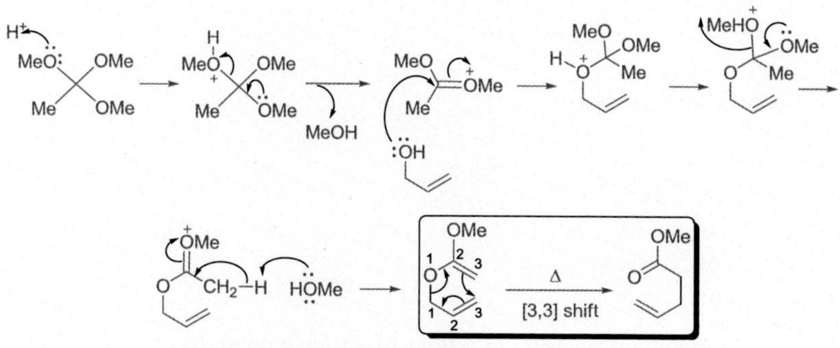

SCHEME 3.25 The Johnson–Claisen rearrangement.

(d) The **Ireland−Claisen (silyl ketene acetal) rearrangement** is the rearrangement of allyl trimethylsilyl ketene acetal, prepared by reaction of allylic ester enolates with trimethylsilyl chloride, to yield γ,δ-unsaturated carboxylic acids (Scheme 3.26).

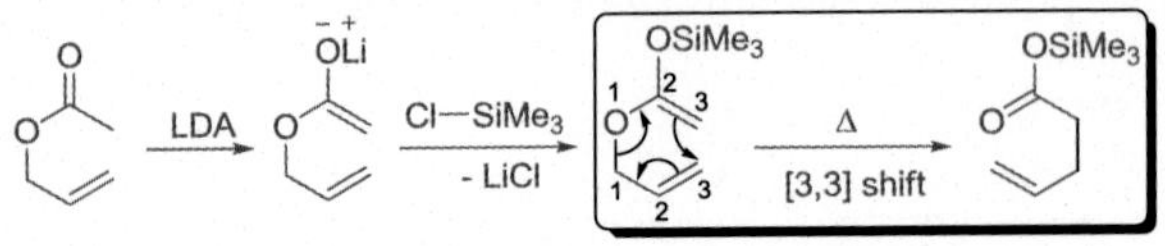

O LDA OLi Cl−SiMe$_3$ − LiCl OSiMe$_3$ 1 2 3 O 1 2 3 Δ [3,3] shift OSiMe$_3$ O H$^+$ OH O

SCHEME 3.26 The Ireland−Claisen rearrangement.

3.4.6.1 *Solved Problems*

Q 1. The product formed in the following three-step reactions is:

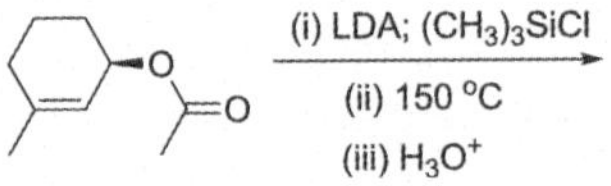

O O (i) LDA; $(CH_3)_3SiCl$ (ii) 150 °C (iii) H_3O^+

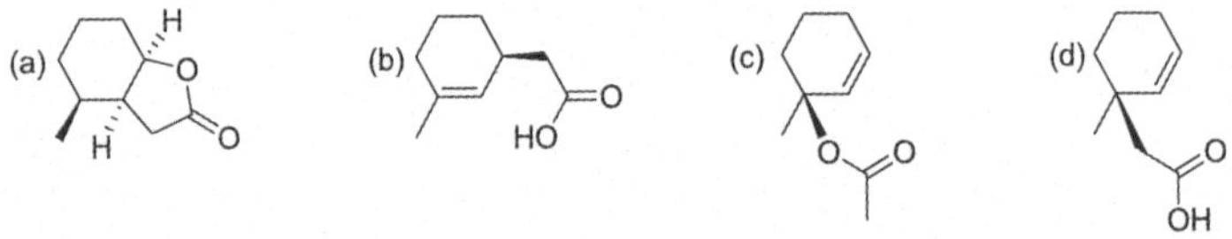

(a) H O O H (b) HO O (c) O O (d) O OH

Sol 1. (d) Abstraction of proton from α-methyl group of ester and subsequent treatment with trimethylsilyl chloride (TMSCl) resulted in the formation of cyclohexenyl silyl ketene acetal, which on heating undergoes [3,3] shift, i.e., Ireland−Claisen (silyl ketene acetal) rearrangement. The final step is the removal of the silyl group by acid hydrolysis to get the free acid.

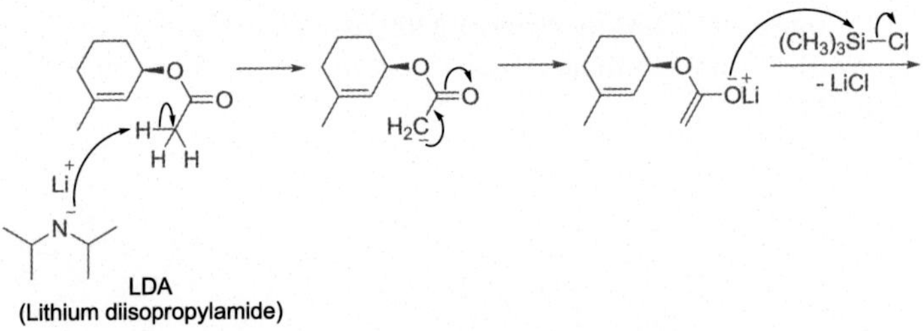

O O H H H Li N H$_2$C O O O OLi $(CH_3)_3Si$−Cl − LiCl

LDA
(Lithium diisopropylamide)

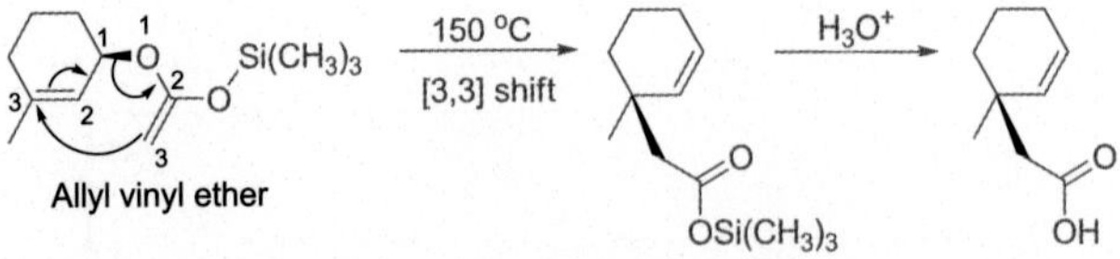

1 1 O 2 O Si(CH$_3$)$_3$ 3 2 3 150 °C [3,3] shift O OSi(CH$_3$)$_3$ H$_3$O$^+$ O OH

Allyl vinyl ether

The rearrangement occurs via a boat-like transition state.

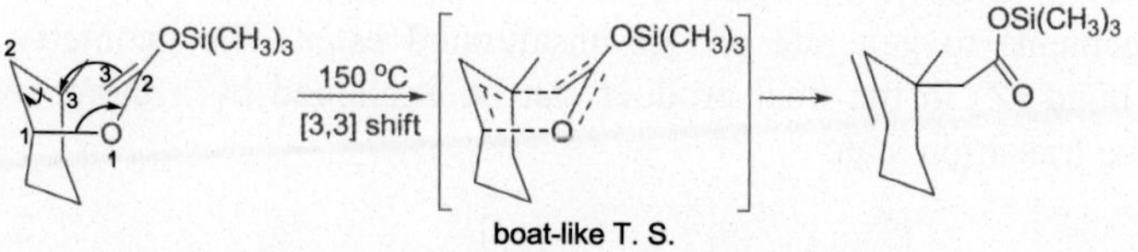

2 OSi(CH3)3 3 2 3 1 O 1 — 150 °C [3,3] shift → [OSi(CH3)3 O] → OSi(CH3)3 O

boat-like T. S.

Q 2. Predict the major product in the following reaction:

Me, Et, O, O, Me — Heat, $-CO_2$ →

(a) Et, Me, Me, O (b) Et, Me, Me, O

(c) Me, Me, Et, O (d) Me, Me, Et, O

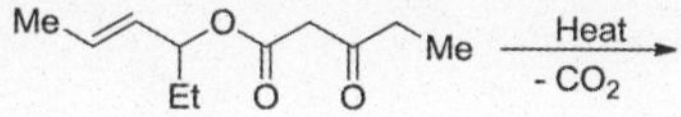

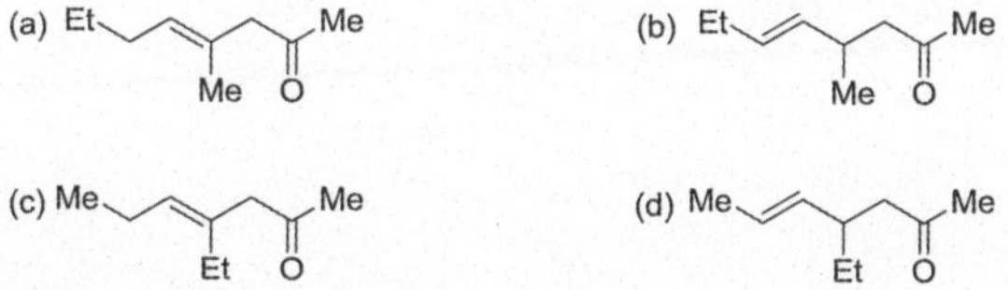

Sol 2. (b) It is an example of Carroll–Claisen rearrangement.

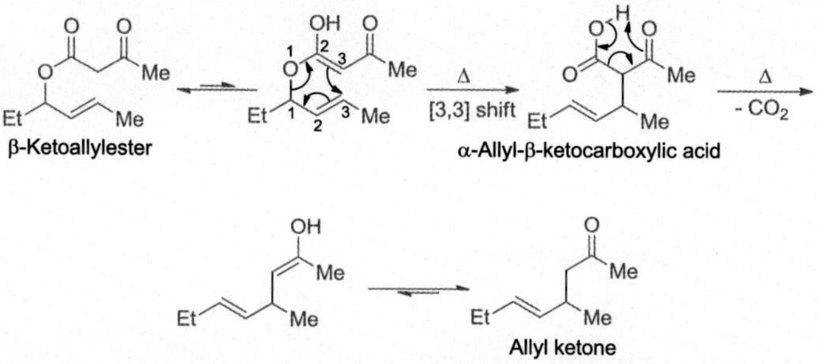

β-Ketoallylester ⇌ (OH, O, 1, 2, 3, Me, O, Et, 1, 2, 3, Me) — Δ, [3,3] shift → α-Allyl-β-ketocarboxylic acid — Δ, $-CO_2$ →

(OH, Me, Et, Me) ⇌ Allyl ketone (O, Me, Et, Me)

Q 3. Identify the product formed in the following reaction:

O — (i) $H_2C{=}CHMgBr$, THF (ii) H_3O^+ (iii) excess $CH_3C(OCH_3)_3$, PTSA, heat →

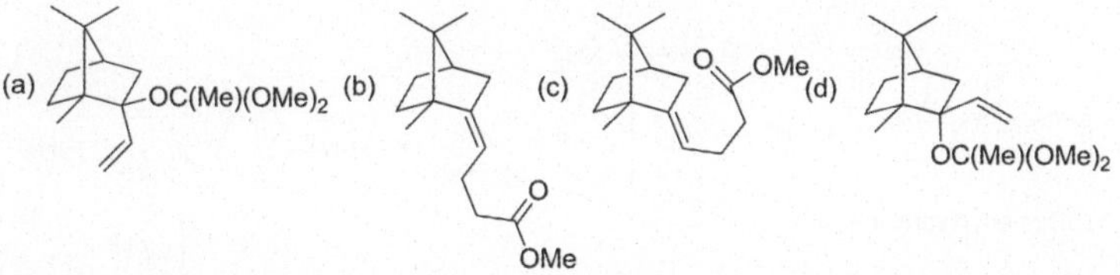

(a) OC(Me)(OMe)$_2$ (b) O, OMe (c) O, OMe (d) OC(Me)(OMe)$_2$

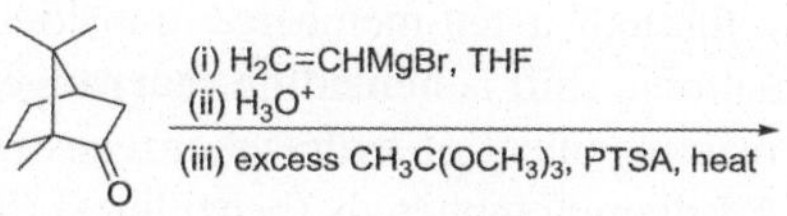

Sol 3. (b) The Grignard reagent attacks from the less hindered face of the carbonyl group to give an allyl alcohol, which is treated with

trimethylorthoacetate in the presence of PTSA to give a ketene acetal, which undergoes a facile [3,3] sigmatropic rearrangement (i.e., Johnson–Claisen rearrangement) to generate an γ,δ-unsaturated ester. The geometry at the double bond (*Z*) in the final product can be explained by a more favorable chair-like transition state.

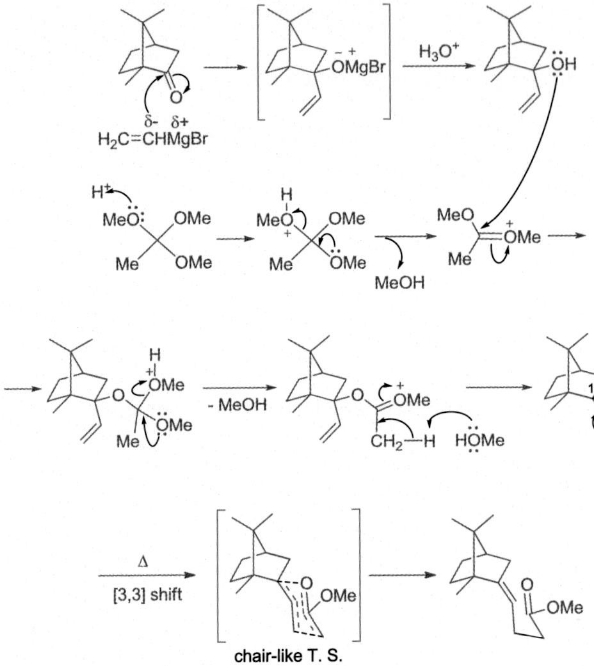

3.5 [5,5] SIGMATROPIC SHIFT

The Woodward–Hoffmann rules predict that [5,5] sigmatropic shifts would proceed suprafacially through a ten-membered transition state. A common example of [5,5] sigmatropic shift is **benzidine rearrangement**. This reaction is an acid-catalyzed rearrangement of hydrazobenzenes (in general, the *N,N*-diarylhydrazines) to 4,4-diaminobiphenyls (benzidines) (Scheme 3.27).

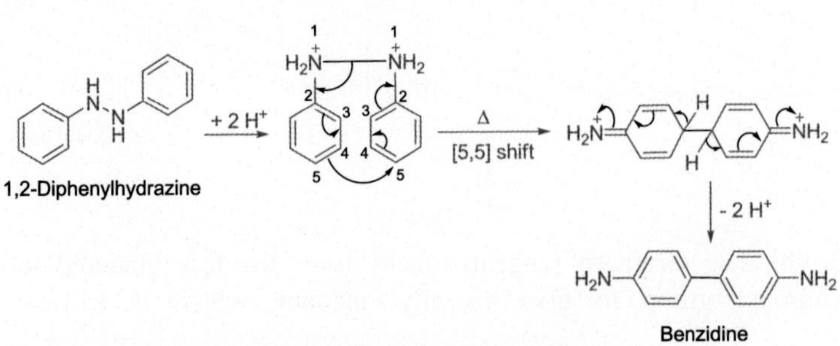

SCHEME 3.27 [5,5] Sigmatropic shift in benzidine rearrangement.

N-Phenyl-*N'*-(2-thiazolyl)hydrazine (**1**) and *N,N'*-bis(2-thiazolyl)hydrazine (**3**) also undergo acid-catalyzed benzidine-type rearrangement into 2-amino-5-(*p*-aminophenyl)thiazole (**2**) and 5,5'-bis(2-aminothiazole) (**4**), respectively (Scheme 3.28).

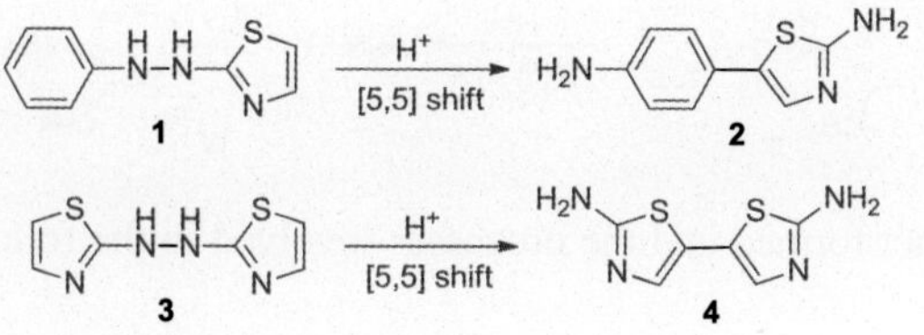

SCHEME 3.28 [5,5] Sigmatropic shift in benzidine-type rearrangement.

3.5.1 Solved Problems

Q 1. The major product obtained on treating a mixture of **I** and **II** in a strongly acidic solution of H_2SO_4 is:

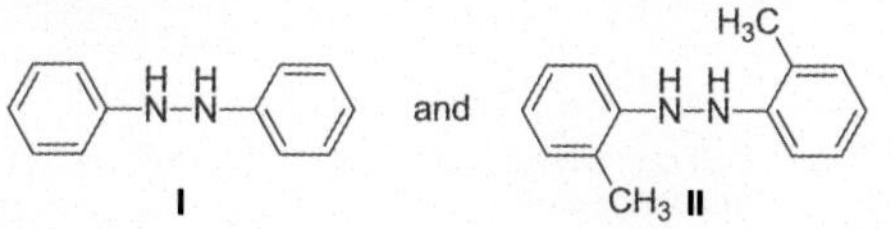

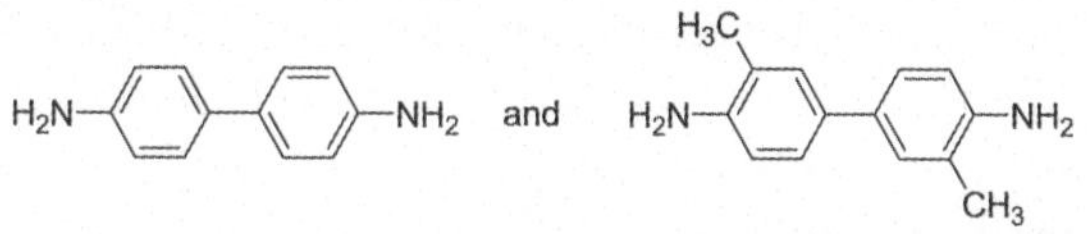

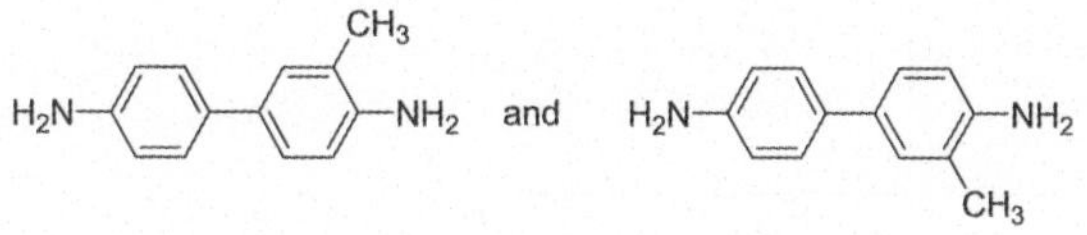

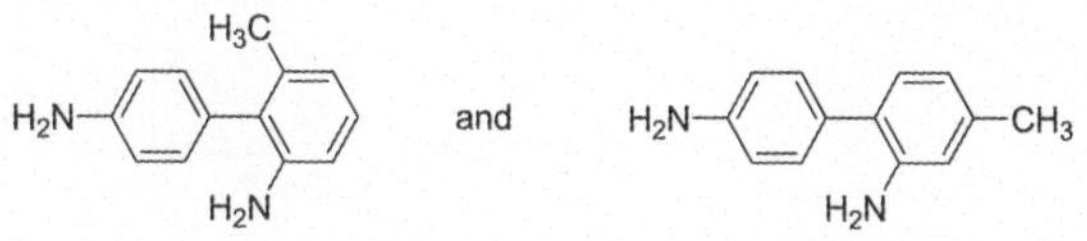

Sol 1. (a) The intramolecular nature of the [5,5] sigmatropic shift can be established by the crossover experiment. When a mixture of **I** and **II** is treated with a strongly acidic solution of H_2SO_4, it yields the same products (**III** and **IV**, respectively) as when they were heated separately. The intramolecularity of the reaction was established by the absence of crossover products.

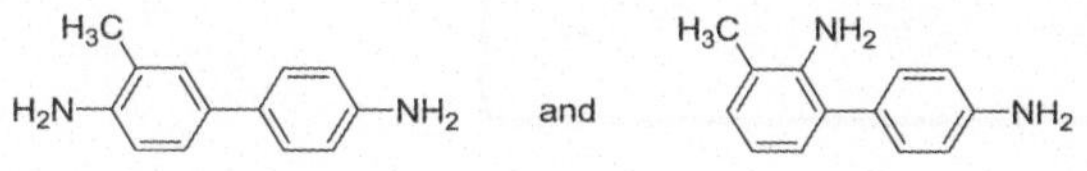

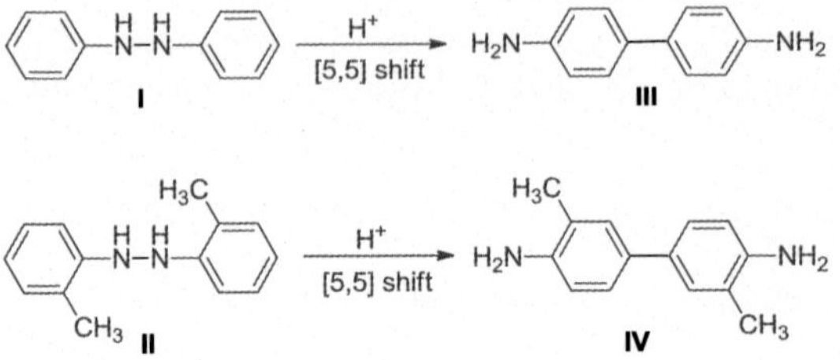

Q 2. The product formed and the processes involved in the following reaction are:

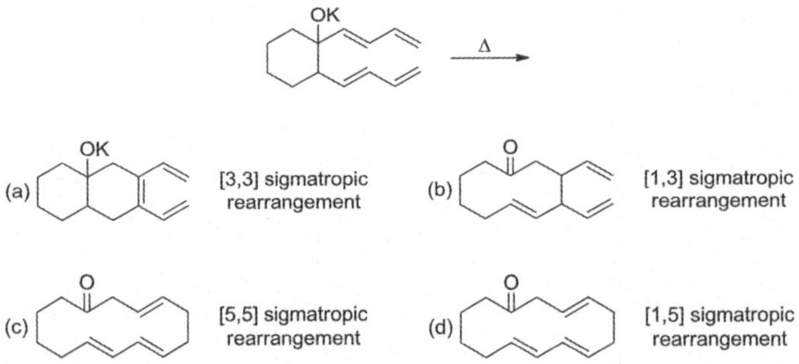

(a) [3,3] sigmatropic rearrangement (b) [1,3] sigmatropic rearrangement

(c) [5,5] sigmatropic rearrangement (d) [1,5] sigmatropic rearrangement

Sol 2. (c) On heating, potassium salt of 1,2-di((*E*)-buta-1,3-dienyl)cyclohexanol transforms into cyclotetradeca-3,7,9-trien-1-one by an oxy [5,5] sigmatropic process. The rearrangement does not proceed by a sequence of [3,3] shifts it is indeed a result of [5,5] shift.

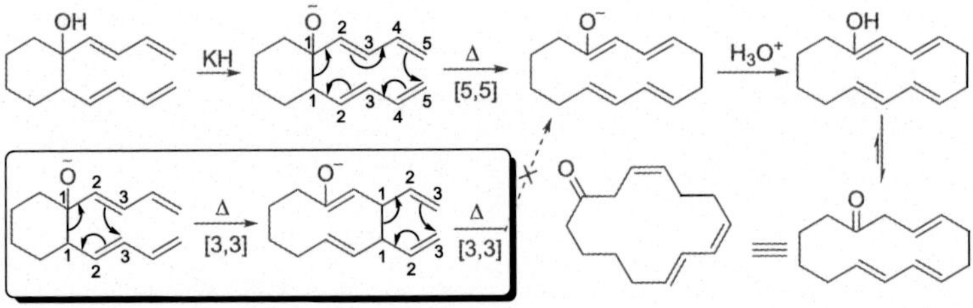

Q 3. Predict the product you would expect to obtain from the following reaction:

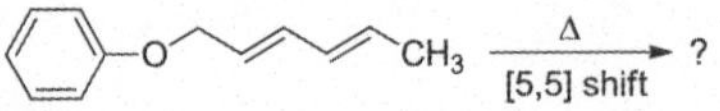

Sol 3. The *trans*, *trans*-hexa-2,4-dienylphenyl ether (**I**) provided on heating at 165 °C, in addition to the *ortho* Claisen rearrangement product **III**, mainly a mixture consisting of 94% 4-(1-methylpenta-2,4-dienyl)-phenol (**II**) and 6% 4-(hexa-2,4-dienyl)-phenol (**IV**). The migration **I** → **II**, proceeding through a

ten-membered transition state, is an example of [5,5] sigmatropic rearrangement and is the major pathway.

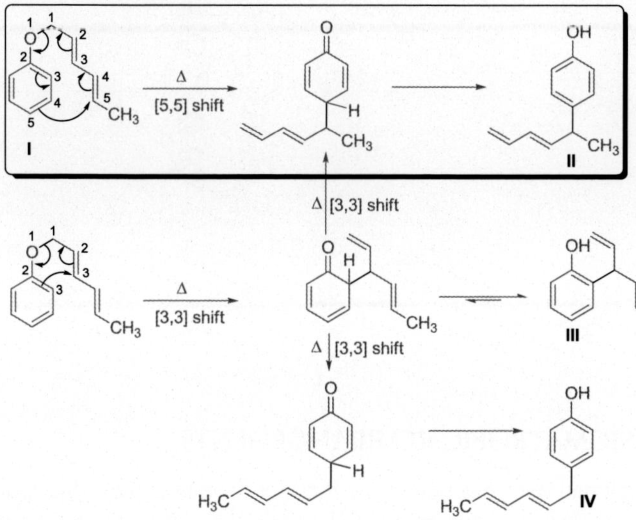

Q 4. Predict the mechanism of the following reactions:

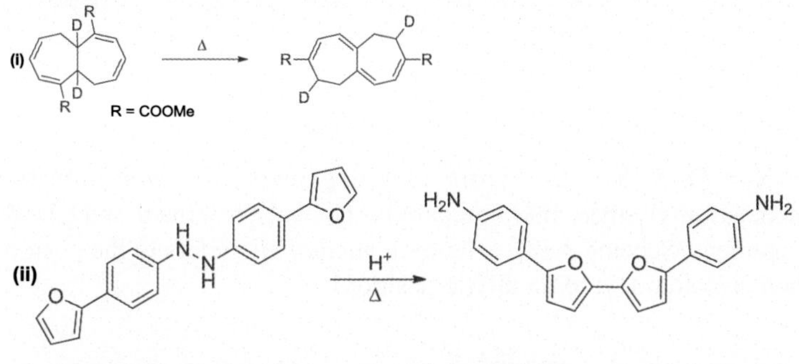

(i) R = COOMe

(ii)

Sol 4. (i) On heating, the 1,6-diester (**I**) undergoes [5,5] shift to give the 3,8-diester (**II**), which reacts immediately by two [1,5] hydrogen shifts, probably via **III**, to give the thermodynamically more stable **IV** with a conjugated octatetraene moiety.

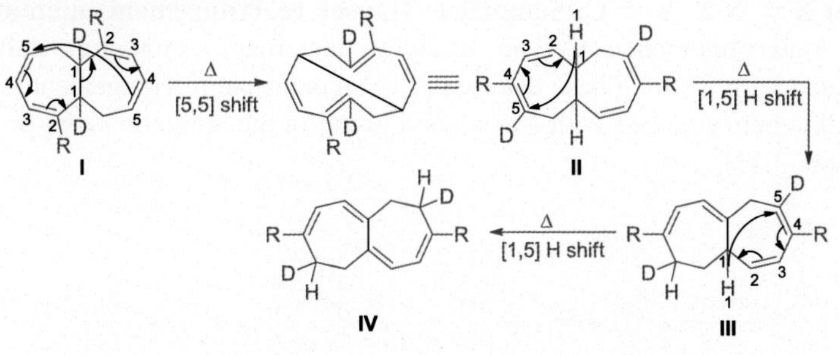

(ii) *Bis*-[4-(2-furyl)phenyl]diazene undergoes acid-catalyzed benzidine-type rearrangement involving [9,9] sigmatropic shift.

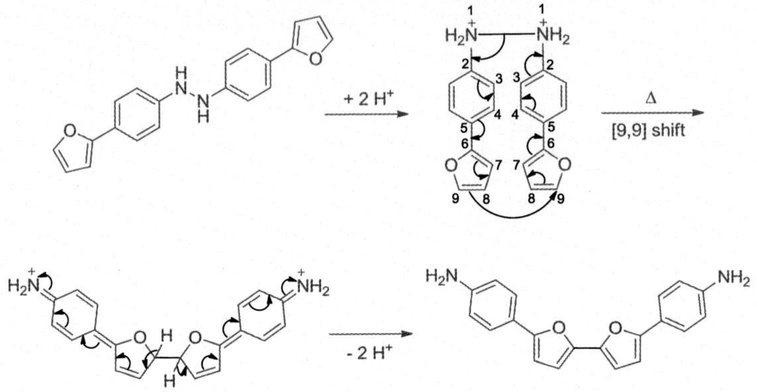

3.6 [2,3] SIGMATROPIC REARRANGEMENTS

The [2,3] sigmatropic reaction is a thermal isomerization that proceeds through a six-electron, five-membered cyclic transition state. A general scheme for [2,3] sigmatropic rearrangements is shown in Scheme 3.29.

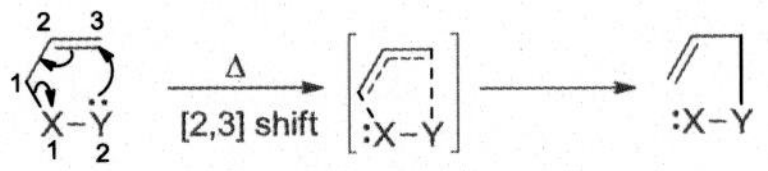

SCHEME 3.29 General scheme for [2,3] sigmatropic rearrangements.

(a) X = O & Y = C; Wittig rearrangement: The transformation of deprotonated allyl ethers into homoallylic alcohols is termed as [2,3] Wittig rearrangement (Scheme 3.30). (The term *homoallylic* refers to the position on a carbon skeleton next to an allylic position).

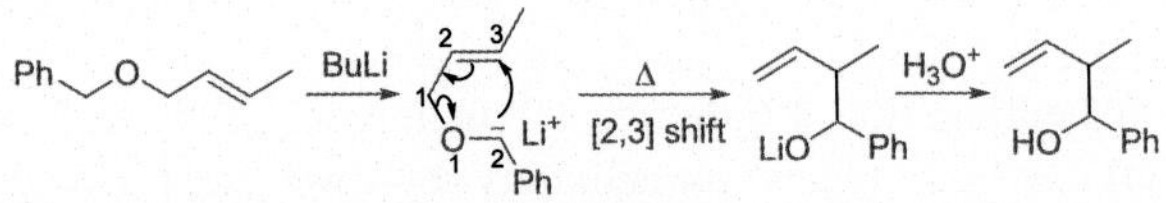

SCHEME 3.30 Wittig rearrangement of allyl ethers.

(b) X = N & Y = C; Sommelet–Hauser rearrangement (ammonium ylide rearrangement): Certain benzylic quaternary ammonium salts on treatment with a strong base undergo [2,3] sigmatropic rearrangement to give N-dialkyl benzyl amine with a new alkyl group in the aromatic *ortho*-position (Scheme 3.31).

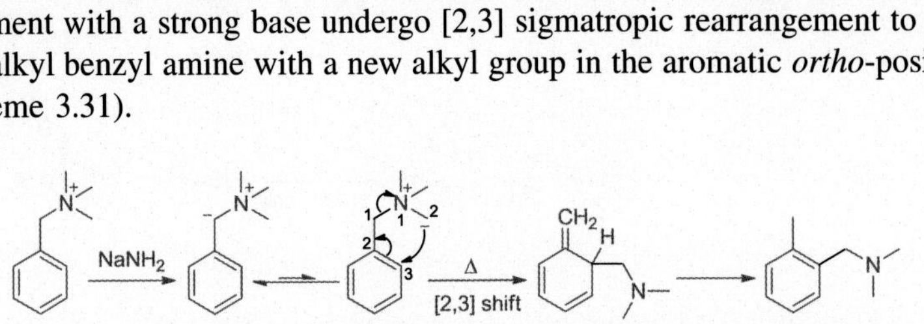

SCHEME 3.31 Sommelet–Hauser rearrangement involving [2,3] sigmatropic shifts.

(c) X = N & Y = C; Sulfonium ylide rearrangement: Sulfur ylide bearing an allylic group are converted on heating to unsaturated sulfides (Scheme 3.32).

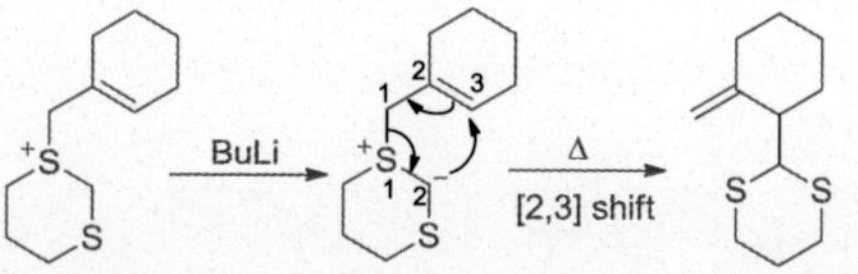

SCHEME 3.32 Sulfonium ylide rearrangement involving [2,3] sigmatropic shifts.

(d) X = S & Y = O; Mislow–Evans rearrangement: It is a [2,3] sigmatropic reaction that allows the formation of allylically rearranged alcohols from allylic sulfoxides (Scheme 3.33).

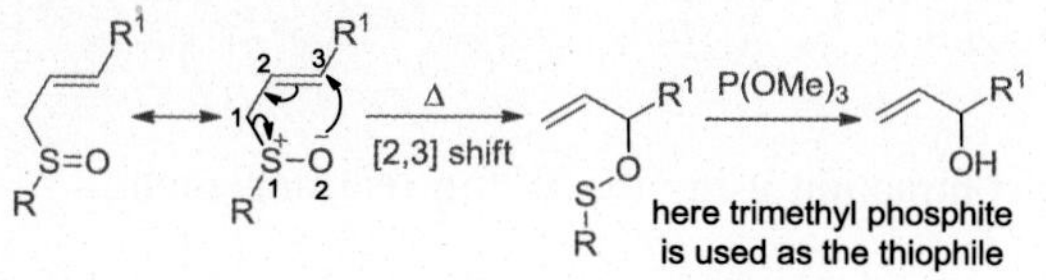

SCHEME 3.33 Mislow–Evans rearrangement of allylic sulfoxides.

(e) X = N & Y = O; Meisenheimer rearrangement: Certain tertiary amine oxides having an allylic or benzilic group rearrange on heating to give substituted hydroxylamines. However, if one of the R groups contains a β-hydrogen, Cope elimination often competes (Scheme 3.34).

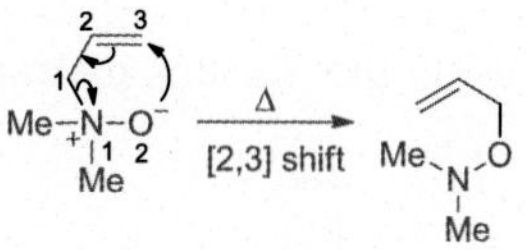

SCHEME 3.34 Meisenheimer rearrangement of tertiary amine oxides.

3.6.1 Solved Problems

Q 1. Give the mechanism of the following reactions:

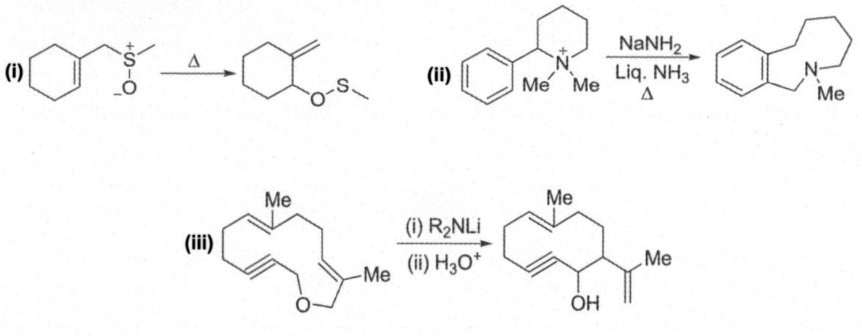

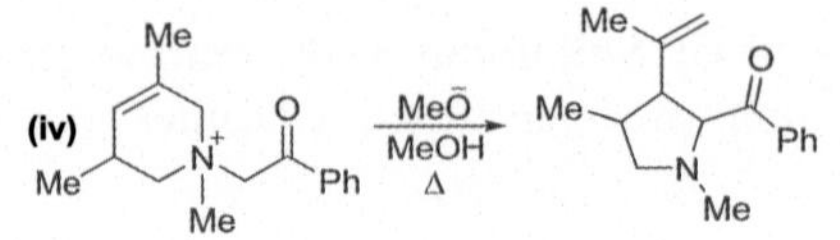

(iv) Me, Me, N+, Me, O, Ph — MeO^- / MeOH / Δ → Me, Me, O, Ph, N, Me

Sol 1.

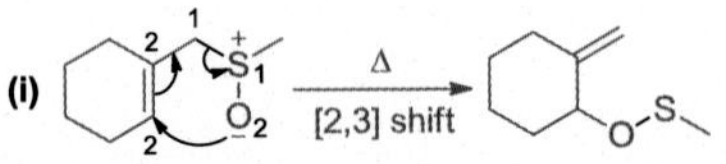

(i) 1, 2, S+ 1, O− 2, 2 — Δ / [2,3] shift → O, S

(ii) A ring expansion using the Sommelet–Hauser rearrangement.

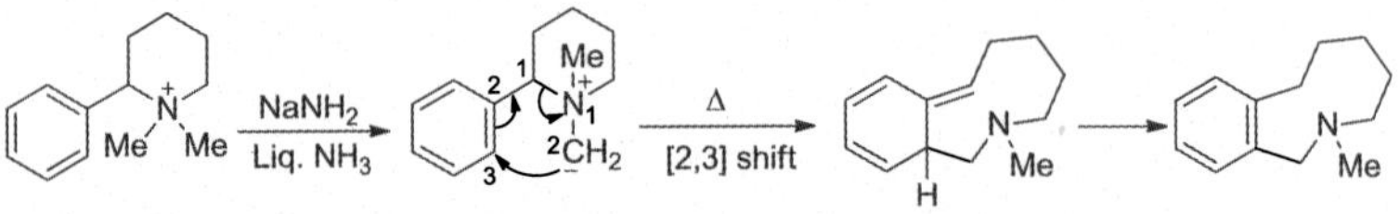

N+, Me, Me — $NaNH_2$ / Liq. NH_3 → Me, 1, 2, N+ 1, $^2CH_2$, 3 — Δ / [2,3] shift → N, Me, H → N, Me

(iii) A ring contraction using the Wittig rearrangement.

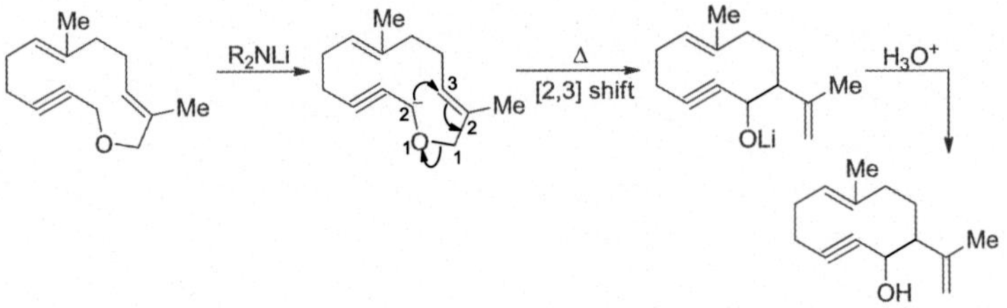

Me, Me, O — R_2NLi → Me, 3, 2, 2, 1 O, 1, Me — Δ / [2,3] shift → Me, Me, OLi — H_3O^+ → Me, Me, OH

(iv) A ring contraction using the Stevens rearrangement.

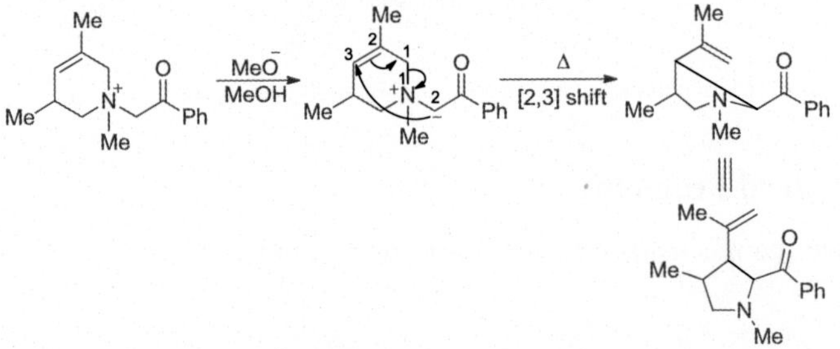

Me, Me, N+, Me, O, Ph — MeO^- / MeOH → Me, 3, 2, 1, 1, +N 2, Me, Me, O, Ph — Δ / [2,3] shift → Me, Me, N, Me, O, Ph ≡ Me, Me, O, Ph, N, Me

Q 2. Pericyclic reaction involved in one of the steps of the following reaction sequence is:

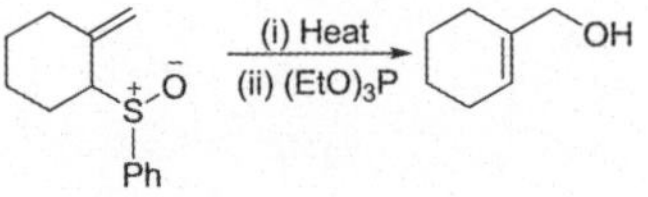

S+, O−, Ph — (i) Heat / (ii) $(EtO)_3P$ → OH

(a) [1,3] Sigmatropic shift
(b) [3,3] Sigmatropic shift
(c) [1,5] Sigmatropic shift
(d) [2,3] Sigmatropic shift

Sol 2. (d)

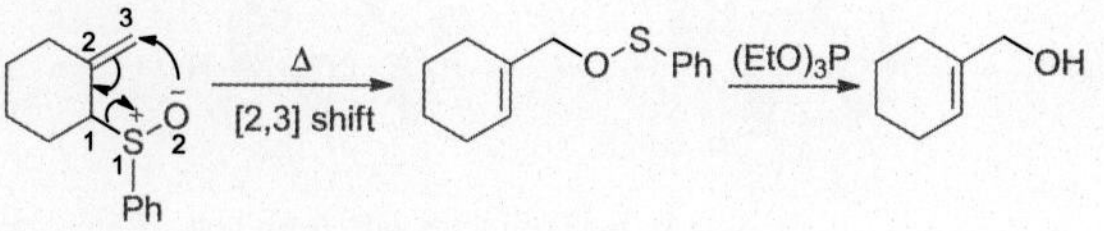

3.7 PERIPATETIC CYCLOPROPANE BRIDGE: WALK REARRANGEMENTS

The migration of a divalent group, which is part of a three-membered ring in a bicyclic molecule, is commonly referred to as a Walk rearrangement. Let us consider the thermolysis of bicyclo[4.1.0]heptadiene (norcaradiene) system. The HOMO of the π-framework of the biradicloid transition state is Ψ_3 with mirror symmetry. Therefore, under thermal conditions, a [1,5] sigmatropic shift constrained to proceed suprafacially must take place with retention at the migrating group. This condition implies that in each shift, substituents on the cyclopropane ring pivot 180°. As a result, the groups A and B will become alternatively *endo* and *exo* with each migration step, giving rise to a mixture of two isomers. On the other hand, if we consider the symmetry forbidden path, i.e., under thermal conditions the [1,5] sigmatropic shift proceeds suprafacially with inversion at the migrating group, then the groups A and B remain throughout *endo* and *exo*, respectively, in a series of [1,5] sigmatropic shifts as shown in Figure 3.16.

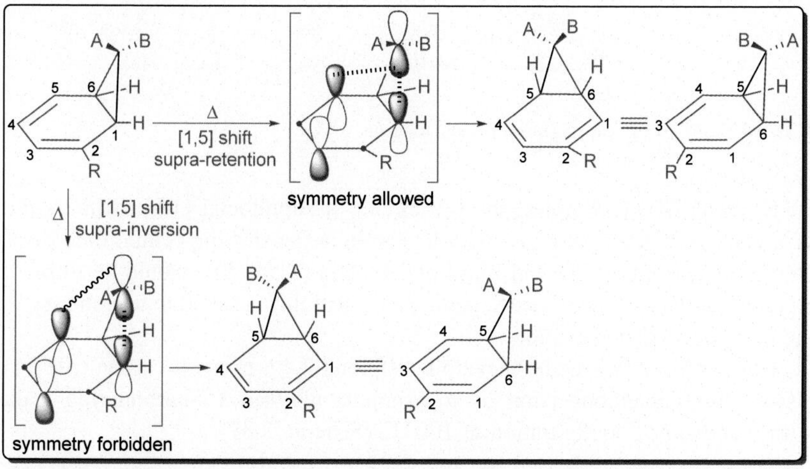

FIGURE 3.16 Analysis of Walk rearrangement by FMO method.

The peripatetic nature of cyclopropane bridge can also be explained by invoking PMO method (Figure 3.17).

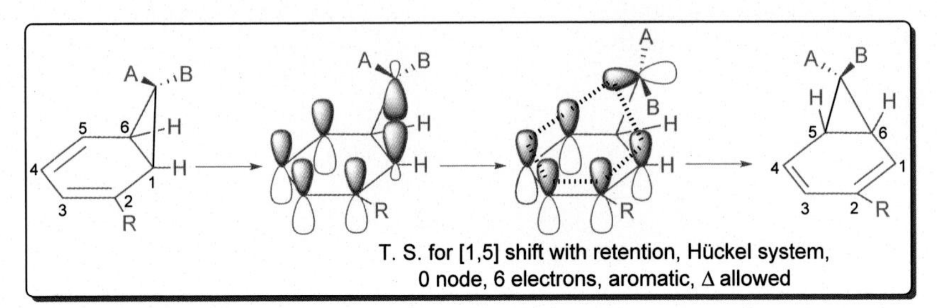

T. S. for [1,5] shift with retention, Hückel system,
0 node, 6 electrons, aromatic, Δ allowed

FIGURE 3.17 Analysis of Walk rearrangement by PMO method.

1,3,5-Cycloheptatrienes undergo [1,5] sigmatropic shifts via norcaradiene intermediates. For example, pyrolysis of an optically active troplidine takes place with no loss of optical activity. This fact can be explained by completing the full circuit of suprafacial [1,5] shifts with retention at the shifting center. It was found that at no stage is a structure produced that is an enantiomer of the other (Scheme 3.35).

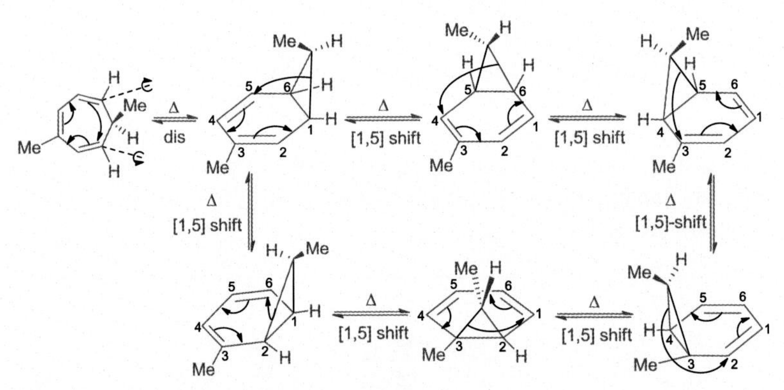

SCHEME 3.35 Walk rearrangement of norcaradiene.

However, if we complete the full circuit of suprafacial [1,5] shifts by the symmetry forbidden path, i.e., with inversion at the shifting center, then three pairs of enantiomers—(**1** and **2**), (**3** and **6**), and (**4** and **5**)—would be formed, resulting in the racemization (Scheme 3.36). But this is found to be contrary to the experimental observation.

Isomers of 2,9-dimethylbicyclo[6.1.0]nona-2,4,6-triene-9-carbonitrile (**1**) also undergo interconversion by a symmetry-allowed 1,7-carbon ring walk with inversion of configuration at 100 °C (Scheme 3.37).

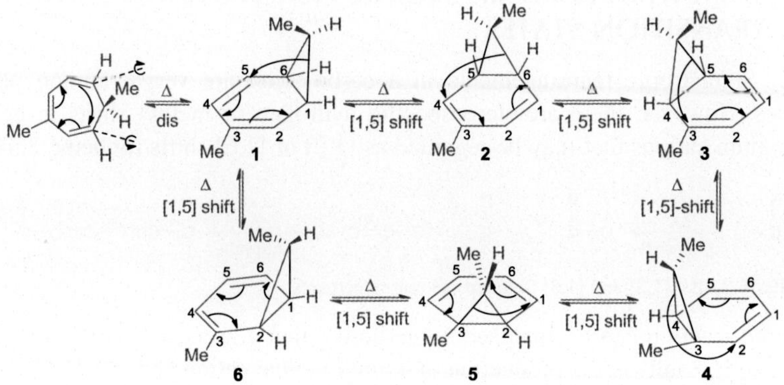

SCHEME 3.36 Walk rearrangement of norcaradiene continued.

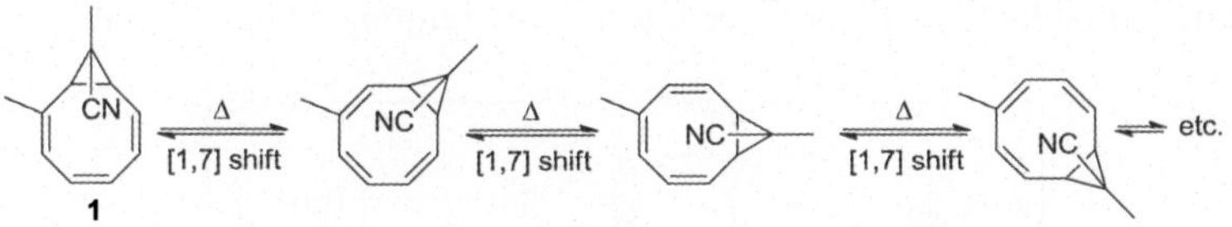

SCHEME 3.37 [1,7] Shifts in the Walk rearrangement of bicyclic nonatriene.

3.7.1 Solved Problem

Q 1. Provide a suitable mechanistic pathway for the following transformations.

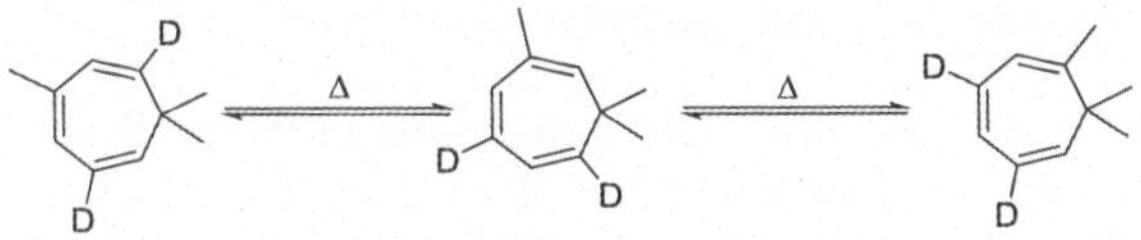

Sol 1. Isomerization of substituted cycloheptatrienes takes place through sequences of [1,5] sigmatropic rearrangements and electrocyclic reactions.

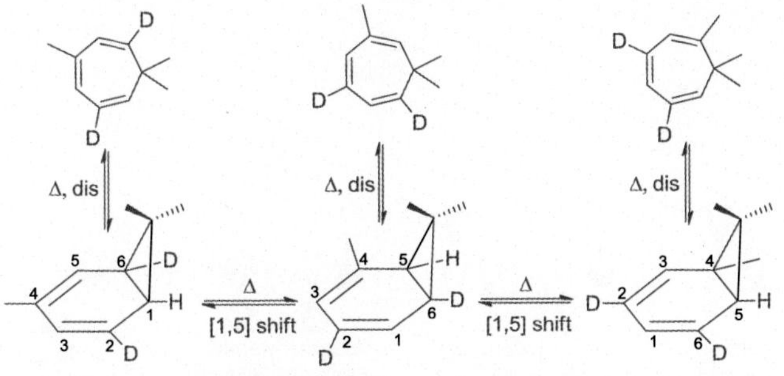

3.8 SIGMATROPIC REARRANGEMENTS INVOLVING IONIC TRANSITION STATES

[1,2] Sigmatropic rearrangements in a carbocation are very common processes. However, there are stereospecific hydrogen or methyl shifts in benzenonium cations that may be regarded as [1,2] or [1,6] shifts (Scheme 3.38).

R–C–C^{+} —Δ, [1,2] shift→ $^{+}$C–C–R R–C–C=C–C=C–C^{+} —Δ, [1,6] shift→ $^{+}$C–C=C–C=C–C–R

SCHEME 3.38 [1,2] and [1,6] Sigmatropic rearrangements.

The reaction can be analyzed by assuming that a homolytic cleavage in a [1,2] shift results in the production of a vinyl radical cation and in a [1,6] shift results in the production of a hexatriene radical cation. In case of a [1,2] shift, the HOMO of the π-framework of the transition state is Ψ_1 and in the [1,6]-shift it is Ψ_3. In either case the HOMO has mirror symmetry, and, therefore, under thermal conditions, the [1,2] or [1,6] sigmatropic shift occurs via a suprafacial pathway without inversion at the migrating center (Figure 3.18).

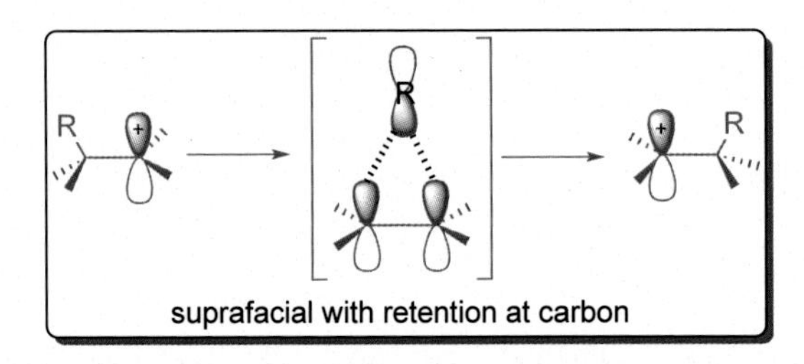

FIGURE 3.18 FMO analysis of [1,2] sigmatropic rearrangement.

A homolytic cleavage in a [1,4] shift results in the production of a butadiene radical cation, whose HOMO is Ψ_2 having C_2 symmetry. Therefore, under thermal conditions, the [1,4] sigmatropic shift occurs via a *suprafacial pathway with inversion at the migrating center* (Figure 3.19).

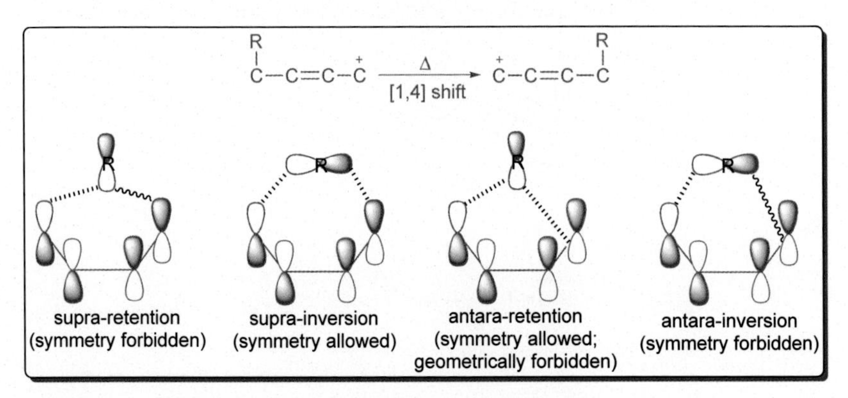

FIGURE 3.19 FMO analysis of [1,4] sigmatropic rearrangements.

Sigmatropic rearrangements of carbocations involving [1,2] and [1,4] shifts can also be analyzed on the basis of PMO method (Figures 3.20 and 3.21).

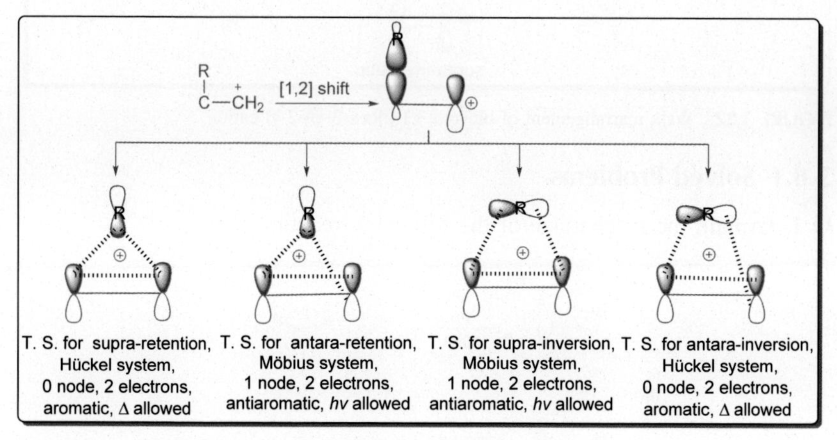

FIGURE 3.20 PMO analysis of [1,2] sigmatropic rearrangements.

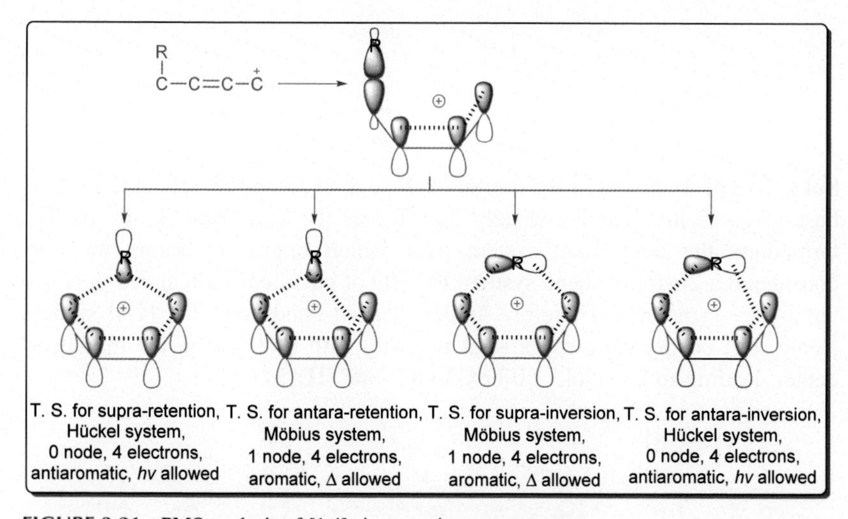

FIGURE 3.21 PMO analysis of [1,4] sigmatropic rearrangements.

The bicyclo[3.1.0]hex-3-en-2-yl cation systems exhibit four electron [1,4] sigmatropic shifts. When a [1,4] sigmatropic shift proceeds suprafacially, it invariably takes place with inversion at the migrating group. As a result, the groups A and B incorporated on the bridge remain throughout *endo* and *exo*, respectively, in a series of [1,4] sigmatropic shifts on thermolysis of a cationic species as shown below in Figure 3.22.

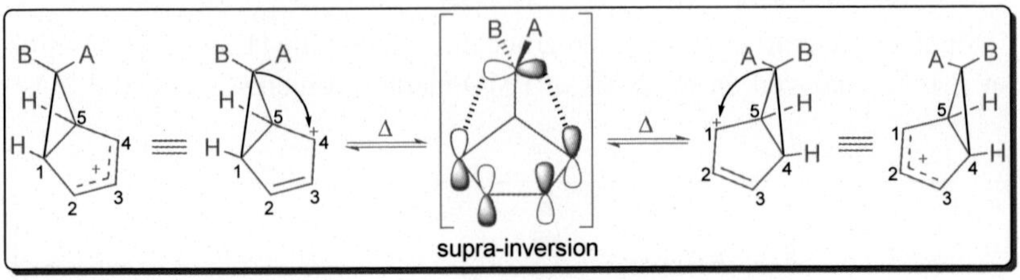

B A H H 5 4 1 2 3 ≡ B A H H 5 4 1 2 3 Δ B A supra-inversion Δ A B H H 5 1 2 3 4 ≡ A B H H 5 1 2 3 4

FIGURE 3.22 Walk rearrangement of bicyclo[3.1.0]hex-3-en-2-yl cation.

3.8.1 Solved Problems

Q 1. Explain the mechanism of the following reactions:

(i) R^1 R^2 Br H H O Br — Zn, heat →; R^1 R^2 Br H H O — KO-t-Bu, t-BuOH, 42 °C → R^1 R^2 O H H

$R^1 = CH_2C_6H_5$
$R^2 = C_6H_5$

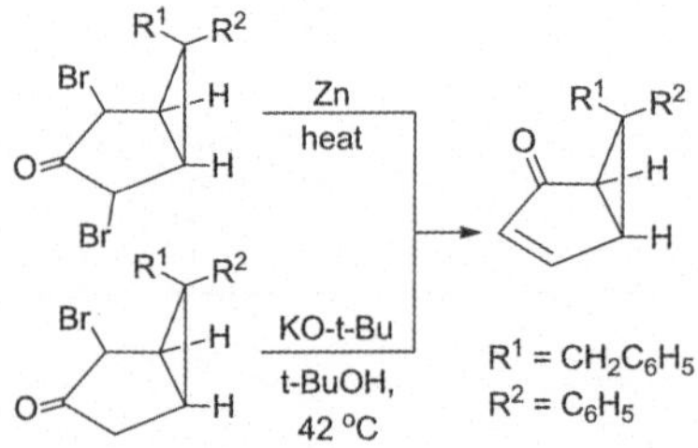

(ii) H HO CH$_3$ * —Δ→ * CH$_3$

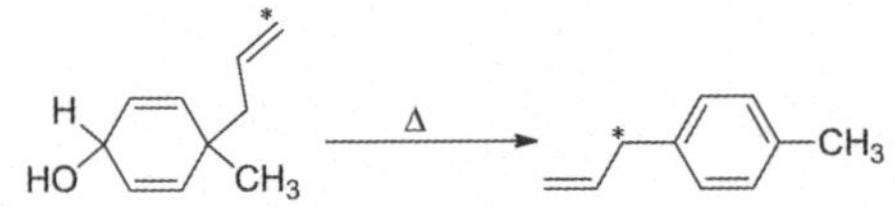

Sol 1. (i) The treatment of the dibromoketone with zinc or bromoketone with a base gives an intermediate usually depicted as the zwitterion (**I**). In this intermediate the conjugated system over which migration occurs may be considered a 2-oxobutadiene system. HOMO of butadiene radical cation is Ψ_2 having C_2 symmetry. Therefore, under thermal conditions, the [1,4] sigmatropic shift occurs via a suprafacial pathway with inversion at the migrating center, leading to bicyclo[3.1.0]hex-3-en-2-one (**II**).

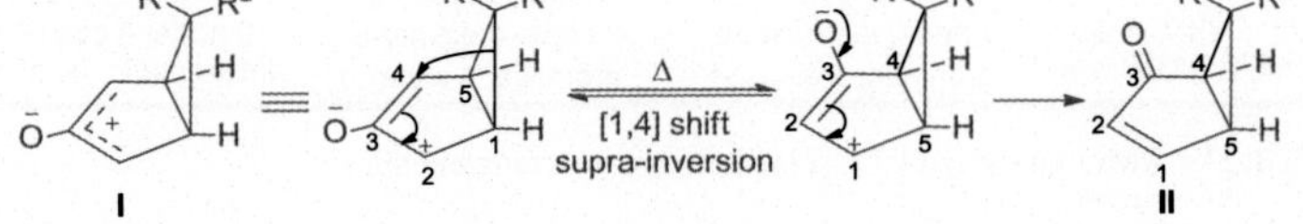

R^1 R^2 H H O$^-$ + (**I**) ≡ R^1 R^2 4 5 H H O$^-$ 3 2 1 —Δ, [1,4] shift, supra-inversion→ R^1 R^2 O 3 4 H 2 5 H 1 → R^1 R^2 O 3 4 H 2 5 H 1 (**II**)

(ii) The reaction is a typical example of a [3,4] cationic sigmatropic shift.

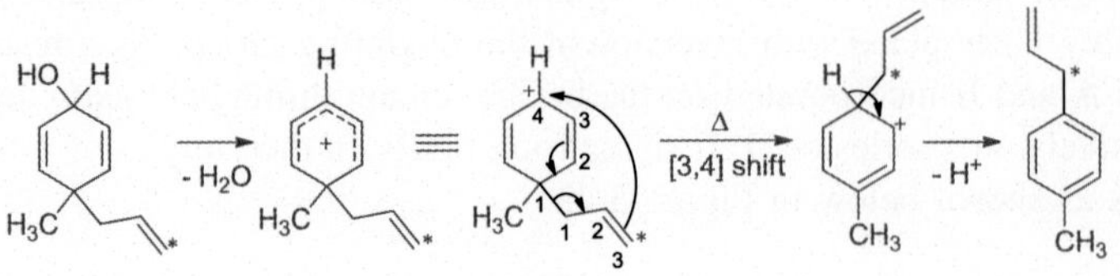

HO H H$_3$C * —- H$_2$O→ H + H$_3$C * ≡ H + 4 3 2 H$_3$C 1 1 2 3 * —Δ, [3,4] shift→ H * + CH$_3$ —- H$^+$→ * CH$_3$

Chapter 4

Cycloaddition Reactions

Chapter Outline

The reaction in which two or more unsaturated systems join together to give a cyclic adduct that possesses lesser bond multiplicity is called a *cycloaddition reaction*. When the joining π-systems belong to same molecule, then it is known as intramolecular cycloaddition reaction. Whereas, in intermolecular cycloaddition reaction, the joining π-systems belong to different molecules. The reaction is said to be *cycloreversion* when reversal of cycloaddition takes place.

Cycloaddition reactions offer a versatile route for the synthesis of cyclic compounds with a high degree of stereoselectivity under thermal and photochemical conditions. These reactions consist of the addition of a system of

Pericyclic Reactions. http://dx.doi.org/10.1016/B978-0-12-803640-2.00004-X
Copyright © 2016 Elsevier Inc. All rights reserved.

p π-electrons to a system of **q** π-electrons to form a cyclic product having [(**p** – 2) + (**q** – 2)] π-electrons. Depending upon the number of π-electrons participating in the reorganization process of electrons, these reactions are termed [$p + q$] or [$p + q + \ldots$] cycloaddition reactions. Some examples of cycloaddition reactions are given in Scheme 4.1.

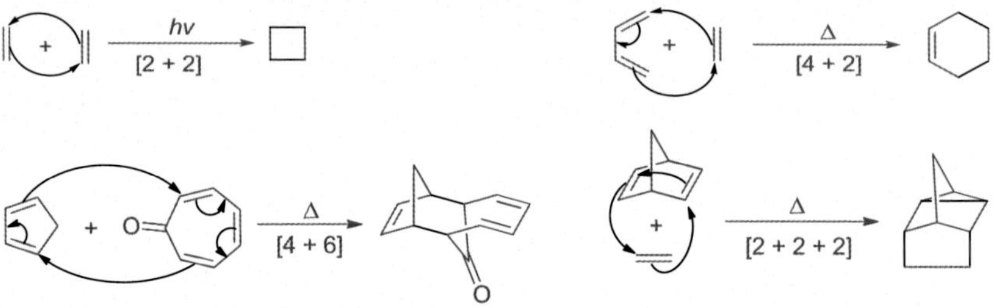

SCHEME 4.1 Examples of cycloaddition reactions.

4.1 STEREOCHEMICAL MODES OF CYCLOADDITION

Since in a typical cycloaddition reaction there is addition of two unsaturated systems, it is logical to expect the addition to occur on the same or the opposite faces of the system involved. Furthermore, as both the π-systems are undergoing addition, it is necessary to specify these modes of addition on each of them especially to discuss and understand the stereochemistry of the product. These different modes have been termed as *suprafacial* (on the same side) and *antarafacial* (on the opposite side) and are shown in Figure 4.1.

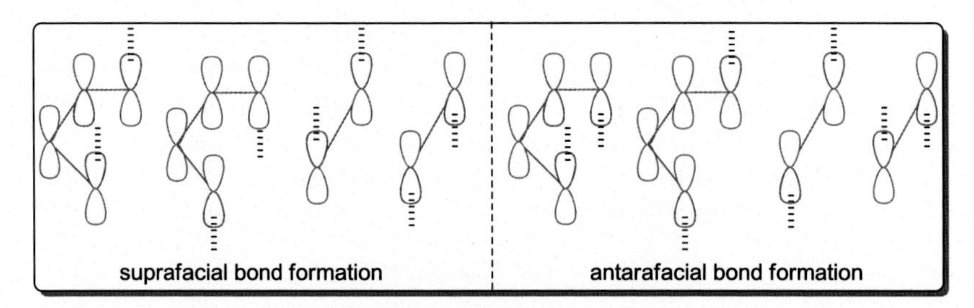

FIGURE 4.1 Suprafacial and antarafacial approaches to a π-bond.

Pericyclic reactions in which p-orbitals at the ends of the π-component of each system overlap and form the new σ-bonds on the same surface are called suprafacial cycloaddition. Almost all pericyclic cycloaddition reactions are suprafacial on both systems and thus the stereochemistry is maintained due to their concerted nature. This specification is usually made by placing a suitable subscript (s or a) after the number referring to the π-component. For example, the most common cycloaddition, Diels–Alder reaction, may be represented as a process involving [π^4s + π^2s] cycloaddition or simply [4s + 2s] cycloaddition. Similarly, cycloaddition between two ethene molecules to give cyclobutane may be represented as a process involving [π^2s + π^2s] cycloaddition or simply [2s + 2s] cycloaddition (Figure 4.2).

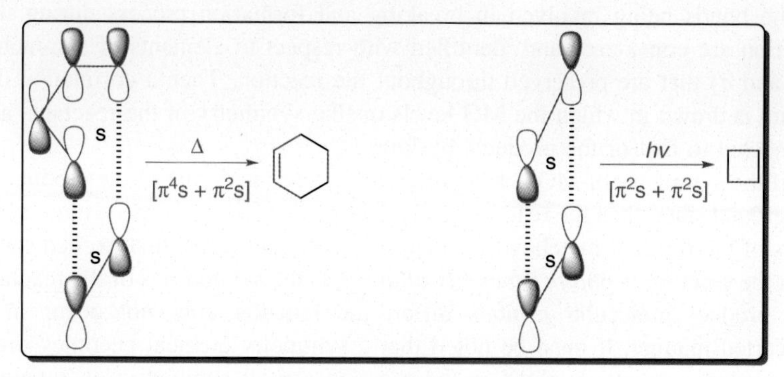

FIGURE 4.2 Representation of suprafacial addition to a π-bond.

Pericyclic reactions in which p-orbitals at the ends of the π-component of each system overlap and form the new σ-bonds on the opposite surface are called antarafacial cycloaddition. In actual practice, straightforward antarafacial attack is rare because it is sterically difficult for one π-system to suffer this type of attack by another π-system, and needs at least one usually long and flexible unsaturated system.

Similarly, in cycloreversion or retro cycloaddition reactions, σ-bonds take part in bond reorganization process. The cycloreversion of Diels–Alder reaction [π^4s + π^2s] and of [π^2s + π^2s] cycloaddition may be designated as [σ^2s + σ^2s + π^2s] and [σ^2s + σ^2s] cycloaddition, respectively. *In suprafacial cycloreversion, either retention or inversion at both the ends of σ-bond takes place. On the other hand, antarafacial process provides retention at one end and inversion at the other end of the σ-bond* (Figure 4.3).

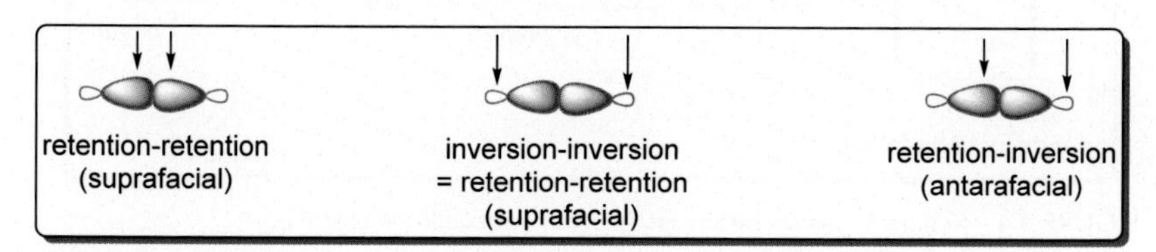

FIGURE 4.3 Suprafacial and antarafacial approaches to a σ-bond.

4.2 FEASIBILITY OF CYCLOADDITION REACTIONS

The feasibility of cycloaddition reactions can be easily predicted on the basis of three methods, namely, orbital symmetry correlation-diagram method, transition state aromaticity method or perturbation molecular orbital (**PMO**) *method and frontier molecular orbital* (**FMO**) *method.*

4.2.1 Orbital Symmetry Correlation-Diagram Method

In this approach, symmetry properties of the various molecular orbitals (MOs) of the bonds being involved in breaking and formation process during the reaction are considered and identified with respect to elements of symmetry (C_2 and σ) that are preserved throughout the reaction. Then a correlation diagram is drawn in which the MO levels of like symmetry of the reactants are connected to that of the products by lines.

If the symmetry of MOs of the reactant matches with that of the product in the ground state, then the reaction is *thermally allowed*, and if the symmetry of MOs of the reactant matches with that of the product in the first excited state, then the reaction is *photochemically allowed*. If the symmetries of the reactant and product molecular orbitals differ, the reaction does not occur in a concerted manner. It must be noted that a symmetry element becomes irrelevant when orbitals involved in the reaction are all symmetric or antisymmetric. In short, during the transformation, symmetry properties of the reactants and products remain conserved.

Let us consider the [π^2s + π^2s] cycloaddition. For a simple example, the approach of the two ethene molecules is considered as parallel. There are two symmetry planes present in the starting materials, transition state and the product. Horizontal (σ_h) plane of symmetry is the mirror plane between the components perpendicular to the *p*-orbitals; vertical (σ_v) plane of symmetry splits the molecules in half perpendicular to the σ-bonds. These planes shall be referred to as 1 and 2, respectively (Figure 4.4). It should be noted that the two planes containing the ethene units are not considered as the symmetry elements for the [π^2s + π^2s] cycloaddition.

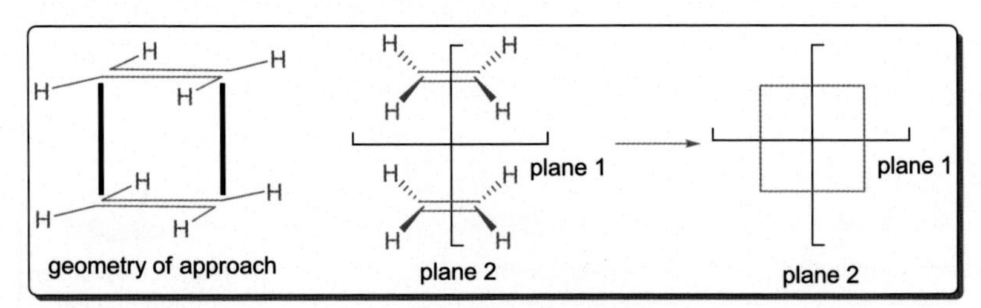

FIGURE 4.4 Symmetry planes for the [π^2s + π^2s] cycloaddition reaction.

In this process, we are primarily concerned with the four π-orbitals of the two ethene molecules and the four σ-orbitals of cyclobutane. The symmetry labels of the combinations with respect to the two mirror planes are considered here. (SS) represents the orbital symmetry where the π-orbitals are symmetric under reflection on both the mirror planes, (AA) represents where they are antisymmetric, and (SA) and (AS) represent levels where the π-orbitals are symmetric under one mirror plane but antisymmetric under the other mirror plane.

Applying the concept of "conservation of orbital symmetry" as advanced by Woodward and Hoffmann, it is shown in the correlation diagram (Figure 4.5) how the levels transform along the course of the reaction. It is

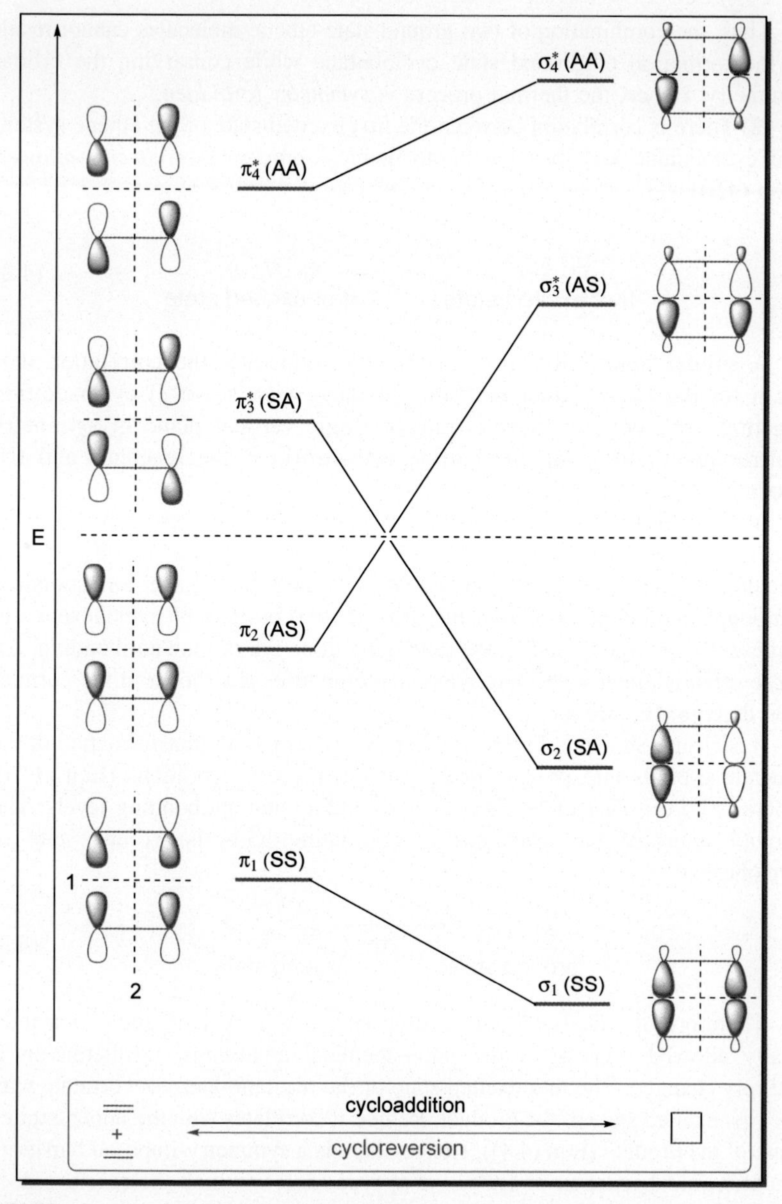

FIGURE 4.5 Correlation diagram for cycloaddition and cycloreversion of ethene-cyclobutane system.

concluded from the diagram that: (1) The ground state orbitals of the ethene correlate with an excited state of cyclobutane as shown in Eqn (4.1).

$$\underset{\text{ground state}}{\pi_1^2\,\pi_2^2} \rightleftharpoons \underset{\text{first excited state}}{\sigma_1^2\,\sigma_3^{*2}} \tag{4.1}$$

Thus, the combination of two ground state ethene molecules cannot result in the formation of ground state cyclobutane while conserving the orbital symmetry. Hence, the thermal process is *symmetry forbidden.*

(2) There is correlation between the first excited state of the ethene system and cyclobutane and thus the photochemical process is *symmetry allowed* (Eqn (4.2)).

$$\underset{\text{first excited state}}{\pi_1^2\,\pi_2^1\,\pi_3^{*1}} \rightleftharpoons \underset{\text{first excited state}}{\sigma_1^2\,\sigma_2^{*1}\,\sigma_3^{*1}} \tag{4.2}$$

A similar approach may be applied to construct the correlation diagram for the Diels–Alder reaction, which is a $[\pi^4s + \pi^2s]$ cycloaddition reaction. In this case, there is only a single vertical plane of symmetry (mirror plane) bisecting the carbon framework of the reactants and the product (Figure 4.6).

In this transformation, we have to consider six orbitals each of the reactants and the product. The ground state orbitals of the reactants are Ψ_1, Ψ_2 (of butadiene), and π (of ethene) while Ψ_3, Ψ_4, and π^* are the corresponding antibonding orbitals. Similarly, the ground state orbitals of cyclohexene are represented by σ_1, σ_2, and π; the remaining three orbitals are antibonding. All these orbitals along with their symmetry properties are shown in the correlation diagram (Figure 4.6).

It is clearly shown in Figure 4.6 that there is a smooth transformation of the reactant orbitals into product orbitals under thermal conditions (Eqn (4.3)). There is no crossing of levels between bonding and antibonding levels. The ground states of the reactants are correlating with the ground state of cyclohexene.

$$\underset{\text{ground state}}{\Psi_1^2\,\pi^2\,\Psi_2^2} \rightleftharpoons \underset{\text{ground state}}{\sigma_1^2\,\sigma_2^2\,\pi^2} \tag{4.3}$$

Therefore, the Diels–Alder reaction ($\pi^4s + \pi^2s$ cycloaddition) is a thermally allowed process. Under photochemical conditions, this situation is entirely changed. The first excited state of the reactant does not correlate with the first excited state of the product. Rather, it correlates with the upper excited state of the product (Eqn (4.4)). Hence, there is a symmetry-imposed barrier to photochemical reaction of $[\pi^4s + \pi^2s]$ type, due to which the reaction does not proceed under photochemical conditions.

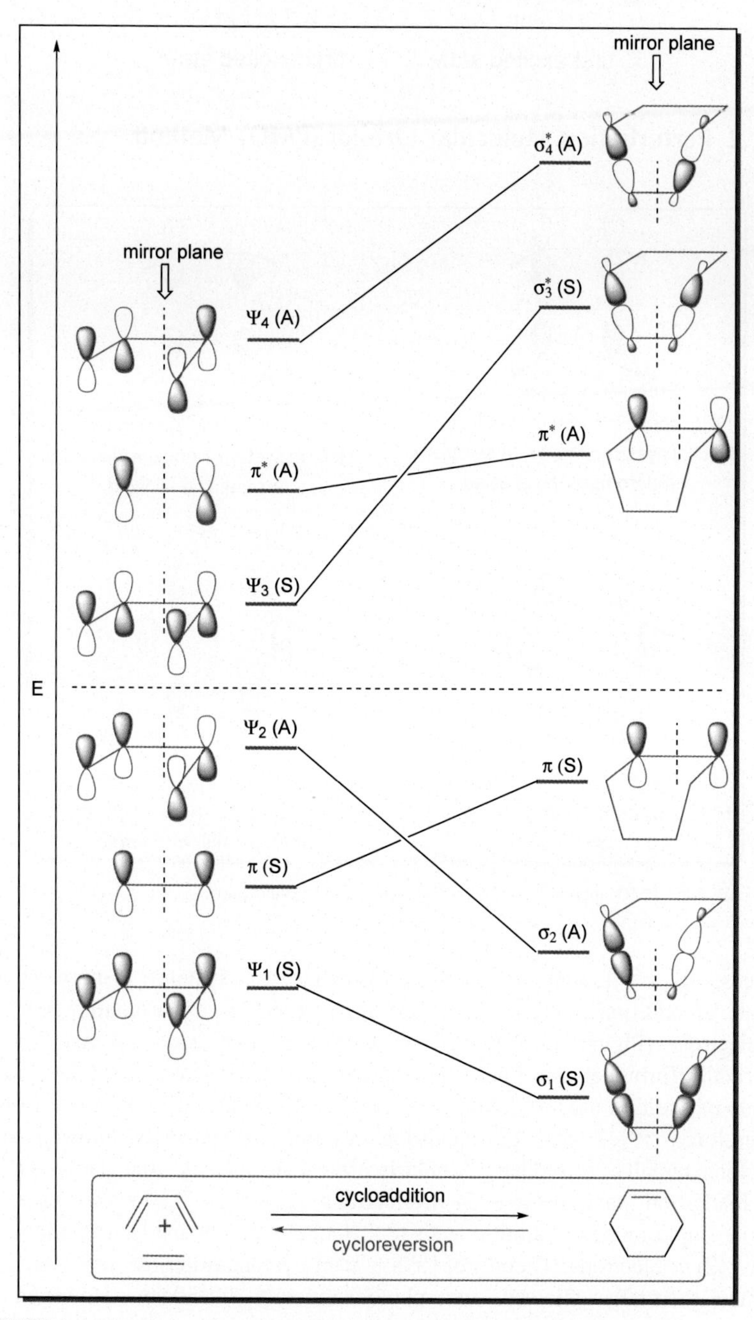

FIGURE 4.6 Correlation diagram for Diels–Alder and retro Diels–Alder reaction.

$$\Psi_1^2 \pi^2 \Psi_2^1 \Psi_3^1 \not\longrightarrow \sigma_1^2 \sigma_2^2 \pi^1 \pi^{*1} \tag{4.4}$$

first excited state → first excited state

4.2.2 Perturbation Molecular Orbital (PMO) Method

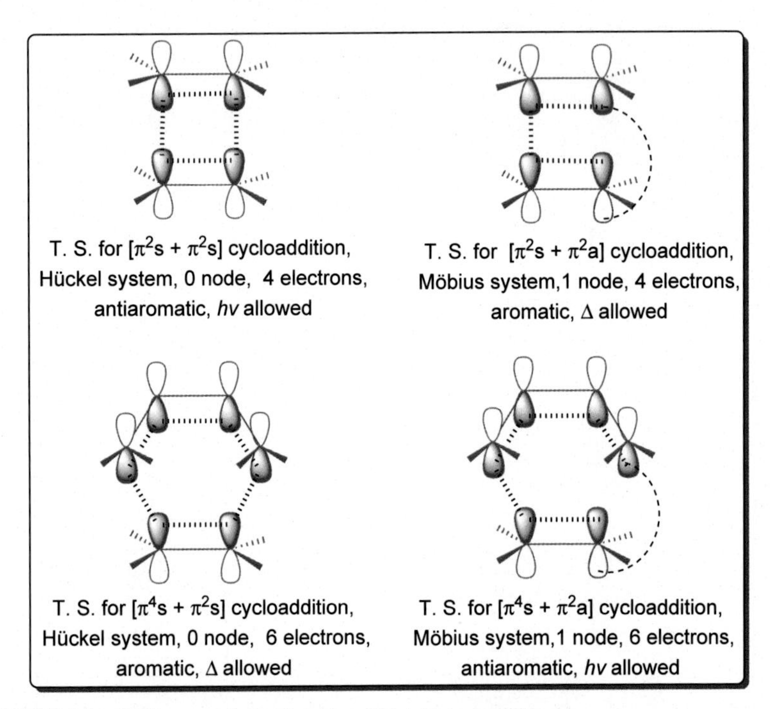

T. S. for [π²s + π²s] cycloaddition,
Hückel system, 0 node, 4 electrons,
antiaromatic, hv allowed
T. S. for [π²s + π²a] cycloaddition,
Möbius system,1 node, 4 electrons,
aromatic, Δ allowed
T. S. for [π⁴s + π²s] cycloaddition,
Hückel system, 0 node, 6 electrons,
aromatic, Δ allowed
T. S. for [π⁴s + π²a] cycloaddition,
Möbius system,1 node, 6 electrons,
antiaromatic, hv allowed

FIGURE 4.7 PMO approach for [2 + 2] and [4 + 2] cycloadditions.

In case of [π^2s + π^2s] cycloaddition (4n π-electron system), a supra–supra mode of addition leads to a Hückel array, which is antiaromatic with 4n π-electrons (Figure 4.7). Therefore, the supra–supra mode of reaction is thermally forbidden and photochemically allowed. However, a supra–antara mode of addition uses a Möbius array, which is aromatic with 4n π-electrons. Therefore, [π^2s + π^2a] cycloaddition reaction is thermally allowed and photochemically forbidden. Similarly, we can analyze the [π^4s + π^2s] cycloaddition having (4n + 2) π-electrons (Figure 4.7). In this case, a supra–supra mode of addition leads to a Hückel array, which is aromatic with (4n + 2) π-electrons. Therefore, [π^4s + π^2s] cycloaddition reaction now becomes thermally allowed and photochemically forbidden. However, a [π^4s + π^2a] cycloaddition uses a Möbius array, which is antiaromatic with (4n + 2) π-electrons. Therefore, the reaction is thermally forbidden and photochemically allowed in this mode.

The selection rules for cycloadditions and cycloreversions by PMO method are summarized in Table 4.1, where n is zero or an integer.

TABLE 4.1 Selection rules for cycloadditions and cycloreversions by PMO method.

No. of electrons p + q	Stereochemical mode of reaction	No. of nodes in the T.S.	Aromaticity of the T.S.	Reaction condition
4n	Supra–supra Antara–antara	0 or Even	Antiaromatic	*hν*
4n	Supra–antara Antara–supra	Odd	Aromatic	Δ
4n + 2	Supra–supra Antara–antara	0 or Even	Aromatic	Δ
4n + 2	Supra–antara Antara–supra	Odd	Antiaromatic	*hν*

4.2.3 Frontier Molecular Orbital Method

According to this method, the feasibility of a cycloaddition reaction depends upon the symmetry properties of the highest occupied molecular orbital (HOMO) of one reactant and the lowest unoccupied molecular orbital (LUMO) of the other. A favorable bonding interaction is possible only when the phases of the lobes of the interacting orbitals in the HOMO and LUMO are the same.

In the $[\pi^2s + \pi^2s]$ cycloaddition of ethene to form cyclobutane, lobes of HOMO in one molecule and that of LUMO in the other do not have the same signs and hence the reaction is thermally forbidden. Irradiation of ethene, however, promotes an electron to the antibonding orbital, which now becomes HOMO and corresponds with the LUMO of the second unexcited ethene molecule. As expected, the combination now proceeds smoothly (Figure 4.8).

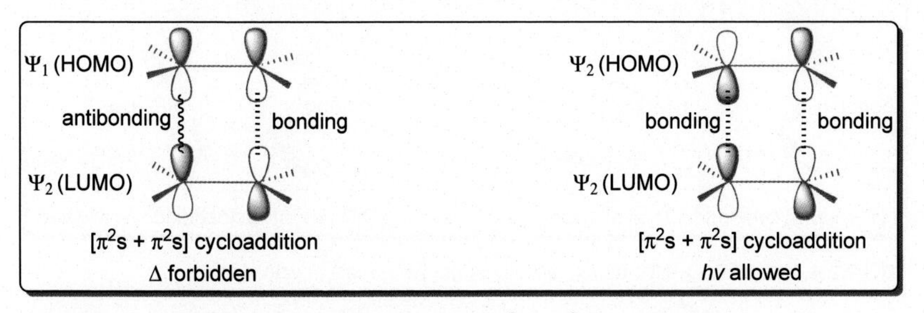

FIGURE 4.8 HOMO–LUMO interactions in the $[\pi^2s + \pi^2s]$ cycloaddition.

An interesting situation arises in the case of $[\pi^2s + \pi^2a]$ cycloaddition, where the ground state interaction allows the reaction to proceed under thermal conditions. The reaction is suprafacial with respect to the HOMO component and antarafacial with respect to the LUMO component. Although this reaction is favorable in terms of overlap of the orbitals, it is not a common reaction because of geometric reasons. The requirement of substituents on one alkene to be oriented directly toward the molecular plane of the second alkene is rather a sterically unfavorable situation (Figure 4.9). It occurs in some very specific cases (for example, ketenes) where steric congestion is reduced.

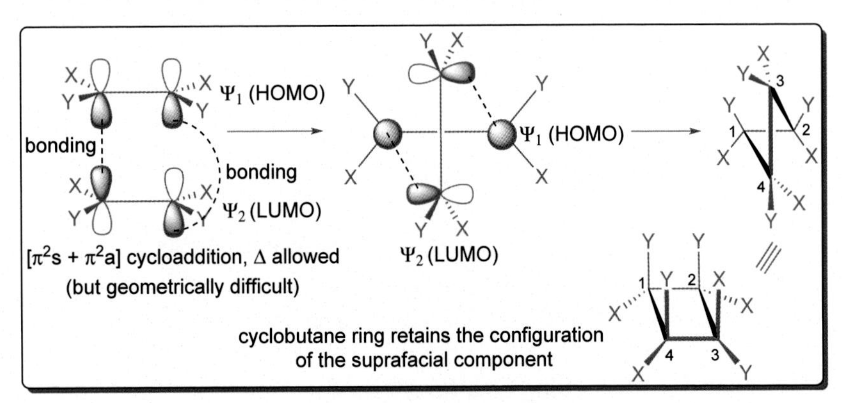

FIGURE 4.9 HOMO–LUMO interactions in the $[\pi^2s + \pi^2a]$ cycloaddition.

The Diels–Alder reaction may also be analyzed by a similar consideration of the molecular orbitals of butadiene and ethene. In this case, there are two possible HOMO–LUMO interactions. Since the phases of the 1,4-lobes of the HOMO of butadiene match with those in the LUMO of ethene, the $[\pi^4s + \pi^2s]$ cycloaddition is thermally allowed. We reach a similar conclusion by considering the symmetry of LUMO of butadiene and the HOMO of ethene (Figure 4.10). However, on energetic grounds the latter interaction will make a smaller contribution than the former.

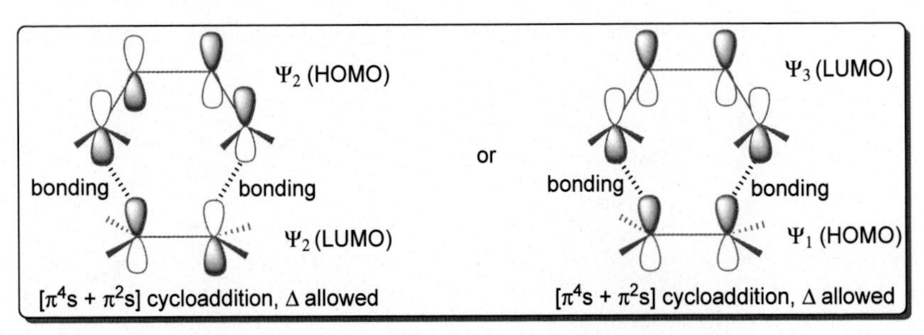

FIGURE 4.10 HOMO–LUMO interactions in the $[\pi^4s + \pi^2s]$ cycloaddition.

On the other hand, a photoinduced [π^4s + π^2s] cycloaddition is symmetry forbidden. The absorption of a photon promotes an electron from the HOMO to the LUMO. In this case, the lower energy gap between HOMO and LUMO is in the diene partner. Thus, a new photochemical HOMO Ψ_3 is produced and this cannot overlap with the LUMO of the dienophile (Figure 4.11).

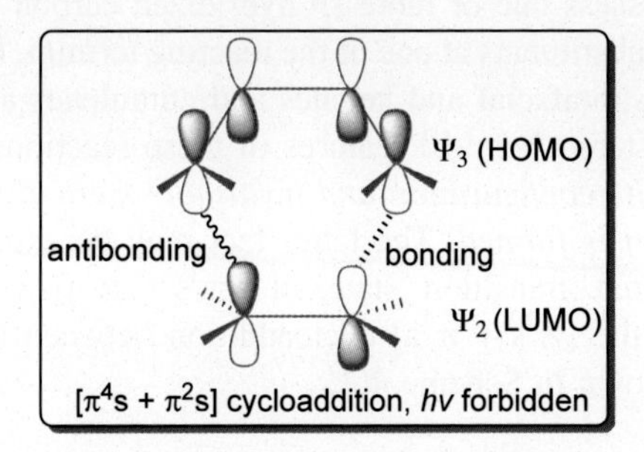

FIGURE 4.11 HOMO–LUMO interactions in the [π^4s + π^2s] cycloaddition continued.

The selection rules for cycloadditions and cycloreversions by the FMO method are summarized in Table 4.2, where n is zero or an integer.

TABLE 4.2 Selection rules for cycloadditions and cycloreversions by FMO method.

No. of electrons p + q	Stereochemical mode of reaction	Reaction condition
4n	Supra–supra Antara–antara	*hv*
4n	Supra–antara Antara–supra	Δ
4n + 2	Supra–supra Antara–antara	Δ
4n + 2	Supra–antara Antara–supra	*hv*

4.3 [2 + 2] CYCLOADDITIONS

We have seen that thermal [π^2s + π^2s] cycloaddition is symmetry forbidden whereas [π^2s + π^2a] cycloaddition is symmetry allowed. However, [π^2s + π^2a] mode of cycloaddition can be expected only when the two double bonds are disposed orthogonally to each other. However, such an orientation is not easily accessible unless some structural features in one of the ethylenic moieties permit it. This explains why cyclobutanes are not readily accessible from simple olefins.

[π^2s + π^2a] Cycloaddition reactions are rather uncommon and occur only in particular cases where steric congestion is reduced. For example, alkenes readily undergo [π^2s + π^2a] cycloadditions with ketenes, cumulenes, or isocyanates, which possess one or more *sp* hybridized carbon atoms and lack a pair of interfering substituents at one of the reacting termini. In these reactions, alkenes act as the suprafacial and ketenes and cumulenes as the antarafacial components. Two stereochemical features of these reactions are noteworthy: *the alkene retains its configuration and invariably sterically more congested [2 + 2] cycloadduct is formed.* The latter fact may be explained by considering the orthogonal transition state of [π^2s + π^2a] cycloaddition. For example, consider the [π^2s + π^2a] cycloaddition between (Z)-but-2-ene and ethoxyketene as shown in Scheme 4.2.

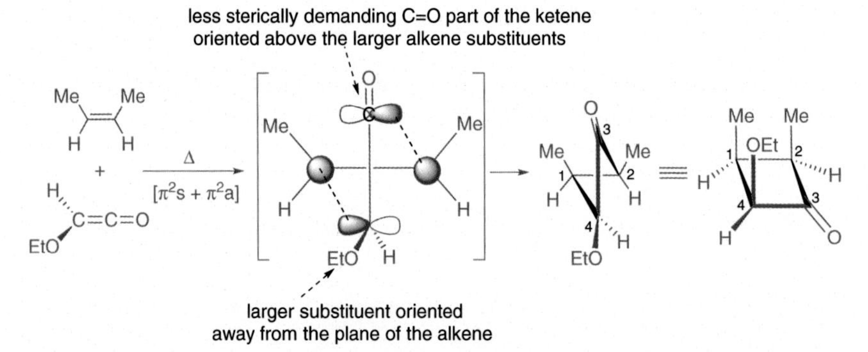

SCHEME 4.2 Thermal [π^2s + π^2a] cycloaddition between (Z)-but-2-ene and ethoxyketene.

In the transition state, the less hindered *sp* carbon is directed toward the more hindered side of the alkene double bond. In this arrangement, the carbonyl group of the ketene and two methyl groups of the *cis*-but-2-ene are on the same side. There is now more steric hindrance in the kinetically controlled product of the reaction with ethoxy and methyl groups being on the same side.

Reaction of *cis*- or *trans*-cyclooctene with dichloroketene takes place in a highly stereospecific manner with retention of configuration for the double-bond substituents (Scheme 4.3). Such a high stereospecificity is entirely

SCHEME 4.3 Stereospecific cycloaddition of dichloroketene to *cis*- or *trans*-cyclooctene.

consistent with a concerted mechanism in which π-bond of the ketenophile is used suprafacially.

Somc mcthods for the generation of ketenes are given below in Scheme 4.4.

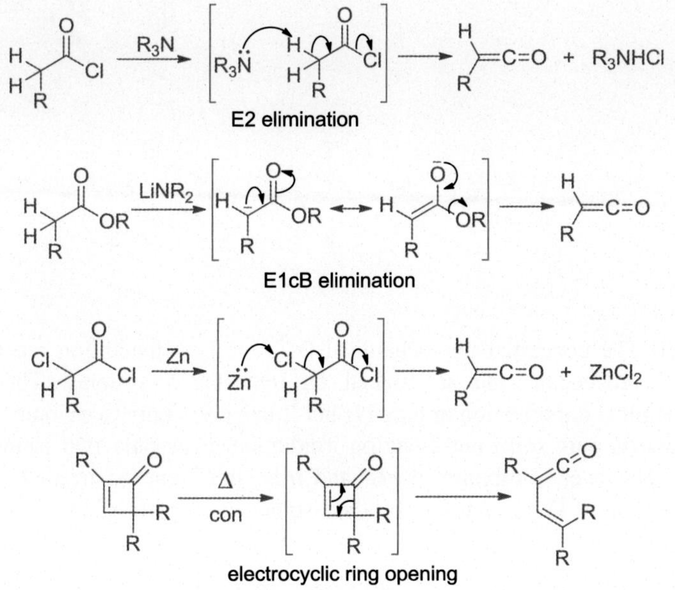

H, H, R–C(=O)Cl —R_3N→ [$R_3\ddot{N}$ H, H, R, O, Cl] → H, R >C=O + R_3NHCl

E2 elimination

H, H, R–C(=O)OR —$LiNR_2$→ [H, R, O, OR ↔ H, R, O, OR] → H, R >C=O

E1cB elimination

Cl, H, R–C(=O)Cl —Zn→ [$\ddot{Zn}$ Cl, H, R, O, Cl] → H, R >C=O + $ZnCl_2$

R, O, R, R —Δ, con→ [R, O, R, R] → R, C=O, R, R

electrocyclic ring opening

SCHEME 4.4 Methods for the generation of ketenes.

4.3.1 Solved Problems

Q 1. Explain the mechanism of following transformations:

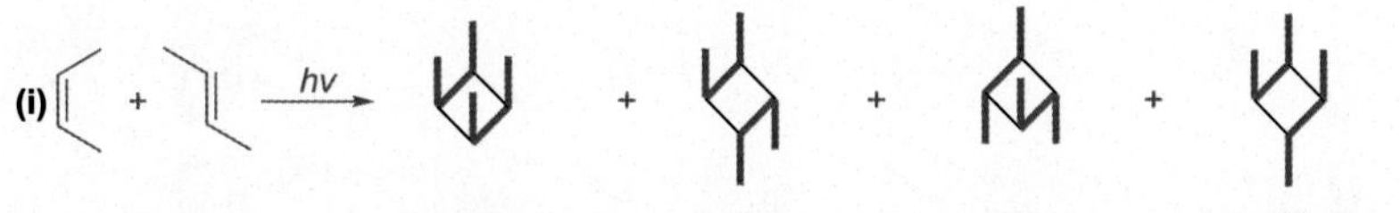

(i) + —*hv*→ I + II + III + IV

I *cis-syn-cis* II *cis-anti-cis* III *trans-anti-trans* IV *cis-anti-trans*

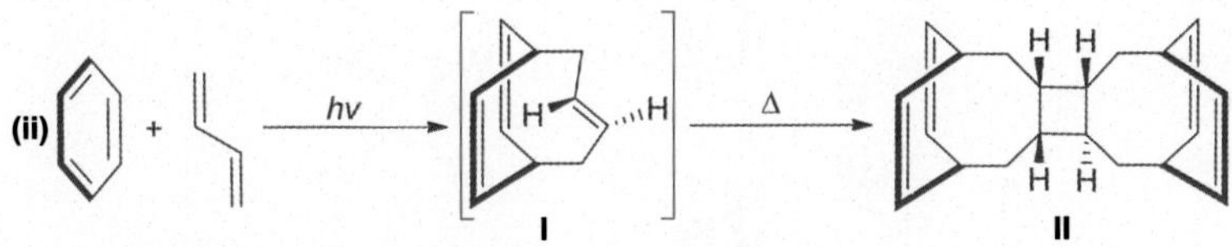

(ii) + —*hv*→ [H, H I] —Δ→ H H H H II

(iii) 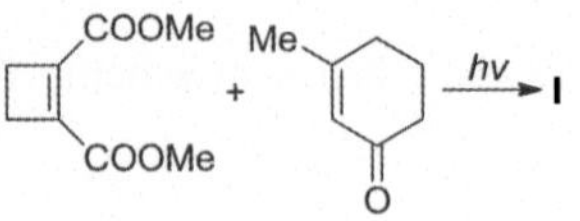 (iv) 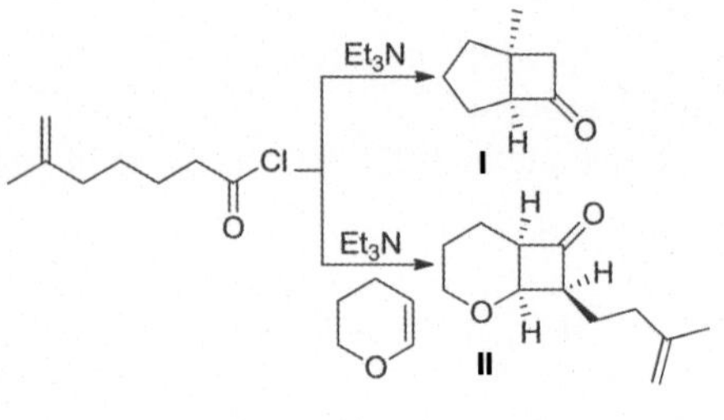 (v) 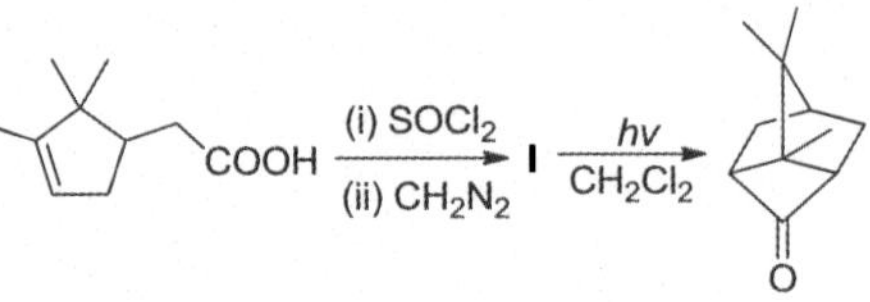 (vi)

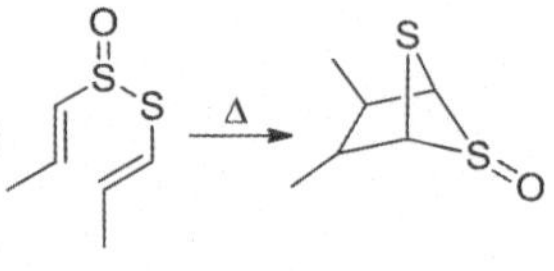

Sol 1. **(i)** The concerted photochemical [$\pi^2 + \pi^2$] cycloaddition reaction is highly stereospecific and suprafacial on both the π-systems. Therefore, photochemical dimerization of pure *cis*-but-2-ene gives only two isomers **I** and **II**. Similarly, pure *trans*-but-2-ene on irradiation gives only two isomers, **II** and **III**. However, a mixture of *cis*- and *trans*-but-2-ene on irradiation also gives the isomer **IV** in addition to other isomers as shown below:

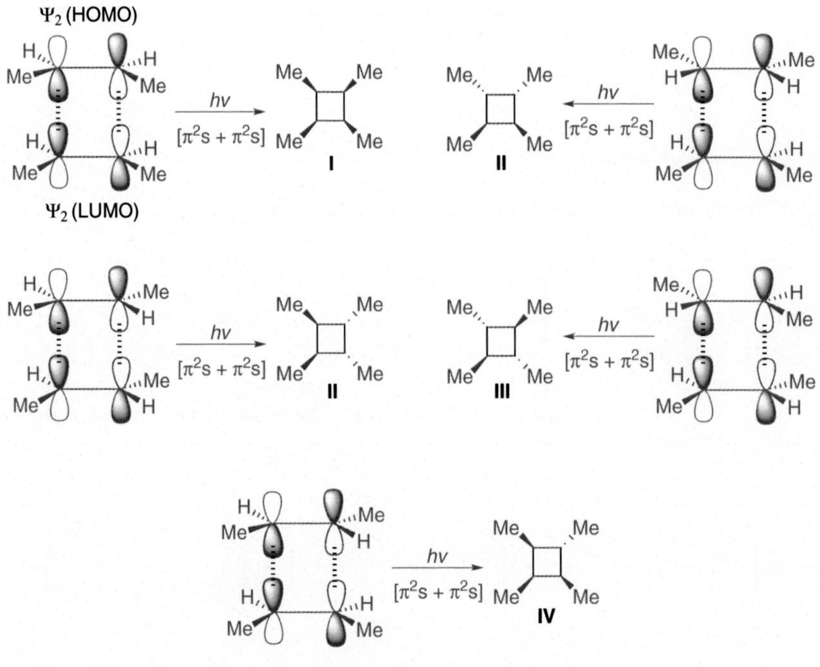

Ψ_2 (HOMO)

Ψ_2 (LUMO)

hv [π^2s + π^2s]

I II III IV

(ii) The first step involves the [π^4s + π^4s] addition of benzene and *transoid* butadiene to give compound **I**. In compound **I**, the two *p*-orbitals are not parallel and coplanar because of the twisting of double bond about its axis. Therefore, FMOs in compound **I** readily overlap to give the dimer **II** by [π^2s + π^2a] addition. One side of the dimer **II** retains the cyclobutaric component in the *trans* configuration while the other side becomes *cis* due to the antara interaction.

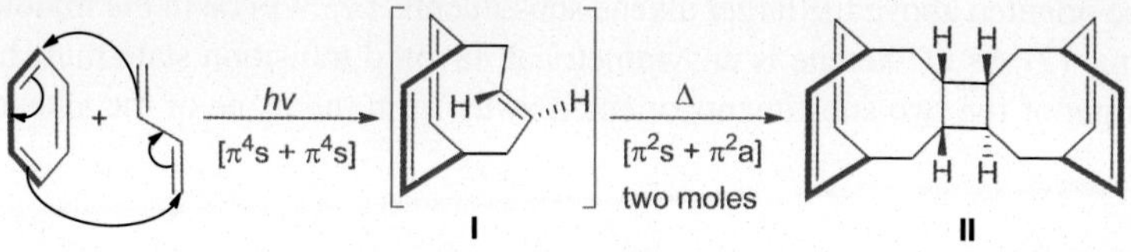

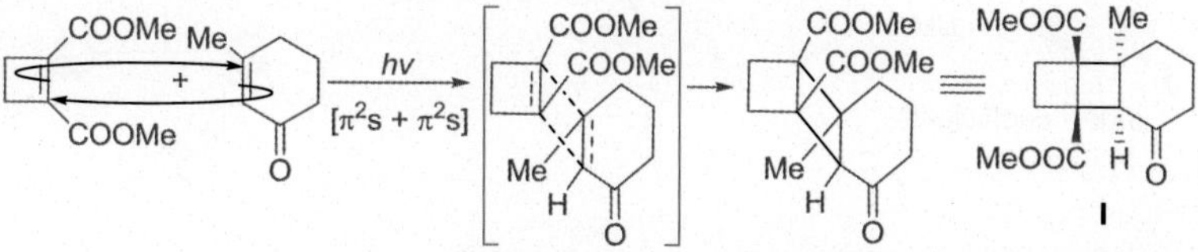

rings arranged in an *exo* manner to minimize repulsions

(iv) When the internal carbon of the olefin is more substituted, the intramolecular cycloaddition of ketenes leads to the formation of bicyclo[n.2.0] alkanones as shown below:

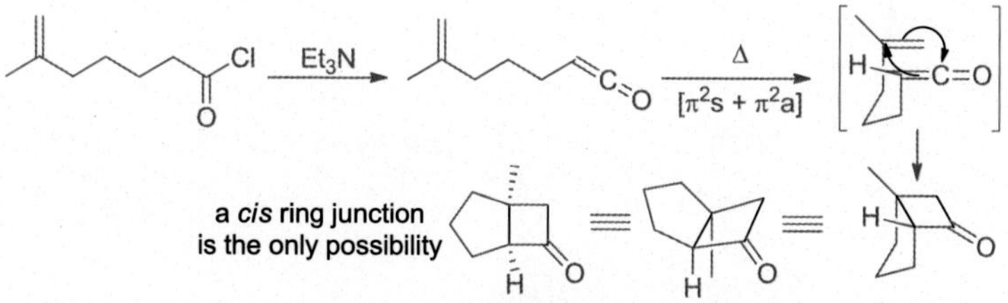

1-Methylbicyclo[3.2.0]heptan-6-one (I)

It may be noted that when the terminal carbon of the olefin is more substituted, the intramolecular cycloaddition of ketenes leads to the formation of bicycle[n.1.1]alkanones as shown below:

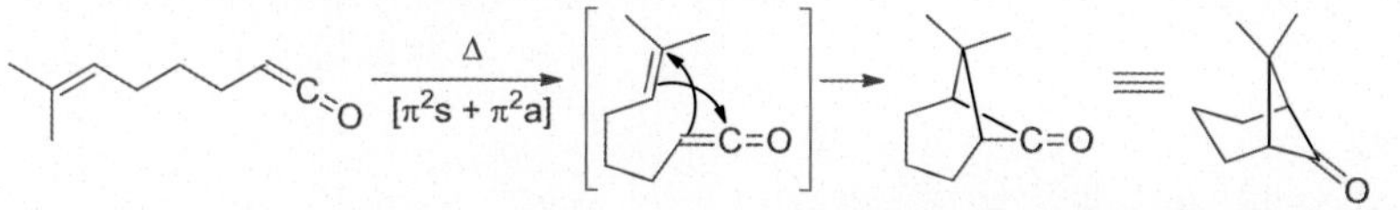

7,7-Dimethylbicyclo[3.1.1]heptan-6-one

On the other hand, 3,4-dihydro-2*H*-pyran is more reactive than the simple alkene (because the electron-donating oxygen makes the enol ether more reactive), therefore, the ketene prefers to react with it instead of undergoing intramolecular addition reaction.

The regiochemistry of the reaction can be explained by considering the interaction of that carbon of alkene (i.e., enol ether) which bears the largest coefficient in its HOMO with the central carbon of the ketene, which bears the largest coefficient in its LUMO.

The olefin retains its *cis*-configuration, i.e., the two *cis*-hydrogens of cyclic alkene must remain *cis* in the product. The stereochemistry at the remaining center comes from the way the two molecules approach each other. (1) As the alkene is unsymmetrical, the less sterically demanding carbonyl part of the ketene will be oriented above the larger alkene substituents, i.e., will be in the middle of the ring. (2) As the ketene is unsymmetrical, favored transition state must have the larger of the two substituents oriented away from the plane of the alkene.

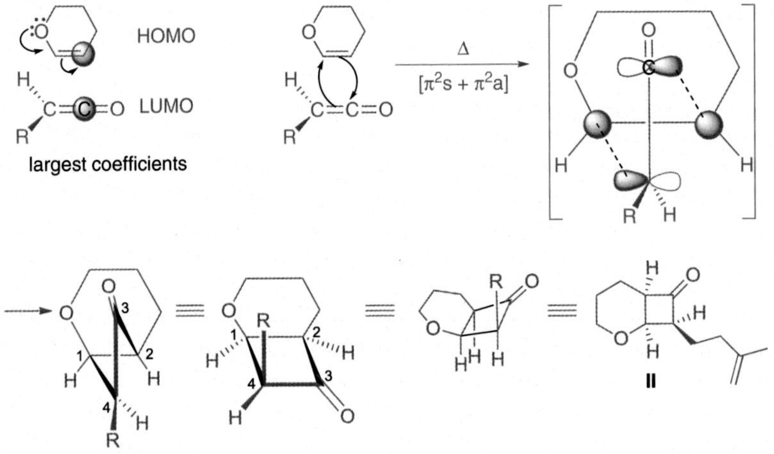

HOMO
LUMO
largest coefficients
Δ
$[\pi^2s + \pi^2a]$
II

(v)
$SOCl_2$
$- SO_2$
$- HCl$
$\bar{C}H_2-\overset{+}{N}\equiv N$
Diazoketones
hv
$- N_2$
α-Ketocarbene intermediate
Ketene intermediate (I)
hv
$[\pi^2s + \pi^2s]$

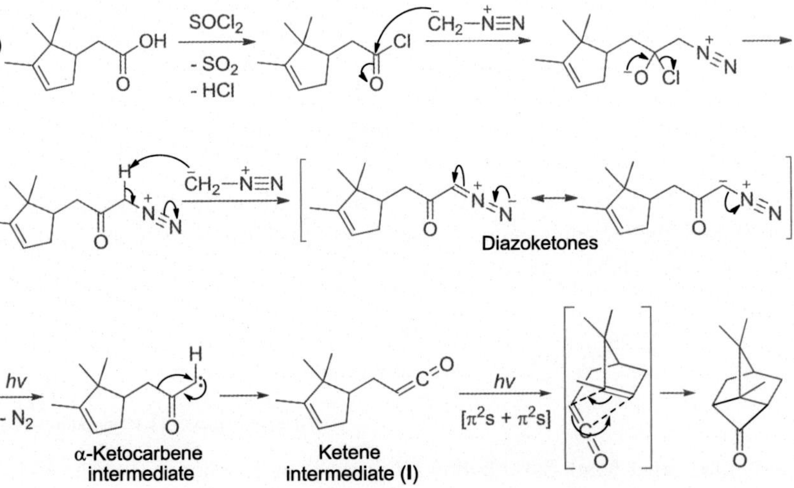

(vi) [3, 3] shift Δ; [2 + 2] Δ

Zwiebelane (a 1,3-dithietane, a compound obtained from the onion)

Q 2. Select the product of the reaction of (*Z*)-(2-methoxyvinyl)benzene with dichloroacetyl chloride in presence of triethylamine.

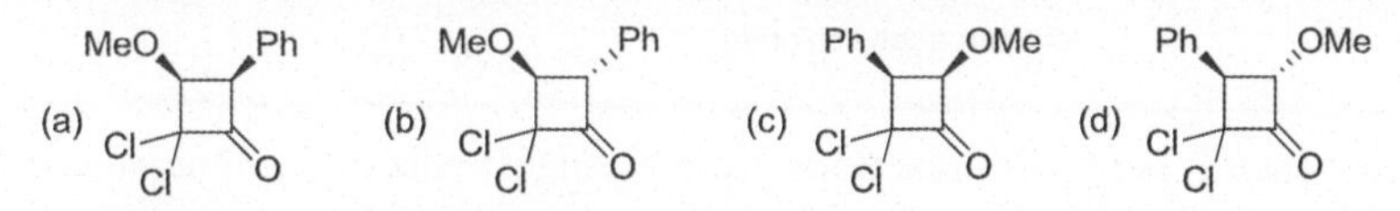

Sol 2. (a) Regiochemistry

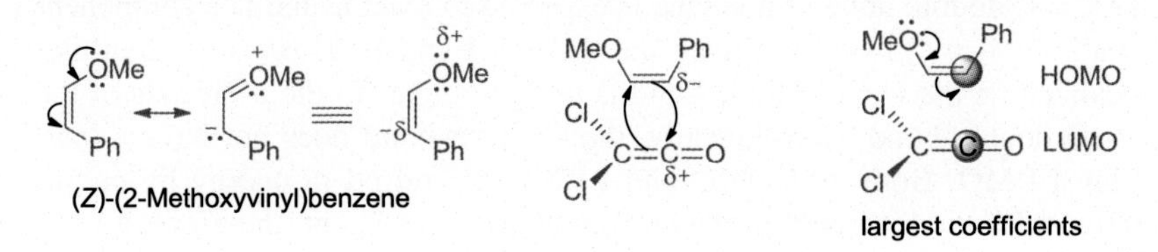

Stereochemistry

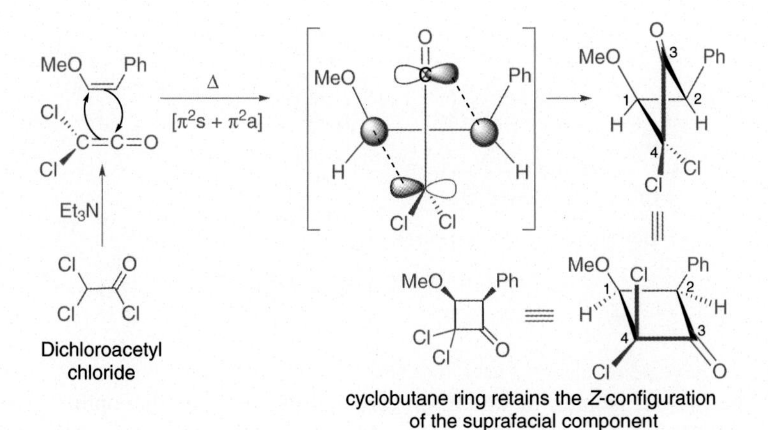

cyclobutane ring retains the *Z*-configuration of the suprafacial component

Q 3. The major product formed in the reaction of cyclopentadiene with a mixture of dichloroacetyl chloride and triethylamine is:

(a) (b) (c) (d)

Sol 3. (b)

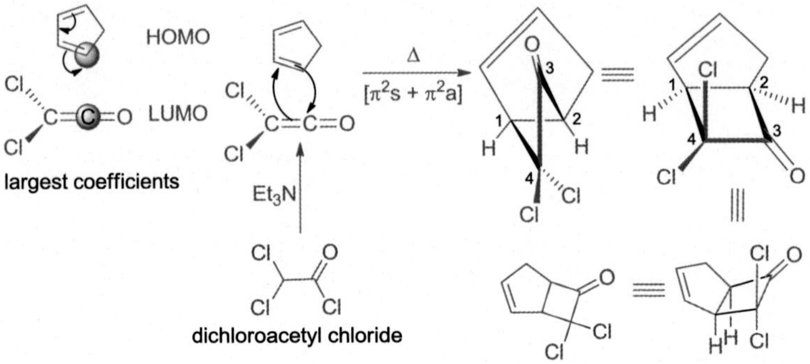

HOMO

LUMO

largest coefficients

$[\pi^2s + \pi^2a]$

Et_3N

dichloroacetyl chloride

In general, pericyclic reactions use the longest part of a conjugated system because the ends of the conjugated systems carry the largest coefficients in the frontier orbitals, which make these reactions proceed faster. Therefore, the C=C double bond of a ketene is expected to react as the π^2s component of a $[\pi^4s + \pi^2s]$ reaction giving the adducts **I** and **II**. However, the orbital localized on the C=C double bond is at right angles to the *p*-orbitals of the C=O double bond. Consequently, the C=C π-bond does not have a low-lying LUMO. Both its HOMO and LUMO are raised in energy by conjugation with the lone pair on the oxygen atom and it is not, therefore, a good dienophile.

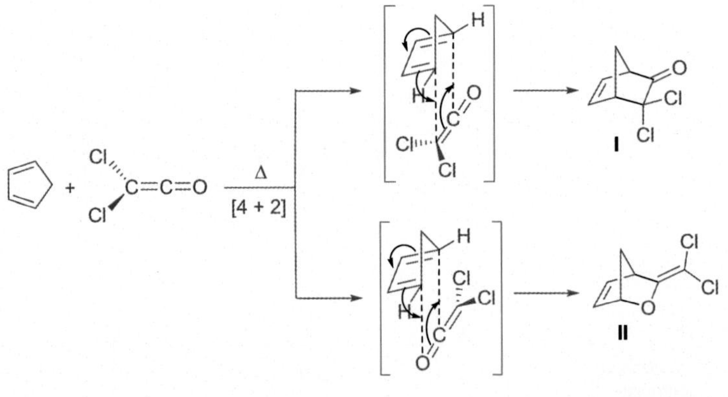

Δ

[4 + 2]

I

II

The [4 + 2] isomer, in which the carbonyl group is the dienophile giving the ether **II**, is not so obviously unfavorable. [4 + 2] Cycloaddition of this type is known for other ketenes as well. A [3,3]-Claisen rearrangement converts this product to the [2 + 2] adduct.

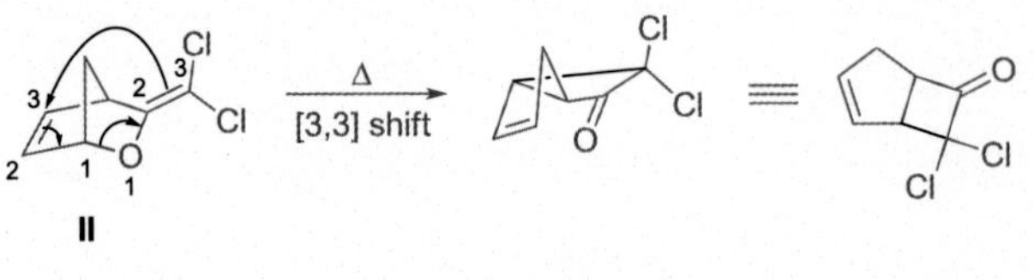

Δ

[3,3] shift

II

Q 4. The major product formed in the following reaction sequence is:

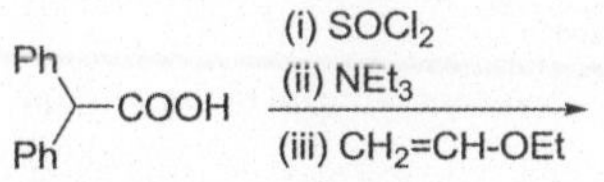

(i) $SOCl_2$
(ii) NEt_3
(iii) CH_2=CH-OEt

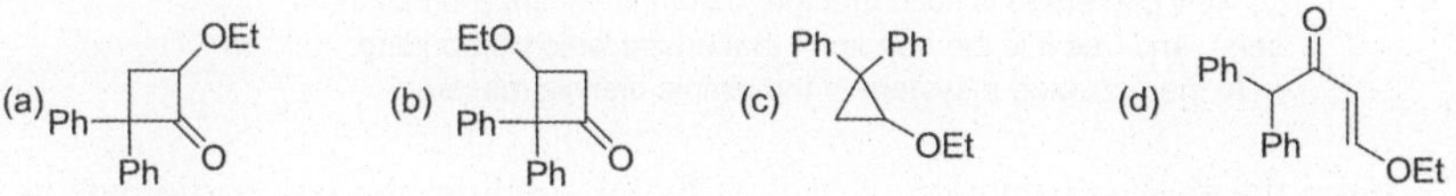

(a) (b) (c) (d)

Sol 4. (b)

$SOCl_2$ NEt_3 Et_3N $=C=O + Et_3NHCl$

E2 elimination Ketene

HOMO LUMO largest coefficients

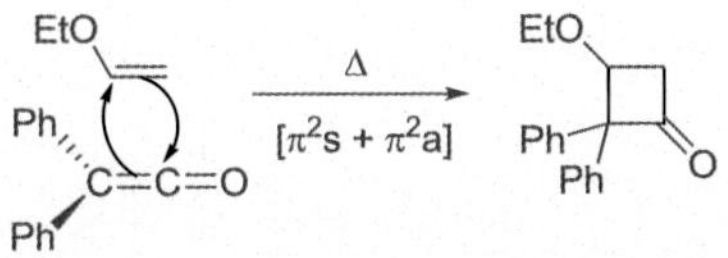

Δ $[\pi^2s + \pi^2a]$

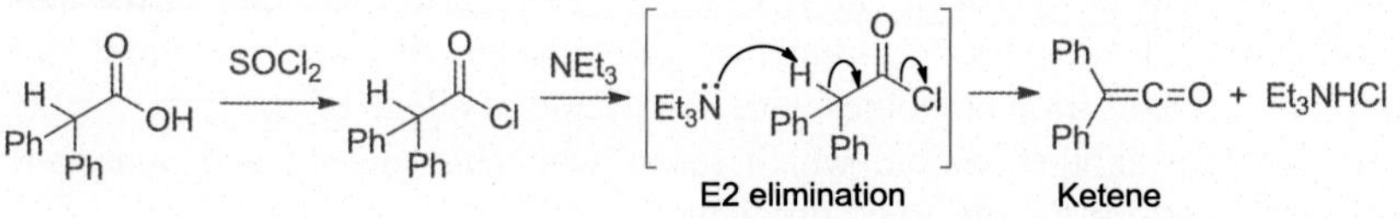

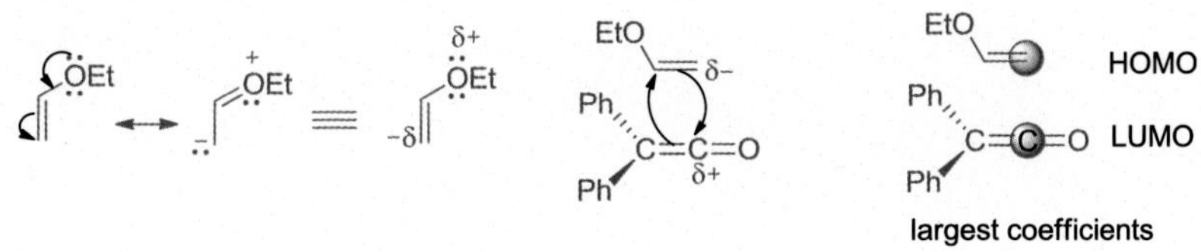

Q 5. What is the special factor in the structure of the ketenes that allows them to act as an antarafacial component in the $[\pi^2s + \pi^2a]$ cycloaddition reactions?

Sol 5. The resonating structure **II** of ketene suggests that we should analyze the combination of a vinyl cation (**III**) with an olefin from the viewpoint of orbital symmetry conservation.

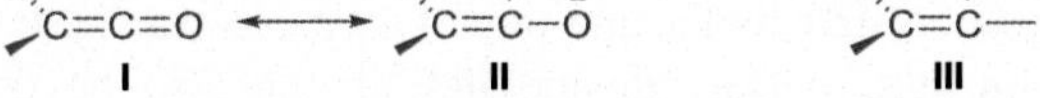

I II III

The presence of the orthogonal vacant *p*-orbitals in the vinyl cation results in the formation of strong bonding interactions with the occupied π-system of the simple olefinic reactant. Thus, the normal symmetry-allowed combination of a cation with an olefin in this particular situation sets the stage for the $[\pi^2s + \pi^2a]$ cycloaddition reaction. However, these secondary interactions are absent from the $[\pi^2s + \pi^2a]$ reaction path for simple olefins.

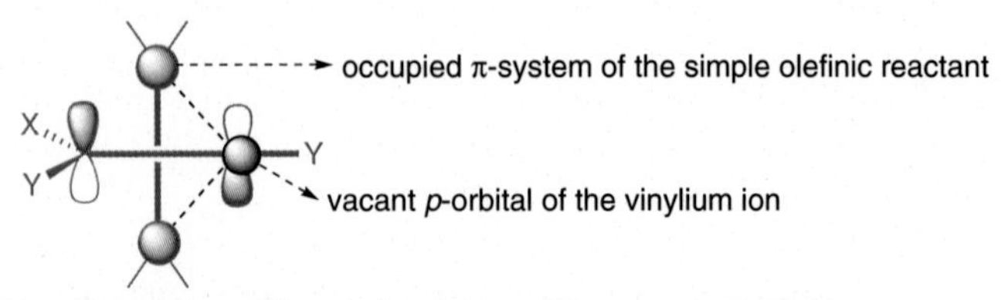

The view presented is such that the top of the vacant p-orbital is seen, and that it is the bottom of that orbital which is bonding to the occupied π-system of the simple olefinic reactant.

From the above discussion, it is clear that a vinylium ion can participate as the π^2a component in the concerted $[\pi^2s + \pi^2a]$ cycloaddition. For example, allene reacts with hydrogen chloride at −70 °C to give a mixture of the stereoisomeric 1,3-dichloro-1,3-dimethylcyclobutanes. In this reaction, an initially formed cation combines with a molecule of allene by the $[\pi^2s + \pi^2a]$ route to give another cation, which reacts with chloride ion and hydrogen chloride to afford the observed products.

$$H_2C{=}C{=}CH_2 \xrightarrow{HCl} H_2C{=}\overset{+}{C}{-}CH_3 \xrightarrow[{[\pi^2s + \pi^2a]}]{H_2C{=}C{=}CH_2} \text{(cyclobutyl cation with } CH_3 \text{ and } {=}CH_2\text{)} \xrightarrow[\text{(ii) } HCl]{\text{(i) } \bar{Cl}} \text{(1,3-dichloro-1,3-dimethylcyclobutane: } Cl, CH_3, H_3C, Cl\text{)}$$

Similarly, treatment of but-2-yne with chlorine in the presence of boron trifluoride gives β-chlorovinylium ion. This ion undergoes $[\pi^2s + \pi^2a]$ combination with but-2-yne to give the cyclobutenyl cation, whose discharge by chloride ion yields 3,4-dichloro-1,2,3,4-tetramethylcyclobutene.

$$Me{-}C{\equiv}C{-}Me \xrightarrow[BF_3]{Cl_2} \overset{Me}{\underset{Cl}{\diagdown}}C{=}\overset{+}{C}{-}Me \xrightarrow[{[\pi^2s + \pi^2a]}]{Me{-}C{\equiv}C{-}Me} \text{(cyclobutenyl cation: Cl, Me, Me, Me, Me)} \xrightarrow{\bar{Cl}} \text{(cyclobutene: Cl, Cl, Me, Me, Me, Me)}$$

4.4 [4 + 2] CYCLOADDITIONS

The Diels–Alder reaction involves the addition of a dienophile, which is an olefinic or acetylinic compound, to the 1,4-position of a conjugated diene system to produce a cyclohexene. Since the reaction forms a cyclic product, via a cyclic transition state by the addition of a 4π-electron system (i.e., diene) to a 2π-electron system (i.e., dienophile), it can also be described as a $[\pi^4s + \pi^2s]$ cycloaddition reaction. Numbers 4 and 2 identify both the number of π-electrons involved in the electronic rearrangement and the number of atoms in the cyclohexene ring. The letter "s" indicates that the reaction takes place suprafacially on both the components. The reaction between buta-1,3-diene and ethene to give cyclohexene is a typical example of Diels–Alder reaction (Scheme 4.5).

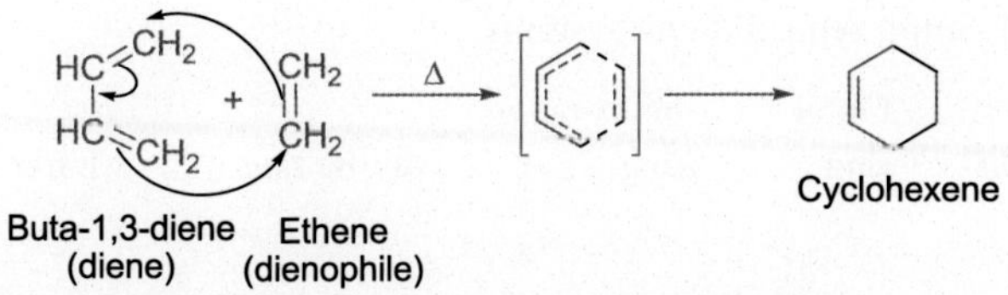

Buta-1,3-diene (diene) Ethene (dienophile) Cyclohexene

SCHEME 4.5 Diels–Alder reaction between buta-1,3-diene and ethene.

The Diels–Alder reaction is a pericyclic reaction, which takes place in a concerted single step by redistribution of bonding electrons. The two reactants simply join through a cyclic transition state in which the two new carbon–carbon σ-bonds are formed at the cost of the two π-bonds of the diene and dienophile. The high *syn* stereospecificity of the reaction, the low solvent effect on the reaction rate, and the large negative values of both activation entropy and activation volume comprise the evidence usually given in favor of a pericyclic Diels–Alder reaction.

4.4.1 Diene and Dienophile

The diene component in Diels–Alder reaction can be either an open chain or a cyclic compound and it can have many different kinds of substituents. A conjugated diene such as buta-1,3-diene can exist in two different planar conformations: *s-cis* conformation and *s-trans* conformation (Scheme 4.6). By "*s-cis*" we mean that the double bonds are *cis* about the single bond (s = single). The *s-trans* conformation is more stable than the *s-cis* conformation because the close proximity of the hydrogens causes some steric strain.

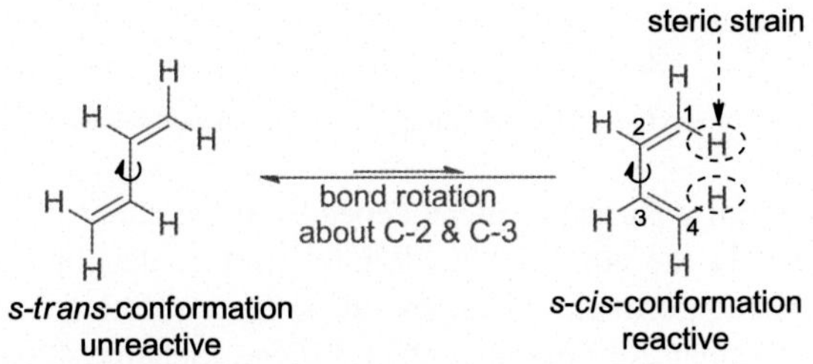

s-trans-conformation unreactive — bond rotation about C-2 & C-3 — *s-cis*-conformation reactive (steric strain)

SCHEME 4.6 Conformations of buta-1,3-diene.

Some typical diene systems are illustrated in Table 4.3.

The diene must be in the *s-cis* conformation to undergo the Diels–Alder reaction, because only in this form are the C-1 and C-4 of the diene close enough to react through a cyclic transition state. In the *s-trans* conformation, C-1 and C-4 of the diene are too far apart to react through a cyclic transition state. Dienes with an enforced coplanar *s-cis* conformation are exceptionally

TABLE 4.3 Some typical diene systems.

Open chain	Outer ring	Inner-outer ring	Across ring	Inner ring

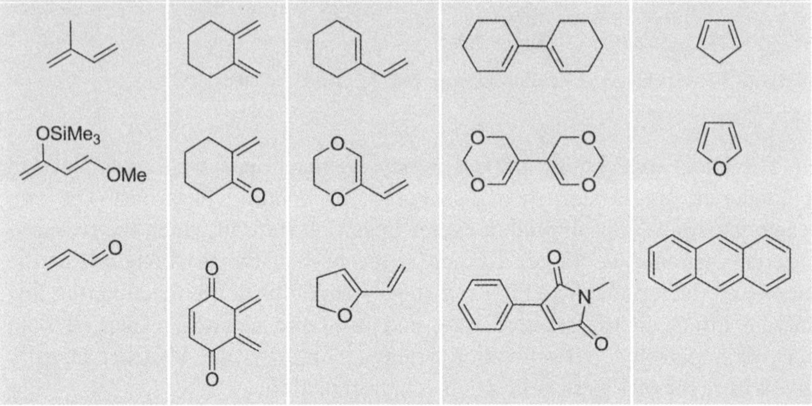

reactive in Diels–Alder reaction. But cyclic dienes that are permanently in the *s-trans* conformation and cannot adopt *s-cis* conformation will not undergo Diels–Alder reaction. The order of reactivity of some conjugated dienes toward maleic anhydride at 30 °C is given in Scheme 4.7.

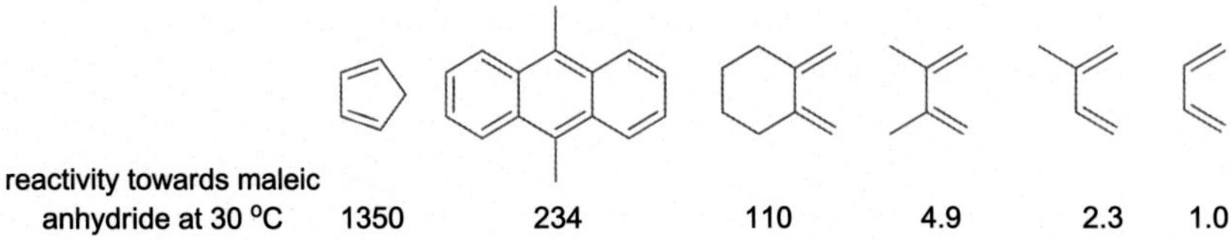

reactivity towards maleic anhydride at 30 °C 1350 234 110 4.9 2.3 1.0

SCHEME 4.7 Order of reactivity of some conjugated dienes toward maleic anhydride at 30 °C.

The reactivity of a particular diene depends on the concentration of the *s-cis* conformation in the equilibrium mixture. Factors that increase the concentration of this conformation make the diene more reactive. As an example of this effect, consider 2,3-dimethylbuta-1,3-diene (Scheme 4.8).

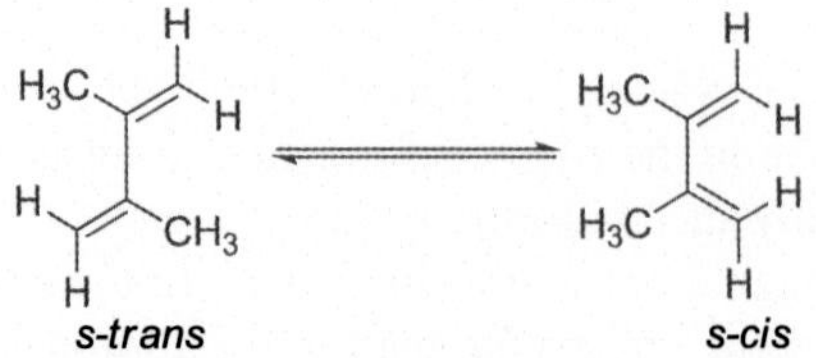

s-trans *s-cis*

SCHEME 4.8 Conformations of 2,3-dimethylbuta-1,3-diene.

The methyl groups on C-2 and C-3 cause the *s-trans* and *s-cis* conformations to have similar amounts of steric strain and, thus, more of the *s-cis* conformer is present at equilibrium for 2,3-dimethylbuta1,3-diene than for buta-1,3-diene itself. Therefore, 2,3-dimethylbuta-1,3-diene reacts with faster rate than buta-1,3-diene in the Diels–Alder reaction. Moreover, presence of the electron-donating substituents (e.g., alkyl groups) further activates dienes.

However, presence of bulky 1-*cis* substituents decreases the equilibrium proportion of the diene in the required *s-cis* conformation, thereby slowing down the reaction. It is thus not surprising that (2*Z*,4*Z*)-hexa-2,4-diene reacts at a slower rate than buta-1,3-diene in the Diels–Alder reaction (Scheme 4.9).

H_3C CH_3 ⇌ CH_3 CH_3

s-trans (more stable) *s-cis* (less stable)

SCHEME 4.9 Conformations of (2*Z*,4*Z*)-hexa-2,4-diene.

Dienophiles are the molecules possessing a double or triple bond. Typical dienophiles are illustrated in Table 4.4.

TABLE 4.4 Some typical dienophile systems.

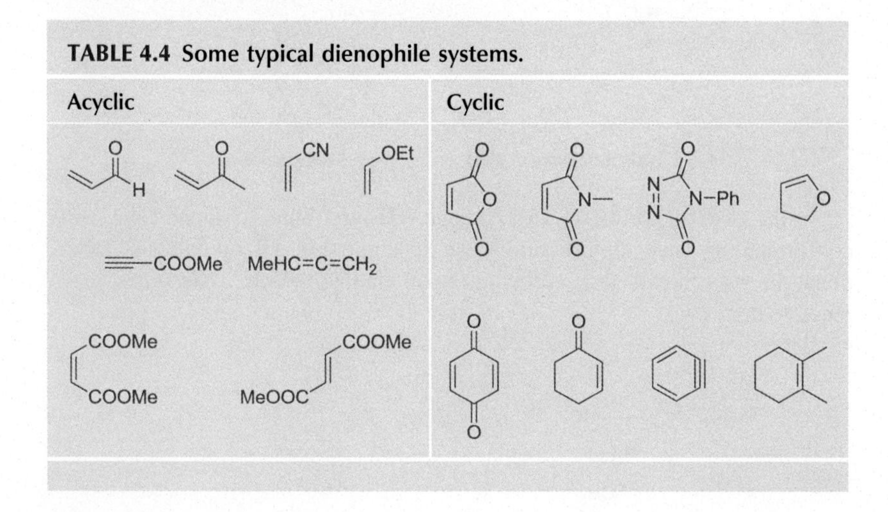

4.4.1.1 Solved Problems

Q 1. The diene that undergoes Diels–Alder reaction with maleic anhydride is:

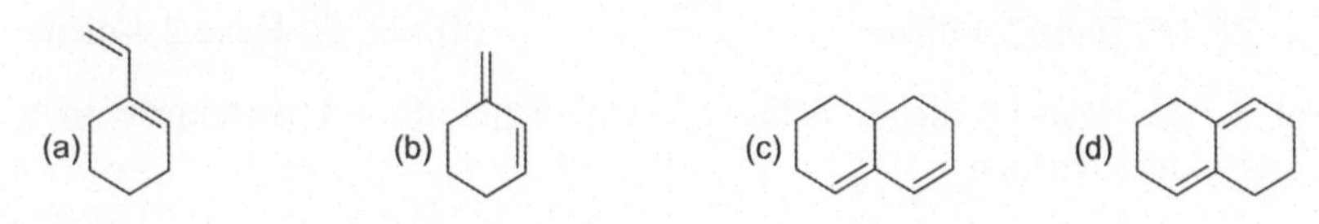

Sol 1. (a) Only the diene (a) can undergo Diels–Alder reaction with maleic anhydride by attaining *s-cis* conformation. The remaining dienes are locked in an *s-trans* conformation and hence do not react.

Q 2. The reactivity of compounds **I–IV** with maleic anhydride (**V**) follows the order:

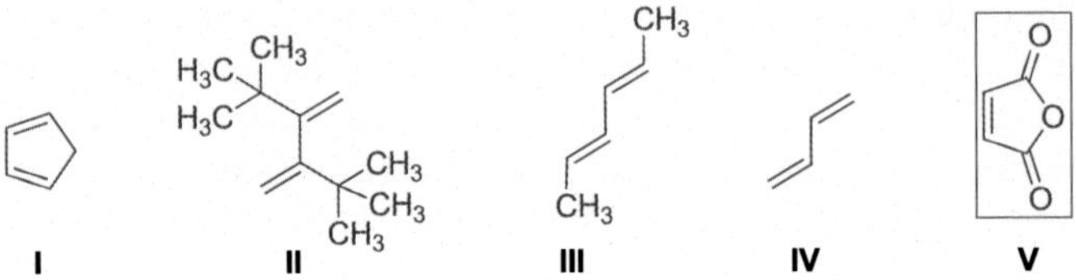

I II III IV V

(a) I < II < III < IV (b) II < IV < III < I (c) II < I < III < IV (d) II < I < IV < III

Sol 2. (b) The diene **I** is most reactive because it has the double bonds locked in an *s-cis* conformation. In case of diene **II**, the very bulky *tert*-butyl groups cause steric strain in the *s-cis* conformation, thereby decreasing its concentration in the equilibrium mixture. Moreover, *s-cis* conformation of **II** is also not planar due to steric strain. Therefore, the diene **II** is least reactive.

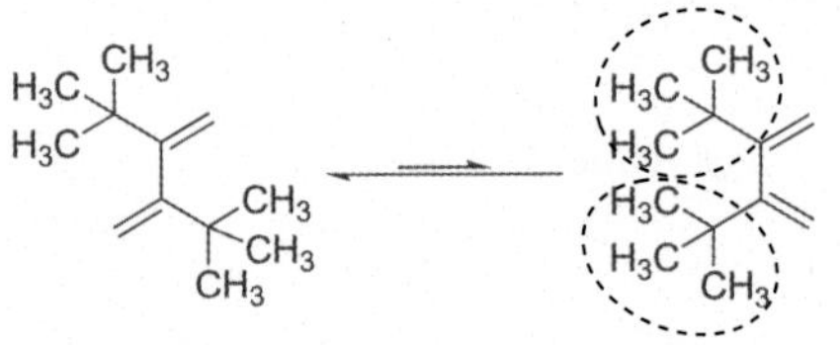

s-*trans* (more stable) s-*cis* (less stable)

In the case of (2*E*,4*E*)-hexa-2,4-diene (**III**) and buta-1,3-diene (**IV**), *s-cis* conformations have almost similar steric strain but **III** contains electron-donating substituents (e.g., alkyl groups), due to which it becomes more reactive than **IV**.

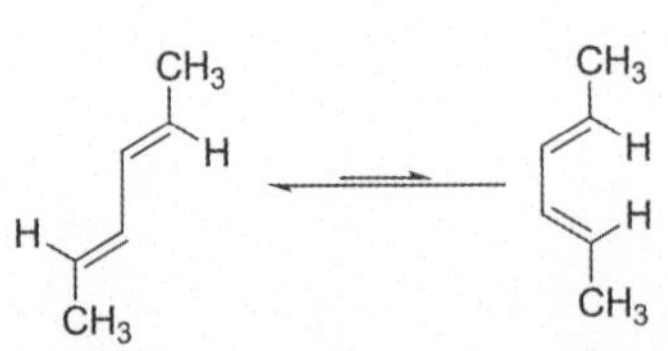

s-*trans* (more stable) s-*cis* (less stable)

Q 3. In a Diels–Alder reaction, the most reactive diene amongst the following is:
(a) (4*E*)-Hexa-1,4-diene (b) (4*Z*)-Hexa-1,4-diene
(c) (2*E*,4*E*)-Hexa-2,4-diene (d) (2*Z*,4*Z*)-Hexa-2,4-diene

Sol 3. (c) Hexa-1,4-dienes being isolated dienes do not participate in a Diels–Alder reaction.

(4*E*)-Hexa-1,4-diene (4*Z*)-Hexa-1,4-diene

Among the remaining two dienes (c) and (d), the former is more reactive due to the fact that in the diene (d) bulky 1-*cis* substituents decrease the equilibrium proportion of the diene that is present in the required *s-cis* conformation, thereby slowing down the reaction.

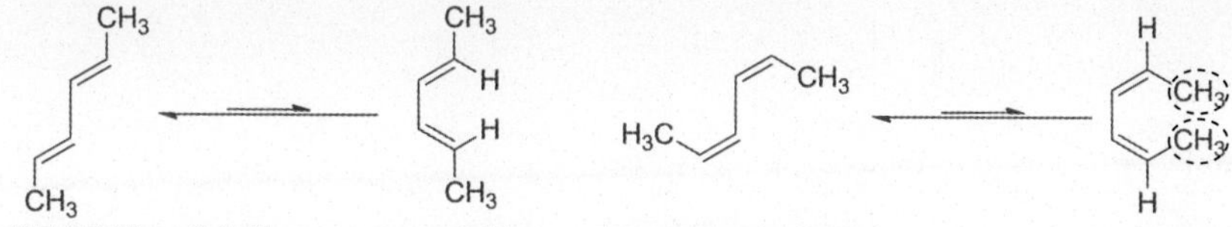

CH_3 CH_3 H H CH_3 CH_3 CH_3 H_3C H CH_3 CH_3 H

(2*E*,4*E*)-Hexa-2,4-diene more stable (2*Z*,4*Z*)-2,4-Hexadiene less stable

Q 4. Which one of the following Diels–Alder reactions is feasible?
(a) Maleic anhydride + ethene (b) Maleic anhydride + cyclopentadiene
(c) Maleic anhydride + norbornadiene (d) Cyclopentadiene + benzene

Sol 4. (b) Cyclopentadiene (diene) readily undergoes Diels–Alder reaction with maleic anhydride (dienophile).

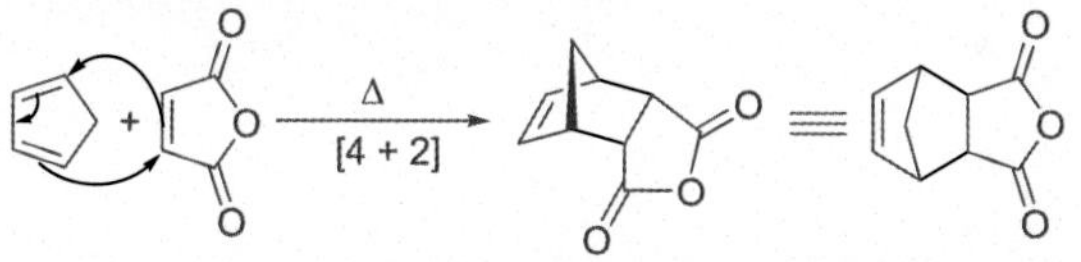

O + O O Δ [4 + 2] O O O ≡ O O O

Q 5. The most appropriate starting materials for the one-step synthesis of compound (I) are:

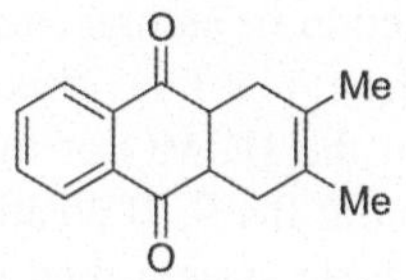

O Me Me O
I

(a) 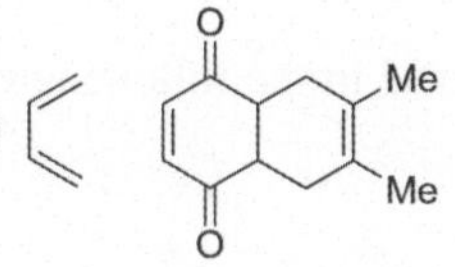 O Me Me O (b) 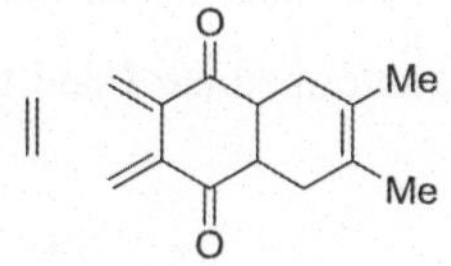 O Me Me O

(c) 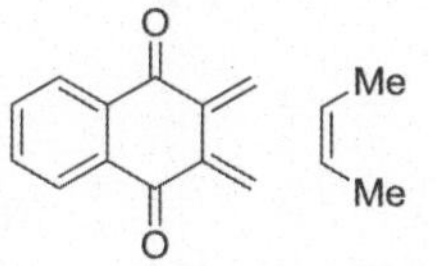 O Me Me O (d) 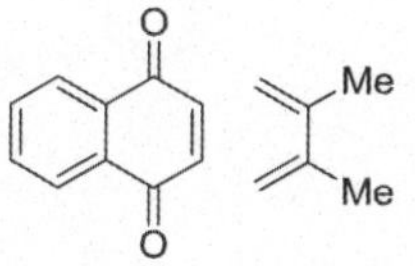 O Me Me O

Sol 5. (d)

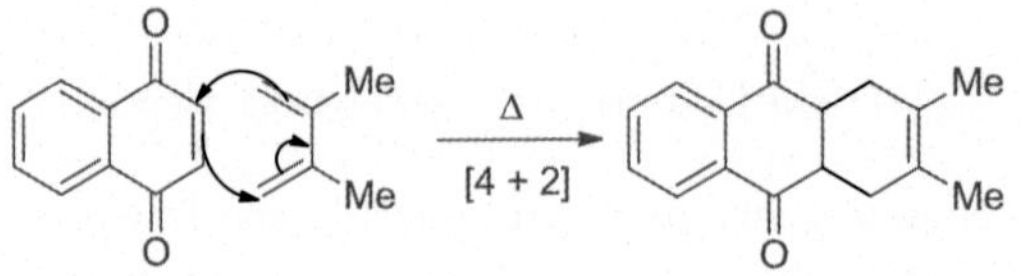

$$\xrightarrow[\text{[4 + 2]}]{\Delta}$$

I

Q 6. The product of the reaction of the benzyne intermediate with anthracene is:

(a) 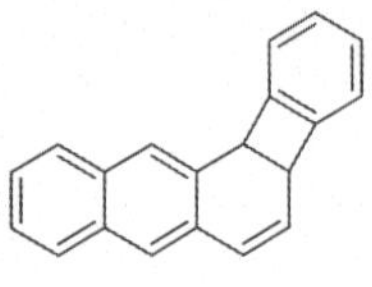 (b) 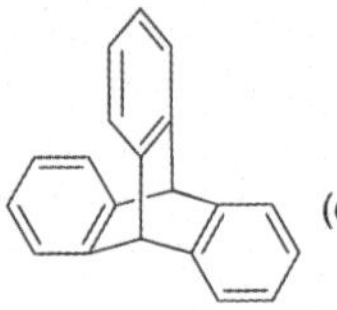 (c) 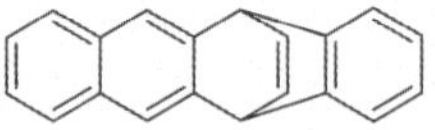 (d)

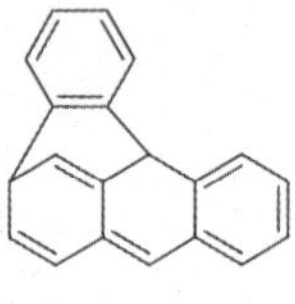

Sol 6. (b) The benzyne can be trapped in a Diels–Alder reaction. It is an unstable electrophilic molecule, which makes it a good dienophile. This is due to its low energy LUMO, which is the π* orbital of the triple bond.

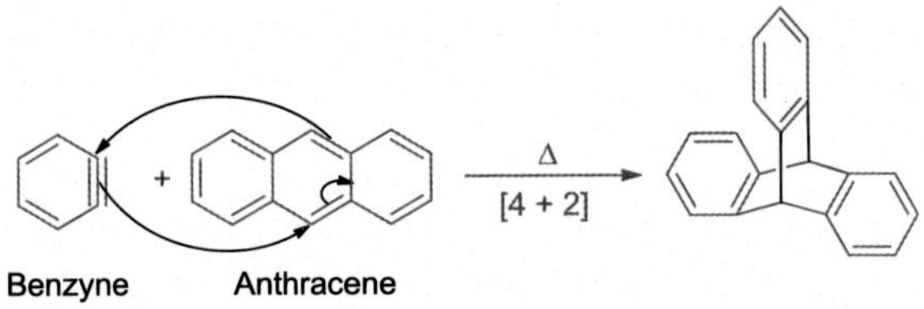

Benzyne + Anthracene $\xrightarrow[\text{[4 + 2]}]{\Delta}$

Site selectivity is the selectivity shown by a reagent toward one site (or more) of a polyfunctional molecule when several sites, in principle, are available. In cycloadditions, site selectivity always involves a pair of sites. For example, the Diels–Alder reaction of anthracene generally takes place across the 9,10 position than across the 1,4 or 3,9a. This may be explained by the fact that the highest coefficients in the HOMO of the anthracene are at the 9,10 position. Furthermore, addition at the 9,10 position creates two isolated benzene rings, which is a more stable system than that of a naphthalene system created by the addition at the 1,4 position.

Q 7. The intermediate **I** and the major product **II** in the following conversion involves:

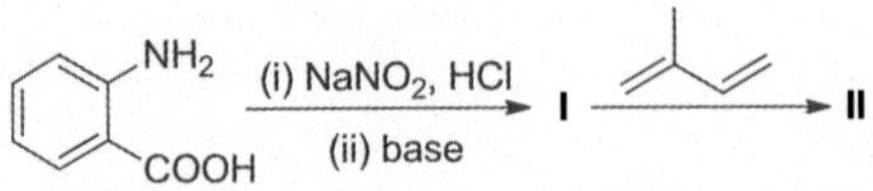

NH_2, COOH $\xrightarrow[\text{(ii) base}]{\text{(i) NaNO}_2\text{, HCl}}$ I $\longrightarrow$ II

(a) Carbocation and COOH

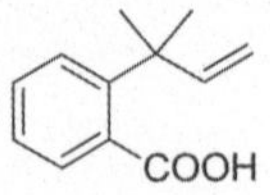

(b) Carbanion and NH_2

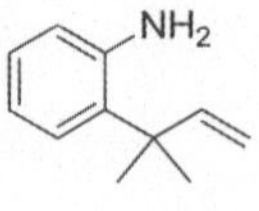

(c) Free radical and 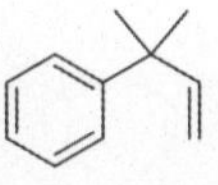 (d) Benzyne and 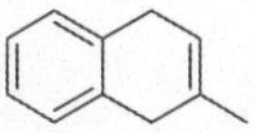

Sol 7. (d) Diazotization of anthranilic acid gives a diazonium salt, which on treatment with a mild base undergoes elimination of CO_2 and N_2 to give benzyne intermediate (**I**). It immediately undergoes Diels–Alder reaction to give **II**.

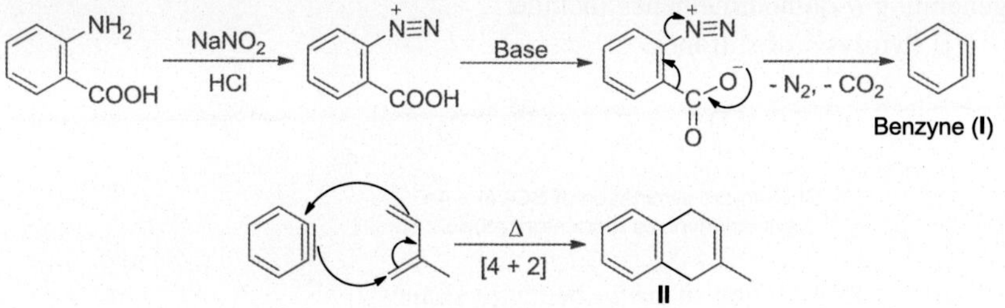
NH2
COOH
NaNO2
HCl
N≡N
COOH
Base
N≡N
C
O
O
- N2, - CO2
Benzyne (I)
Δ
[4 + 2]
II

Q 8. The major product **I** formed in the following reaction is:

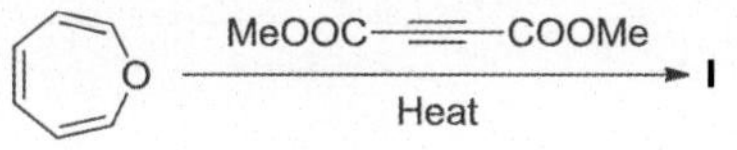
O
MeOOC—≡—COOMe
Heat
I

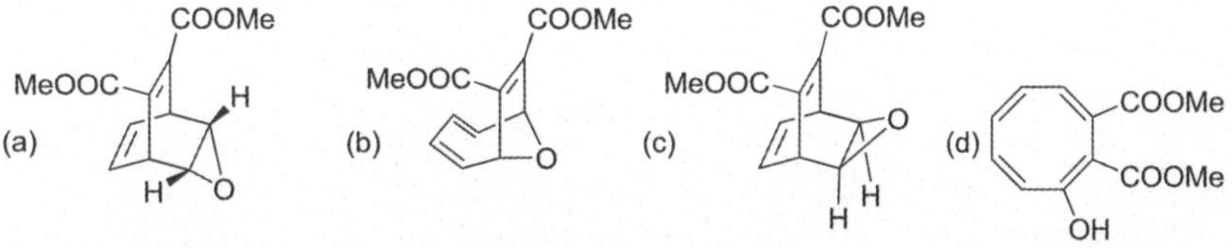
(a)
COOMe
MeOOC
H
H
O
(b)
COOMe
MeOOC
O
(c)
COOMe
MeOOC
O
H
H
(d)
COOMe
COOMe
OH

Sol 8. (a) On heating, oxapine undergoes 6π-electron disrotatory ring closure to give its valence isomer, which readily undergoes [4 + 2] cycloaddition with dimethyl acetylene dicarboxylate to give **I**.

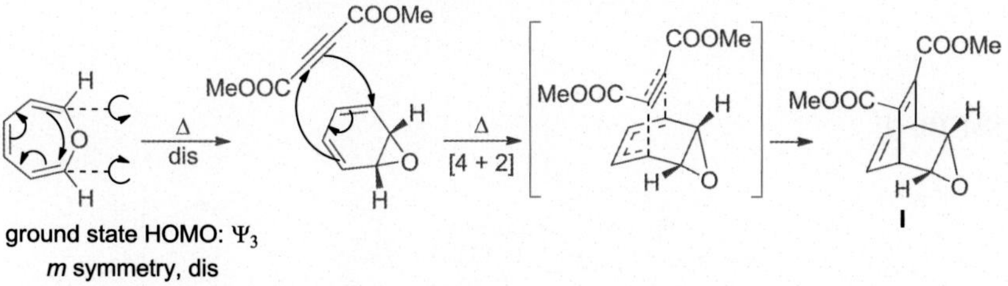
H
O
H
Δ
dis
ground state HOMO: Ψ3
m symmetry, dis
COOMe
MeOOC
H
O
H
Δ
[4 + 2]
COOMe
MeOOC
H
H
O
COOMe
MeOOC
H
H
O
I

Q 9. Identify the product in the following Diels–Alder reaction.

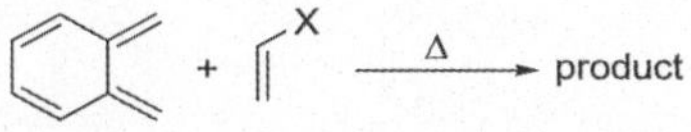
+
X
Δ
product

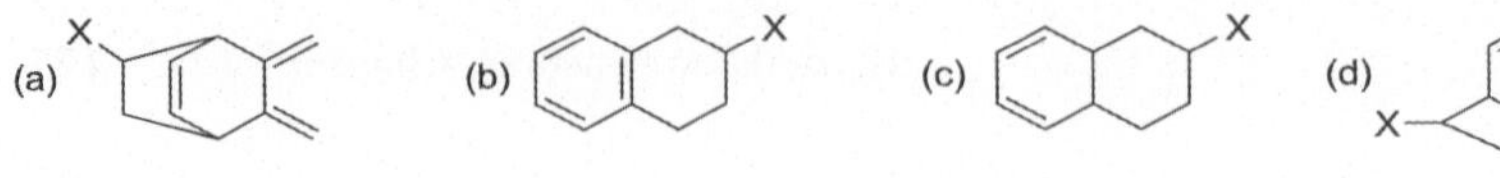

(a) (b) (c) (d)

Sol 9. (b)

[4 + 2]

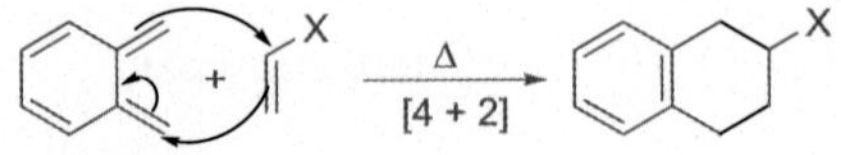

o-Quinodimethane

Unstable (and thus highly reactive) dienes, of which perhaps the most synthetically useful are *o*-quinodimethanes, can be generated in situ. A strong driving force for the [4 + 2] cycloaddition of such species is a result of the establishment (or reestablishment) of aromaticity. Common methods for generating *o*-quinodimethanes include:

(i) Pyrolysis of sulfones.

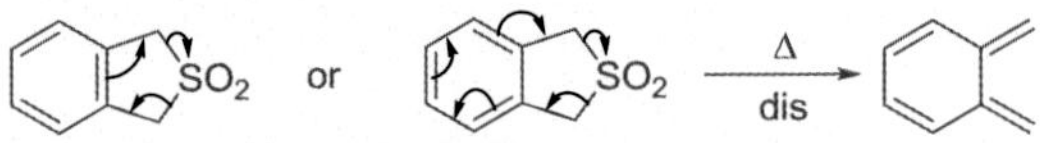

SO_2 or SO_2 Δ dis

Cheletropic elimination of SO_2 in a 4n system involves disrotatory mode

(ii) 1,4-Elimination of *ortho* benzylic silanes.

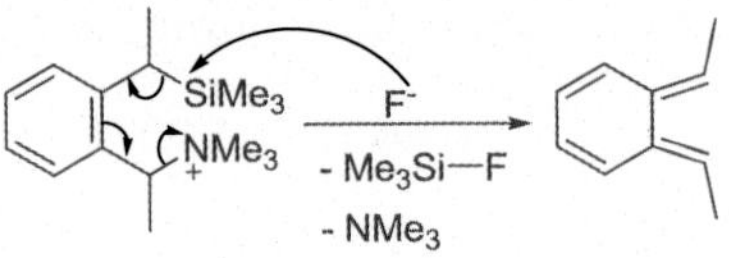

$SiMe_3$ $\overset{+}{N}Me_3$ F^- - Me_3Si—F - NMe_3

Q 10. Explain the mechanisms of the following transformations:

(i) $\overset{+}{N}_2$ $CO\bar{O}$ + (i) Heat (ii) H_3O^+ OH

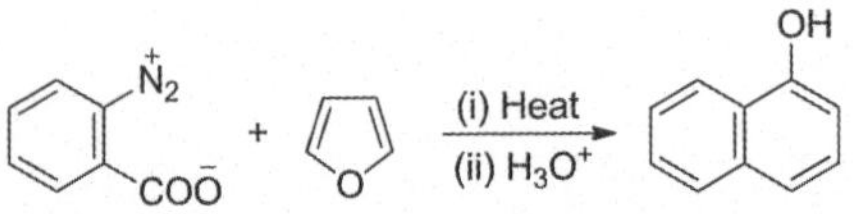

(ii) + COOMe COOMe Δ COOMe COOMe major + MeOOC COOMe minor

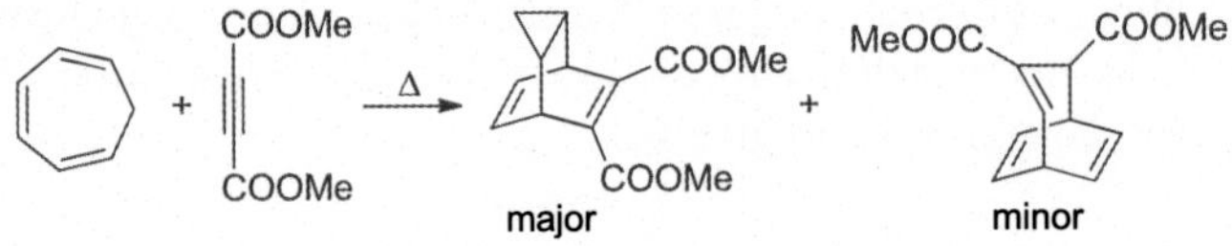

Sol 10. (i)

$\overset{+}{N}$≡N $-N_2$, $-CO_2$ Δ [4 + 2] H_3O^+

HO H OH

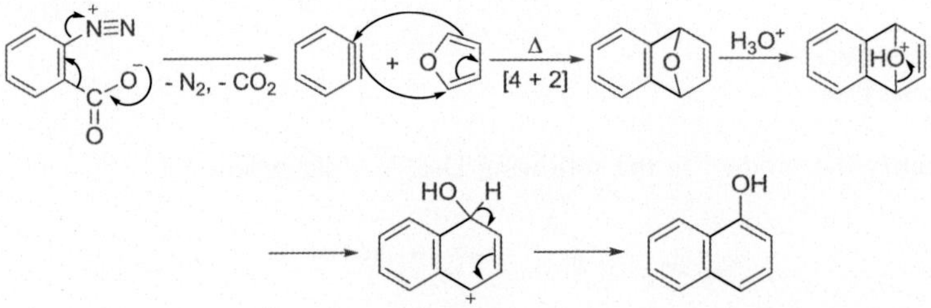

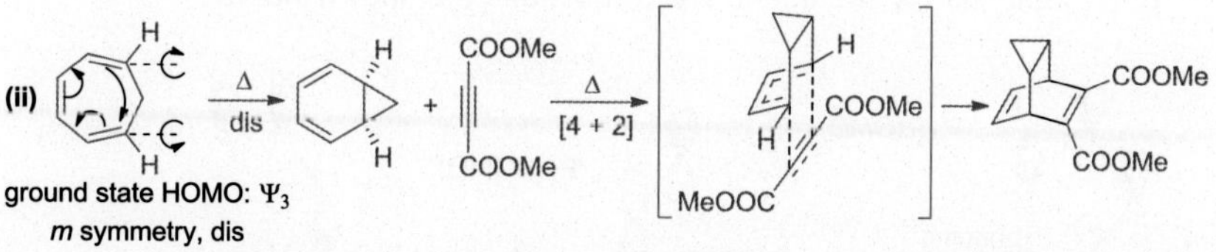

(ii)

ground state HOMO: Ψ_3
m symmetry, dis

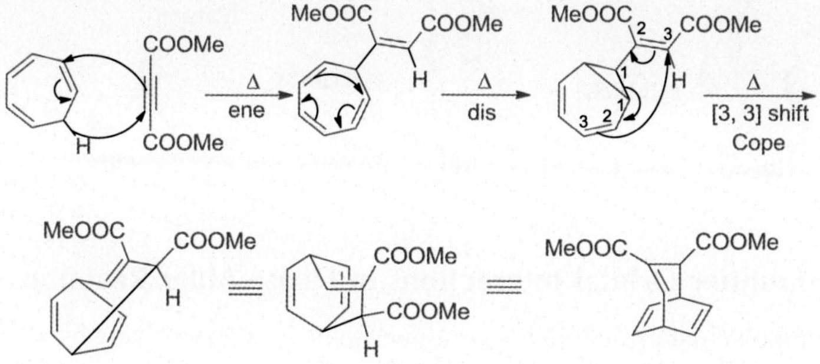

Q 11. Indicate condition (thermal/photochemical) as well as feasibility for carrying out the following transformations. Justify your answer.

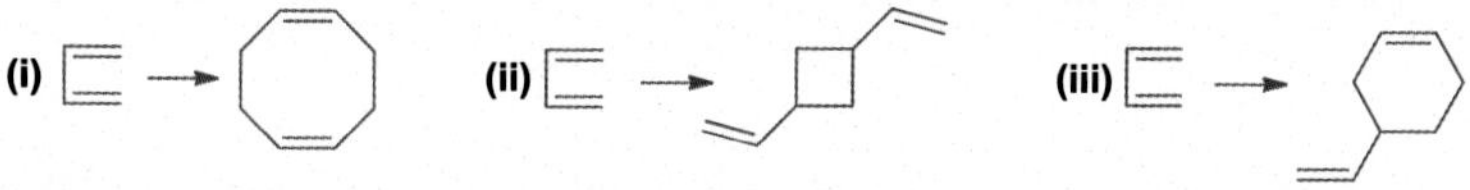

(i) (ii) (iii)

Sol 11.

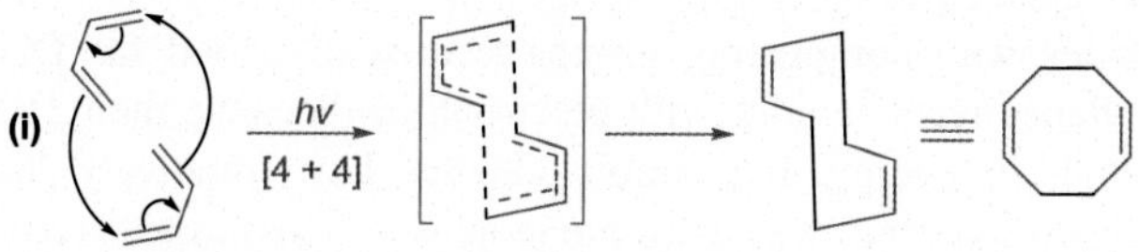

(i) (ii) (iii)

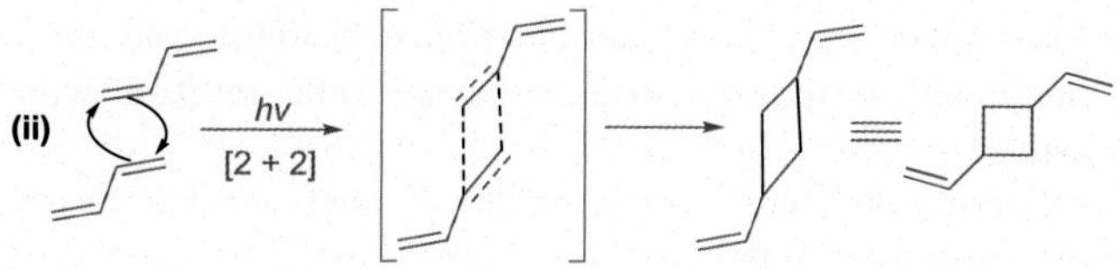

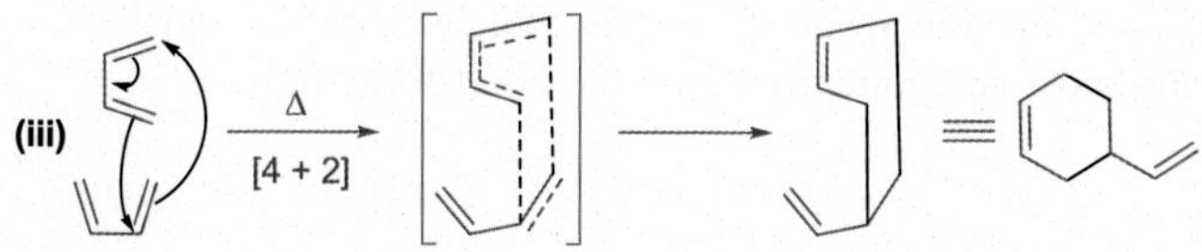

FMO analysis for the above three reactions are shown below:

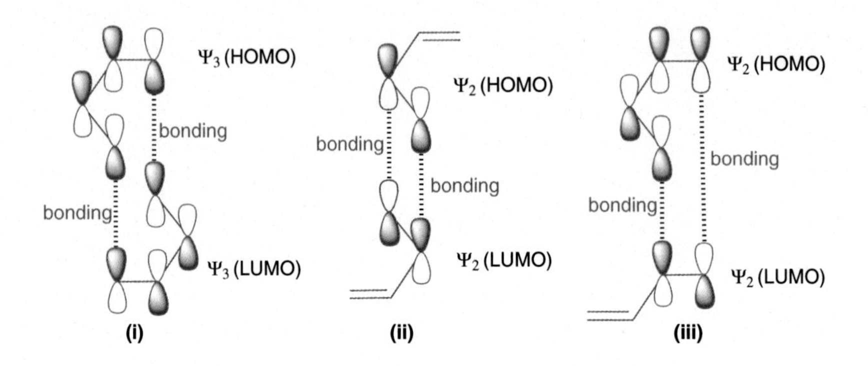

4.4.2 Frontier Orbital Interactions in Diels–Alder Reaction

Frontier molecular orbital theory continues to be used extensively by synthetic organic chemists for the prediction of the reactivity and selectivity of many organic reactions. As in the Diels–Alder reaction, predictions of reactivity and selectivity are normally based on the strength of a single FMO interaction between the diene and the dienophile, the so-called "dominant" interaction. The dominant interaction is usually taken to be the one involving the two frontier orbitals having the smallest energy gap between them. When the $HOMO_{diene}$–$LUMO_{dienophile}$ energy gap is least, the reaction is called a *normal Diels–Alder cycloaddition*; when the $HOMO_{dienophile}$–$LUMO_{diene}$ energy gap is smallest, the reaction is called an *inverse electron demand Diels–Alder cycloaddition* (Figure 4.12).

Dienophiles with conjugating groups are usually good for Diels–Alder reactions. Dienes react rapidly with electrophiles because their HOMOs are relatively high in energy, but simple alkenes have relatively high-energy LUMOs and do not react well with nucleophiles. The most effective modification is to lower the alkene LUMO energy by conjugating the double bond with an electron-withdrawing group such as carbonyl or nitro. This type of Diels–Alder reaction, involving an electron-rich diene and an electron-deficient dienophile, is referred to as a Diels–Alder reaction with *normal electron demand.*

Highly electron-deficient dienes undergo Diels–Alder reaction with electron-rich dienophiles in the inverse electron demand Diels–Alder (DA_{INV}) reaction. In the DA_{INV} reaction, the $LUMO_{diene}$ and $HOMO_{dienophile}$ are closer in energy than the $HOMO_{diene}$ and $LUMO_{dienophile}$. Thus, the $LUMO_{diene}$ and $HOMO_{dienophile}$ are the frontier orbitals that interact the most strongly and result in the most energetically favorable bond formation.

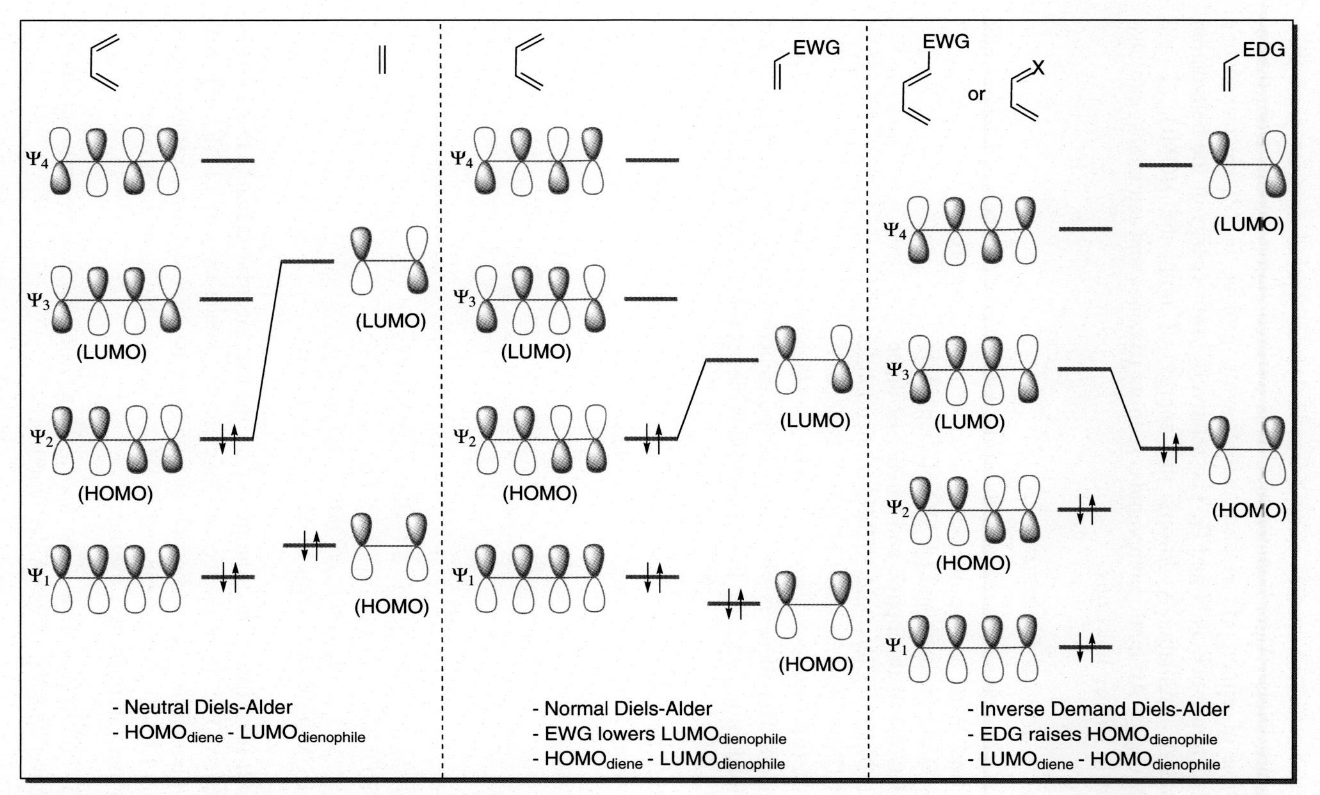

FIGURE 4.12 Frontier molecular orbital interactions in Diels–Alder reactions.

The dienes used in DA_{INV} reaction are relatively electron-deficient species, compared to the standard Diels–Alder, where the diene is electron rich. These electron-deficient species have lower molecular orbital energies than their standard Diels–Alder counterparts. This lowered energy results from the inclusion of either: (1) electron-withdrawing groups, or (2) electronegative heteroatoms such as N and O. The relatively lower energy of the heteroatom *p*-orbitals dramatically lowers the energy of both the $HOMO_{diene}$ and $LUMO_{diene}$. Aromatic compounds, such as triazines and tetrazines, can also react in DA_{INV} reactions.

4.4.2.1 Solved Problems

Q 1. Cyclopentadiene reacts with acrylic ester to give products of Diels–Alder reaction. What are the interacting frontier molecular orbitals?
(a) HOMO of diene and LUMO of dienophile
(b) HOMO of dienophile and LUMO of diene
(c) HOMO of diene and HOMO of dienophile
(d) LUMO of diene and LUMO of dienophile

Sol 1. (a) This is an example of Diels–Alder reaction with normal electron demand. The diene is electron-rich and hence will use its HOMO for the cycloaddition, whereas, the dienophile is electron-deficient and hence will use its LUMO.

Q 2. In the following Diels–Alder reaction, interaction between which of the following frontier molecular orbitals takes place?

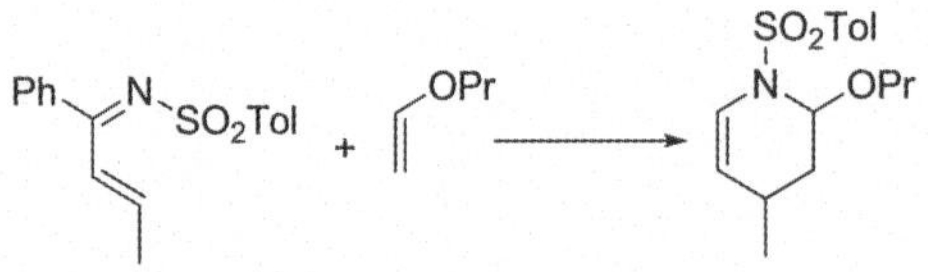

Ph, N, SO_2Tol + OPr → SO_2Tol, N, OPr

(a) HOMO of diene with HOMO of dienophile
(b) HOMO of diene with LUMO of dienophile
(c) LUMO of diene with HOMO of dienophile
(d) LUMO of diene with LUMO of dienophile

Sol 2. (c) This is an example of an inverse electron demand Diels–Alder reaction. The diene is electron-deficient and hence uses its LUMO, whereas the dienophile is electron-rich and will use its HOMO in the cycloaddition. Here, LUMO of the diene and HOMO of dienophile have the smallest energy gap and can have best overlap.

Q 3. Pick a pair of a diene and a dienophile that will undergo an inverse electron demand Diels–Alder reaction.

OMe OMe COOMe
TMSO NO2 COOMe NMe2
I II III IV A B C D

Sol 3. IV (electron-deficient diene) and **C** (electron-rich dienophile) will undergo an inverse electron demand Diels–Alder reaction.

Q 4. An adduct **I** is formed by the Diels–Alder reaction of cyclopentadiene with ethyne. When **I** is treated with hexachlorocyclopentadiene (**II**), it forms the well-known pesticide, Aldrin (**III**). Epoxidation of **III** with peracid produces another pesticide, Dieldrin (**IV**). Draw structures of **I**, **II**, **III**, and **IV** and explain how these products are formed.

Sol 4. The normal electron demand Diels–Alder reaction of cyclopentadiene with ethyne gives norbornadiene (**I**). Aldrin (**III**) is produced by combining hexachlorocyclopentadiene with norbornadiene in an inverse electron demand Diels–Alder reaction. Epoxidation of **III** with peracid produces another pesticide, Dieldrin (**IV**).

Δ [4 + 2] Δ [4 + 2]

Norbornadiene (I)
electron-rich dienophile

Hexachlorocyclopentadiene (II)
electron-deficient diene

Aldrin (III)

RCO_3H (per acid)
epoxidation at the more electron-rich double bond

Dieldrin (IV)

Q 5. Explain the mechanism of the following reactions:

(i) COOMe, NMe_2, Δ, $-N_2$ (ii) COOMe, Me_2N, OMe, Δ, $-N_2$

Sol 5. (i) This reaction is an example of inverse electron demand Diels–Alder reaction. In the first step, the reaction sites of the diene and the dienophile react in a [4 + 2] cycloaddition as indicated by the arrows. The formation of gaseous nitrogen makes this reaction irreversible and produces the 1,2,4-triazine derivative as an exclusive reaction product.

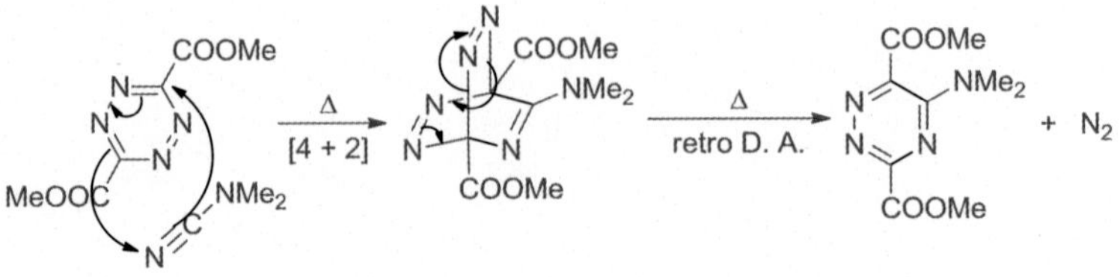

(ii) The first step of this reaction involves an inverse electron demand Diels–Alder reaction. The Diels–Alder adduct then undergoes a retro Diels–Alder reaction resulting in the cleavage of the C–N bonds and giving N_2 as a by-product. An elimination reaction (probably E1) then gives the observed product.

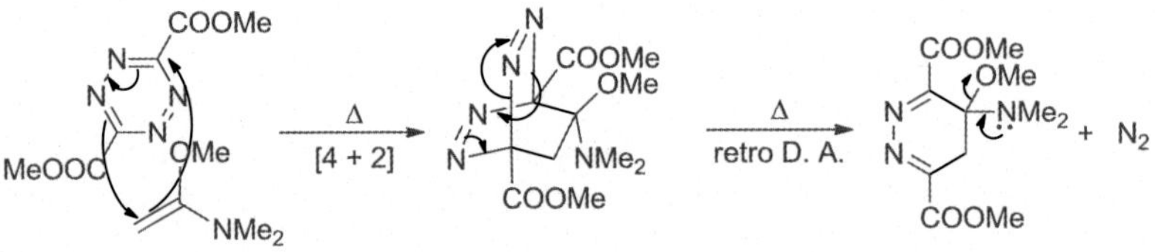

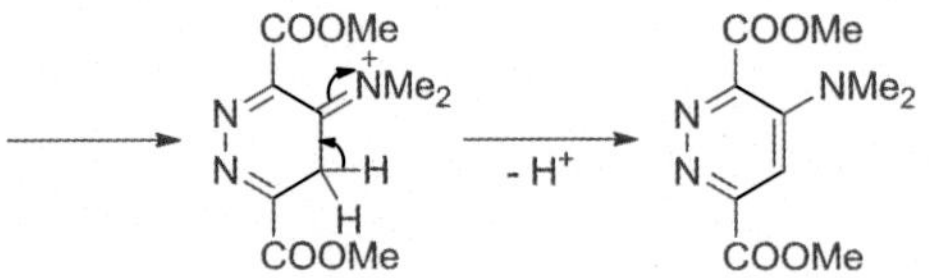

4.4.3 Stereochemistry of Diels–Alder Reaction

(i) The "*cis*" principle: The addition of a dienophile to the diene component is purely a *cis* addition. The relative positions of the substituents in the dienophile are retained in the adduct. The *cis* addition can be readily explained by the synchronous formation of the bonds between the two components in a one-step reaction. *cis-* or *trans-*Dienophiles react with dienes to give 1:1 adducts in which the *cis-* or *trans-*arrangement of the substituents in the dienophile is retained exhibiting the stereoselective nature of the Diels–Alder reaction. An illustrative example is shown for the reactions of the isomeric dimethyl maleate and dimethyl fumarate with buta-1,3-diene in Scheme 4.10.

The *cis* principle also applies for the substituents in the diene component, thereby exhibiting its stereospecificity with regard to the diene. The relative configuration of the substituents in the 1 and 4 positions of butadiene is retained in the adduct. *trans, trans-*1,4-Disubstituted dienes give rise to adducts in which the 1 and 4 substituents are *cis* to each other. *cis, trans-*1,4-Disubstituted dienes give adducts with *trans* substituents. For example, Diels–Alder reaction of *trans, trans-*1,4-diphenylbutadiene and maleic

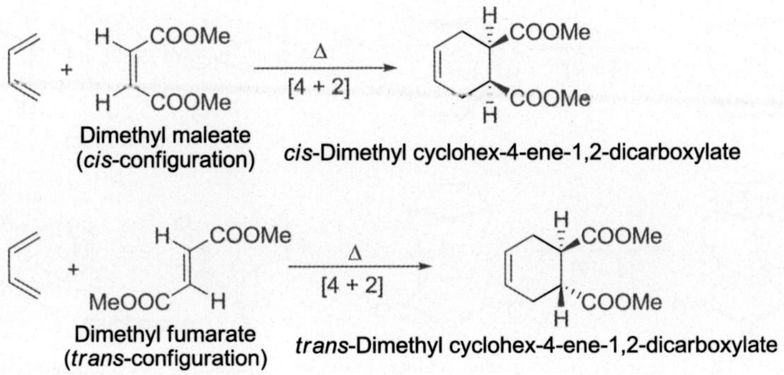

Dimethyl maleate (*cis*-configuration) → Δ [4 + 2] → *cis*-Dimethyl cyclohex-4-ene-1,2-dicarboxylate

Dimethyl fumarate (*trans*-configuration) → Δ [4 + 2] → *trans*-Dimethyl cyclohex-4-ene-1,2-dicarboxylate

SCHEME 4.10 Diels—Alder reactions of isomeric dimethyl maleate and dimethyl fumarate with buta-1,3-diene.

anhydride gives 1:1 adduct in which the phenyl groups are *cis* to each other (Scheme 4.11).

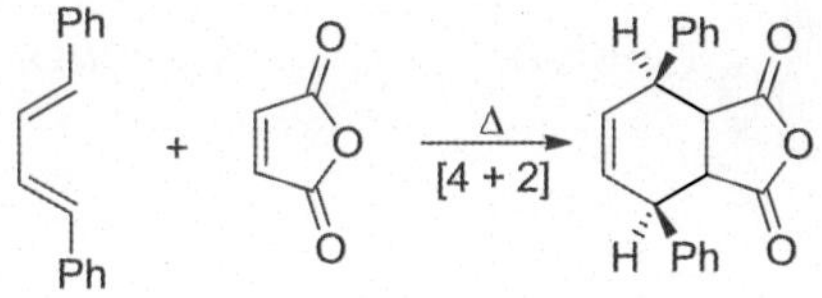

SCHEME 4.11 Diels—Alder reaction of *trans, trans*-1,4-diphenylbutadiene and maleic anhydride.

(ii) The "*endo*" rule: Another stereochemical point of significance is that in many Diels—Alder reactions there is the possibility of two alternative modes of cycloaddition, the *exo* and the *endo*. Diels—Alder reaction generally proceeds selectively via the transition state in which the most powerful electron-withdrawing group on the dienophile is *endo*, i.e., below the diene, as opposed to pointing away from it. This generalization is known as the *endo rule*. For example, in the reaction of maleic anhydride with cyclopentadiene, two modes of addition are theoretically possible, leading to the formation of an *exo* adduct or an *endo* adduct, respectively. Actually, the *endo* configuration is produced exclusively (Scheme 4.12).

Despite the fact that the *exo* adduct is thermodynamically more stable than the *endo* adduct, it is often found in the Diels—Alder reaction that the *endo* adduct is the major, if not the sole, product. This is because in *endo* adduct, additional stabilization of the transition state can occur (which results in faster rate of reaction) through secondary interaction of those lobes of HOMO and of the LUMO that are not involved directly in bond formation, provided these are of the same phase. However, such interactions are not possible in the transition

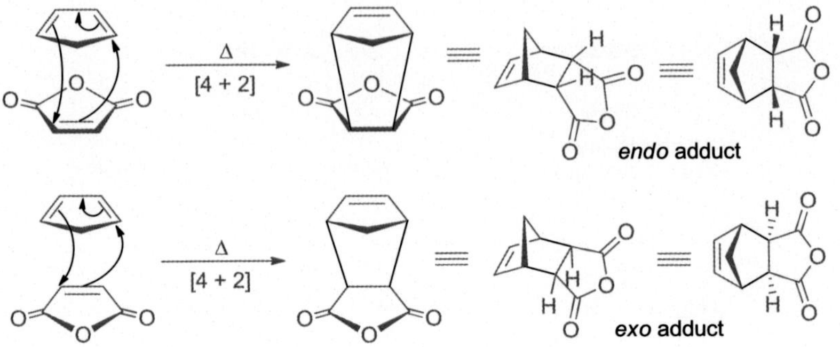

SCHEME 4.12 Formation of *exo* or *endo* adducts in the [4 + 2] reaction of cyclopentadiene and maleic anhydride.

state for *exo* addition because the relevant sets of centers are now too far from each other (Figure 4.13). Thus the *endo* adduct is the kinetically controlled product.

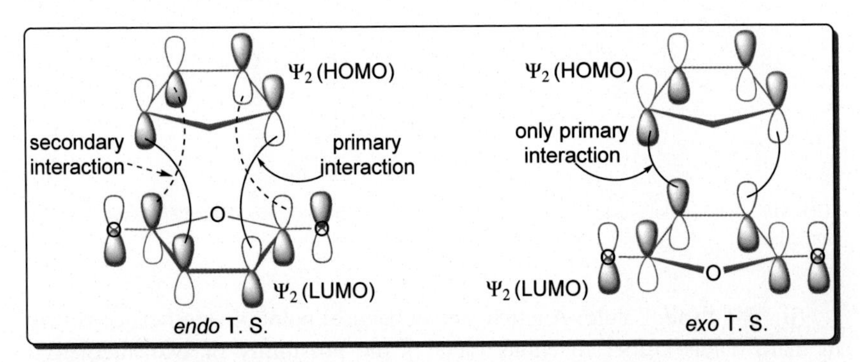

FIGURE 4.13 Orbital interactions in Diels–Alder reaction of cyclopentadiene and maleic anhydride.

However, establishment of an equilibrium between the higher energy *endo* and lower energy *exo* results in the predominant formation of the *exo* product (thermodynamic control) (Scheme 4.13). Cycloaddition of furan with maleic anhydride, for instance, provides only the *exo* product. This may be ascribed to

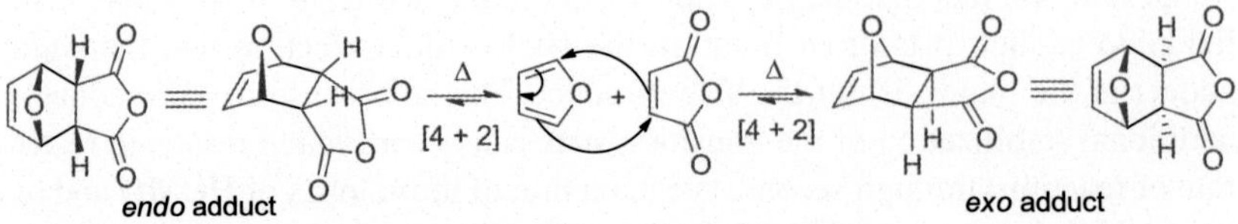

SCHEME 4.13 Diels–Alder reaction of furan and maleic anhydride.

the lower energy of furan (an aromatic compound) allowing the retro Diels–Alder reaction of the *endo* adduct (Figure 4.14).

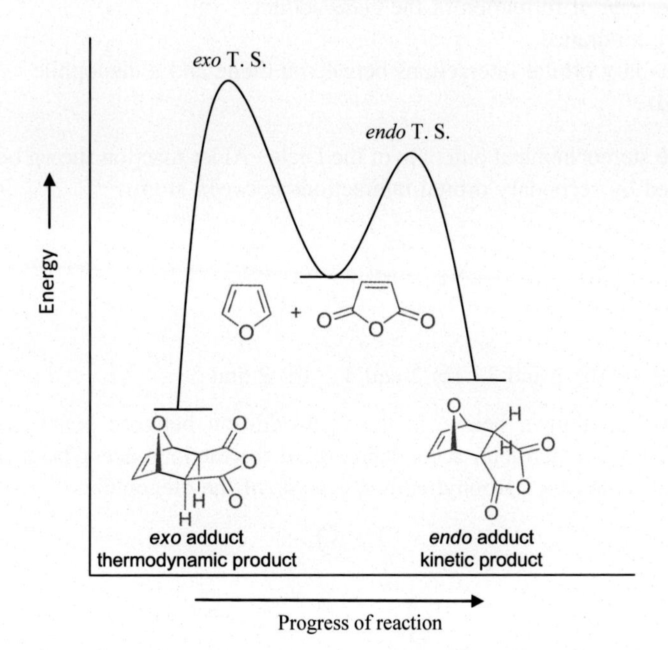

FIGURE 4.14 Energy profile diagram for *exo* and *endo* products.

4.4.3.1 Solved Problems

Q 1. Which of the following statements is true about the Diels–Alder reaction?

(a) Nonconcerted reaction
(b) Stereoselective
(c) Concerted, but not stereospecific reaction
(d) Concerted and stereospecific reaction

Sol 1. (d) Diels–Alder reaction is concerted and stereospecific. *Stereospecificity* is the characteristic of a reaction that leads to the formation of different stereoisomeric products from different stereoisomeric reactants or that operates on only one (or a subset) of the stereoisomers.

The Diels–Alder reaction involves a stereospecific *cis* addition (suprafacial) with respect to both the diene and dienophile. Existing stereochemical relationships in the dienophile (*cis* or *trans*) and the diene (*trans–trans* or *cis–trans*) are retained in the product.

Q 2. Diels–Alder reaction normally yields *endo*-adduct as a major product. This is due to:
(a) Higher stability of the product
(b) Faster rate of formation of the *endo*-adduct
(c) Steric hindrance
(d) Secondary orbital interactions between a diene and a dienophile
Sol 2. (d)

Q 3. The stereochemical outcome of the Diels–Alder reaction shown below is controlled by secondary orbital interactions between atoms:

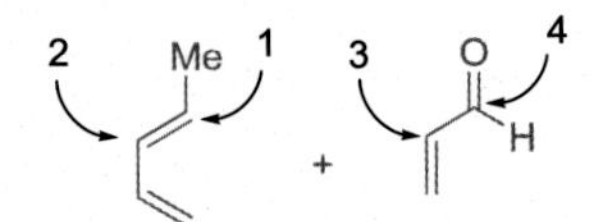

(a) 1 and 4 (b) 1 and 3 (c) 2 and 4 (d) 2 and 3

Sol 3. (c) As shown below, in the cycloaddition between penta-1,3-diene (piperylene) and acrolein, secondary orbital interactions occur between C-2 of the diene and the carbonyl carbon (i.e., 4) of the dienophile.

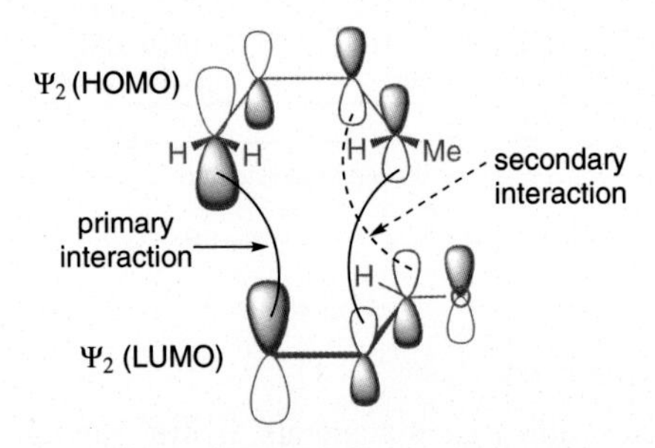

Q 4. Buta-1,3-diene on heating with maleic anhydride gives:

(a) (b) (c) (d)

Sol 4. (b)

Δ [4 + 2]

endo T. S.

endo adduct

Q 5. For the synthesis of:

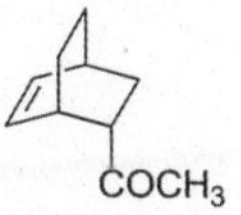

$COCH_3$

using Diels–Alder reaction, the reactants required are:

(a)  and $COCH_3$ (b) and $COCH_3$

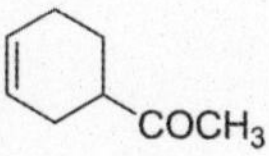

(c) and $COCH_3$ (d) and $COCH_3$

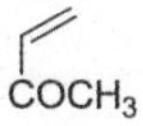

Sol 5. (d)

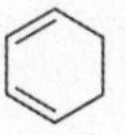

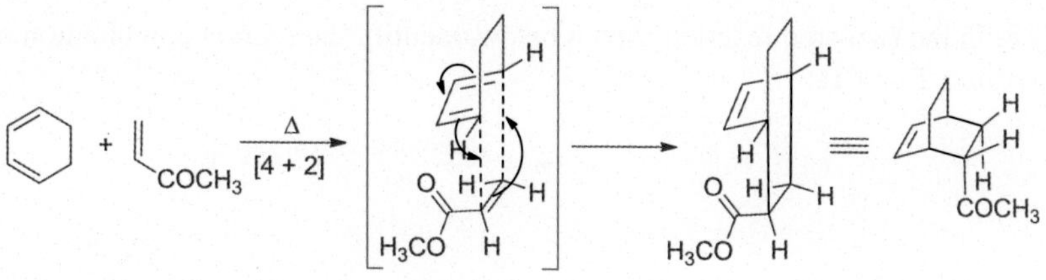

+ $COCH_3$ [4 + 2] H H H O H H_3CO H *endo* T. S. H H O H H H_3CO H ≡ H H H $COCH_3$ *endo* adduct

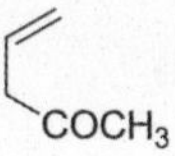

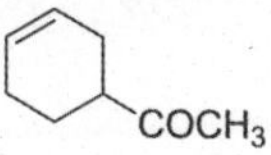

Q 6. The major product formed in the reaction given below is:

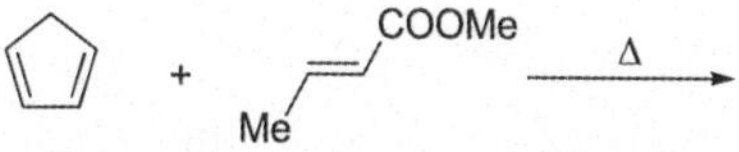

+ COOMe Me Δ

(a) 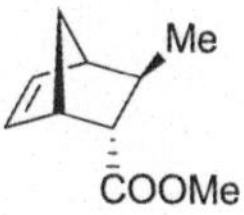 Me COOMe (b) 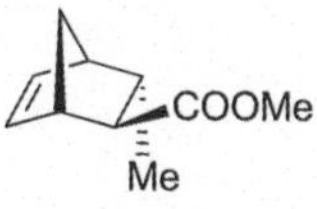 COOMe Me (c)  Me COOMe (d)  Me COOMe

Sol 6. (a)

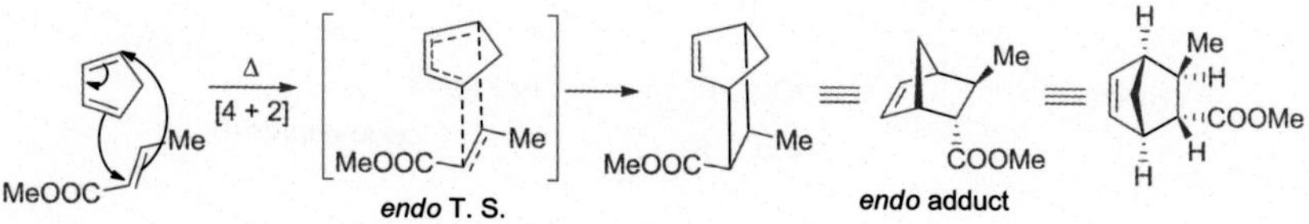

Me MeOOC Δ [4 + 2] MeOOC Me *endo* T. S. MeOOC Me ≡ Me COOMe *endo* adduct ≡ H Me H COOMe H H

Q 7. The structures of the major products **I** and **II** in the following transformations are:

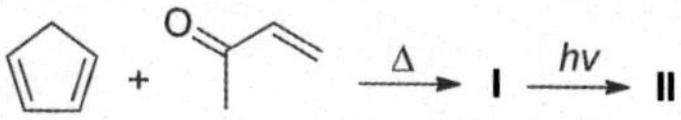

+ O Δ **I** *hv* **II**

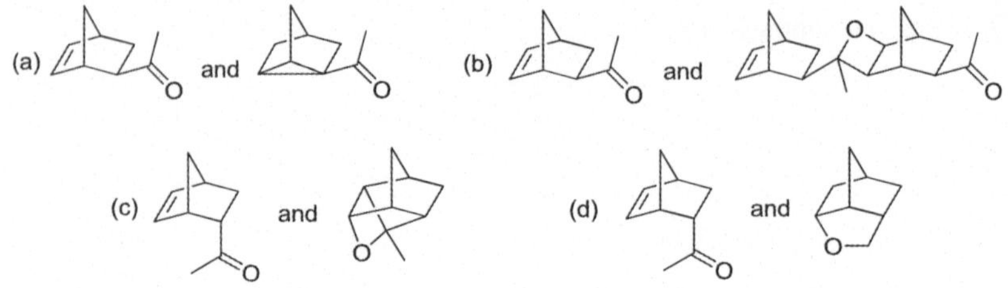

(a) and (b) and

(c) and (d) and

Sol 7. (d)

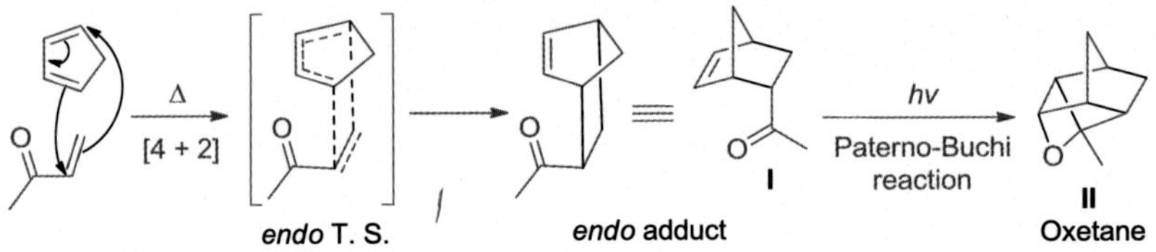

Δ [4 + 2] *endo* T. S. *endo* adduct I *hv* Paterno-Buchi reaction II Oxetane

Q 8. In the two-step reaction shown below, identify the correct combination of products **I** and **II**.

COOH Δ I N-bromosuccinimide II

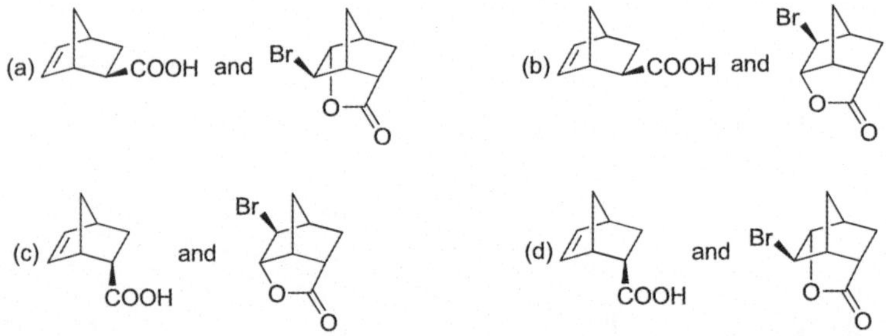

(a) COOH and Br (b) COOH and Br

(c) COOH and Br (d) COOH and Br

Sol 8. (c)

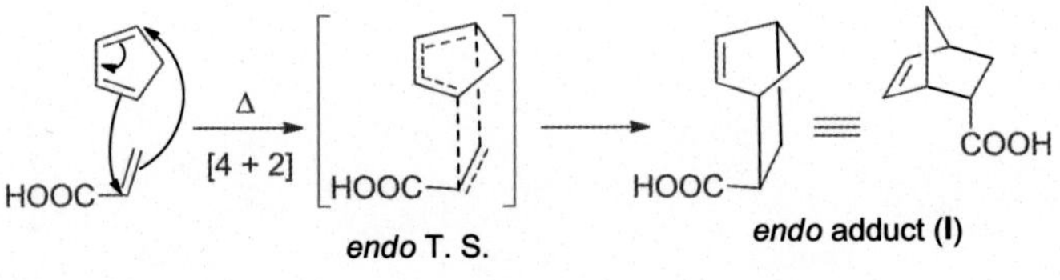

Δ [4 + 2] HOOC *endo* T. S. HOOC COOH *endo* adduct (I)

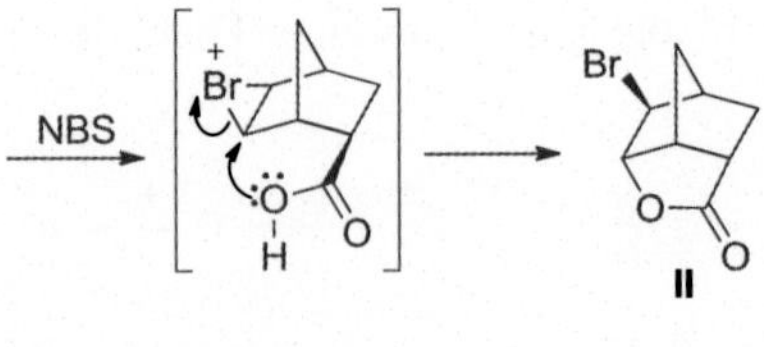

NBS Br H O Br O II

Q 9. In the reaction sequence, the major products **I** and **II**, respectively, are:

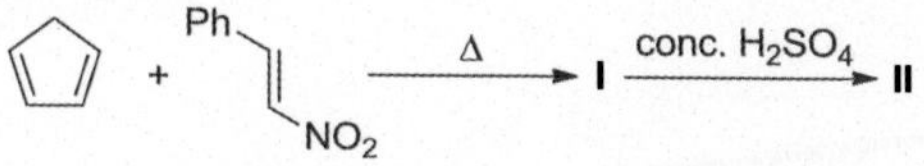

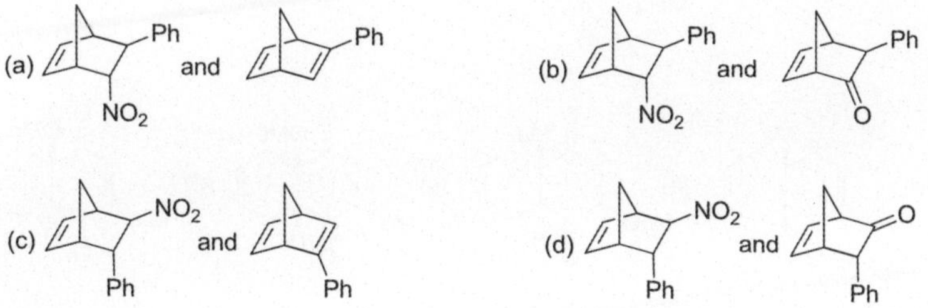

Sol 9. (b) The Diels–Alder reaction of cyclopentadiene with nitrostyrene gives adduct **I**. The nitro group of the adduct **I** is then converted to a carbonyl group by application of the Nef reaction.

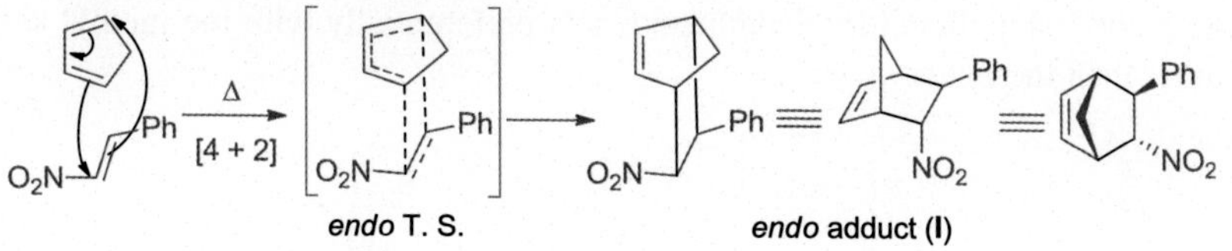

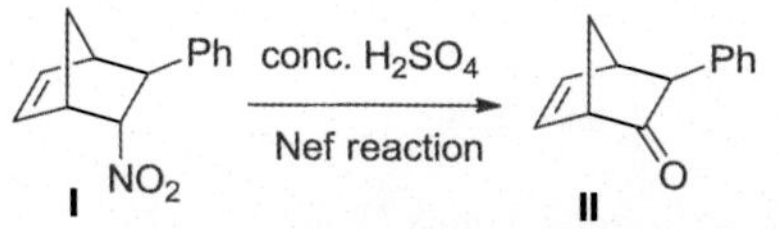

endo T. S.

endo adduct (I)

However, the reaction of nitrostyrene derivatives with cyclopentadiene in the presence of tin tetrachloride affords exclusively a nitronate. In this reaction, the nitrostyrene derivative behaves as a diene whereas cyclic diene behaves as a dienophile.

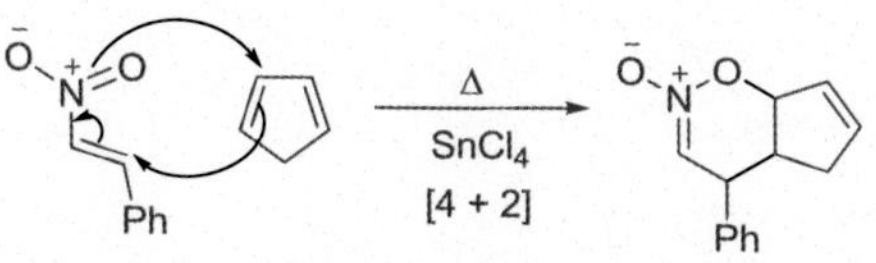

Since these reactions are inverse electron demand cycloadditions, the relevant frontier orbital interactions are between the LUMO of the diene and the HOMO of the dienophile. The accelerating effect of a Lewis acid can be understood since complexation of the nitroalkene should decrease the HOMO–LUMO gap by lowering the LUMO of the diene. Thus, the peri-selectivity of the reaction is controlled by the complexation of the nitroalkene to the Lewis acid.

Q 10. Predict the [4 + 2] Diels–Alder cycloaddition product with the right stereochemistry.

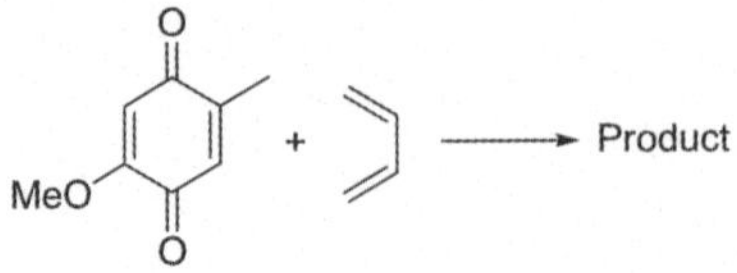

MeO + → Product

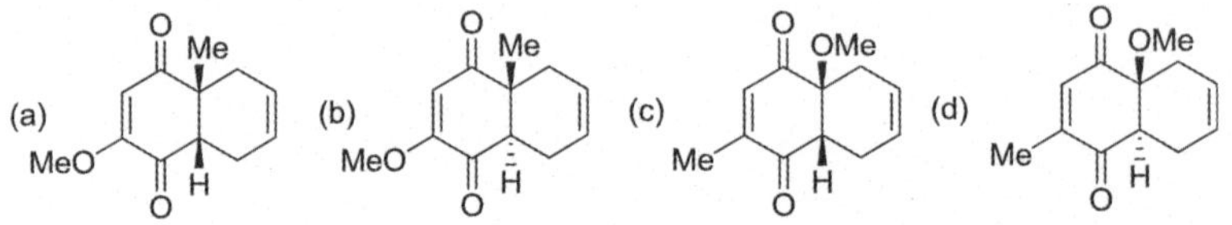

(a) Me, MeO, H (b) Me, MeO, H (c) OMe, Me, H (d) OMe, Me, H

Sol 10. (a) The Diels–Alder reaction of 2-methoxy-5-methylbenzoquinone and buta-1,3-diene occurs preferentially at the double bond bearing the methyl group rather than at the more electron-rich vinyl ether. The $HOMO_{diene}$–$LUMO_{dienophile}$ energy difference (ΔE) for the methoxy side is larger than that for the methyl side, therefore, butadiene reacts preferentially with the methyl side rather than the methoxy side.

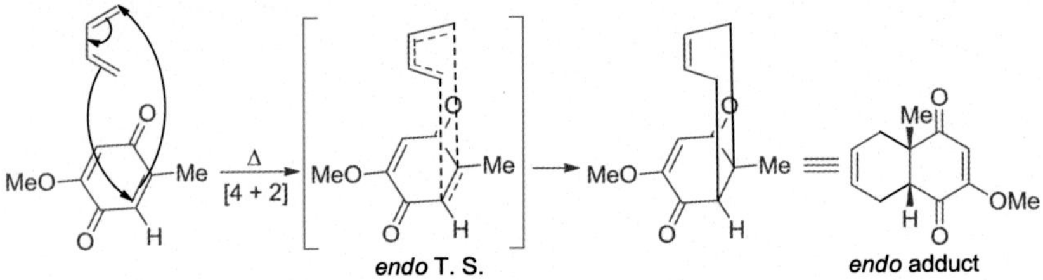

MeO, Me, H —Δ [4 + 2]→ [MeO, Me, H *endo* T. S.] → MeO, Me, H ≡ Me, OMe, H *endo* adduct

Q 11. In the reaction sequence **I** and **II** are:

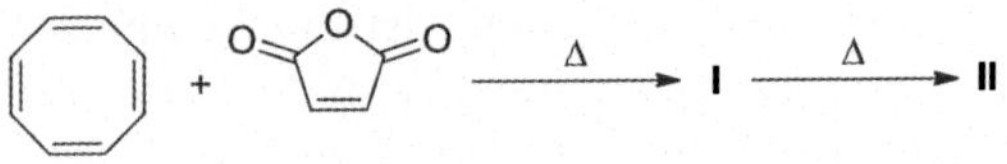

+ —Δ→ I —Δ→ II

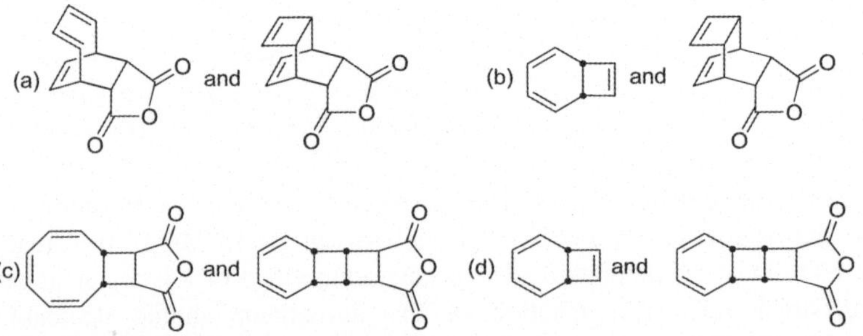

(a) and (b) and

(c) and (d) and

Sol 11. (b) The given reaction involves a 6π-electron disrotatory electrocyclic ring closure of cyclooctatetraene to give the *cis*-isomer **I**, which then undergoes [4 + 2] cycloaddition with maleic anhydride to give **II**.

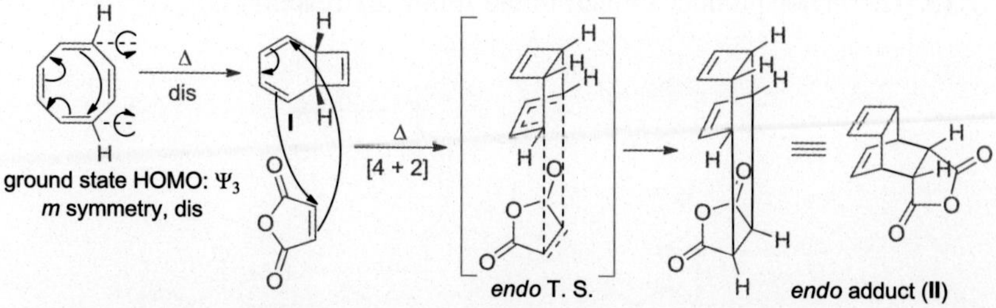

ground state HOMO: Ψ_3
m symmetry, dis

endo T. S.

endo adduct (**II**)

It should be noted that cyclooctatetraene is a tub-shaped molecule, therefore, the double bonds in it are not conjugated enough to react with maleic anhydride. As a result, the Diels–Alder reaction products shown in option (a) will not be obtained.

Q 12. The major products **I** and **II** in the following reaction sequence are:

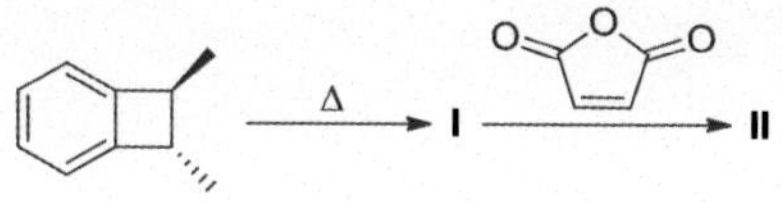

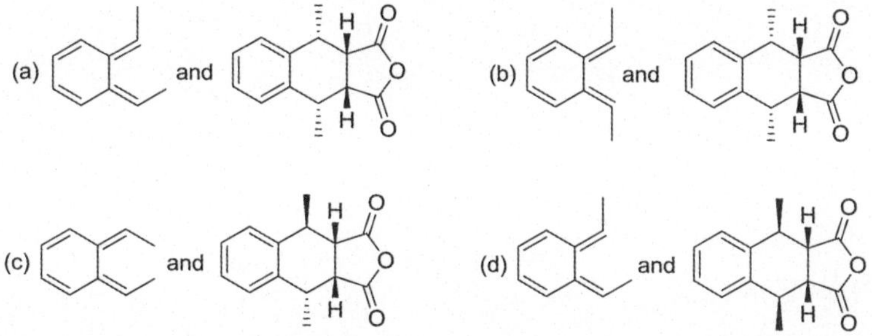

(a) and (b) and

(c) and (d) and

Sol 12. (b) The reaction proceeds via thermal 4π-conrotatory ring opening reaction of benzocyclobutane to *o*-quinodimethane (**I**) followed by *endo* Diels–Alder reaction to give **II**.

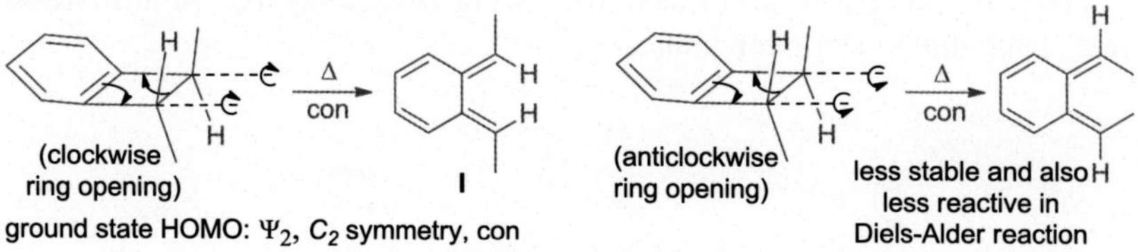

(clockwise ring opening)
ground state HOMO: Ψ_2, C_2 symmetry, con

(anticlockwise ring opening)
less stable and also less reactive in Diels-Alder reaction

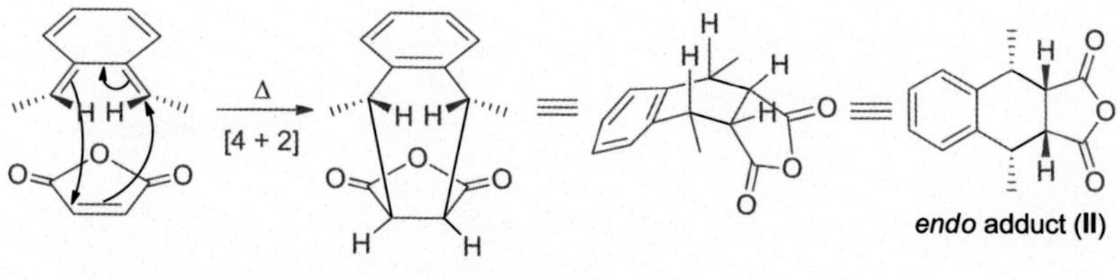

endo adduct (**II**)

Q 13. The major product formed in the following reaction is:

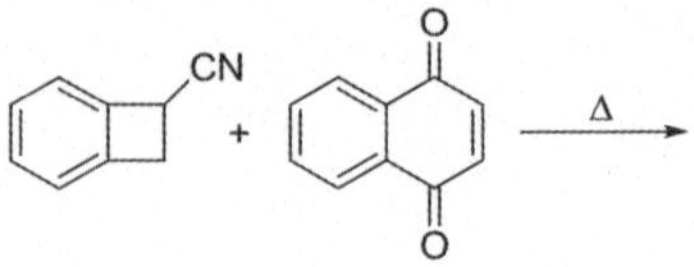

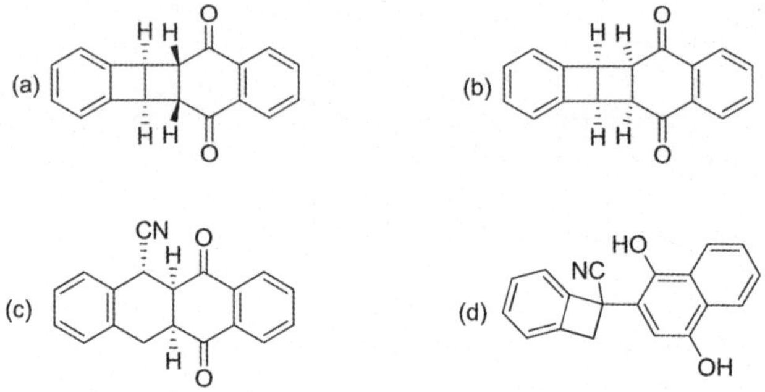

(a)

(b)

(c)

(d)

Sol 13. (c)

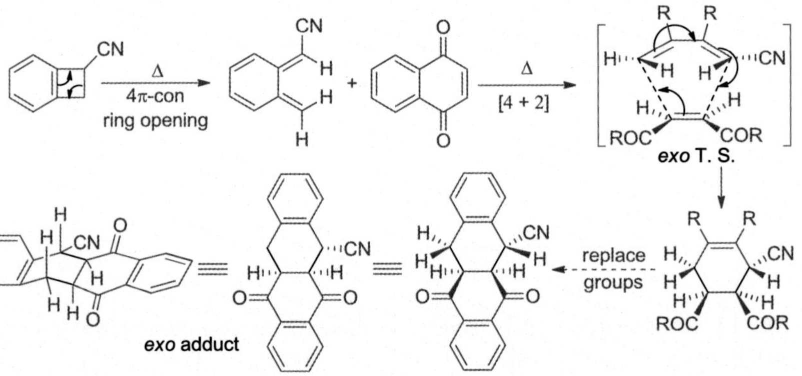

The *endo* adduct is not formed, because the transition state in this case is less stable due to steric hindrance.

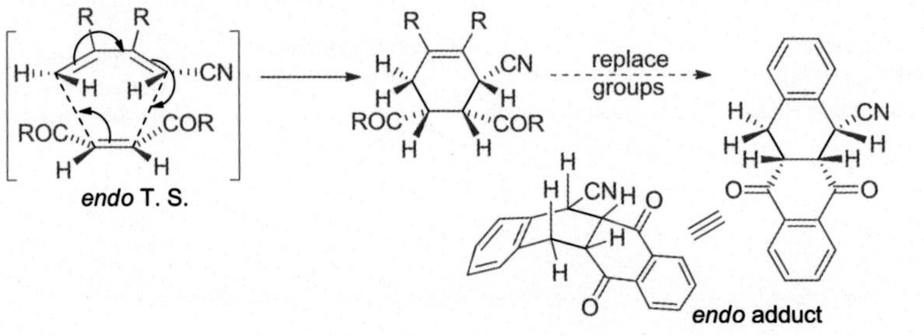

Q 14. What are the products formed in the following reaction? Provide a mechanistic pathway for their formation.

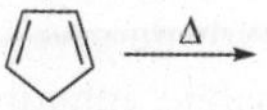

$\xrightarrow{\Delta}$

Sol 14. The double bonds of the cyclopentadiene are held in the *s-cis* conformation. This makes the cyclopentadiene so reactive in the Diels–Alder reaction that it dimerizes at room temperature. One molecule acts as the diene and the other as the dienophile to form dicyclopentadiene. The dicyclopentadiene formed has the ring of dienophile in an *endo* orientation to the cyclopentadiene ring that acts as the diene. Usually, substituents on the dienophile are found to be *endo* in the adduct if the substituents contain the π-bonds due to favorable secondary interactions.

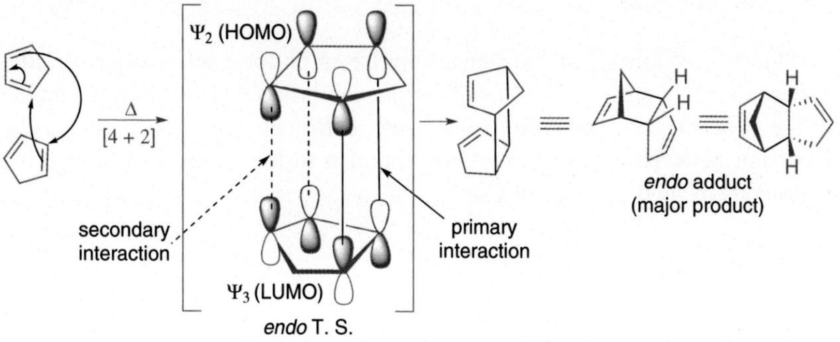

Δ [4 + 2]
Ψ$_2$ (HOMO)
Ψ$_3$ (LUMO)
secondary interaction
primary interaction
endo T. S.
H H H H
endo adduct (major product)

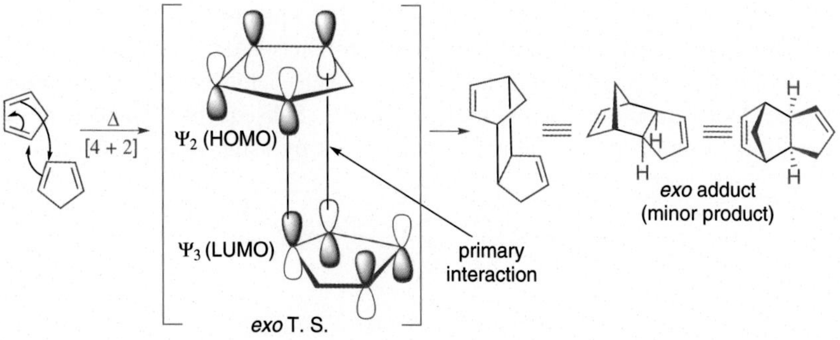

Δ [4 + 2]
Ψ$_2$ (HOMO)
Ψ$_3$ (LUMO)
primary interaction
exo T. S.
H H H H
exo adduct (minor product)

Q 15. Explain why the following [4 + 2] cycloaddition reaction does not occur to produce cantharidin isolated from dried beetles (*Cantharis vesicatoria*).

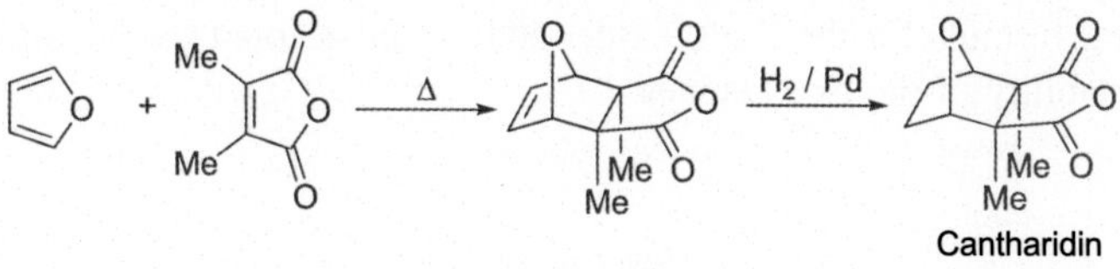

Me Me O O O
Δ
Me Me O O O
H_2 / Pd
Me Me O O O
Cantharidin

Sol 15. The majority of Diels–Alder reactions yield *endo* products. In most of the cases, the *exo* product is thermodynamically more stable, but the *endo* adduct forms much more rapidly (kinetic control). However, [4 + 2] cycloaddition of furan and maleic anhydride gives predominantly *exo* adduct. Such a stereochemistry of the furan-maleic anhydride adduct is due to the fact that the initially formed *endo* compound readily reverses into the reactants whereas the *exo* cycloaddition gives a relatively stable adduct that is the product of the thermodynamic control.

endo adduct
kinetic product
(less stable, minor)

[4 + 2] Δ [4 + 2] Δ

exo adduct
thermodynamic product
(more stable, major)

However, synthesis of the cantharidin based on the Diels–Alder reaction between the two reactants also involves an equilibrium that is very unfavorable to the product. The instability of the *exo* adduct in this case is due to the repulsion between the methyl groups and also to the repulsion between the hydrogens located at C-5 and C-6.

This observation is supported by the fact that when the natural cantharidin is dehydrogenated it readily undergoes a retro Diels–Alder reaction.

Synthesis of cantharidin has indeed been achieved by performing Diels–Alder reaction under high pressure. Subsequent hydrogenation/desulfurization of the *exo* adduct is carried out by using Raney nickel.

Δ, 15 kbar; H_2 / Ni, - H_2S

Cantharidin

Q 16. Cyclopenta-2,4-dienone is unstable and cannot be isolated. All attempts to prepare this molecule give only a Diels–Alder adduct. Interestingly, cyclohepta-2,4,6-trienone is quite stable and can be readily isolated. Draw a structural formula for the Diels–Alder adduct and account for the differences in the stability of the two ketones.

2 Cyclopenta-2,4-dienone → Diels-Alder adduct

Cyclohepta-2,4,6-trienone

Sol 16. A major contributing structure of cyclopenta-2,4-dienone has only four π-electrons (antiaromatic) and, therefore, is extremely unstable and highly reactive. It readily undergoes Diels–Alder reaction to give an *endo* adduct. On the other hand, a major contributing structure of cyclohepta-2,4,6-trienone has only six π-electrons (aromatic) and, therefore, is extremely stable and unreactive.

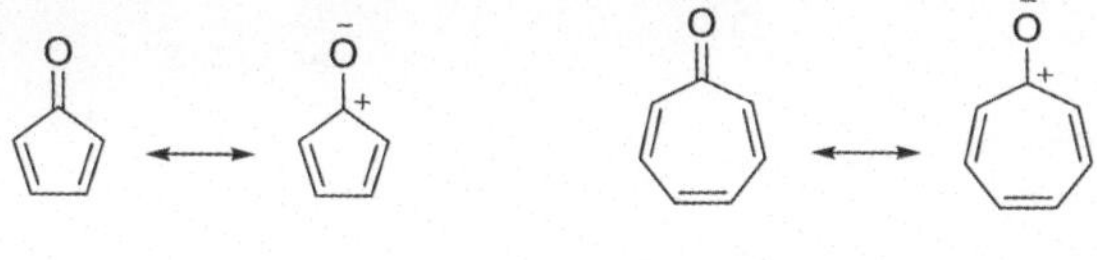

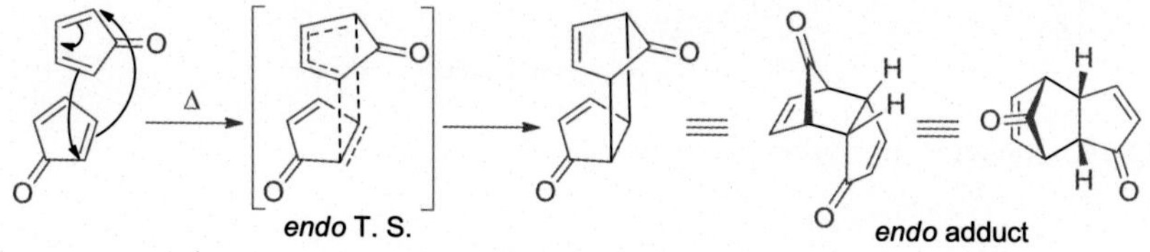

endo T. S.

endo adduct

Q 17. What stereochemistry would you expect for the product of the Diels–Alder reaction between (2*E*,4*E*)-hexadiene and ethene? What stereochemistry would you expect if (2*E*,4*Z*)-hexadiene were used instead?

Sol 17.

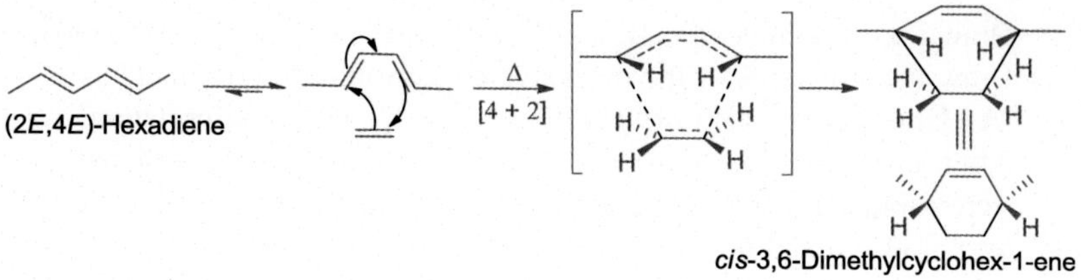

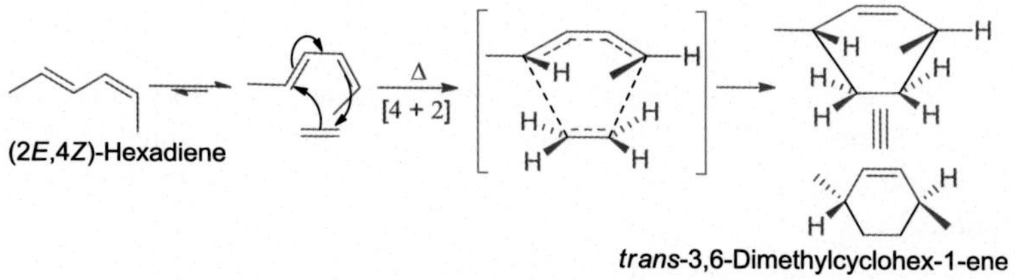

(2*E*,4*E*)-Hexadiene

[4 + 2]

cis-3,6-Dimethylcyclohex-1-ene

(2*E*,4*Z*)-Hexadiene

[4 + 2]

trans-3,6-Dimethylcyclohex-1-ene

Q 18. Explain the mechanism of the following transformation:

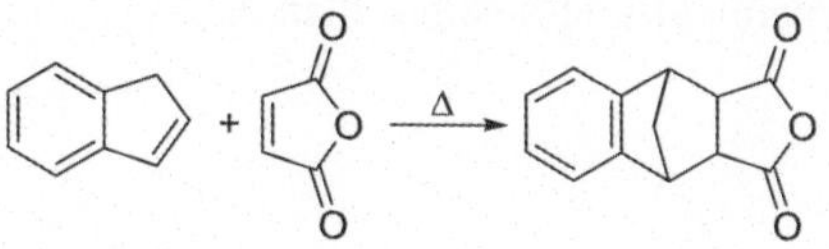

Sol 18.

[1,9] H shift

[4 + 2]

endo T. S.

endo adduct

Q 19. Consider the following sequence of reactions:

I ($C_{11}H_{10}O_2$)

II ($C_{11}H_{10}O_2$)

III ($C_{16}H_{16}O_2$)

(i) Draw the structures of **I, II,** and **III** with stereochemical information.
(ii) Draw the structures of the other five conceivable stereoisomers (**A–E**) of **III** and also explain the reasons for their nonformation.
(iii) After a prolonged heating (15 h, 120 °C) of the originally isolated stereoisomer **III** (melting point mp: 157 °C), two new stereoisomers **IV** (mp: 153 °C) and **V** (mp: 163 °C) are obtained.

III ⇌ IV + V

10% 20% 70%

Decide whether the following questions concerning the Diels–Alder reaction are true or false. (Hint: You do not need to know which of the five stereoisomers **A–E** correspond to either **IV** or **V** in order to answer these questions.)

(a) The Diels–Alder reaction is reversible.
(b) The formation of **III** in the original reaction is thermodynamically controlled.
(c) **III** is thermodynamically more stable than **IV**.
(d) **IV** is thermodynamically less stable than **V**.

Sol 19. (i) Cyclopentadiene undergoes double [4 + 2] cycloaddition with quinone to give a diadduct. Both the steps proceed selectively via the *endo*-transition state to yield the *cis-anti-cis* diadduct **III**.

Δ [4 + 2] *endo* T. S. monoadduct I

Δ [4 + 2] *endo* T. S. diadduct III

The initially formed *endo* adduct **I** undergoes photochemical [2 + 2] cycloaddition to give the cage compound **II**.

I hv [2 + 2] II

(ii) The structures of other five conceivable stereoisomers of **III** are:

A (*endo, endo*) **B** (*endo, exo*) **C** (*endo, exo*) **D** (*exo, exo*) **E** (*exo, exo*)

The Diels–Alder reaction gives products with an *endo* stereochemistry. This *endo* configuration is characterized by the two H atoms and the CH_2 bridge of the bicyclic system being on the same side of the ring. Only structures **A** and **III** of the six stereoisomers have an *endo*, *endo* stereochemistry. All other isomers have at least one *exo* configuration. In structure **A**, the three rings form a U-shaped molecule that is sterically more hindered than structure **III**, which has a zigzag structure.

(iii) (a) True (b) False (c) False (d) True

Q 20. Write the structure of **I**, **II**, **III**, etc. in the following reaction sequences:

(i) 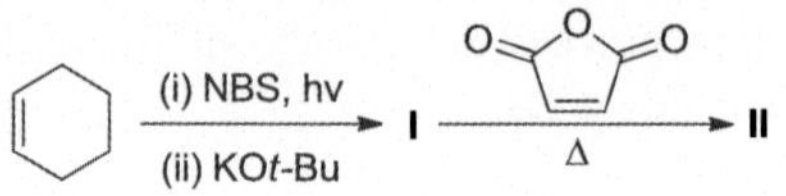 (i) NBS, hv; (ii) KO*t*-Bu → **I**; Δ → **II**

(ii) 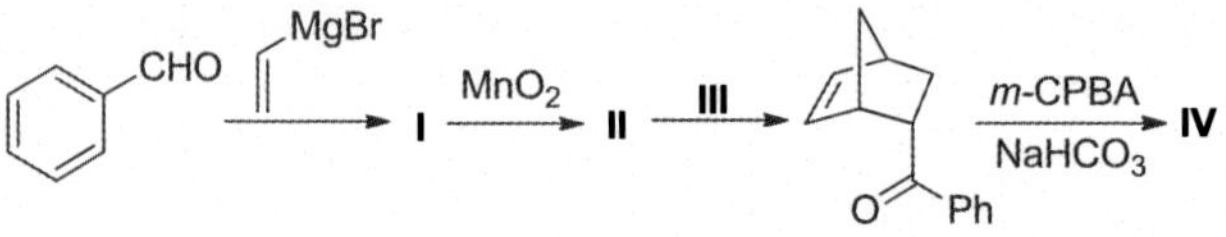 CHO; MgBr → **I**; MnO_2 → **II**; **III** → ; *m*-CPBA, $NaHCO_3$ → **IV**

O Ph

(iii) CH_3, CH_2Br, CH_2Br, CH_3; Δ, Zn / Ag → **I**; $CH_2{=}CH{-}\overset{O}{\overset{\|}{C}}{-}CH_3$ → **II**

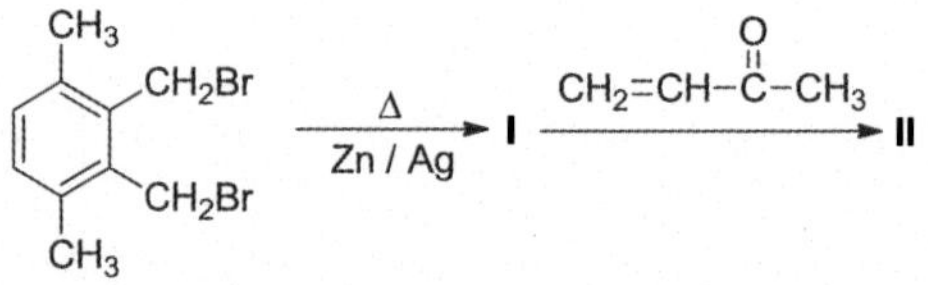

Sol 20.

(i) 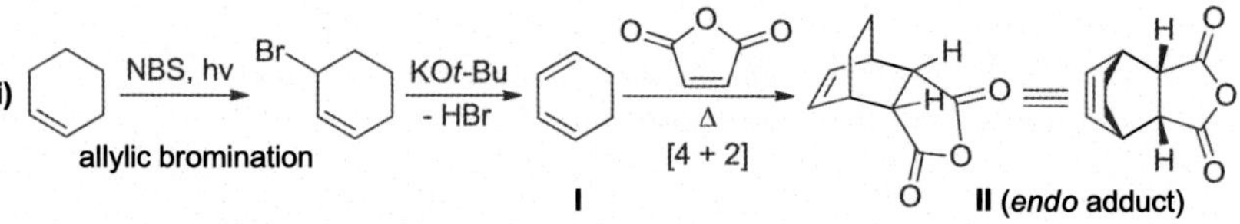 NBS, hv → Br; allylic bromination; KO*t*-Bu, - HBr → **I**; Δ, [4 + 2] → H, H, O ≡ H, O, O, O, H, O; O; **II** (*endo* adduct)

(ii) O, Ph, H; MgBr → OH, Ph **I**; MnO_2 → O, Ph **II**; MnO_2: a selective oxidant for allylic and benzylic alcohols; Δ, [4 + 2] → O, Ph, *endo* T. S.

→ O, Ph ≡ O, Ph; **IV** (*endo* adduct); *m*-CPBA, $NaHCO_3$ → O, O, Ph, O

Baeyer-Villiger to give ester and epoxidation in one step (*exo* epoxidation on less hindered face)

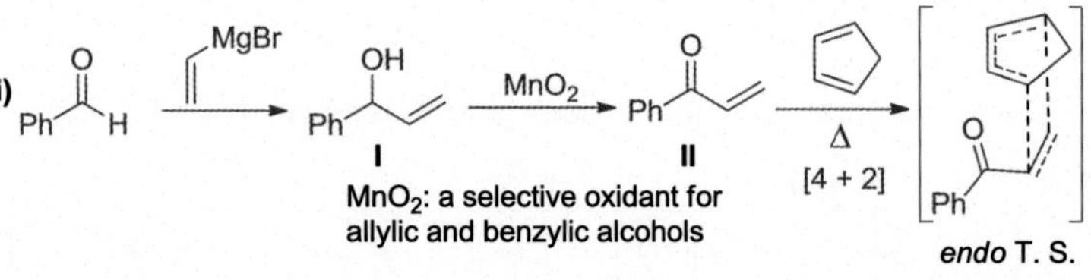

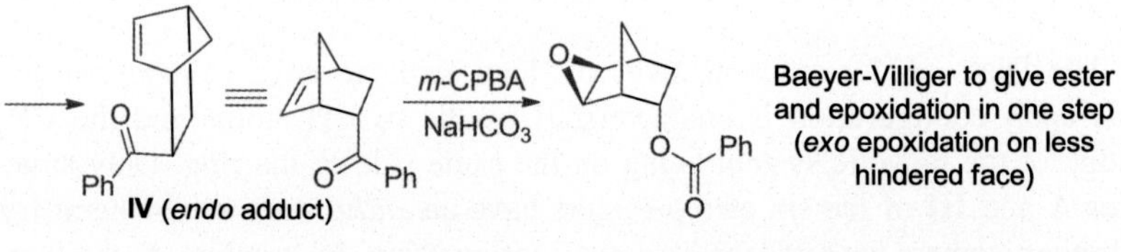

(iii) *o-bis*-(Bromomethyl)benzene can be converted to *o*-quinodimethane (**I**) with reductants such as zinc, nickel, chromous ion, and tri-n-butylstannide. Intermediate **I** then undergoes Diels–Alder reaction to give **II**.

4.4.4 Regiochemistry of Diels–Alder Reaction

Constitutionally homogeneous cycloadducts are obtained through symmetrically substituted dienes and dienophiles. In contrast, when an unsymmetrical diene and an unsymmetrical dienophile combine in a Diels–Alder reaction, it may afford two constitutionally isomeric cycloadducts. The regioisomeric behavior of Diels–Alder reaction can be interpreted through the application of FMO theory on the basis of the orbital coefficients of the atoms forming the σ-bonds. *The Diels–Alder reaction proceeds in such a way that there is an interaction between the largest coefficient in the diene HOMO (at the diene terminus) and the largest coefficient in the dienophile LUMO (at the dienophile terminus)* (Figure 4.15).

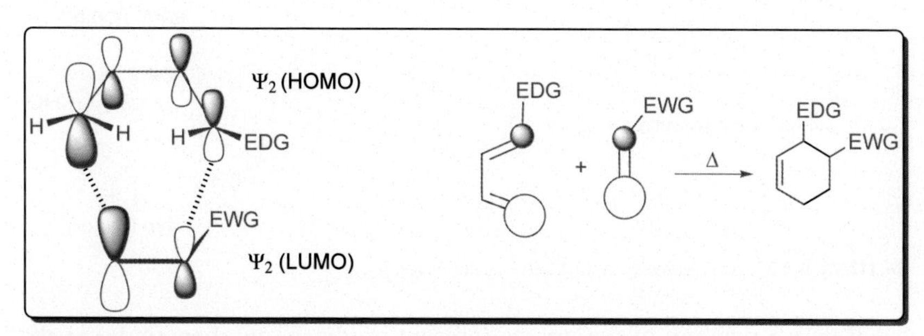

FIGURE 4.15 Orbital interactions in a Diels–Alder reaction resulting in the formation of a regioisomer.

Let us consider an extreme case of mechanism where due to resonance a negative charge is developed on one end of the diene and a positive charge is developed on one end of the dienophile. An electron-donating group (EDG) increases the reactivity of the dienes due to increase in the electron density in the molecule. In case the EDG is located on C-1, there will be more charge density on C-4, and if it is located on C-2, there will be more charge density on C-1. The reactive dienophiles possess an electron-withdrawing group (EWG). An EWG on C-1 will reduce the electron density on C-2 (Scheme 4.14).

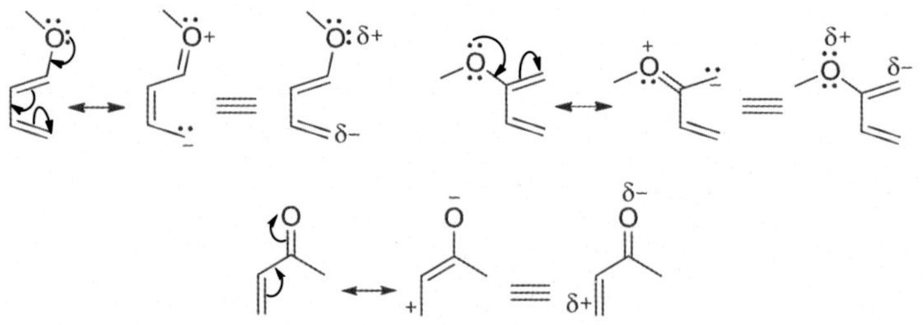

SCHEME 4.14 Analysis of the substituents resonance effects.

The regiochemistry of the Diels—Alder reaction can then easily be predicted by a careful match of the charges. If an EDG is located at C-1 of a diene, then the dienophile will approach with the EWG group next to the EDG group. This is called a 1-2 or "*ortho*" arrangement of the substituents (Scheme 4.15).

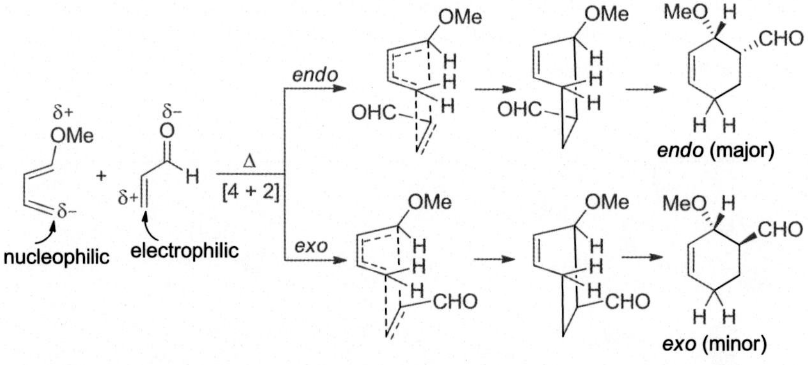

SCHEME 4.15 Regioselectivity in Diels—Alder reaction.

Relative position of the groups (present on diene and dienophile) in the product depends on the stereochemistry of the Diels—Alder reaction. The groups on the dienophile that are *endo* in the transition state will become *cis* to the groups on the outer rim of the diene (in the *s-cis* conformation). Similarly, the groups on the dienophile that are *exo* in the transition state will become *trans* to the groups on the outer rim of the diene (in the *s-cis* conformation).

If the EDG group is located at C-2, the dienophile will flip over resulting in a 1-4 or "*para*" arrangement of the substituents (Scheme 4.16).

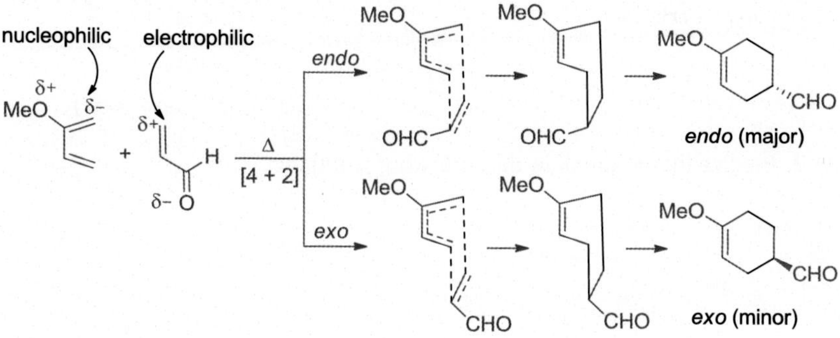

nucleophilic electrophilic

MeO δ+ δ− + δ+ H δ− O Δ [4 + 2]

endo: MeO ... OHC → MeO ... OHC → MeO ... CHO — *endo* (major)

exo: MeO ... CHO → MeO ... CHO → MeO ... CHO — *exo* (minor)

SCHEME 4.16 Regioselectivity in Diels–Alder reaction continued.

As the charges in the diene and dienophile do not match, a 1-3 or "meta" arrangement is not observed in both the cases (Scheme 4.17).

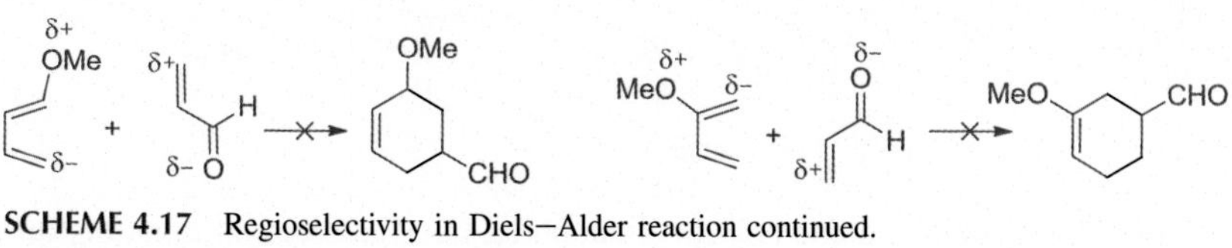

δ+ OMe δ− + δ+ H δ− O ⟶✗ OMe CHO   δ+ MeO δ− + δ− O H δ+ ⟶✗ MeO CHO

SCHEME 4.17 Regioselectivity in Diels–Alder reaction continued.

In conclusion, it can be said that nucleophilicity of dienes and electrophilicity of dienophiles play significant roles in facilitating Diels–Alder reactions. One can also say that the reaction proceeds in a way so as to put the most electron-donating substituent on the diene component and the most electron-withdrawing substituent on the dienophile moiety either *ortho* or *para* to one another in the product.

4.4.4.1 Solved Problems

Q 1. 1-Methoxy buta-1,3-diene and methyl acrylate readily undergo Diels–Alder reaction yielding methyl *cis*-2-methoxycyclohexene carboxylate as the only (or major) product. Explain the stereo- and region-isomerism in the reaction.

Sol 1.

δ+ OMe δ− + δ− O OMe δ+ Δ [4 + 2]

nucleophilic electrophilic

endo: OMe H H MeOOC H → OMe H H MeOOC H → MeO H COOMe H H

endo (major)

exo: OMe H H H COOMe → OMe H H H COOMe → MeO H COOMe H H

exo (minor)

Q 2. Predict the products in the following reactions:

(i) MeO + CN Δ →

(ii) OMe + O Δ →

(iii) O + O Δ →

Sol 2.

(i) δ+ MeO δ− + δ− CN δ+ Δ [4 + 2] → [MeO NC] → MeO NC → MeO CN

endo T. S.

(ii) The reactants, diene and the dienophile, have an electron-donating and an electron-withdrawing group, respectively. As the methoxy group is located at C-1 of diene, the dienophile will approach with the carbonyl group next to the methoxy group to give a 1-2 or "*ortho*" arrangement of the substituents. These groups will be mutually *cis* because the carbonyl directs the product to be *endo*. In the dienophile, carbonyl and methyl groups are having a *trans* geometry that is reflected in the product as well.

δ+ OMe δ− δ− O δ+ Δ [4 + 2] → [OMe H H H O H] → OMe H H H O H ≡ MeO H H O H H ≡ OMe H O

endo T. S.

(iii) This reaction is an example of inverse electron demand Diels–Alder reaction. The regiochemistry of this reaction can also be determined by charge control method.

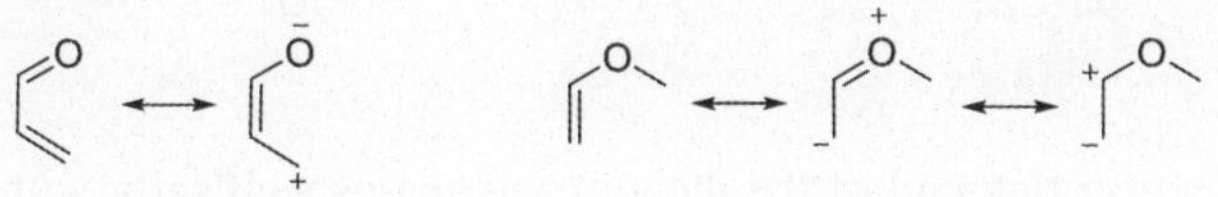

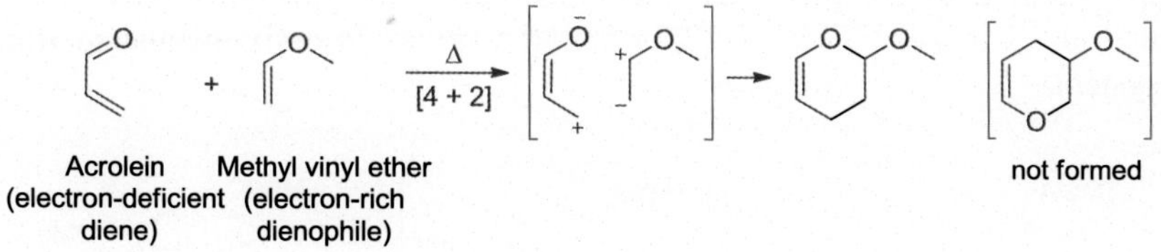

Acrolein (electron-deficient diene) + Methyl vinyl ether (electron-rich dienophile) —Δ, [4 + 2]→ ... not formed

Q 3. Explain the mechanism of the following reactions:

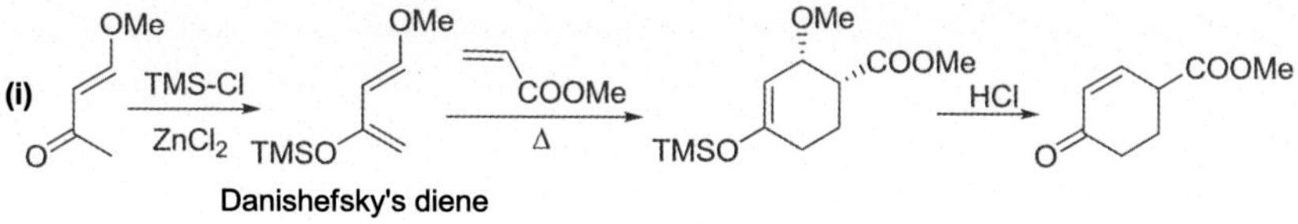

(i) TMS-Cl, $ZnCl_2$ → Danishefsky's diene; COOMe, Δ; HCl

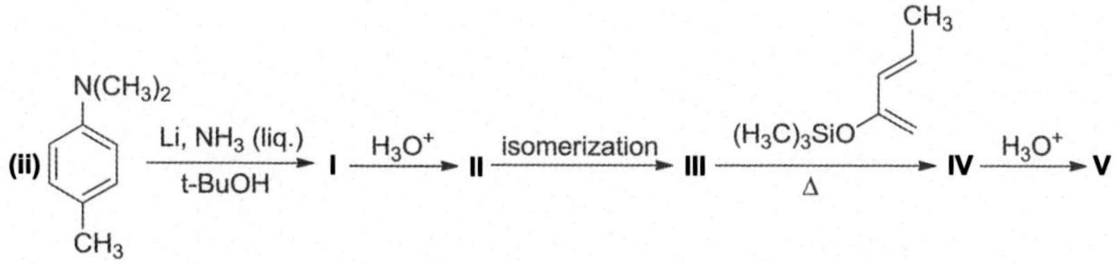

(ii) $N(CH_3)_2$, CH_3; Li, NH_3 (liq.), t-BuOH → I —H_3O^+→ II —isomerization→ III —$(H_3C)_3SiO$ (diene with CH_3), Δ→ IV —H_3O^+→ V

Sol 3. (i) Danishefsky's diene is obtained from α, β-unsaturated ketones via treatment with trimethyl silyl chloride and zinc chloride. The zinc chloride is a Lewis acid that activates the oxygen for reaction with the silicon. The mechanism of this transformation is as follows:

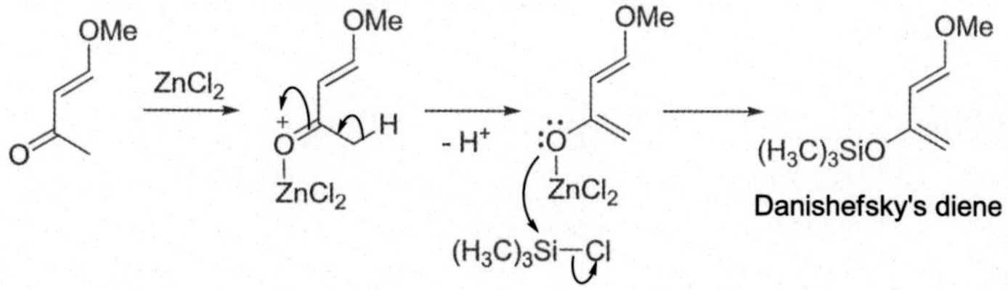

OMe —$ZnCl_2$→ ... —$-H^+$→ ... $(H_3C)_3Si$–Cl → $(H_3C)_3SiO$ ... Danishefsky's diene

This diene is highly electron rich as both the oxygen atoms of the substituents donate electrons toward one end of the conjugated system and thus make the C-4 highly electron rich.

The electron-rich end of the diene reacts regiospecifically with the more electron-deficient part of the dienophile to give the 1,2-product. After the addition is complete, the functionalities are easily removed by acid hydrolysis to generate an α, β-unsaturated ketone in a six-membered ring system.

(ii) In the first step, Birch reduction of *N,N*,4-trimethylaniline results in the rapid isomerization of the initially formed kinetic product: 1,4-cyclohexadienamine to a conjugated 1,3-cyclohexadienamine (**I**), even in absence of an acid. Hydrolysis of the enamine (**I**) gives a β,γ-unsaturated ketone (**II**), which further undergoes isomerization to a conjugated ketone (**III**) under acidic conditions.

Finally, the conjugated ketone (**III**) undergoes Diels–Alder reaction with 2-trimethylsilyloxypenta-1,3-diene to give an adduct (**IV**), which is hydrolyzed to remove a trimethylsilyloxy group to give **V**.

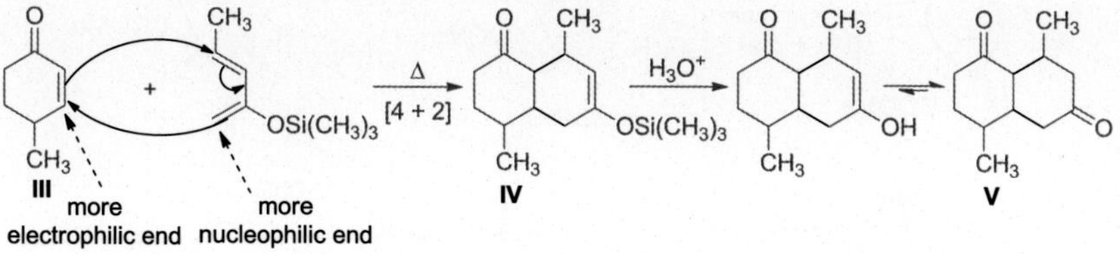

III + OSi(CH$_3$)$_3$ diene → Δ [4 + 2] → IV → H_3O^+ → V

more electrophilic end   more nucleophilic end

Q 4. Which is a major product formed in the following reaction?

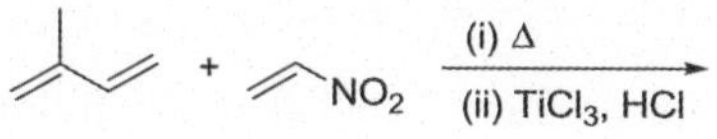

(i) Δ (ii) $TiCl_3$, HCl

(a) 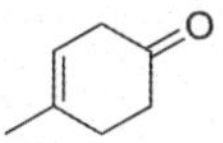 (b) 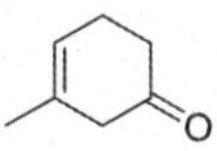 (c) 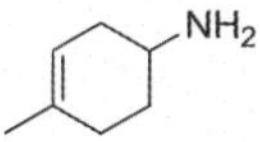 NH_2 (d) 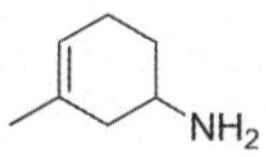 NH_2

Sol 4. (a)

δ− NO_2 δ+ δ+ δ− Δ [4 + 2] *endo* T. S. O_2N NO_2 $TiCl_3$, HCl Nef reaction

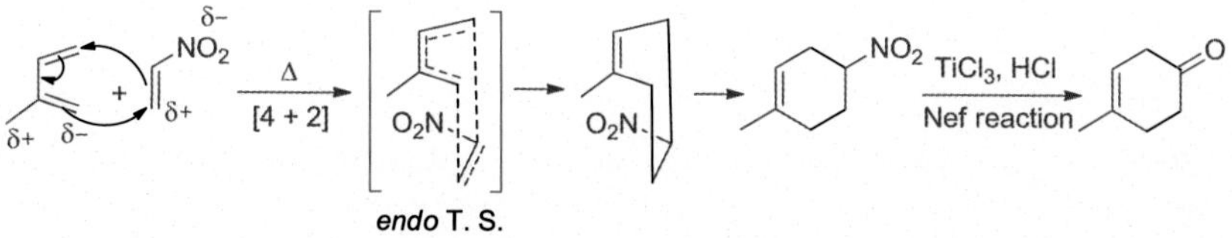

Q 5. What are the major products **I** and **II** formed in the following reaction?

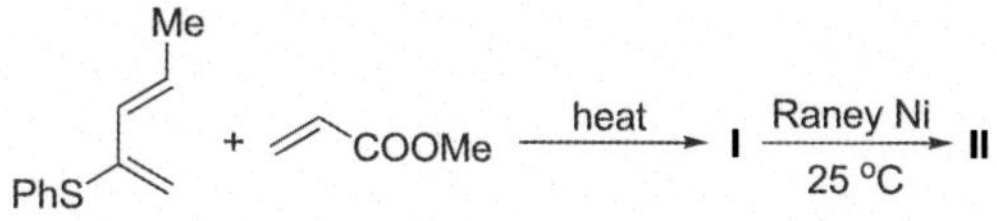

Me PhS + COOMe heat → **I** Raney Ni 25 °C → **II**

(a) 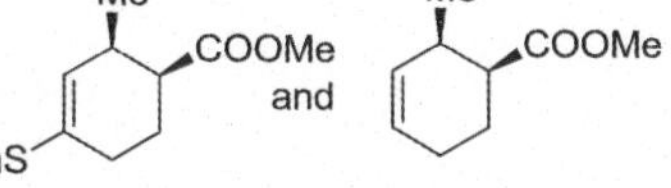 Me PhS COOMe and Me COOMe (b) 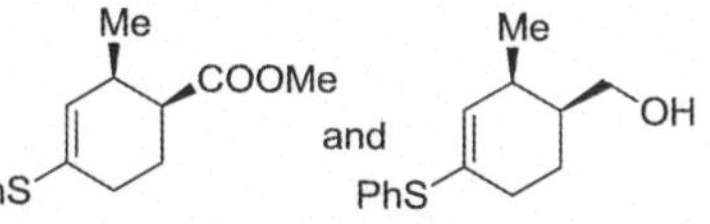 Me PhS COOMe and Me PhS OH

(c) 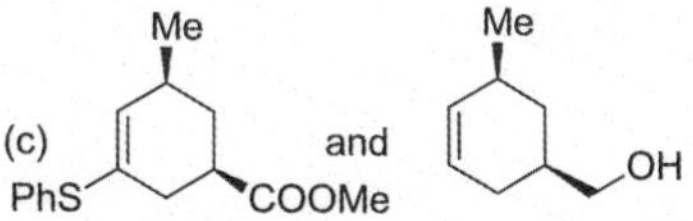 Me PhS COOMe and Me OH (d) 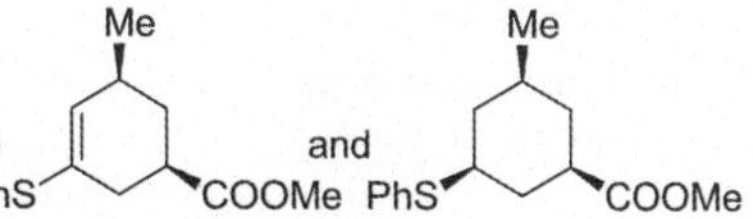 Me PhS COOMe and Me PhS COOMe

Sol 5. (a)

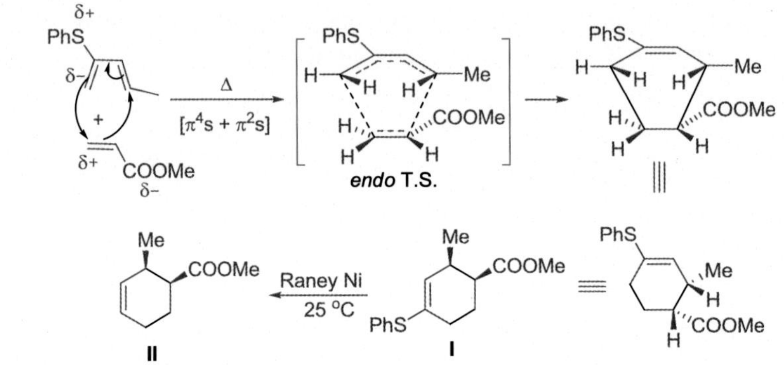

δ+ PhS δ− + δ+ COOMe δ− Δ [$\pi^4s + \pi^2s$] PhS H Me H H H COOMe H H *endo* T.S. PhS H Me H H H COOMe H H

Me COOMe II Raney Ni 25 °C Me COOMe PhS I ≡ PhS Me H COOMe H

Q 6. What are the major products **I** and **II** in the following reactions?

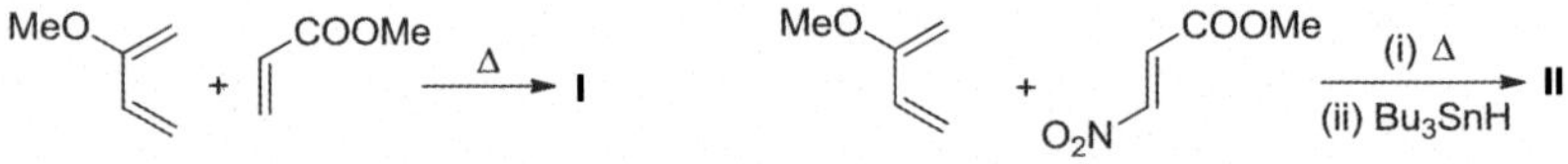

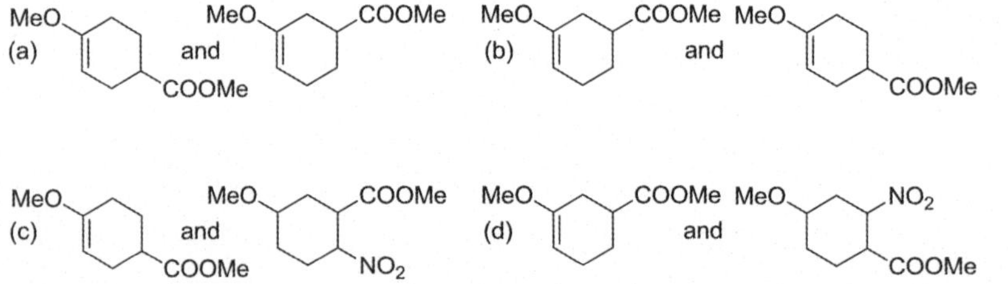

MeO + COOMe Δ → I MeO + COOMe O_2N (i) Δ (ii) Bu_3SnH → II

(a) MeO COOMe and MeO COOMe (b) MeO COOMe and MeO COOMe

(c) MeO COOMe and MeO COOMe NO_2 (d) MeO COOMe and MeO NO_2 COOMe

Sol 6. (a)

OMe δ− H H H H MeOOC H δ+ H Δ [4 + 2] OMe H H H H MeOOC H H ≡ OMe H H MeOOC H H H H ≡ MeO COOMe I

OMe δ− H H H H O_2N H δ+ COOMe Δ [4 + 2] OMe H H H H O_2N H COOMe ≡ OMe H H H H O_2N H H COOMe

NO$_2$ group dominates the direction of reactiion

MeO COOMe II ← Bu_3SnH removal of NO_2 group ← MeO COOMe NO_2

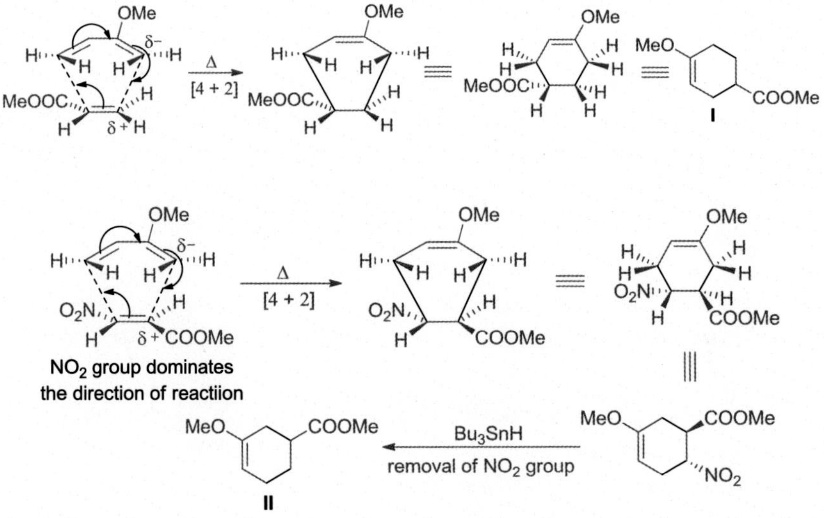

Q 7. Identify the major products **I** and **II** in the following reaction.

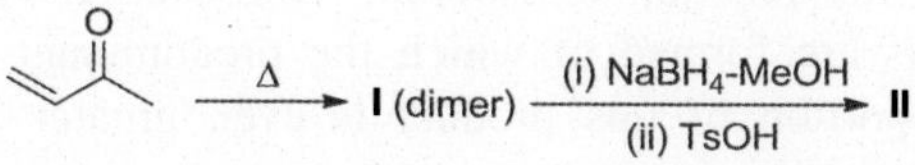

Δ → I (dimer) → (i) $NaBH_4$-MeOH (ii) TsOH → II

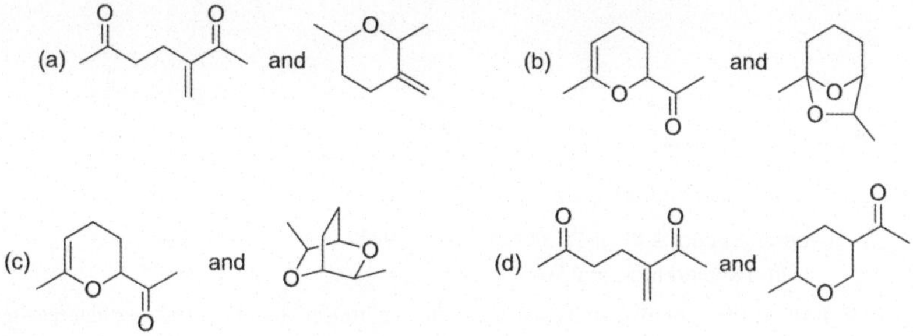

(a) and (b) and

(c) and (d) and

Sol 7. (b) But-3-en-2-one undergoes dimerization by the Diels–Alder reaction in which the oxygen atom is the part of the diene system. The regioselectivity of dimerization of but-3-en-2-one has been a problem. The fact that the reaction apparently involves bonding between two formally electrophilic carbons has been explained by FMO theory, which also accounts for the observed regioselectivity.

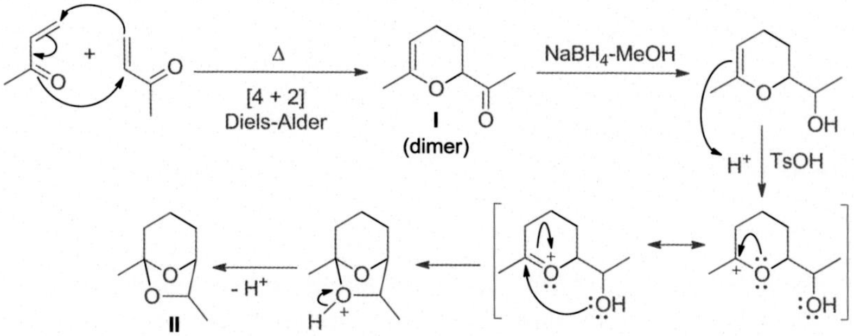

Δ [4 + 2] Diels-Alder → I (dimer) → $NaBH_4$-MeOH → OH → H^+ TsOH → :OH ↔ :OH → → $-H^+$ → II

The transition state for the formation of **I** is less stable on account of polar forces. The "polarity" model also predicts that the dimerization should occur through resonance forms shown below:

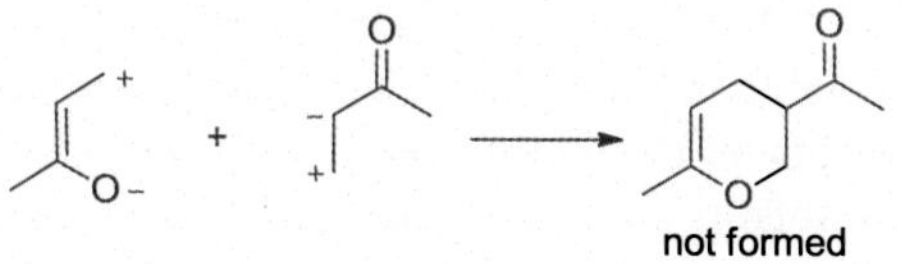

→ not formed

4.4.5 Lewis Acid Catalyzed Diels–Alder Reaction

Diels–Alder reactions are usually accelerated by Lewis acid catalysts such as aluminum chloride, boron trifluoride and tin(IV) chloride. In the presence of a catalyst, an increase in the regio- and stereoselectivity is often observed. For example, in the reaction of isoprene with methyl vinyl ketone, two structural isomers are formed of which the predominant one is the *para* adduct. The proportion of this product is even greater in the catalyzed reaction (Scheme 4.18).

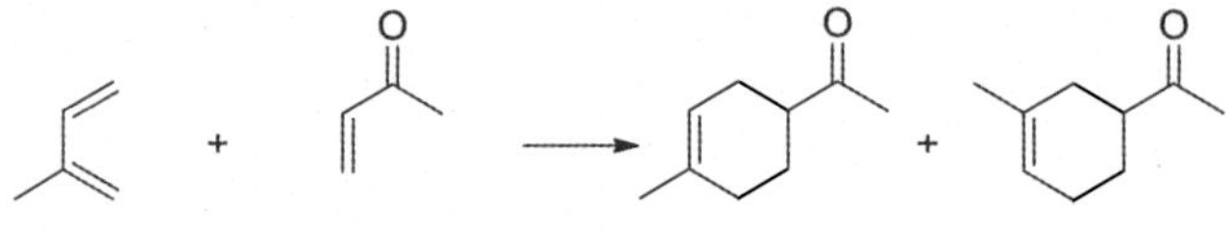

heat in a sealed tube at 120 °C 71% 29%

keep in contact with $SnCl_4.5H_2O$ in a sealed tube at 0 °C 93% 7%

SCHEME 4.18 Effect of Lewis acid catalysis on the regioselectivity of the Diels–Alder reaction.

Similarly, in the reaction of cyclopentadiene with acrylic acid, the proportion of *endo* adduct was found to increase noticeably in the presence of aluminum chloride etherate (Scheme 4.19).

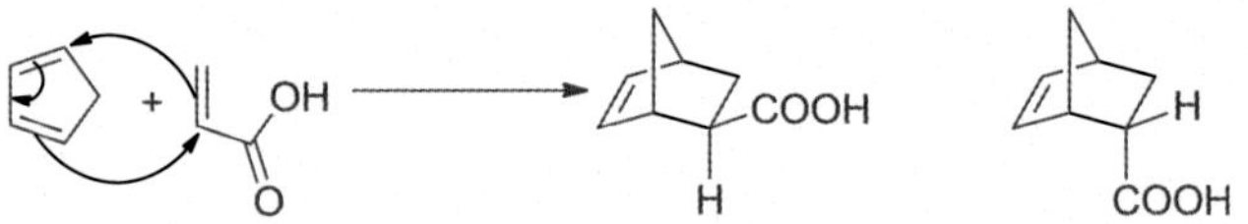

OH O COOH H H COOH

endo adduct *exo* adduct

0 °C, no catalyst 84% 16%

0 °C, $AlCl_3.Et_2O$ 93% 7%

SCHEME 4.19 Effect of Lewis acid catalysis on the *endo/exo* selectivity of the Diels–Alder reaction.

These effects are ascribed to complex formation between the Lewis acid and the polar group of the dienophile, which brings about changes in energies and orbital coefficients of the frontier orbitals of the dienophile.

Rate Enhancement: Addition of a Lewis acid results in the reversible complexation to Lewis basic sites, and the electron-withdrawing ability of the activating substituents increase on such a complexation to the dienophile. Lowering of the LUMO and a decrease in the $HOMO_{diene}$–$LUMO_{dienophile}$ gap help in a more favorable transition state and thus a faster reaction. The effect of Lewis acid catalysts on the interaction of the frontier orbitals of butadiene and methyl vinyl ketone is shown diagrammatically in Figure 4.16.

Regioselectivity: An increase in the regioselectivity in the catalyzed reactions can also be explained on the basis of FMO interactions. More effective overlap of the orbitals in the transition state of a Diels–Alder reaction takes place when the reacting compounds are oriented in such a way that the atom with the largest coefficient in the dienophile interacts preferentially with the

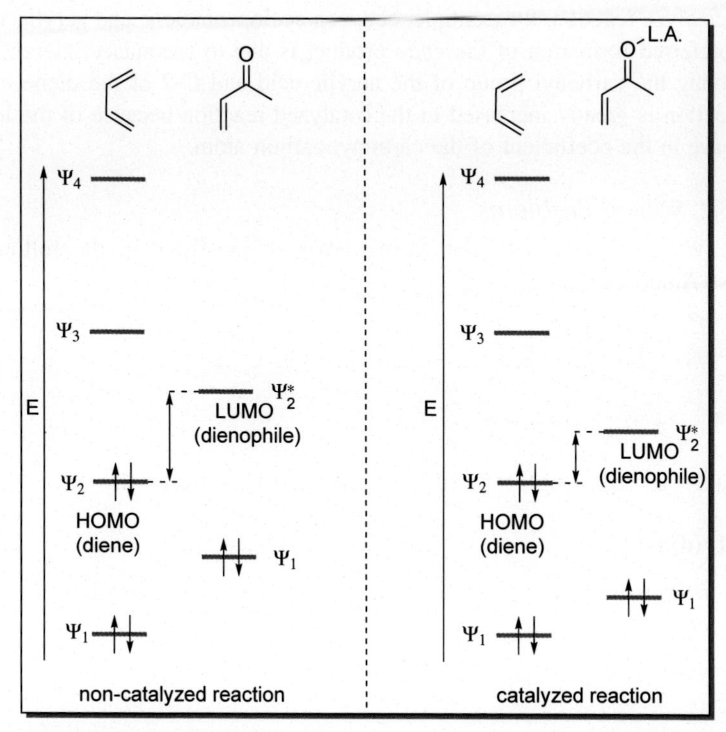

FIGURE 4.16 Effect of Lewis acid catalysts on the interaction of the frontier orbitals of butadiene and methyl vinyl ketone.

one having the larger coefficient in the diene. For example, in the reaction of isoprene with methyl vinyl ketone, the T. S.-**I** is favored over T. S.-**II** and the ratio of *para* to *meta* products formed is 71:29 (Figure 4.17).

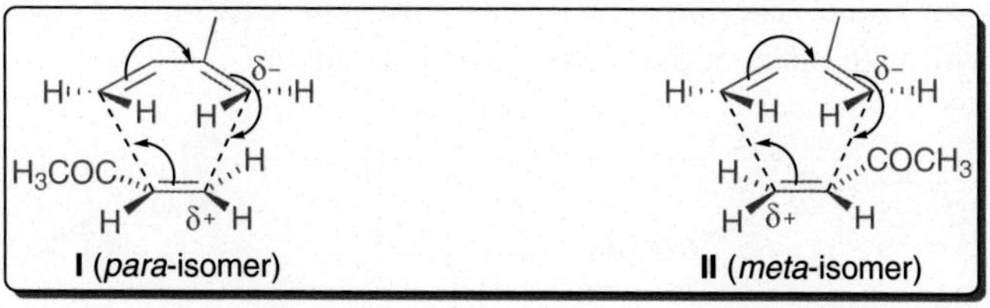

FIGURE 4.17 Transition states for the formation of *para* and *meta* products in the reaction of isoprene with methyl vinyl ketone.

However, in the catalyzed reaction, the interaction of the catalyst with the carbonyl group of the ketone increases the difference between the coefficients at C-2 and C-3 with the result that the reaction becomes more selective; T. S.-**I** is favored even more and the *para* to *meta* ratio now rises to is 93:7.

Stereoselectivity: The increase in *endo*:*exo* ratio in catalyzed reaction is also ascribed to increased secondary orbital interactions. In a typical non-catalyze reaction, for example, between cyclopentadiene and acrylic acid, the preferred formation of the *endo* product is due to secondary interactions involving the carbonyl group of the acrylic acid and C-2 of the diene. This interaction is greatly increased in the catalyzed reaction because of the large increase in the coefficient of the carbonyl carbon atom.

4.4.5.1 Solved Problems

Q 1. What would be the decreasing order of reactivity in the following Diels–Alder reactions?

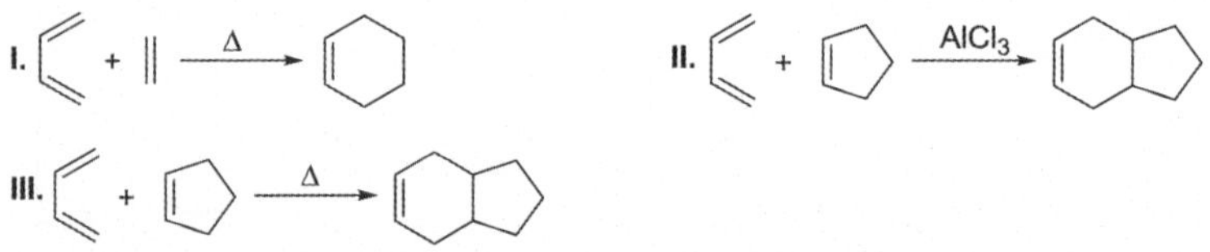
I.
Δ
II.
$AlCl_3$
III.
Δ

(a) I > II > III (b) II > I > III (c) III > II > I (d) II > III > I

Sol 1. (d)

Q 2. What is the major product formed in the following reaction?

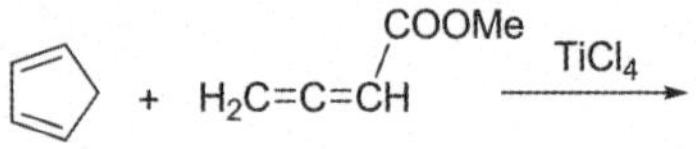
COOMe
$H_2C=C=CH$
$TiCl_4$

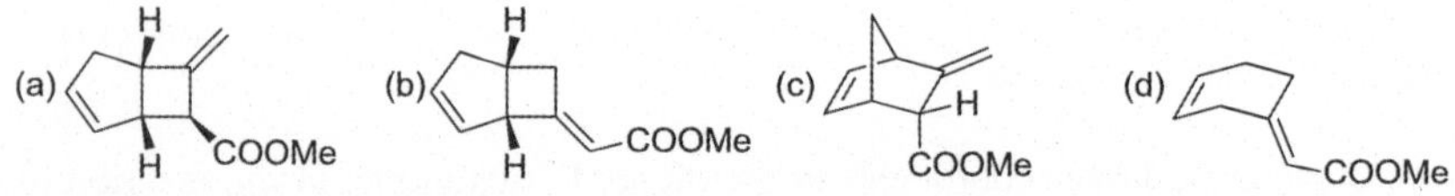
(a)
H
H
COOMe
(b)
H
H
COOMe
(c)
H
COOMe
(d)
COOMe

Sol 2. (c) Allenes usually participate in Diels–Alder reactions as electron-deficient dienophiles. LUMO energy of the allene is lowered by an electron-withdrawing group. The largest LUMO coefficient rests on C-2. Diels–Alder reaction will take place at the internal C=C bond.

H
H
C
C
C
H
COOMe
2
1
H
H
H
COOMe

structure of allene LUMO

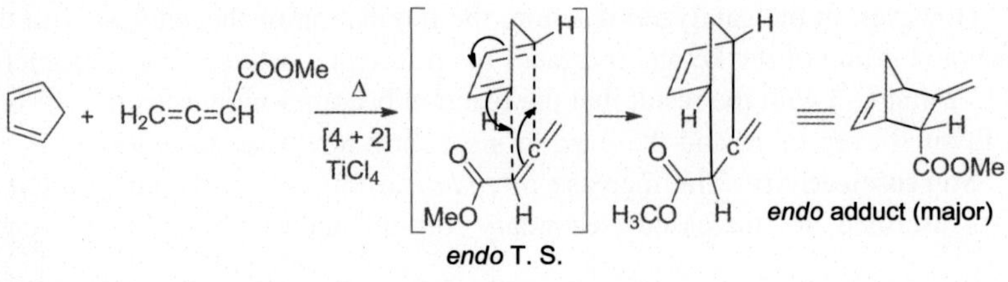
COOMe
$H_2C=C=CH$
Δ
[4 + 2]
$TiCl_4$
H
O
MeO
H
endo T. S.
H
H
O
H_3CO
H
H
COOMe
endo adduct (major)

Allenic esters generally show *endo* selectivity. Lewis acids increase efficiency of the reaction as well as *endo/exo* selectivity. Other electron-withdrawing groups also show preference for the *endo* product.

4.4.6 Retro Diels–Alder Reaction

The Diels–Alder reaction is a reversible reaction and the direction of cycloaddition is favored because two π-bonds are replaced by two σ-bonds. The cycloreversion also occurs more readily when the diene and/or dienophile are particularly stable molecules (i.e., formation of an aromatic ring, nitrogen, carbon dioxide, ethene, ethyne, nitriles, etc.) or when one of them can be easily removed or consumed in a subsequent reaction (Scheme 4.20).

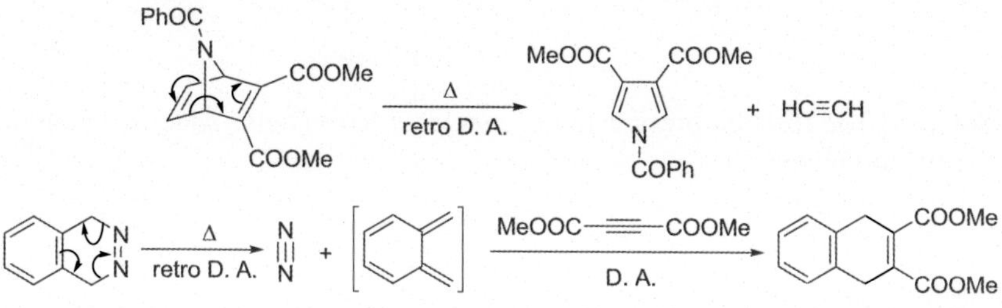

PhOC N COOMe COOMe Δ retro D. A. MeOOC COOMe N COPh + HC≡CH

N N Δ retro D. A. N≡N + MeOOC—≡—COOMe D. A. COOMe COOMe

SCHEME 4.20 Examples of retro Diels–Alder reactions.

The retro Diels–Alder reaction usually requires higher temperatures in order to overcome the high activation barrier of the cycloreversion. Moreover, the strategy of using a retro Diels–Alder reaction is often used in organic synthesis to mask a diene fragment or to protect a double bond.

4.4.6.1 Solved Problems

Q 1. Explain the mechanism of following transformations:

(i) 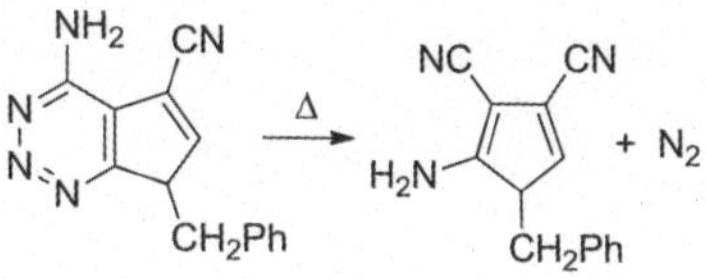 NH_2 CN N N N CH_2Ph Δ NC CN H_2N CH_2Ph + N_2

(ii) 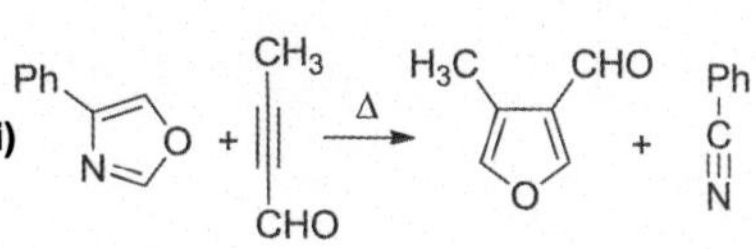 Ph O N + CH_3 CHO Δ H_3C CHO O + Ph C≡N

(iii) 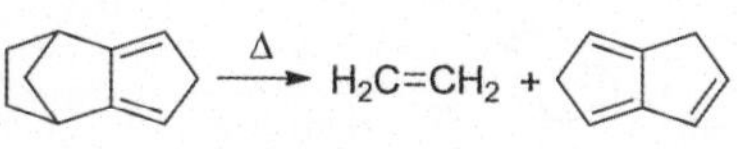 Δ $H_2C{=}CH_2$ +

(iv) 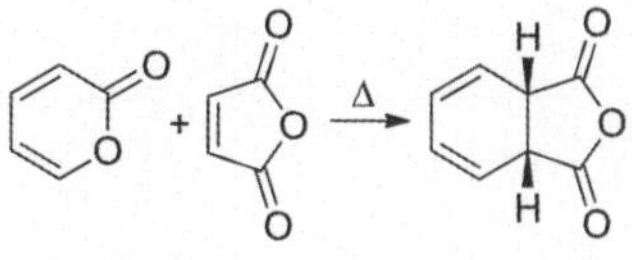 O O + O O O Δ H O O O H

(v) 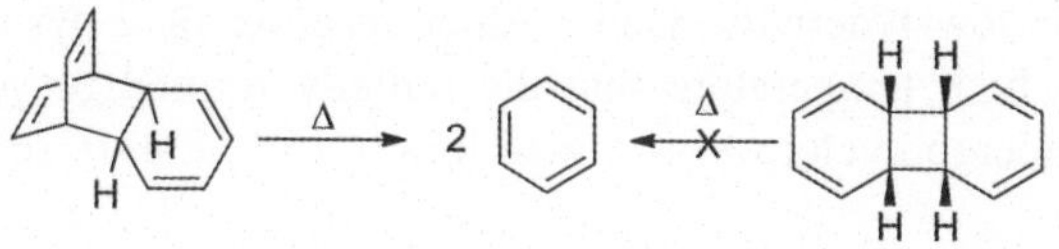 H H Δ 2 ← Δ H H H H

(vi)

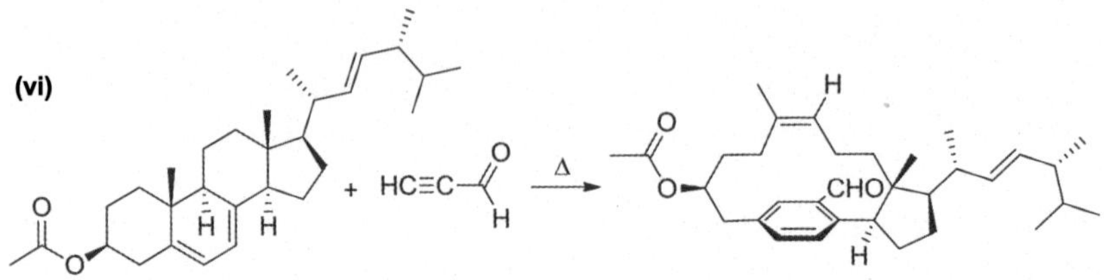

Δ

(vii)

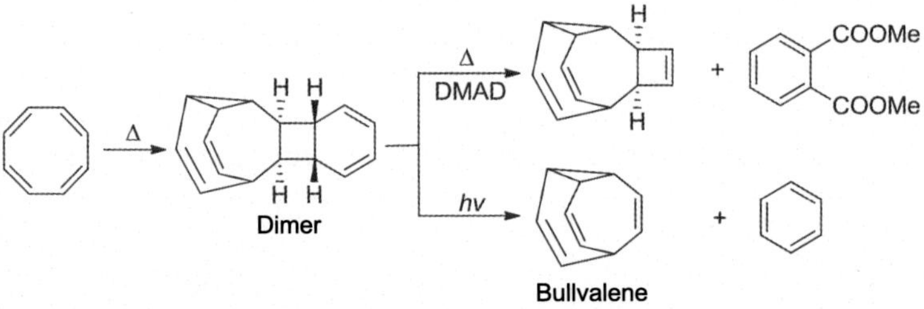

Dimer

DMAD

hv

Bullvalene

Sol 1. **(i)** The reaction presumably occurs via a retro Diels–Alder reaction of the imino tautomer.

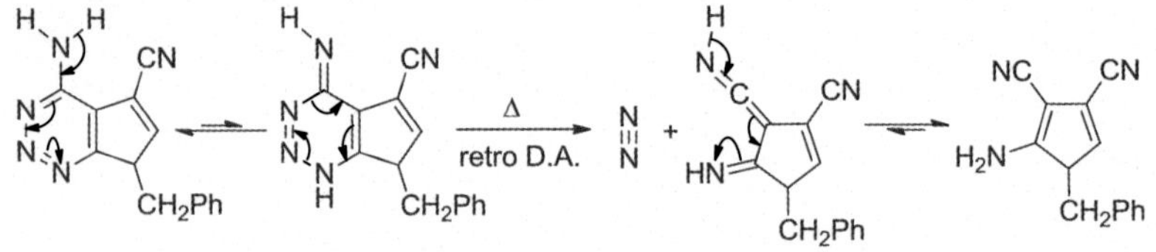

retro D.A.

(ii)

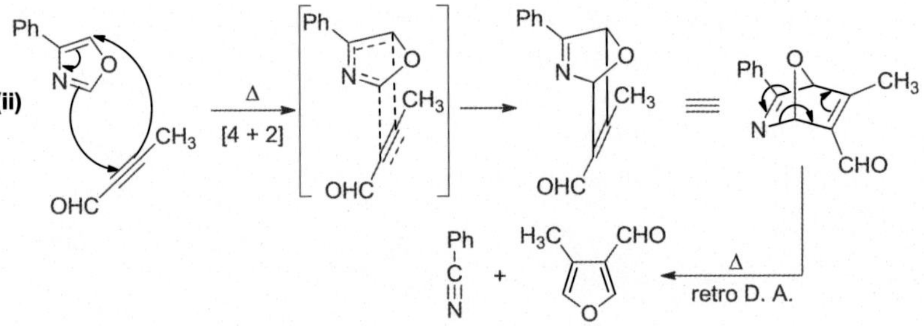

[4 + 2]

retro D. A.

(iii)

[1,5] H shift

retro D. A.

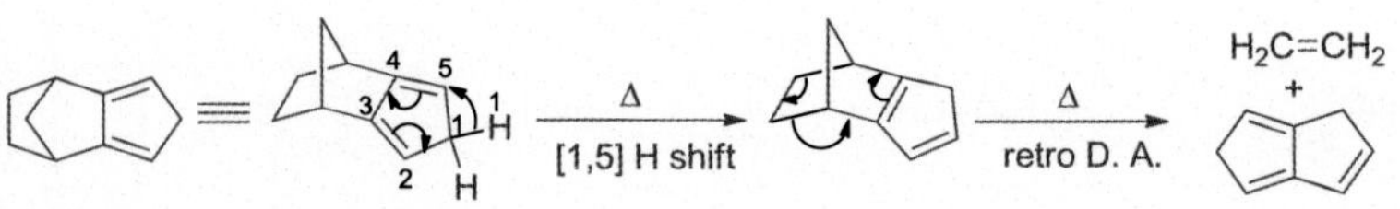

(iv) Because of their low reactivity, Diels–Alder reaction of 2-pyrones usually requires such a high temperature that the initially formed bicyclic lactone adducts often undergo cycloreversion with loss of carbon dioxide.

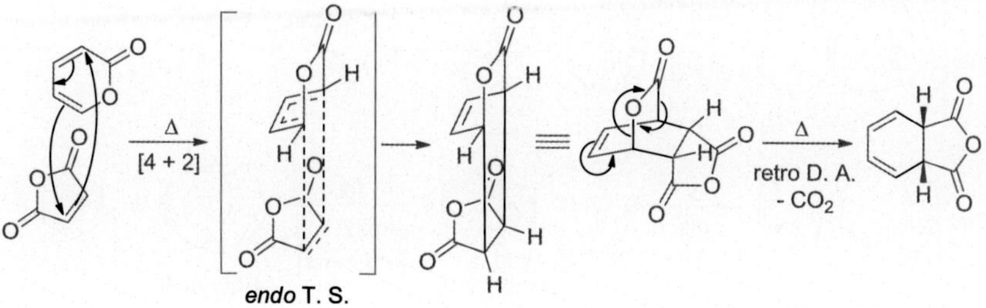

(v) Retro [4 + 2] cycloaddition can take place under thermal conditions by a symmetry allowed suprafacial pathway. Whereas, retro [2 + 2] cycloaddition cannot occur by a concerted process because the antarafacial geometry required for the thermal reaction is not possible for a four π-electron system.

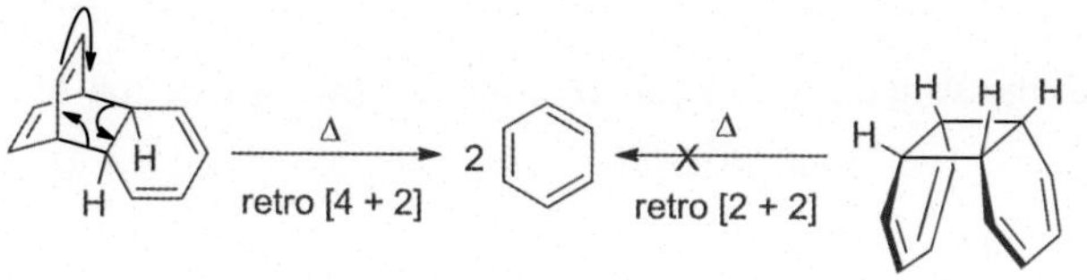

(vi) The ring system of steroids is usually broken by a two-step process: a Diels–Alder followed by a retro Diels–Alder to generate a 14-membered *ansa* compound (benzene derivatives having *para* or *meta* positions bridged by a chain [commonly 10–12 atoms long] are called *ansa* compounds [Latin *ansa*, handle]).

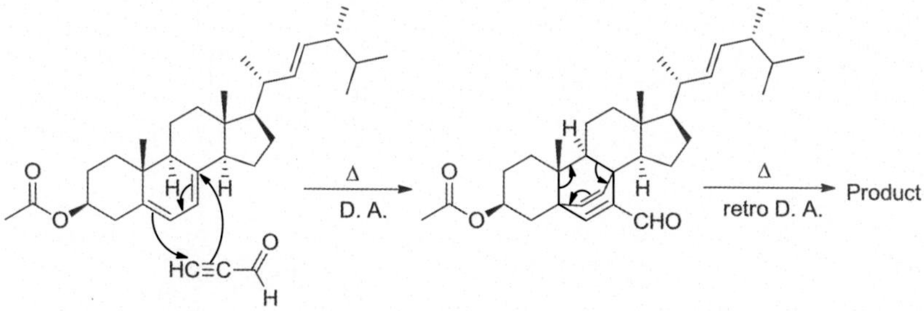

(vii) On heating, cyclooctatetraene undergoes intermolecular Diels–Alder reaction to give an adduct, which then undergoes intramolecular Diels–Alder reaction followed by a six-electron shift and disrotatory electrocyclization to give the dimer. The dimer has a *cis*-butadiene moiety, and hence it readily undergoes thermally allowed Diels–Alder reaction with dimethyl acetylene dicarboxylate (DMAD) to give the adduct, which then undergoes thermally allowed retro Diels–Alder reaction to give the final products. Under photochemical conditions, the dimer undergoes symmetry-allowed cycloreversion of [σ^2s + σ^2s] type to give bullvalene and benzene.

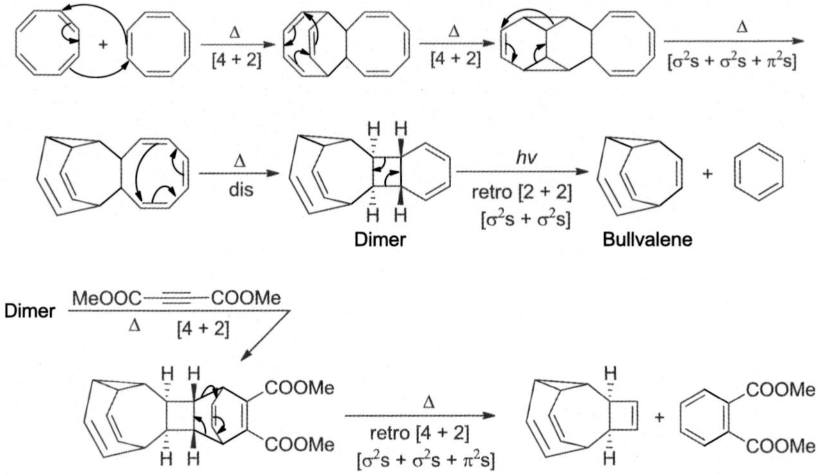

Dimer
Bullvalene

Q 2. Write the structures of **I** and **II** in the following reaction sequence:

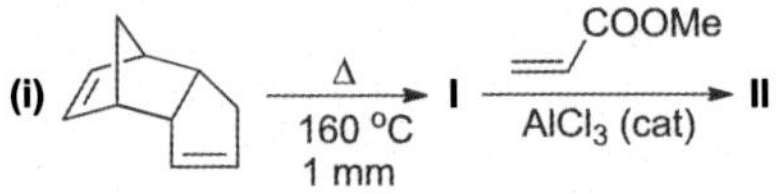

(i) Δ, 160 °C, 1 mm → I; COOMe, $AlCl_3$ (cat) → II

(ii) Furan + Dimethyl acetylenedicarboxylate —Δ→ I —Δ→ Acetylene + II

Sol 2.

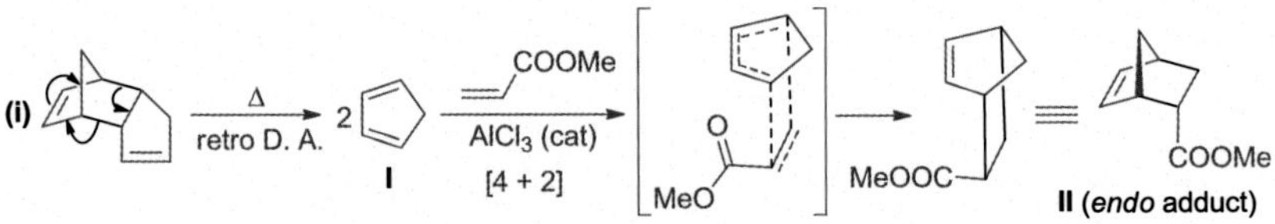

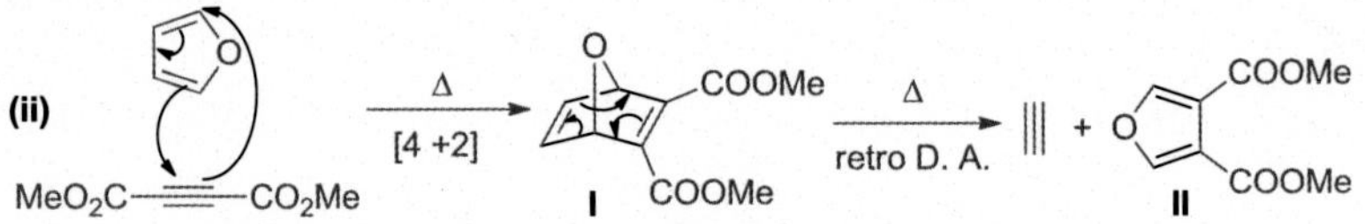

(i) retro D. A. → 2 (I); $AlCl_3$ (cat), [4 + 2] → II (*endo* adduct)

(ii) [4 +2] → I; retro D. A. → ||| + II

4.4.7 Intramolecular Diels–Alder Reactions

The intramolecular Diels–Alder (IMDA) reactions have a number of advantages over intermolecular reactions. They offer one of the most important synthetic routes for the stereoselective construction of polycyclic compounds. The presence of the diene and the dienophile together with a tether reduces the degree of freedom in the starting material. The length of the chain connecting the diene and the dienophile has a pronounced effect on the rate of the reaction by considering the entropic advantages. Furthermore, the nature of the connecting tether is also important in influencing the rate of reaction.

There are two types of IMDA reactions: Type-I and Type-II. *Type-I reactions are those in which the diene is attached to the dienophile by a tether from its terminus while in Type-II reactions, the tether is attached to an internal diene position* (Figure 4.18). In general, both Type-I and Type-II IMDA reactions only occur if the tether contains three or more atoms. This is due to the high level of strain involved in the transition states of reactions of precursors with one or two atoms in the connecting chains.

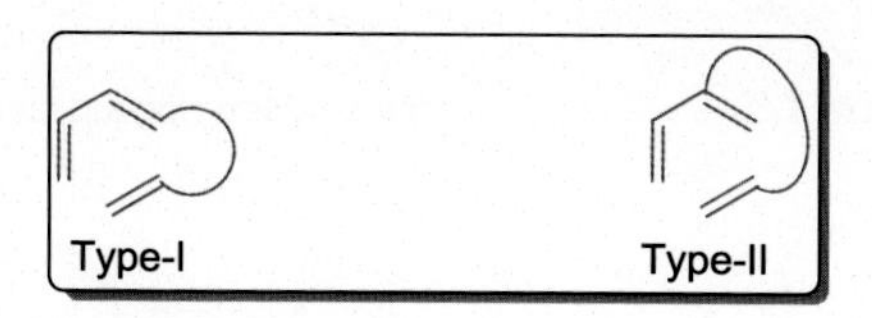

FIGURE 4.18 Types of IMDA substrates.

In IMDA reactions, regioselectivity is higher than in the intermolecular reactions because of the constraints imposed by the tether. Although the possibility of two regioisomers does exist for Type-I IMDA reactions, the majority of cases give the fused isomer exclusively, irrespective of substituent effects, due to the higher degree of strain in the transition state leading to the bridged product. Bridged products become more likely when the length of the tether is 10 or more atoms (Figure 4.19).

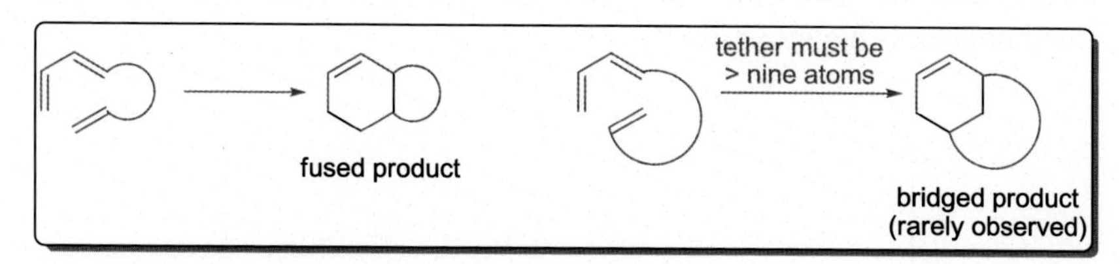

FIGURE 4.19 Regioselectivity in Type-I IMDA reactions.

The nature of the transition state determines whether the newly formed ring junction is *cis*-fused or *trans*-fused. The *endo* rule applicable to intermolecular Diels–Alder reaction cannot be relied upon to predict the stereochemical outcome of the IMDA reaction. The IMDA reaction of *Z*- and *E*-dienes can give either *cis*- or *trans*-fused products depending on orientation of the diene relative to the dienophile in the transition state (i.e., whether the addition is via an *exo* or *endo* mode). In this case, the terms *exo* and *endo* refer to the spatial position of the tether group. When the bulk of the tether group is oriented toward the developing cyclohexene ring, the situation is referred to as an *endo* transition state. When the bulk of the tether group is facing away from the forming cyclohexene ring, this is an *exo* transition state. The terms *Z*- and *E*-diene are used to denote the stereochemistry of the double bond immediately attached to the tether. An *E*-diene will give a *trans*-fused bicyclic adduct from an *exo*-transition state and a *cis*-fused bicyclic adduct from an *endo*-transition state (Figure 4.20).

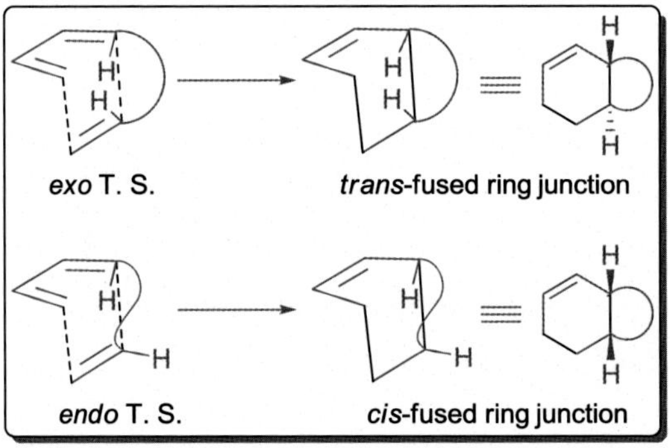

FIGURE 4.20 *Exo*- and *endo*-transition states of an *E*-diene leading to the formation of *trans*- and *cis*-fused bicyclic adducts.

On the other hand, a *cis*-fused bicyclic adduct is obtained from the *exo*-transition state of a Z-diene, and a *trans*-fused bicyclic adduct is the product of the *endo*-transition state (Figure 4.21). Z-dienes with three or four atom tethers produce only *cis*-fused adducts as the *endo*-transition state leading to the *trans*-fused product is highly strained.

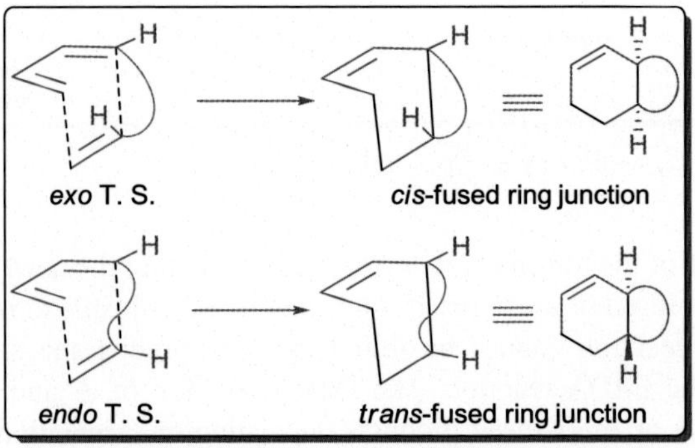

FIGURE 4.21 *Exo*- and *endo*-transition states of Z-diene leading to the formation of *cis*- and *trans*-fused bicyclic adducts.

Type-II IMDA reactions can only form bridged products. Again, the possibility of two regioisomers exists, but tethers containing fewer than five atoms produce *meta*-bridged cycloadducts exclusively (Figure 4.22).

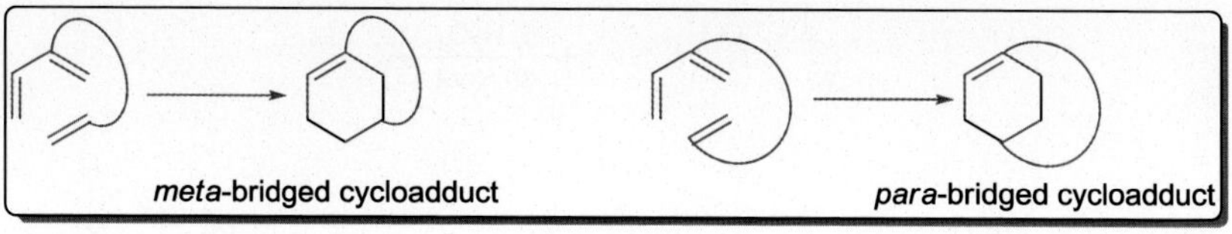

meta-bridged cycloadduct

para-bridged cycloadduct

FIGURE 4.22 Regioselectivity in Type-II IMDA reactions.

Some examples of Type-I and Type-II IMDA reactions are given in Scheme 4.21.

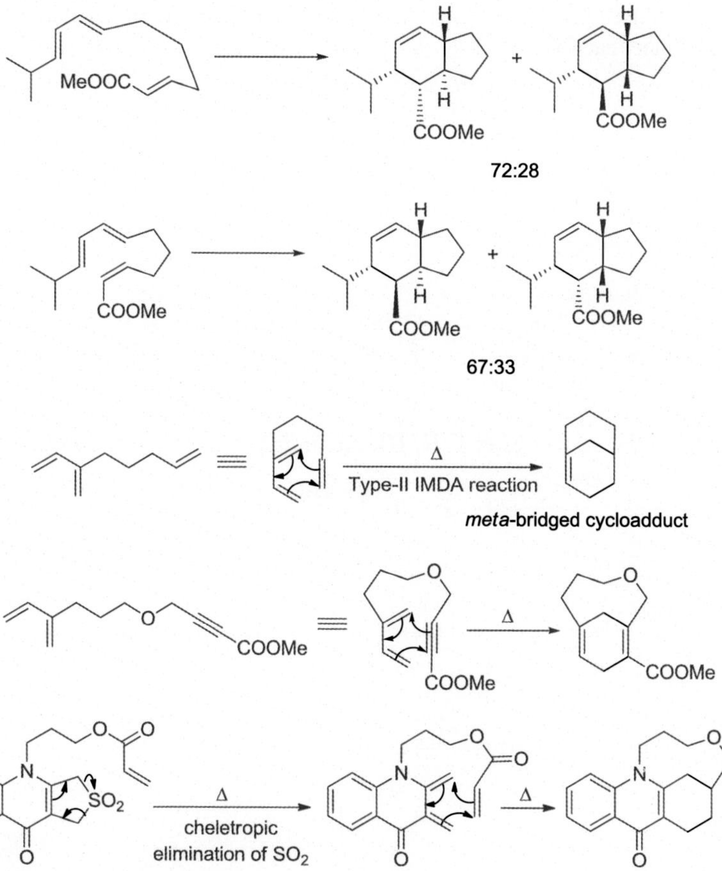

MeOOC

H H H H COOMe COOMe

72:28

COOMe

H H H H COOMe COOMe

67:33

Δ

Type-II IMDA reaction

meta-bridged cycloadduct

O COOMe O COOMe Δ O COOMe

N O O SO$_2$ O

Δ

cheletropic elimination of SO$_2$

Δ

SCHEME 4.21 Examples of Type-I and Type-II IMDA reactions.

4.4.7.1 Solved Problems

Q 1. The major product formed in the following reaction sequence is:

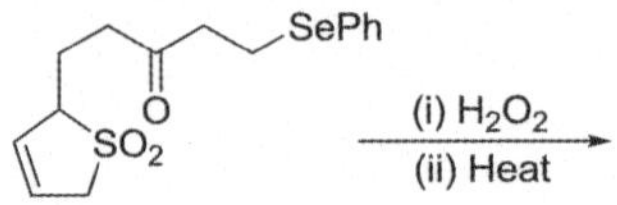
SePh
O
SO2
(i) H2O2
(ii) Heat

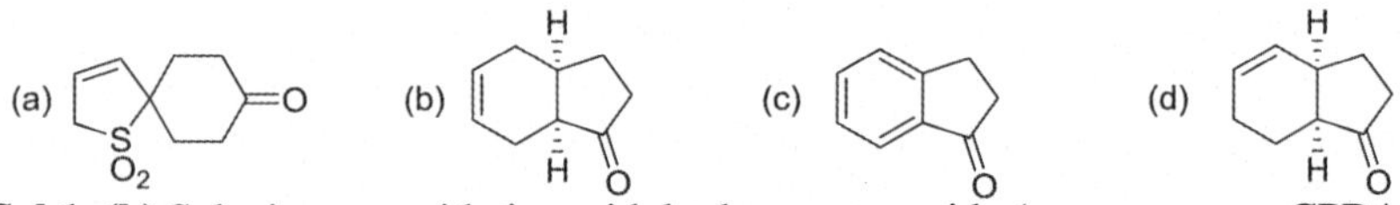
(a)
(b)
(c)
(d)

Sol 1. (b) Selenites on oxidation with hydrogen peroxide (or ozone or *m*-CPBA) give selenoxides. The latter in the presence of a β-hydrogen undergoes an intramolecular *syn* elimination to leave behind an alkene and selenenic acid. The diene component is generated in situ by thermal sulfur dioxide extrusion from sulfolene in a reverse reaction and is then trapped by the dienophile in a Diels–Alder reaction.

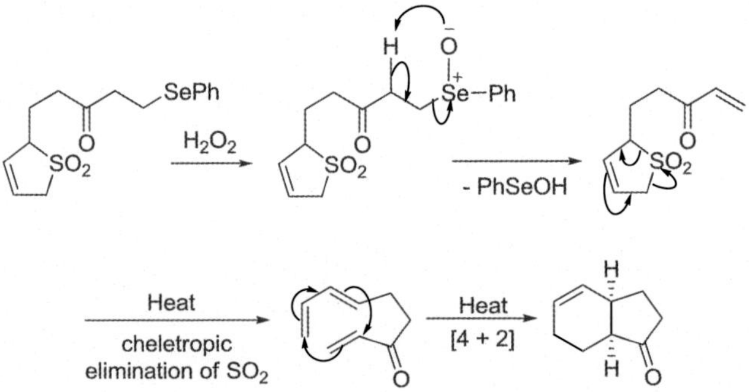
SePh
H2O2
Se–Ph
- PhSeOH
SO2
Heat
cheletropic
elimination of SO2
Heat
[4 + 2]

Q 2. Write the structures of **I**, **II**, **III**, etc. in the following reaction sequence:

(i)

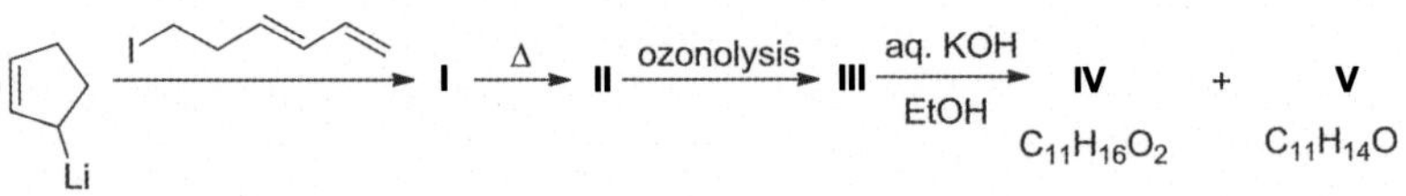
Li
I
Δ
II
ozonolysis
III
aq. KOH
EtOH
IV
+
V
C11H16O2
C11H14O

(ii)

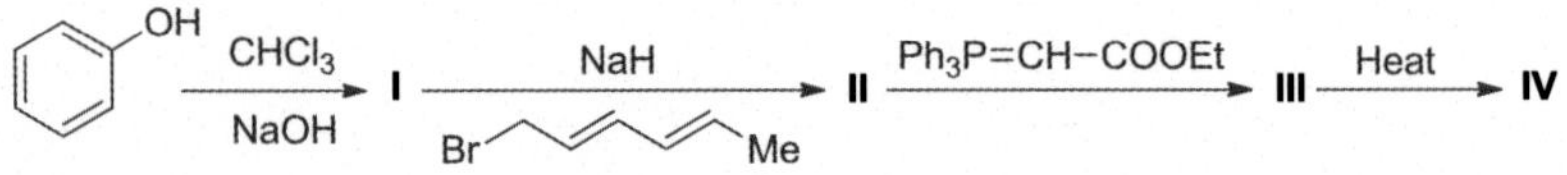
OH
CHCl3
NaOH
I
NaH
Br
Me
II
Ph3P=CH–COOEt
III
Heat
IV

(iii)

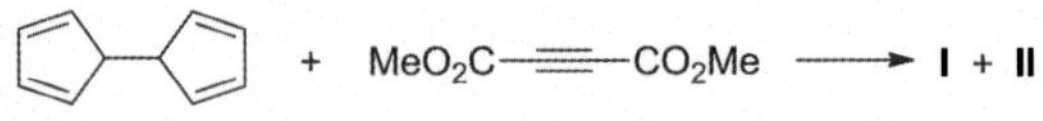
+
MeO2C–≡–CO2Me
I + II

Sol 2.

(i)

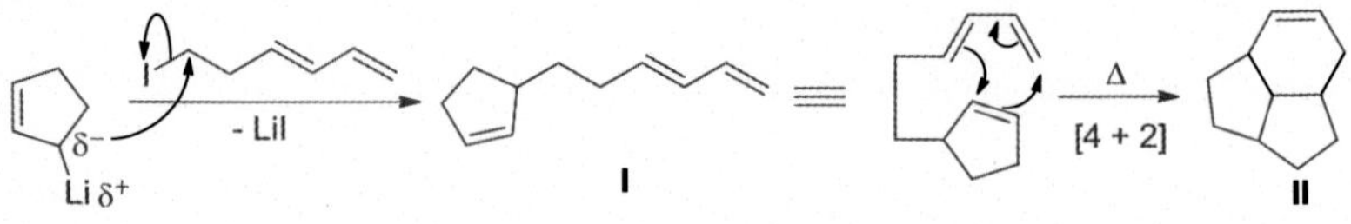
δ−
Li δ+
- LiI
I
Δ
[4 + 2]
II

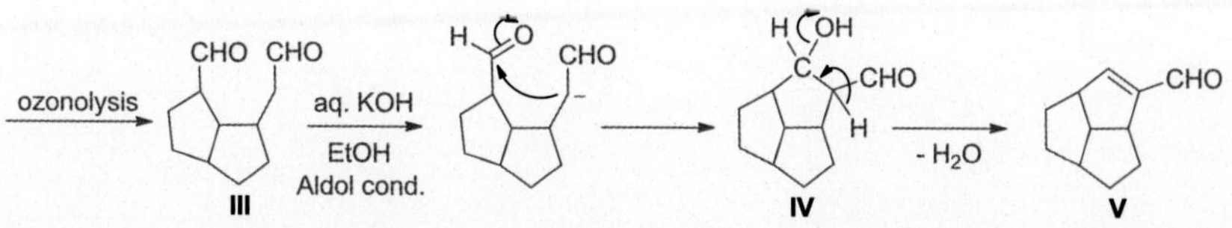

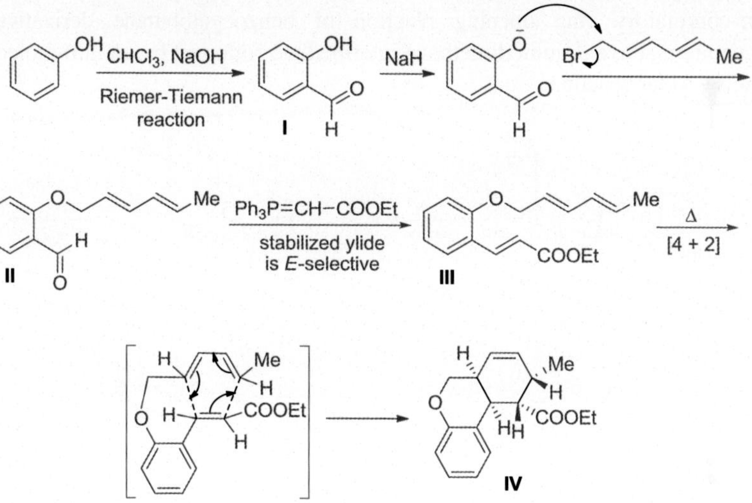

(iii) Intermolecular Diels–Alder reaction of the bicyclic *bis*-diene with dimethyl acetylenedicarboxylate leads to the formation of a bimolecular adduct, which can then undergo intramolecular cycloaddition to give the bridged tetracyclic products in appreciable yield. This reaction is termed as a "domino" Diels–Alder reaction.

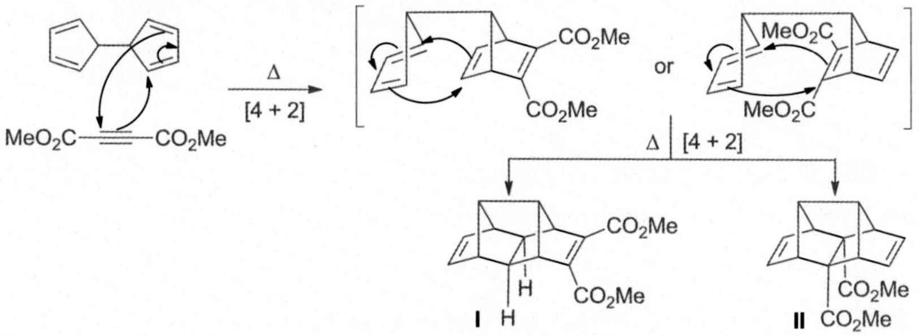

Q 3. Write the mechanism of the following reactions:

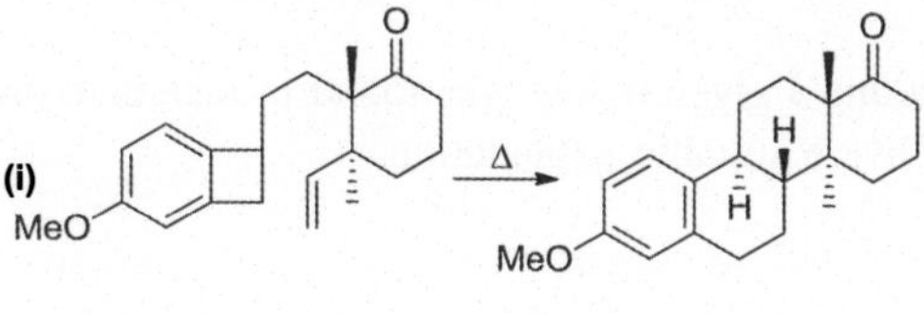

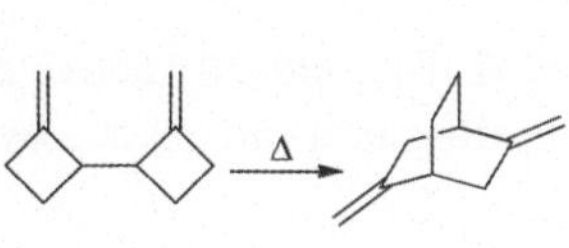

(iii) 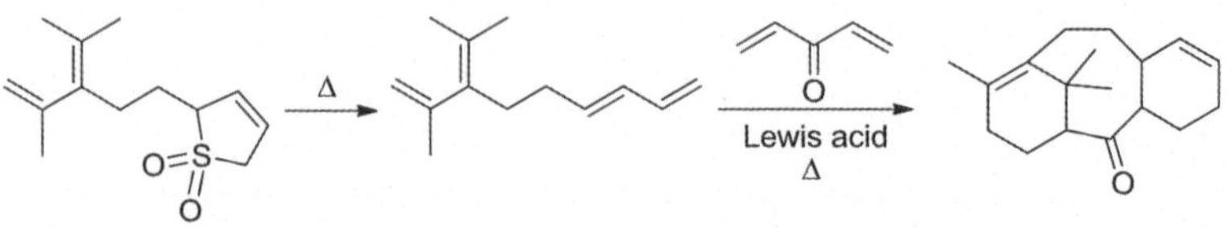 Lewis acid Δ

Sol 3. (i) A stereoselective total synthesis of estrone proceeds via thermal 4π-conrotatory ring opening reaction of benzocyclobutane derivative to generate reactive *o*-quinodimethane intermediate followed by an intramolecular Diels–Alder reaction.

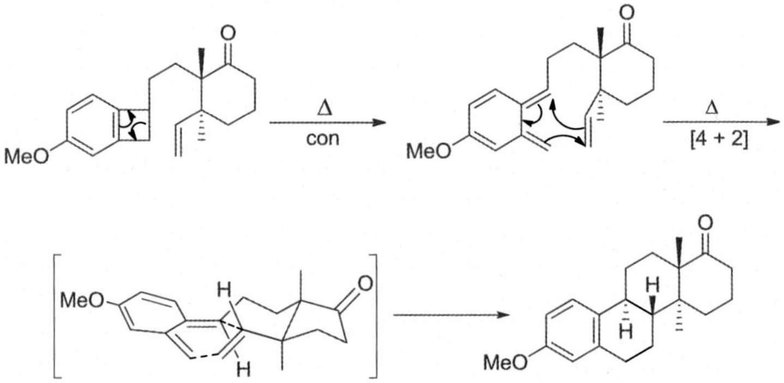

MeO Δ con MeO Δ [4 + 2]

MeO H H MeO H H

(ii) 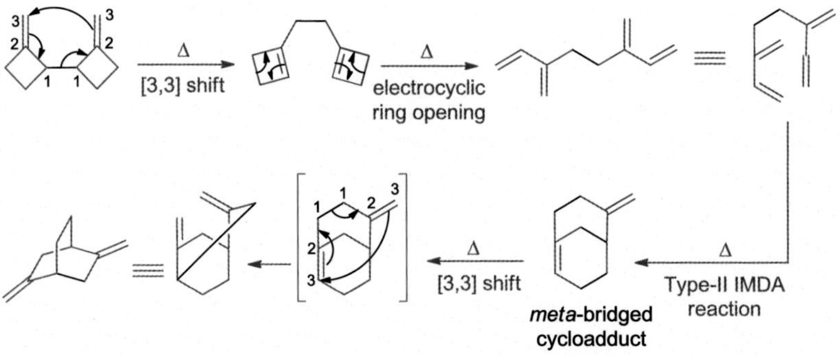 Δ [3,3] shift Δ electrocyclic ring opening ≡

Δ Type-II IMDA reaction *meta*-bridged cycloadduct Δ [3,3] shift

(iii) The tandem Diels–Alder reaction leads to a taxane nucleus found in the natural compound paclitaxel (taxol), an antitumor agent.

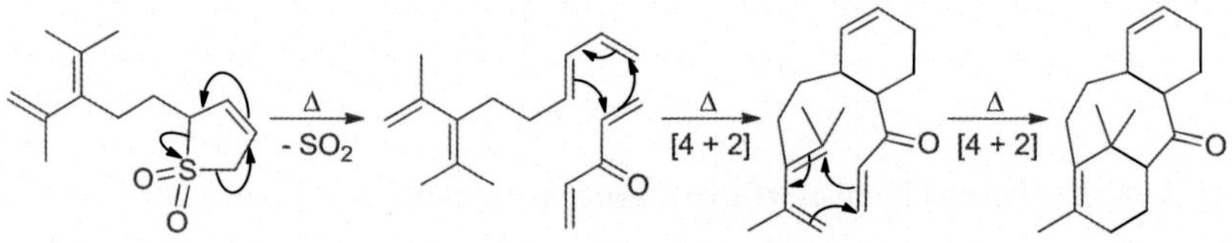

Δ $-SO_2$ Δ [4 + 2] Δ [4 + 2]

Q 4. The optically active compound **I** given below was found to racemize on heating in a microwave oven. Give a suitable explanation.

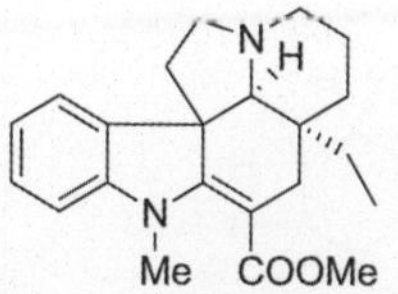

Sol 4. Type-I IMDA reactions also exhibit facial selectivity. There are two *exo* and two *endo* modes of addition, as the dienophile can approach the diene from above or below. The two different approaches of the dienophile to diene are illustrated below by the *exo*-transition state of a *Z*-diene. As discussed above, the *exo*-transition state of a *Z*-diene produces a *cis*-fused ring junction.

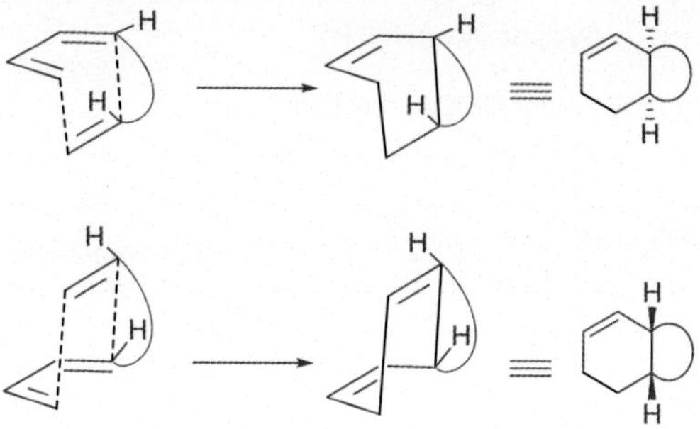

An optically active compound such as **I** on heating undergoes retro Diels–Alder reaction to give the compound **II**. In the compound **II**, the dienophile can attack the diene on both the faces to give either **I** or **III**, which are a pair of enantiomers. Thus, a racemic mixture is obtained.

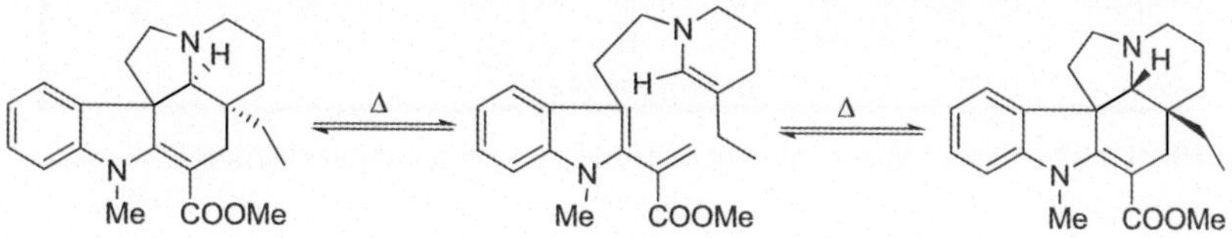

4.4.8 Dehydro-Diels–Alder Reactions

In a typical Diels–Alder reaction involving buta-1,3-diene and ethene, cyclohexene having 10 hydrogens is obtained. However, products with a fewer number of hydrogens—8, 6, and 4—can also be obtained depending upon the nature of the diene and dienophile. These situations can be represented as a, b, c, and d in Scheme 4.22.

HDDA can also be an intramolecular process. As shown in Scheme 4.23, the reactant has a conjugated diyne component and tethered alkynyl diynophile.

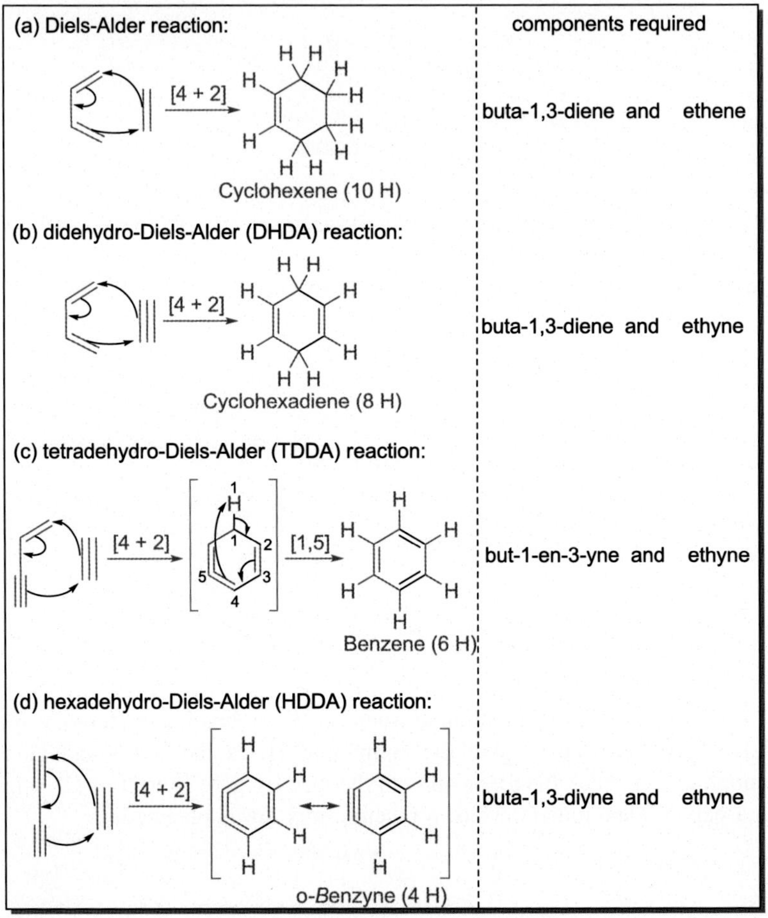

(a) Diels-Alder reaction:	components required
[4 + 2] → Cyclohexene (10 H)	buta-1,3-diene and ethene
(b) didehydro-Diels-Alder (DHDA) reaction:	
[4 + 2] → Cyclohexadiene (8 H)	buta-1,3-diene and ethyne
(c) tetradehydro-Diels-Alder (TDDA) reaction:	
[4 + 2] → [1,5] → Benzene (6 H)	but-1-en-3-yne and ethyne
(d) hexadehydro-Diels-Alder (HDDA) reaction:	
[4 + 2] → o-*Benzyne* (4 H)	buta-1,3-diyne and ethyne

SCHEME 4.22 General representation of Diels-Alder (a) and dehydro-Diels-Alder (b-d) reactions.

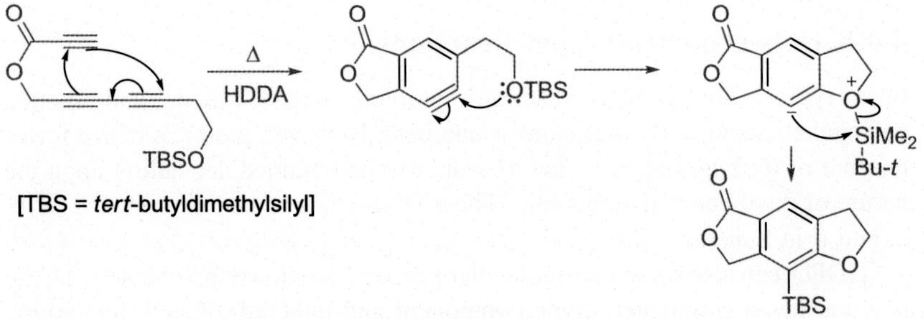

TBSO
Δ
HDDA
:OTBS
$SiMe_2$
Bu-*t*
TBS

[TBS = *tert*-butyldimethylsilyl]

SCHEME 4.23 An intramolecular HDDA reaction.

Therefore, it readily undergoes intramolecular hexadehydro-Diels−Alder (HDDA) reaction to give the benzyne intermediate, which is rapidly trapped by the nucleophilic oxygen of the β-silyloxyethyl group. Further migration of the silyl group from oxygen to carbon within zwitterion results in the formation of the final product.

4.4.8.1 Solved Problems

Q 1. Give the product(s) and mechanism of the following reactions:

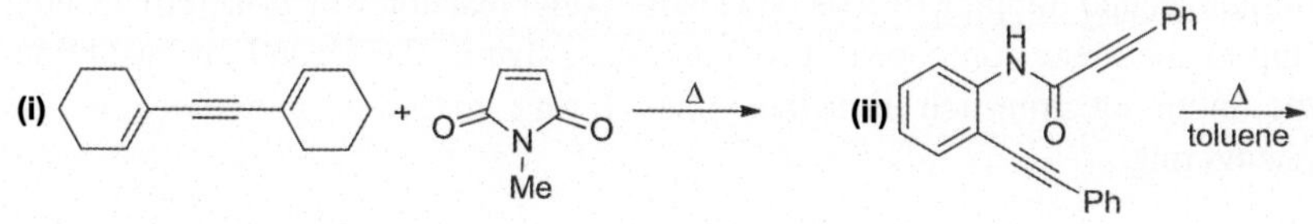

(i) + Δ (ii) Δ toluene

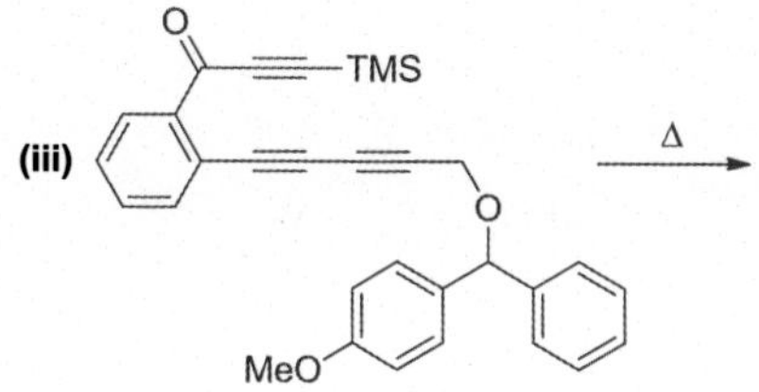

(iii) Δ

Sol 1. (i) Didehydro-Diels−Alder (DHDA) reaction of 1,2-di(cyclohex-1-en-1-yl) ethyne and *N*-methylmaleimide gives a cyclic allene intermediate that further undergoes second Diels−Alder reaction with another molecule of *N*-methylmaleimide to give the observed product.

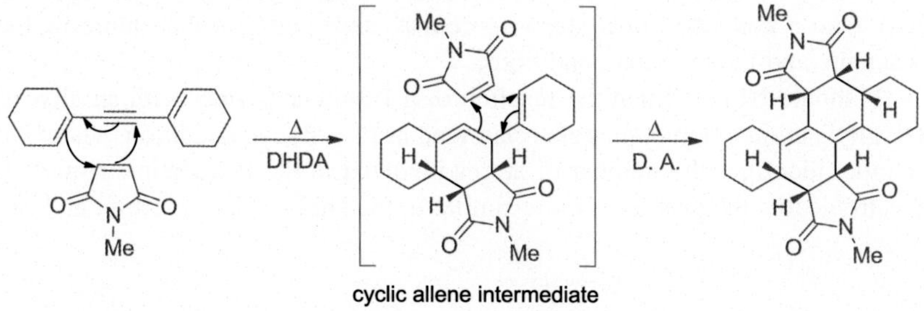

cyclic allene intermediate

(ii) The product is a mixture of two isomeric compounds obtained by the intramolecular tetradehydro-Diels−Alder (TDDA) reaction in which two phenyl acetylene components of the reactant behave as the diene or the dienophile in alternate manner.

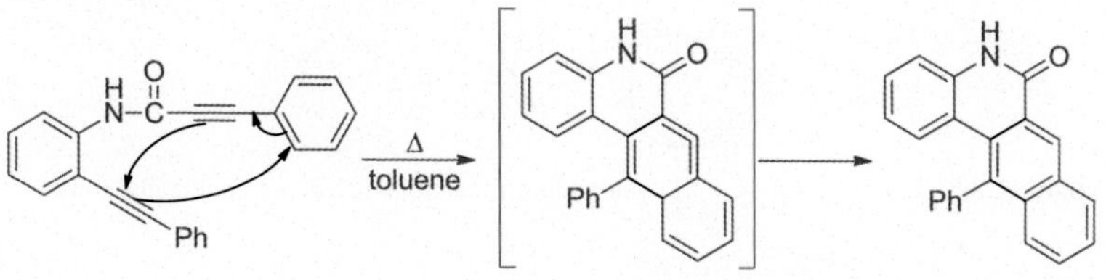

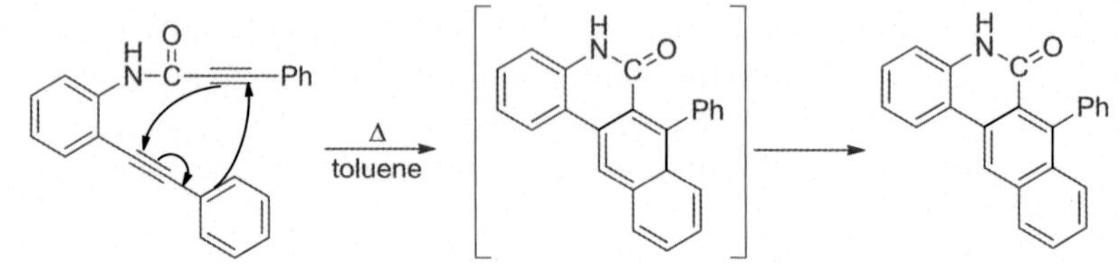

(iii) In this case, the HDDA generated benzyne intermediate undergoes intramolecular trapping by a second Diels–Alder reaction with *p*-methoxyphenyl ring as the diene component. It is interesting that in the second cycloaddition, the more electron-rich *p*-methoxyphenyl ring reacts in preference to the phenyl ring.

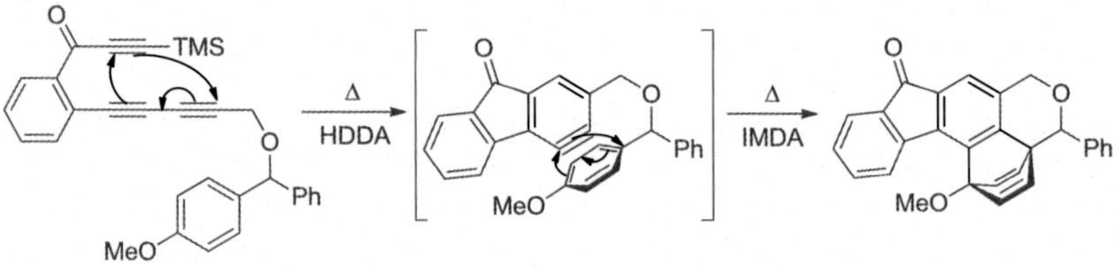

4.4.9 [4 + 2] Cycloaddition Reactions with Allyl Cations and Allyl Anions

Conjugated ions like allyl cations, allyl anions, and pentadienyl cation are also capable of undergoing [4 + 2] cycloadditions (Scheme 4.24). These reactions may take a concerted or a stepwise course, depending on the nature of the reactants and the reaction conditions.

It should be noted that the reaction of a 1,3-diene partner with an allyl or oxyallyl cation is sometimes also classified as [4 + 3] cycloaddition, where the number identifies the number of atoms involved in the two chains. However, electronically, the process is quite similar to the Diels–Alder reaction and can

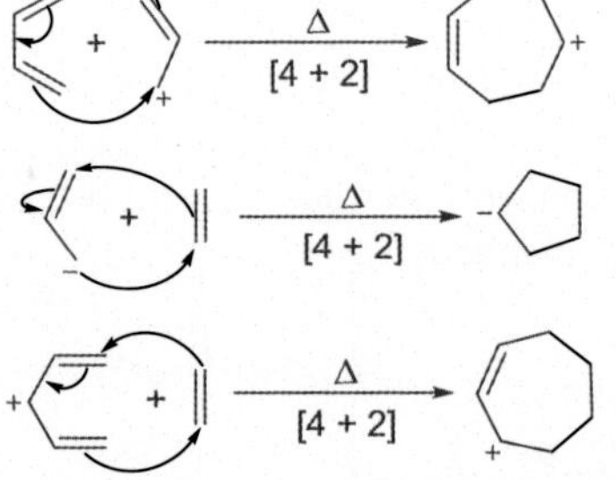

SCHEME 4.24 Examples of [4 + 2] cycloaddition reactions involving conjugated ions.

be viewed as a [4π (4C) + 2π (3C)] cycloaddition in which the allyl or oxyallyl cation participates as the reactive 2π-component.

4.4.9.1 *Solved Problems*

Q 1. Give the mechanism of the following reactions:

(i) + Me I $\xrightarrow[\Delta]{CF_3COOAg}$ Me

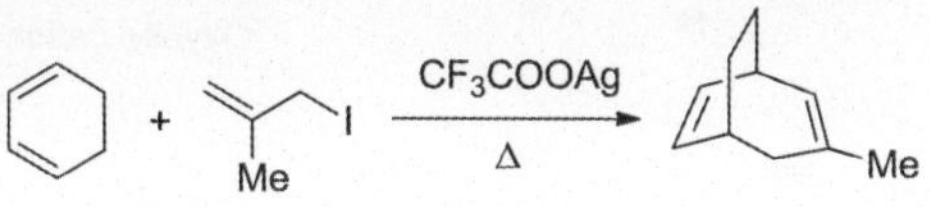

(ii) Me Br O Br Me + O $\xrightarrow{Zn/Cu}$ Me O =O Me + Me O =O Me
48% 5%

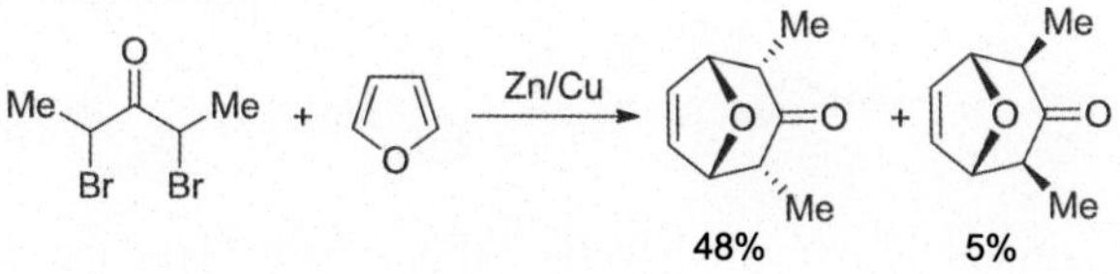

(iii) Ph + Ph Ph $\xrightarrow[(ii)\ H^+]{(i)\ LDA,\ THF}$ Ph Ph Ph

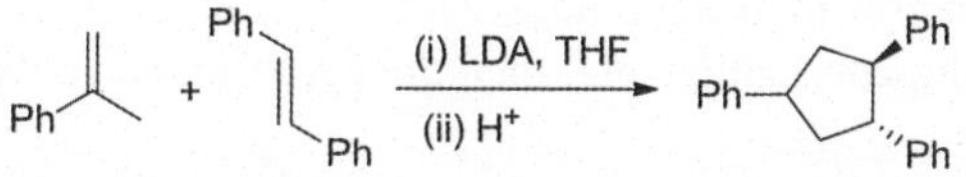

(iv) MeO OMe OMe O + $\xrightarrow{SnCl_4}$ O OMe OMe +

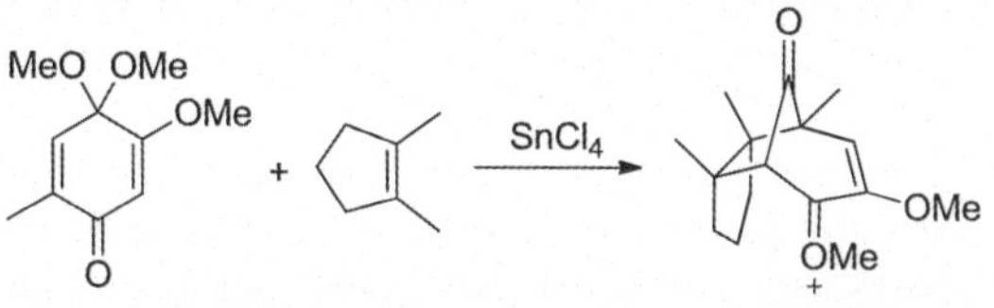

(v) Me_3Si O O Me Me CF_3 $\xrightarrow{ZnCl_2}$ Me Me

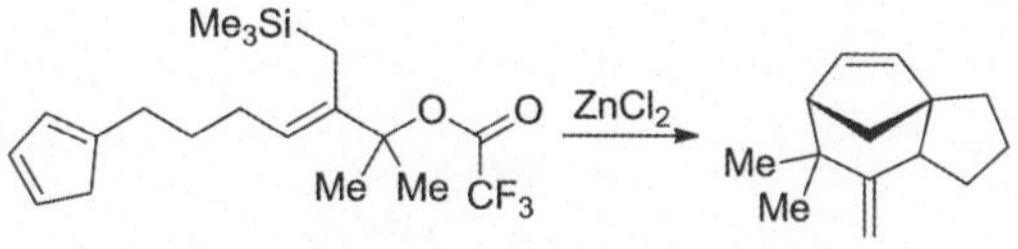

Sol 1. **(i)** Reaction of allyl halide and silver trifluoroacetate generates the methylallyl cation, which then reacts with cyclohexadiene to give the seven-membered ring cation. Like other tertiary cations, it loses a neighboring proton to give the alkene, i.e., 3-methylbicyclo[3.2.2]nona-2,6-diene.

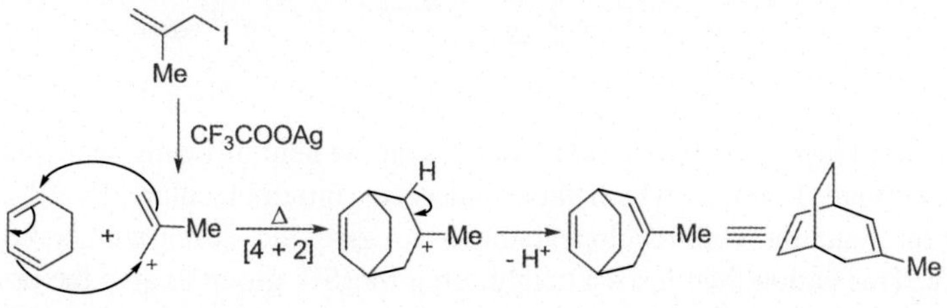

Me I $\xrightarrow{CF_3COOAg}$ + Me + $\xrightarrow[[4+2]]{\Delta}$ H + Me $\xrightarrow{-H^+}$ Me ≡ Me

(ii) Allyl cations can undergo cycloaddition with the dienes. However, oxyallyl cations are the most common reactive species because of their greater stability as compared to simple allyl cations. A common method for producing oxyallyl cation intermediates involves two-electron reduction of appropriate α,α′-dihaloketones with reducing agents such as zinc(0) and iron(0). Halogenated metal enolates are formed initially and the oxyallyl cations are then generated by a subsequent El elimination of the halide.

O Me Me Br Br M - Br⁻ O M Me Me Br - Br⁻ O M Me Me O M Me Me

Oxyallyl cation

Me O Me Δ [4 + 2] Me O Me + Me O Me

48% 5%

(iii) Treatment of α-methylstyrene with a base gives the corresponding allyl anion. This anion then undergoes cycloaddition with stilbene, present in situ, to give cyclopentyl anion, and then the cyclopentane after protonation.

Ph LDA THF Ph Ph + Ph Δ [4 + 2] Ph Ph Ph H^+ Ph Ph Ph

(iv) Ionization of the acetal in the presence of tin (IV) chloride gives a highly stabilized pentadienyl cation system, which then undergoes intramolecular $[4+2]$ cycloaddition with 1,2-dimethyl cyclopentene to give the adduct.

O OMe OMe OMe $SnCl_4$ O OMe OMe [4 + 2] O OMe OMe

O OMe OMe O OMe OMe

(v) Cleavage of the carboxylate bond with the help of Lewis acid generates the methylallyl cation, which then undergoes intramolecular $[4+2]$ cycloaddition reaction with cyclopentadiene to give the seven-membered ring cation. This cation then loses a neighboring Me_3Si^+ group to give the product.

Me_3Si O O Me Me CF_3 $ZnCl_2$ Me_3Si $ZnCl_2$ O O+ Me Me CF_3

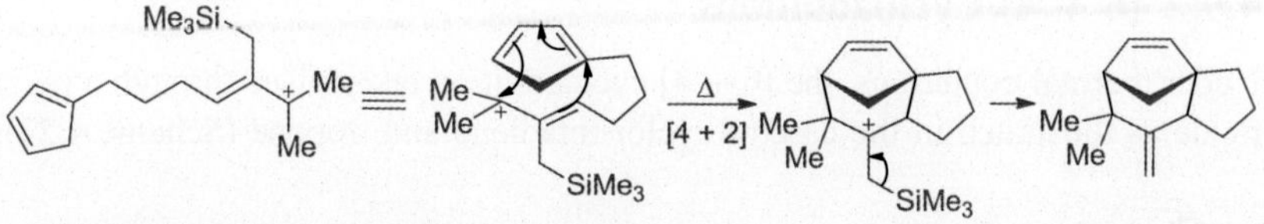

4.5 HIGHER CYCLOADDITIONS

The cycloaddition reactions that involve more than six π-electrons are generally referred to as higher cycloadditions. These reactions provide a convenient synthesis of medium-sized rings, which are otherwise difficult to obtain by other methods.

4.5.1 [4 + 4] Cycloadditions

According to the Woodward–Hoffman rules, a [π^4s + π^4s] cycloaddition reaction is symmetry allowed under photochemical conditions. For example, due to the proximity of the two *s-cis*-1,3-dienes in the same molecule, *cis*-9,10-dihydronaphthalene on photolysis undergoes an intramolecular [4 + 4] cycloaddition (Scheme 4.25).

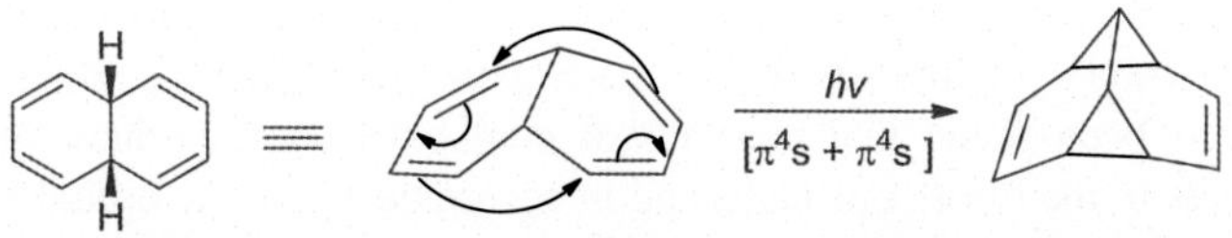

SCHEME 4.25 Intramolecular [4 + 4] cycloaddition of *cis*-9,10-dihydronaphthalene.

Anthracene undergoes photochemical [π^4s + π^4s] cycloaddition to yield the corresponding dimer. Similarly, photo-[4 + 4]-dimerization of 4,6-dimethyl-2H-pyran-2-one gives a mixture of *trans* and *cis* adducts. This mixture on heating undergoes extrusion of carbon dioxide to give (1Z,3Z,5Z,7Z)-1,3,5,7-tetramethylcycloocta-1,3,5,7-tetraene (Scheme 4.26).

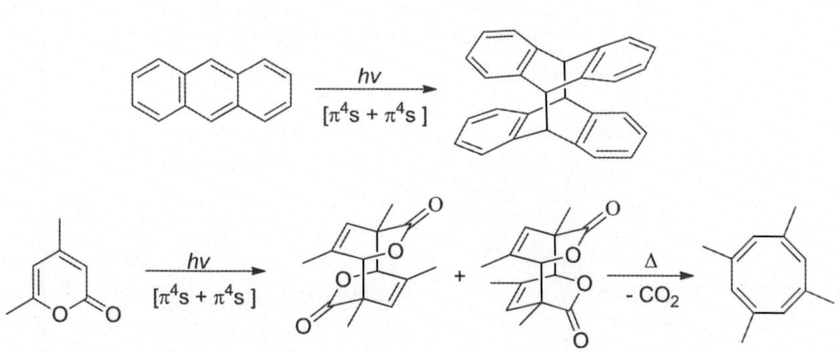

SCHEME 4.26 Photodimerization of anthracene and 4,6-dimethyl-2*H*-pyran-2-one.

4.5.2 [6 + 4] Cycloadditions

Under thermal conditions, the [6 + 4] cycloaddition takes place through a (s, s) mode as illustrated in the case of cyclopentadiene and tropone (Scheme 4.27).

SCHEME 4.27 $[\pi^6s + \pi^4s]$ Cycloaddition of cyclopentadiene and tropone.

In this reaction, formation of *exo* adduct predominates. This is in contrast to the Diels–Alder reaction, which gives *endo* adduct as the major product. Moreover, the [6 + 4] cycloaddition takes place in preference to a Diels–Alder [4 + 2] cycloaddition (both are thermally allowed). Such a situation is known as *periselectivity* and is explained by the fact that the coefficients of the frontier molecular orbitals of the LUMO of the tropone are highest at atoms C-2 and C-7. It has been found that the ends of conjugated systems have the largest coefficients in the frontier orbitals, and in accordance with the orbital symmetry rules, pericyclic reactions make use of the longest part of such systems. However, such reactions have to be permissible by the geometry of the molecule.

The frontier molecular orbitals involved in reaction of tropone illustrate that a repulsive secondary orbital interaction destabilizes the *endo* transition state leading to the complete selectivity for *exo* products. The frontier orbitals have a repulsive interaction (wavy lines) between C-3 and C-4 on the tropone and C-2 on the diene (and between C-5 and C-6 on the tropone and C-3 on the diene) in the *endo* transition state (Figure 4.23).

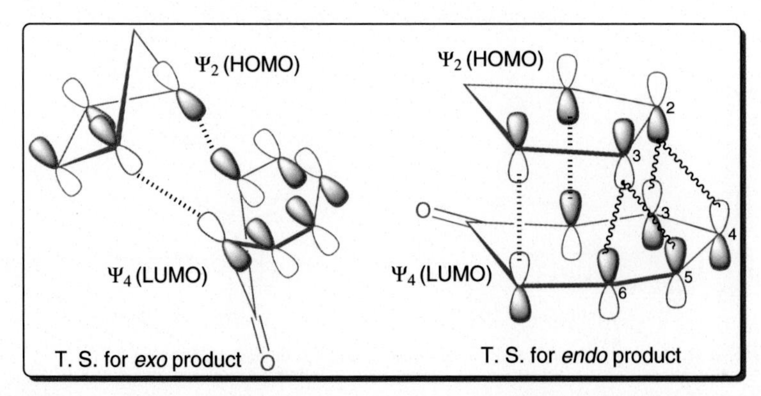

FIGURE 4.23 HOMO–LUMO interactions for *endo* and *exo* approaches of cyclopentadiene and tropone.

Similarly, *N*-ethoxycarbonylazepine dimerizes on heating by a $[\pi^6s + \pi^4s]$ cycloaddition (Scheme 4.28).

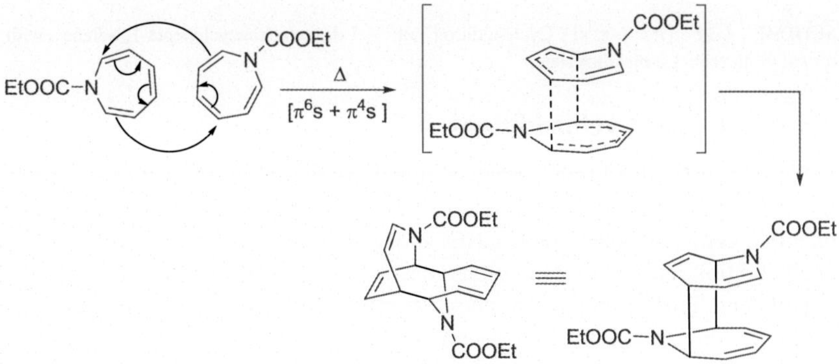

SCHEME 4.28 $[\pi^6s + \pi^4s]$ Dimerization of *N*-ethoxycarbonylazepine.

4.5.3 [8 + 2] Cycloadditions

Indolizine undergoes thermally allowed $[\pi^8s + \pi^2s]$ cycloaddition with dimethyl acetylenedicarboxylate to give a cycloadduct, which readily undergoes aromatization in the presence of a dehydrogenating catalyst (Scheme 4.29).

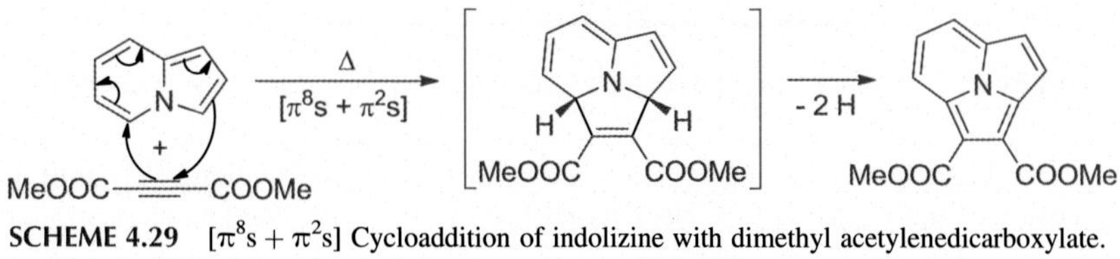

SCHEME 4.29 $[\pi^8s + \pi^2s]$ Cycloaddition of indolizine with dimethyl acetylenedicarboxylate.

In a similar fashion, heptafulvene, i.e., 7-methylenecyclohepta-1,3,5-triene, adds to the acetylenic ester to give a dihydroazulene derivative (Scheme 4.30).

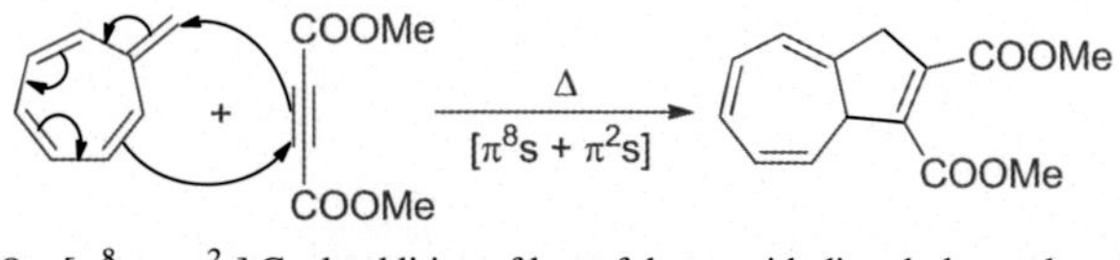

SCHEME 4.30 $[\pi^8s + \pi^2s]$ Cycloaddition of heptafulvene with dimethyl acetylenedicarboxylate.

5,7-Dimethylenecyclohepta-1,3-diene undergoes [8 + 2] cycloaddition with dimethyl diazene-1,2-dicarboxylate to give the bridged bicyclic adduct (Scheme 4.31).

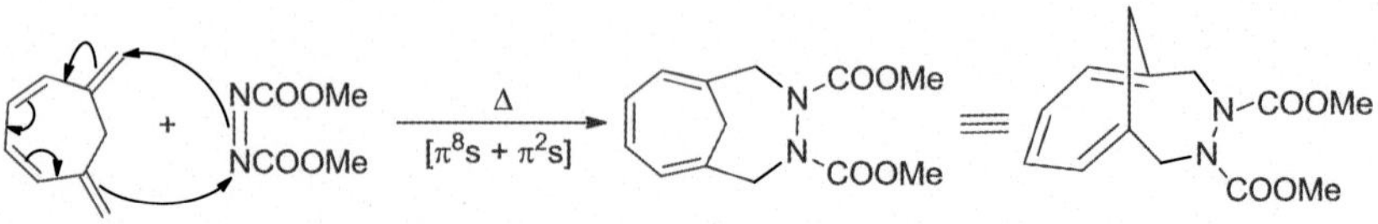

SCHEME 4.31 $[\pi^8s + \pi^2s]$ Cycloaddition of 5,7-dimethylenecyclohepta-1,3-diene with dimethyl diazene-1,2-dicarboxylate.

4.5.4 [14 + 2] Cycloadditions

$[\pi^{14}s + \pi^2s]$ Cycloadditions involve an antiaromatic transition state, so it has to be $[\pi^{14}a + \pi^2s]$ for the reaction to be a thermally allowed process. For example, tetracyanoethylene (TCNE) adds to heptafulvalene in a $[\pi^{14}a + \pi^2s]$ cycloaddition. Heptafulvalene undergoes antarafacial attack, i.e., acts as $\pi^{14}a$ component. This is possible because heptafulvalene is flexible enough not to lose the conjugation through seven double bonds. As heptafulvalene uses opposite lobe for overlapping with TCNE, the two hydrogen atoms in the adduct are *trans* to each other (Scheme 4.32).

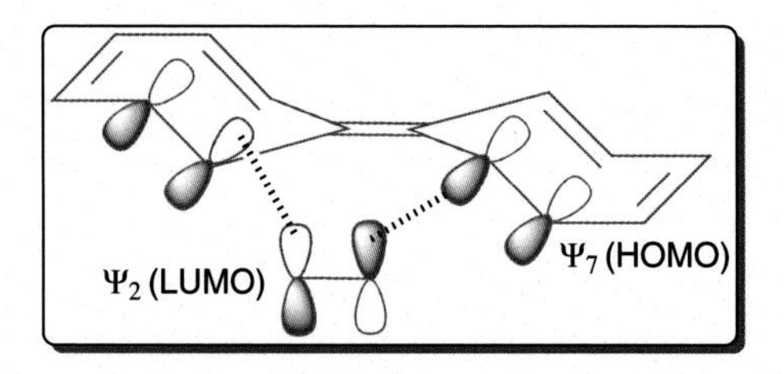

FIGURE 4.24 Transition state for thermal cycloaddition of heptafulvalene to tetracyanoethylene.

Overlapping takes place between the HOMO of heptafulvalene (Ψ_7 with m symmetry) and LUMO of TCNE (Ψ_2 with C_2 symmetry). Models confirm that heptafulvalene is not flat and hence opposite lobes at the two ends are close together and can easily overlap with the π-lobes of TCNE (Figure 4.24).

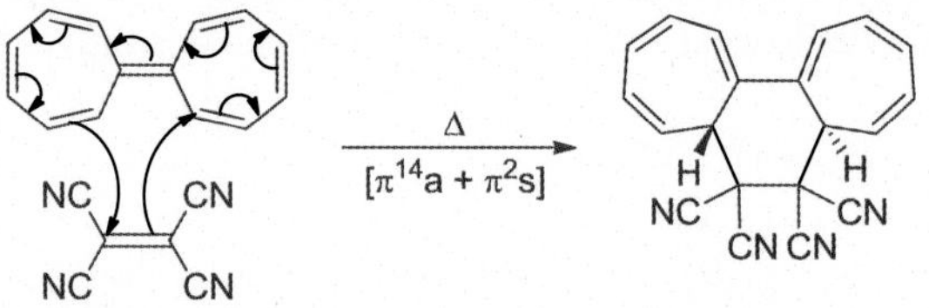

SCHEME 4.32 Thermal cycloaddition of heptafulvalene to tetracyanoethylene.

As heptafulvalene is not flat, it is apparent that the conformational requirements of the $[\pi^{14}a + \pi^2s]$ reaction allow it to compete successfully with the various alternative $[\pi^4s + \pi^2s]$ and $[\pi^8s + \pi^2s]$ cycloaddition modes.

4.5.5 Solved Problems

Q 1. Explain the mechanism of the following transformations:

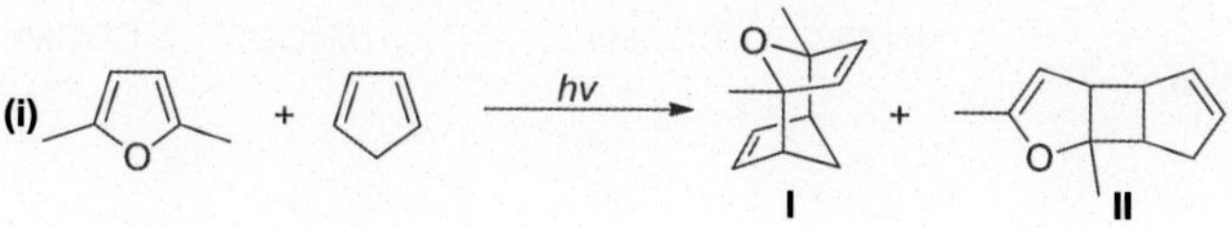

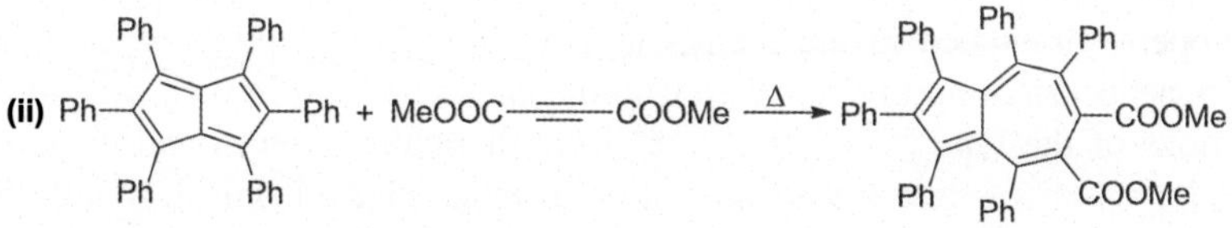

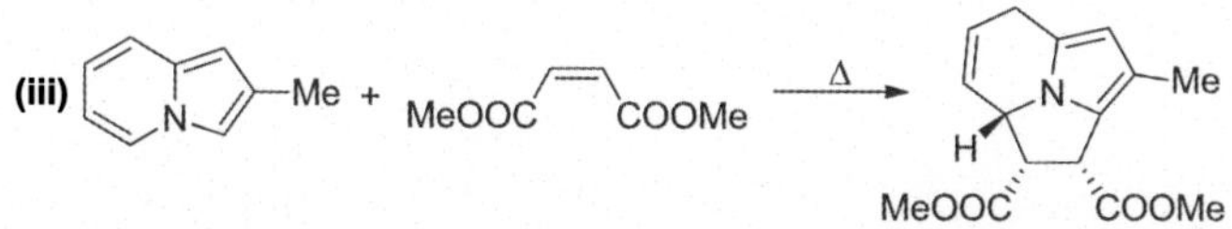

(i) [2,5-dimethylfuran + cyclopentadiene →(hv) I + II]

(ii) [hexaphenylpentalene + MeOOC–C≡C–COOMe →(Δ) azulene derivative]

(iii) [2-methylindolizine + MeOOC–CH=CH–COOMe →(Δ) product]

Sol 1. (i) Irradiation of a mixture of 2,5-dimethylfuran and cyclopentadiene gives a mixture of *trans* and *cis* [4 + 4] adducts. **II** is formed by the thermally allowed [3,3] shift of *cis* [4 + 4] adduct that was not observed.

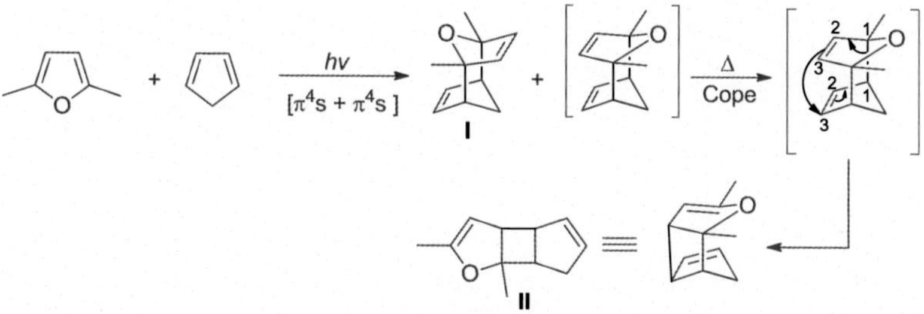

[Scheme: hv, $[\pi^4s + \pi^4s]$ → I + [cis adduct] →(Δ, Cope) → II]

(ii) Hexaphenylpentalene undergoes [8 + 2] cycloaddition with dimethyl acetylenedicarboxylate to give an unstable adduct, which ultimately gives an azulene derivative.

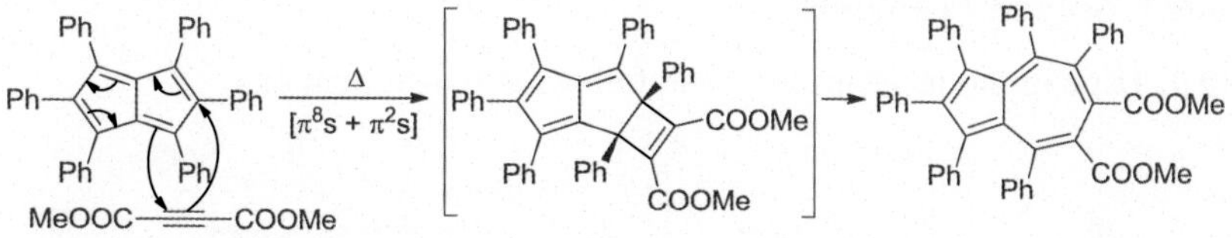

[Scheme: Δ, $[\pi^8s + \pi^2s]$ → unstable adduct → azulene derivative]

(iii) The product of the reaction between 2-methylindolizine and dimethyl maleate may be obtained by an initial [8 + 2] cycloaddition followed by [1,5] hydrogen shift.

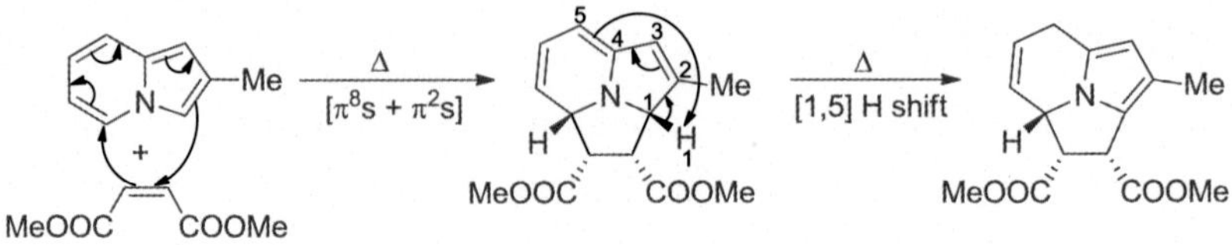

4.6 CYCLOADDITION OF MULTIPLE COMPONENTS

If properly orientated in the molecular structure, cycloaddition of multiple components can occur in a bi- or even unimolecular manner. In such situations, reactions of the type $[\pi^2s + \pi^2s + \pi^2s]$ are thermally allowed and those of type $[\pi^2s + \pi^2s + \pi^2s + \pi^2s]$ are allowed under photochemical conditions (Scheme 4.33).

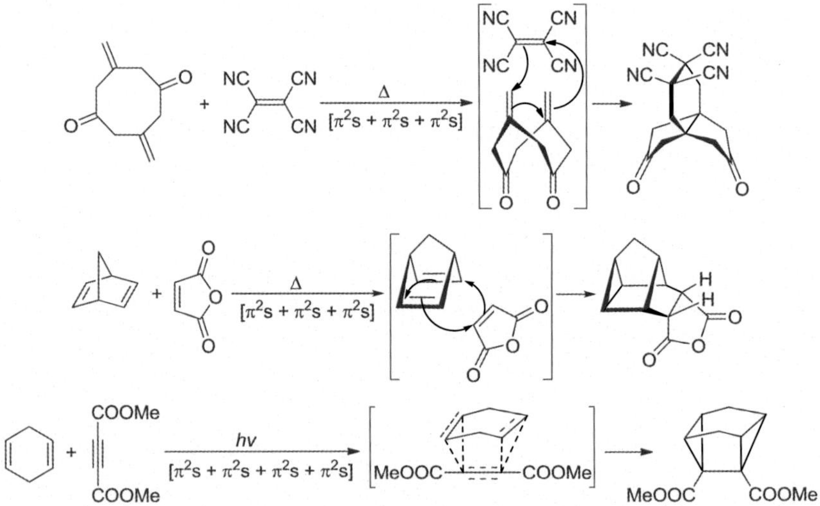

SCHEME 4.33 Examples of reactions involving multiple cycloadditions.

4.6.1 Solved Problems

Q 1. Explain the mechanism of the following transformations:

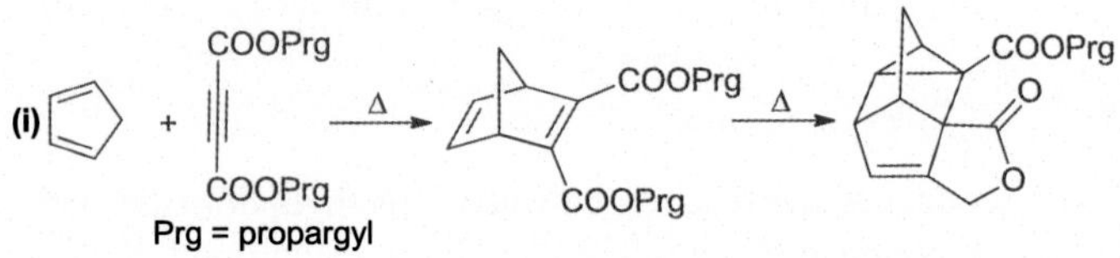

(i) Prg = propargyl

(ii) + R—≡—R (excess) —Δ→ + ; R = COOMe

Sol 1. **(i)** [π^4s + π^2s] Cycloaddition of cyclopentadiene and di(prop-2-yn-1-yl) but-2-ynedioate at room temperature gives norbornadiene dicarboxylate (**I**), which on further heating undergoes intramolecular [π^2s + π^2s + π^2s] cycloaddition in a concerted manner to give **II**.

COOPrg; Δ [π^4s + π^2s]; COOPrg; PrgOOC; COOPrg

Δ [π^2s + π^2s + π^2s]; COOPrg; O

(ii) Δ [π^2s + π^2s + π^2s]; Δ retro D. A.; R R

H; R; group transfer (transfer of two *cis*-hydrogens); H R; H R

Chapter 5

Cheletropic Reactions and 1,3-Dipolar Cycloadditions

Chapter Outline

5.1 CHELETROPIC REACTIONS

Cheletropic reactions *are a special group of concerted cycloadditions or cycloreversions in which two σ-bonds are made or broken* on the same atom of the ("monocentric") species. There is formal loss of one π-bond in the substrate and an increase in coordination number of the relevant atom of the species.

Pericyclic Reactions. http://dx.doi.org/10.1016/B978-0-12-803640-2.00005-1
Copyright © 2016 Elsevier Inc. All rights reserved.

5.1.1 [2 + 2] Cheletropic Reactions

An important example of cheletropic reactions is the reversible insertion of a singlet carbene into a carbon–carbon double bond to give a cyclopropane derivative. Only singlet carbenes will be considered here, as the pericyclic selection rules cannot be applied to triplet states. Singlet carbenes add thermally to alkenes in a concerted manner, therefore, the geometry of the alkene is preserved in the product, i.e., the reaction is *stereospecific*. Hence, in the present context, the reaction is described as suprafacial on the olefin. Therefore, the reaction of *E*-but-2-ene and *Z*-but-2-ene with singlet carbene gives *trans*- and *cis*-1,2-dimethylcyclopropane, respectively (Scheme 5.1).

SCHEME 5.1 Reaction of singlet carbene with diastereomeric but-2-enes.

In singlet carbenes, the central carbon atom is in a state of sp^2 hybridization. Of the three sp^2-hybrid orbitals, two are used in forming two single bonds with the monovalent atoms or groups attached to carbon. The third sp^2 hybrid orbital contains the unshared pair of electrons and acts as the highest occupied molecular orbital (HOMO), while the unhybridized *p*-orbital is empty and acts as the lowest unoccupied molecular orbital (LUMO) (Figure 5.1).

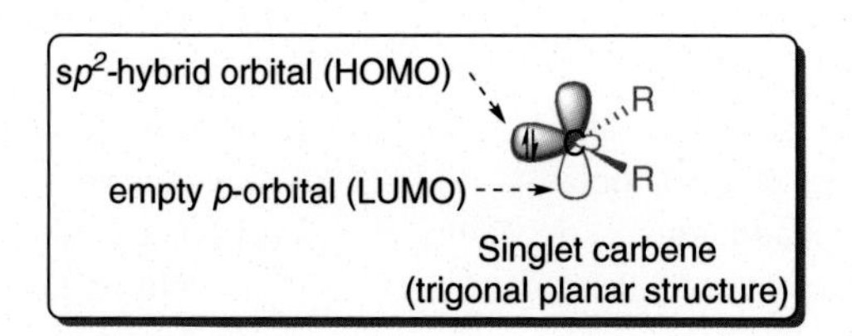

FIGURE 5.1 Structure of singlet carbene.

Applying FMO method, there are two possibilities in which the carbene can approach the olefin: linear and nonlinear. In the linear approach, the plane of bonds of two substituents on carbene is perpendicular to that of the C–C bond of the olefin (Figure 5.2). However, this linear approach is ruled out as it creates an antibonding interaction.

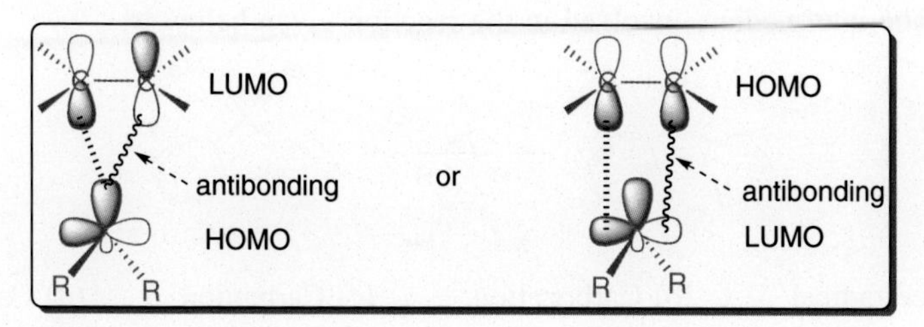

FIGURE 5.2 HOMO–LUMO interactions in the linear approach of singlet carbene suprafacial to the π-system.

However, in the nonlinear approach of carbene, the plane of bonds of two substituents on carbene is parallel to that of the C–C bond of the π-system. As shown in Figure 5.3, in this approach the HOMO–LUMO interactions are oriented suprafacially to the $(4n + 2)$ π-system and are bonding. As the electrons reorganize themselves into new bonds, the plane of bonds of the two substituents on carbene and that of the C–C bond of the olefins become at right angles to each other. However, there is no way to prove that carbene approaches linearly or nonlinearly.

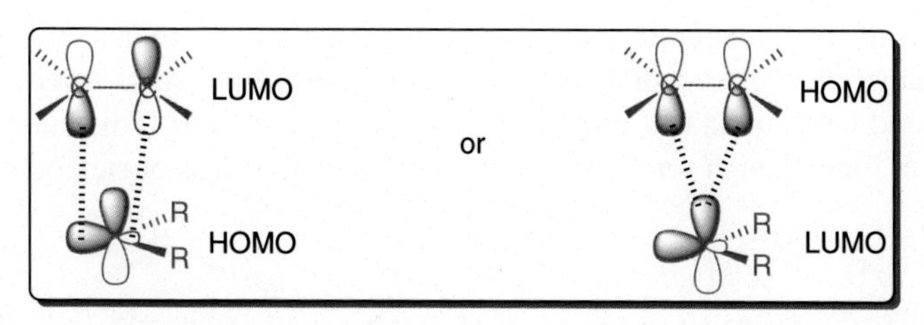

FIGURE 5.3 HOMO–LUMO interactions in the nonlinear approach of singlet carbene suprafacial to the π-system.

5.1.1.1 Solved Problems

Q 1. The major product formed in the following reaction is:

Me, Me, H, H (Z-alkene) + $:CH_2$ (singlet) ⟶

(a) cyclopropane with Me, Me, H, H (b) cyclopropane with Me, H, H, Me (c) 50:50 mixture of (a) and (b) (d) Me, CH_2CH_3, H, H (alkene)

Sol 1. (a) For singlet carbenes, the alkene insertion reaction is stereospecific. Therefore, the reaction of *Z*-but-2-ene with singlet carbene gives *cis*-1,2-dimethylcyclopropane.

Q 2. The intermediate involved in the reaction given below is:

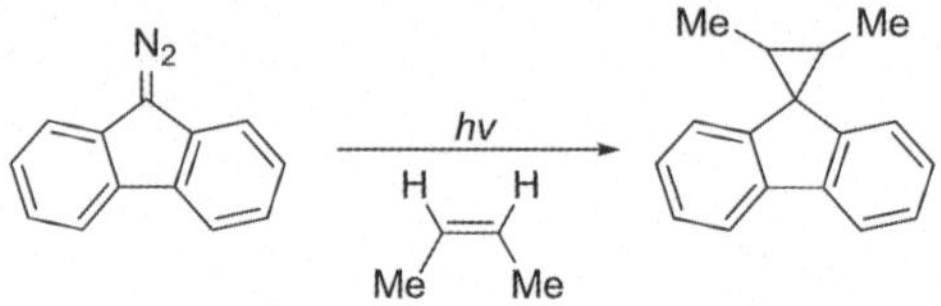

(a) Free radical (b) Carbocation (c) Carbanion (d) Carbene

Sol 2. (d) Though carbenes formed thermally from diazoalkanes must initially be singlet, photochemistry is one way to provide the more stable triplet species. For triplet carbenes, the alkene insertion reaction is nonstereospecific. Therefore, the reaction of *Z*-but-2-ene with triplet carbene gives a mixture of *cis/trans*-1,2-dimethylcyclopropane derivatives.

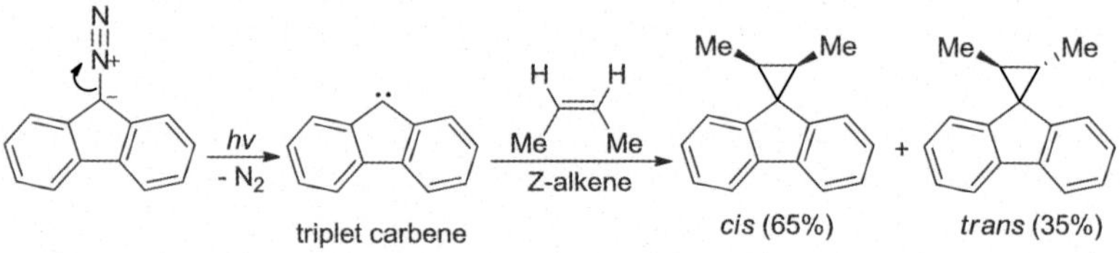

Alkenes react with a triplet carbene (i.e., diradical) in a stepwise fashion. A concerted reaction is not possible for triplet carbenes because of the spins of the electrons involved. After the carbene adds to the alkene in a radical reaction, the diradical (triplet) intermediate must wait until one of the spins inverts so that the second C—C bond can be formed with paired electrons. This intermediate also lives long enough for C—C bond rotation and thus loss of stereochemistry.

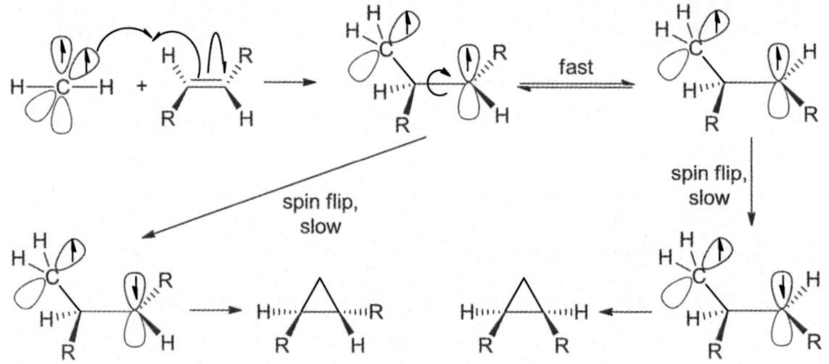

Q 3. In the following reaction **I** and **II**, respectively, are:

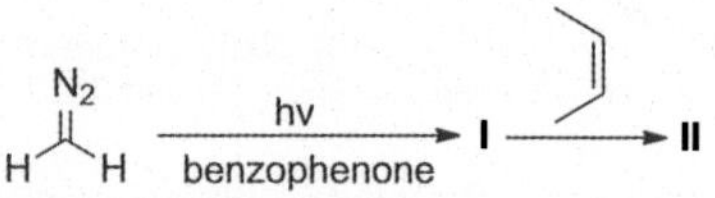

(a) ^{1}:CH_2 and *cis*-1,2-dimethylcyclopropane
(b) ^{3}:CH_2 and *cis*-1,2-dimethylcyclopropane
(c) ^{1}:CH_2 and a mixture of *cis/trans*-1,2-dimethylcyclopropane
(d) ^{3}:CH_2 and a mixture of *cis/trans*-1,2-dimethylcyclopropane

Sol 3. (d) Diazo compounds liberate the free carbenes under photochemical conditions that are in singlet states. This process can be directed toward the formation of a triplet carbene using a triplet sensitizer like benzophenone. For triplet carbenes, the alkene insertion reaction is nonstereospecific. Therefore, the reaction of *Z*-but-2-ene with triplet carbene gives a mixture of *cis*/*trans*-1,2-dimethylcyclopropane.

Q 4. In the following reaction, the reactive intermediate **I** and the product **II** are:

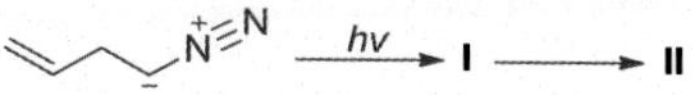

$\xrightarrow{hv}$ I ⟶ II

(a) Carbene and 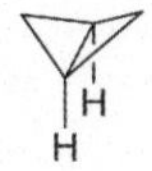 (b) Radical and

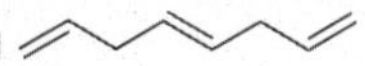

(c) Carbene and 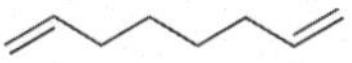 (d) Radical and

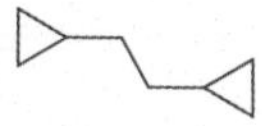

Sol 4. (a)

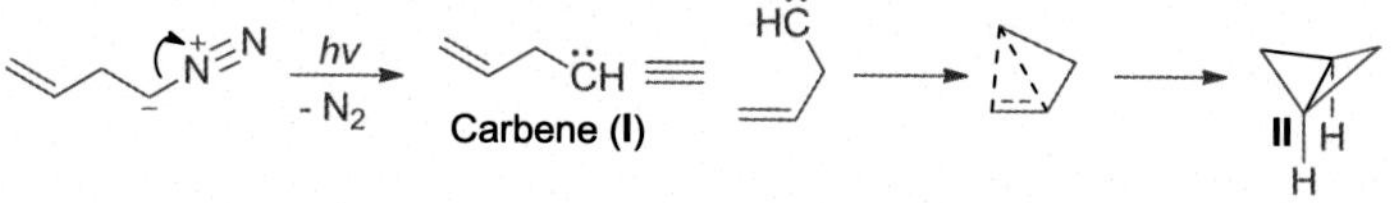

$\xrightarrow[-N_2]{hv}$ Carbene (I) ⟶ ⟶ II

Q 5. The major product in the following reaction is:

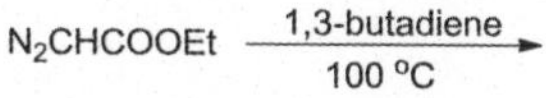

$N_2CHCOOEt \xrightarrow[100\ ^{o}C]{1,3\text{-butadiene}}$

(a) OEt (b) COOEt (c) COOEt (d) 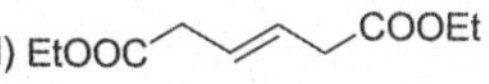 EtOOC COOEt

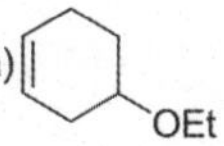

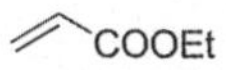

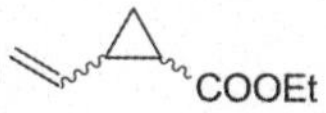

Sol 5. (c) Heating of ethyl diazoacetate furnishes a carbene, mostly in the singlet state. The generated alkene undergoes a concerted cycloaddition to a double bond to give a cyclopropane ring. This step is stereospecific with respect to the geometry of the double bond. However, a mixture of *cis* and *trans* forms is formed due to two different possible approaches of the carbene.

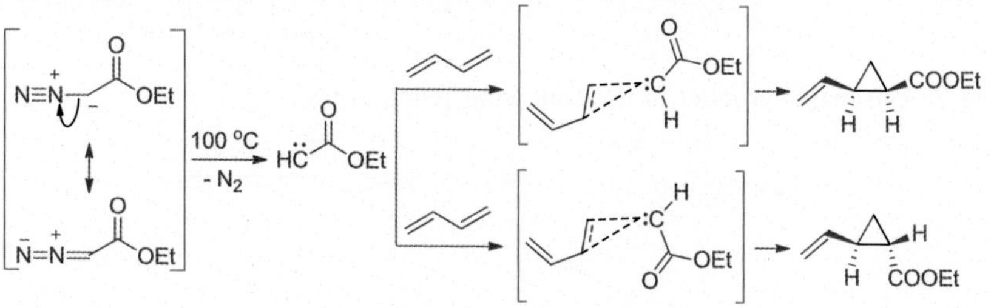

$\xrightarrow[-N_2]{100\ ^{o}C}$ HC̈ OEt

COOEt

H COOEt

It should be noted that the reaction is stopped after addition of carbene to only one double bond. This is due to the fact that dienes are more nucleophilic than alkenes because of the presence of high energy HOMO.

Q 6. In the following reaction, the reagent **I** and the major product **II** are:

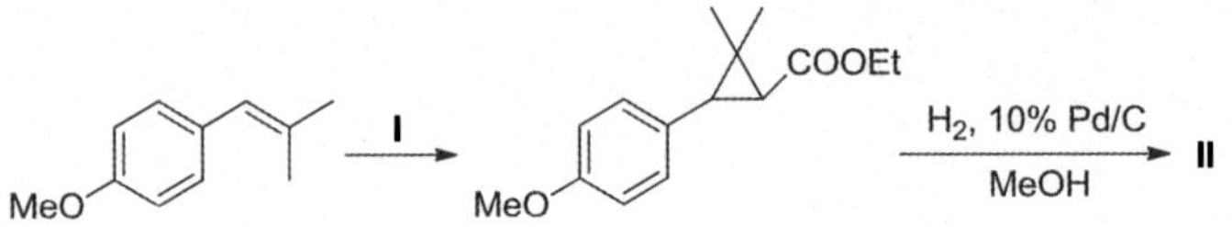

MeO, COOEt, I, H_2, 10% Pd/C, MeOH, II

I	II
(a) $N_2CHCOOEt$, $Cu(acac)_2$	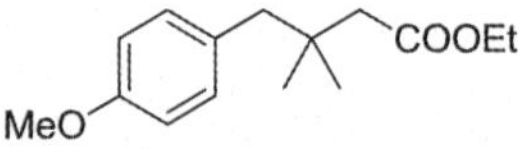 MeO, COOEt
(b) $N_2CHCOOEt$, $Cu(acac)_2$	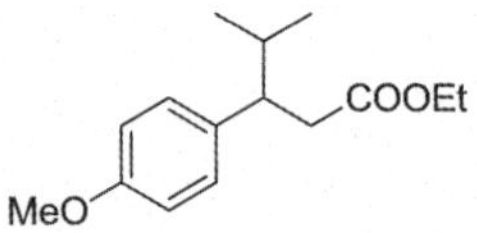 MeO, COOEt
(c) NaH, $Me_2S^+(=O)CH_2COOEt$ Br^- 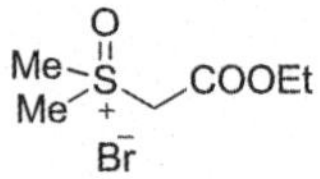	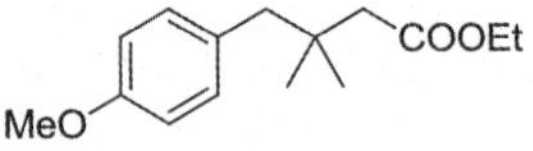 MeO, COOEt
(d) NaH, $Me_2S^+(=O)CH_2COOEt$ Br^- 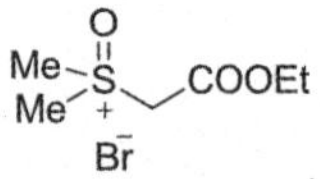	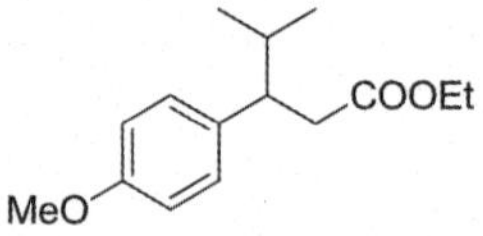 MeO, COOEt

Sol 6. (a)

$\bar{N}=\overset{+}{N}=CHCOOEt$ $\xrightarrow[- N_2]{Cu(acac)_2}$ $\ddot{H}C$–COOEt $\xrightarrow[\text{addition of carbene to double bond}]{\text{MeO-aryl alkene}}$ cyclopropane (MeO, COOEt) $\xrightarrow[MeOH]{H_2,\ 10\%\ Pd/C}$ MeO, COOEt

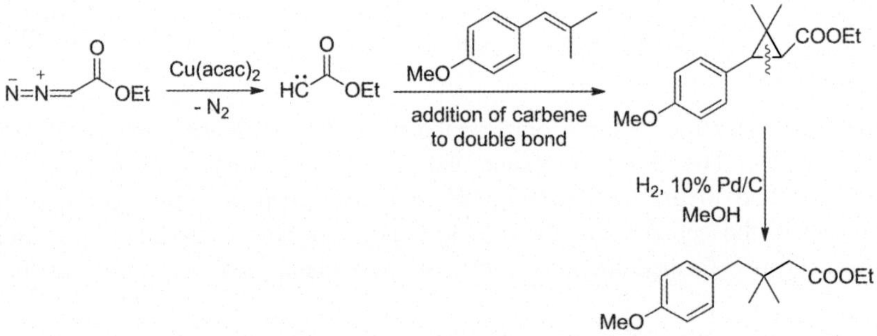

Q 7. The major product in the following reaction is:

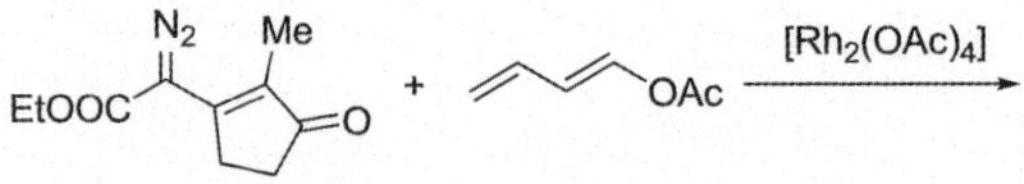

EtOOC–C(N_2)–(Me-cyclopentenone) + diene–OAc $\xrightarrow{[Rh_2(OAc)_4]}$

(a) 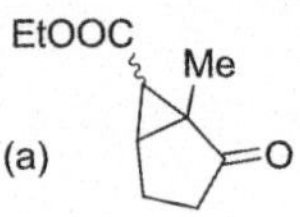 EtOOC, Me, O (b) 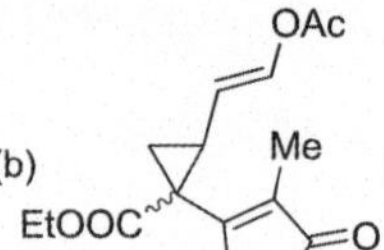 OAc, Me, EtOOC, O (c) 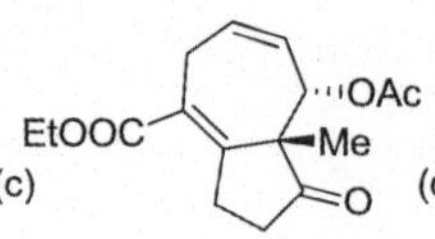 EtOOC, OAc, Me, O (d) 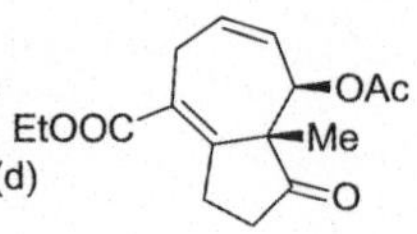 EtOOC, OAc, Me, O

Sol 7. (c) The rhodium-carbenoid addition to less hindered double bond furnishes a cyclopropane intermediate in which both the alkene groups are *cis* to each other, which allows the subsequent Cope rearrangement of the divinyl cyclopropane. The rearrangement proceeds through the boat-like transition state and has the methyl and acetate groups *trans* to each other.

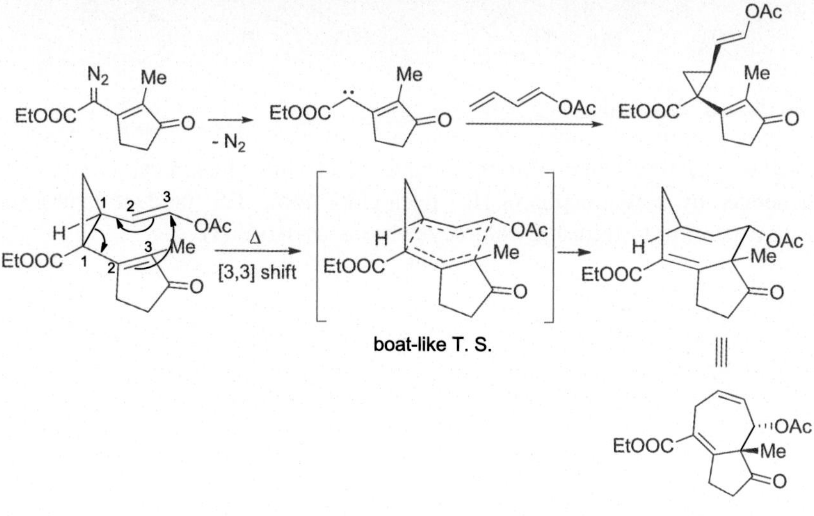

boat-like T. S.

Q 8. The required reagent **I** and the major product **II** formed in the following reaction sequence are:

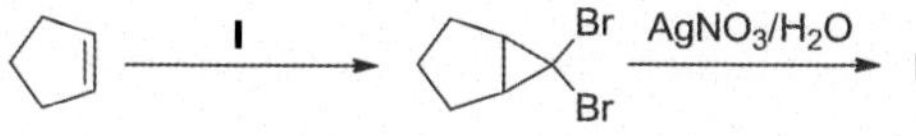

(a) CH_2Br_2/KO^tBu and

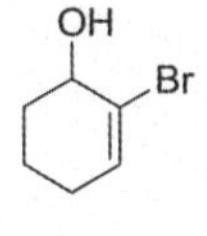

(b) CH_2Br_2/KO^tBu and

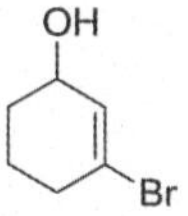

(c) $CHBr_3/KO^tBu$ and

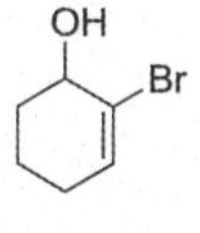

(d) $CHBr_3/KO^tBu$ and

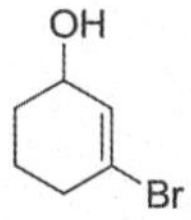

Sol 8. (c) Dibromocarbene, generated from $CHBr_3$ and base by E1cB mechanism, is added to the double bond to give a cyclopropane derivative that undergoes subsequent ring expansion through the carbocation intermediate when treated with $AgNO_3$.

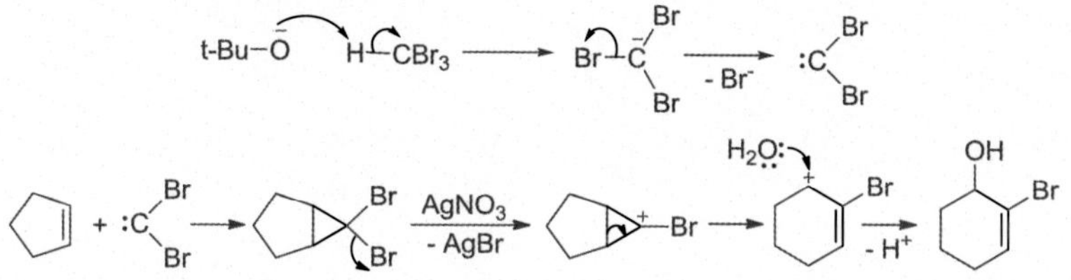

Q 9. The major product **I** in the following reaction sequence is:

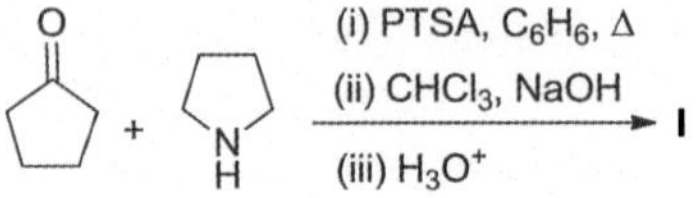

(i) PTSA, C_6H_6, Δ
(ii) $CHCl_3$, NaOH
(iii) H_3O^+

(a) 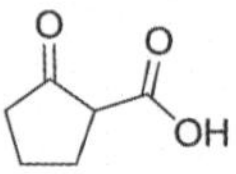 (b) 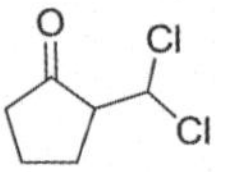 (c) 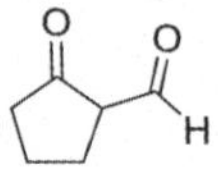 (d)

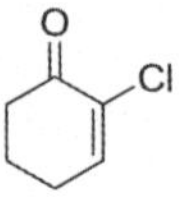

Sol 9. (d) Aldehydes and ketones having α-protons react with secondary amines to form **enamines**. The enamine formed in the first step undergoes cyclopropanation reaction at the double bond with dichlorocarbene that is generated by α-elimination of HCl from chloroform. This product undergoes ring expansion to furnish 2-chlorocyclohex-2-en-1-one (**I**).

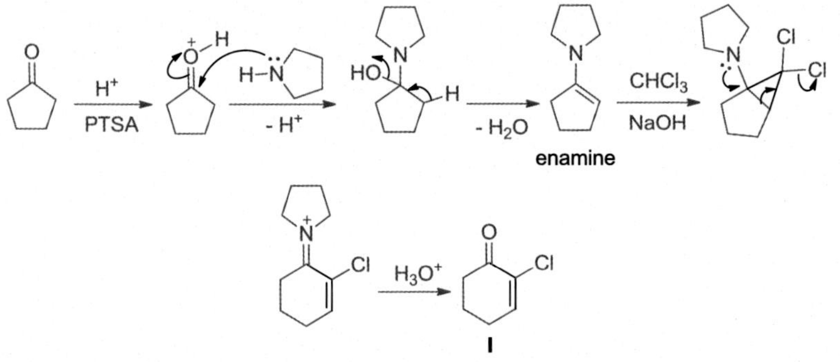

H^+ PTSA; $- H^+$; $- H_2O$; enamine; $CHCl_3$ NaOH; H_3O^+; I

Q 10. In the following reaction, the intermediate and the major product **I** are:

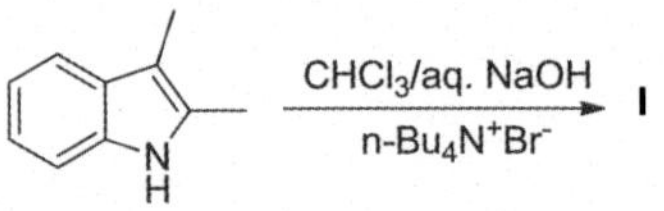

$CHCl_3$/aq. NaOH
n-$Bu_4N^+Br^-$

(a) :CHCl and

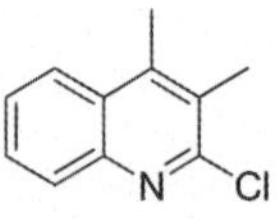

(b) :CCl_2 and

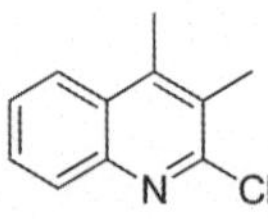

(c) :CHCl and

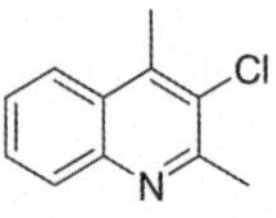

(d) :CCl_2 and

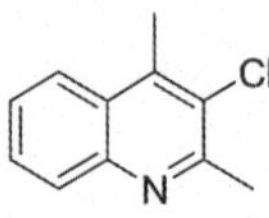

Sol 10. (d)

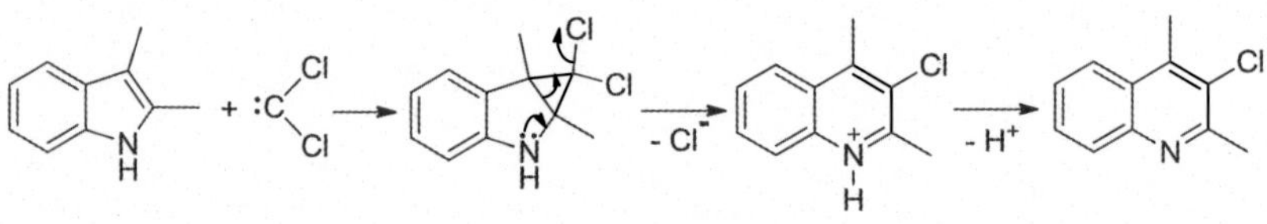

$- Cl^-$; $- H^+$

Q 11. Give the mechanism of the following reaction.

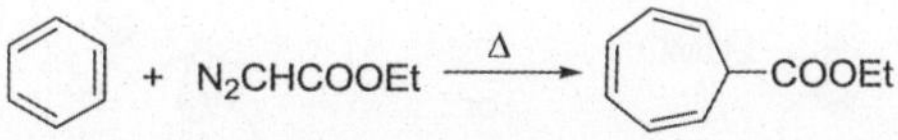

$$C_6H_6 + N_2CHCOOEt \xrightarrow{\Delta} \text{cycloheptatriene-COOEt}$$

Sol 11. Carbenes substituted with electron-withdrawing carbonyl groups are more powerful electrophiles than simple carbenes and will even add to the double bonds of benzene. The product is not stable but immediately undergoes electrocyclic ring opening.

Electrocyclic ring opening

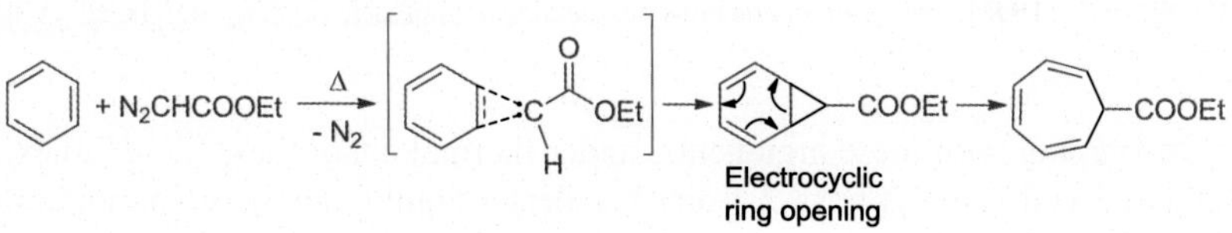

5.1.2 [4 + 2] Cheletropic Reactions

The irreversible extrusion of nitrogen from the cyclic diazene and the easy loss of carbon monoxide from norbornadienone are important examples of six-electron cheletropic reactions (Scheme 5.2). The driving force for these reactions is often the entropic benefit of gaseous evolution (e.g., N_2 or CO).

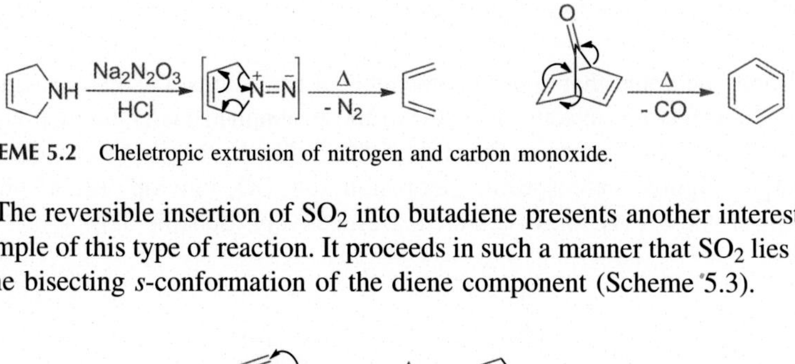

SCHEME 5.2 Cheletropic extrusion of nitrogen and carbon monoxide.

The reversible insertion of SO_2 into butadiene presents another interesting example of this type of reaction. It proceeds in such a manner that SO_2 lies in a plane bisecting *s*-conformation of the diene component (Scheme 5.3).

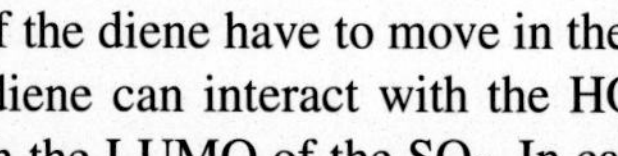

SCHEME 5.3 Addition of SO_2 to buta-1,3-diene.

The HOMO−LUMO interactions in the linear approach of SO_2 molecule suprafacial to the π-system are bonding (Figure 5.4). In the transition state, the terminal carbon atoms of the diene have to move in the **disrotatory** manner so that the LUMO of the diene can interact with the HOMO of the SO_2 or the HOMO of the diene with the LUMO of the SO_2. In case of SO_2 molecule, the HOMO is that which has a lone pair of electrons in the plane having the atoms, while the LUMO represents the *p*-orbital perpendicular to the plane.

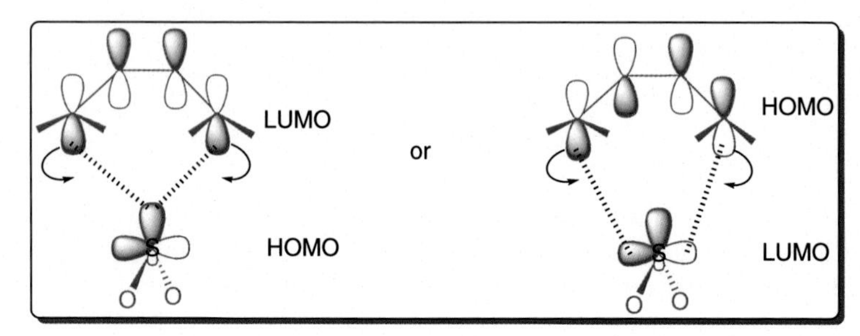

FIGURE 5.4 HOMO–LUMO interactions in the linear approach of SO_2 suprafacial to the π-system.

In keeping with these arguments, under thermal conditions, (*2E,4E*)-hexa-2,4-diene and (*2Z,4E*)-hexa-2,4-diene undergo highly stereospecific addition of SO_2 to give *cis*- and *trans*-2,5-dimethyl-2,5-dihydrothiophene 1,1-dioxide (*cis*- and *trans*-2,5-dimethylsulfolene), respectively (Scheme 5.4).

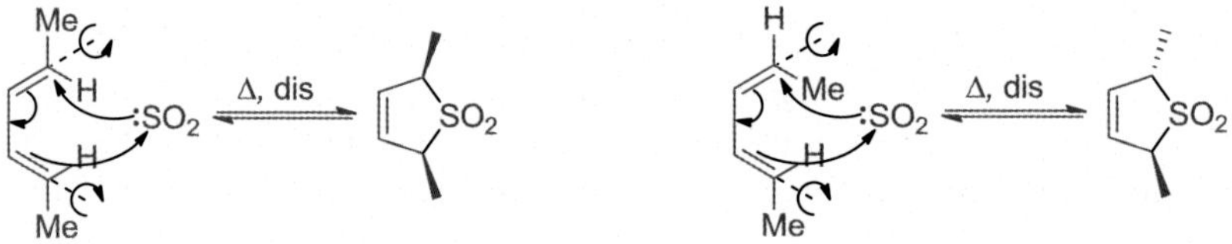

SCHEME 5.4 Stereochemistry of cheletropic addition of SO_2 to diastereomeric hexa-2,4-dienes.

It may be noted that under photochemical conditions, the stereochemistry of the products are exactly opposite to those obtained under thermal conditions. Thus, *cis*- and *trans*-2,5-dimethyl-2,5-dihydrothiophene 1,1-dioxides undergo highly stereospecific extrusion of SO_2 giving (2*Z*,4*E*)-hexa-2,4-diene and (2*E*,4*E*)-hexa-2,4-diene, respectively (Scheme 5.5).

SCHEME 5.5 Stereospecific extrusion of SO_2 from diastereomeric 2,7-dimethyl-2,7-dihydrothiophene 1,1-dioxides.

The reaction by nonlinear approach of SO_2 antarafacial to the 4n π-system is also symmetry allowed and takes place in a conrotatory fashion (Figure 5.5).

However, from the structures of the products of cheletropic reaction of SO_2 with a 4n π-system, it appears that the reagent approaches the π-system in a linear fashion.

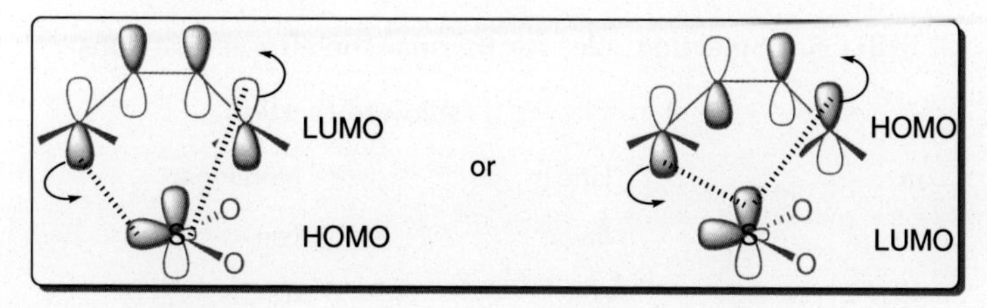

FIGURE 5.5 HOMO–LUMO interactions in the nonlinear approach of SO_2 antarafacial to the π-system.

5.1.3 [6 + 2] Cheletropic Reactions

With trienes, the extrusion process dominates and the ring opens in a conrotatory fashion. This indicates that in this reaction, one of the components is acting in an antarafacial manner. For example, *cis*- and *trans*-2,7-dimethyl-2,7-dihydrothiepine 1,1-dioxides undergo highly stereospecific extrusion of SO_2 giving (2*Z*,4*Z*,6*E*)-octa-2,4,6-triene and (2*E*,4*Z*,6*E*)-octa-2,4,6-triene, respectively (Scheme 5.6).

SCHEME 5.6 Stereospecific extrusion of SO_2 from diastereomeric 2,7-dimethyl-2,7-dihydrothiepine 1,1-dioxides.

The formation of the above products can be explained by considering the HOMO–LUMO interactions in the linear approach of SO_2 antarafacial to the (4n + 2) π-system (Figure 5.6). As the HOMO–LUMO interactions are bonding, the reaction is thermally allowed in a conrotatory fashion. By the principle of microscopic reversibility, the same conclusion applies to the extrusion of SO_2.

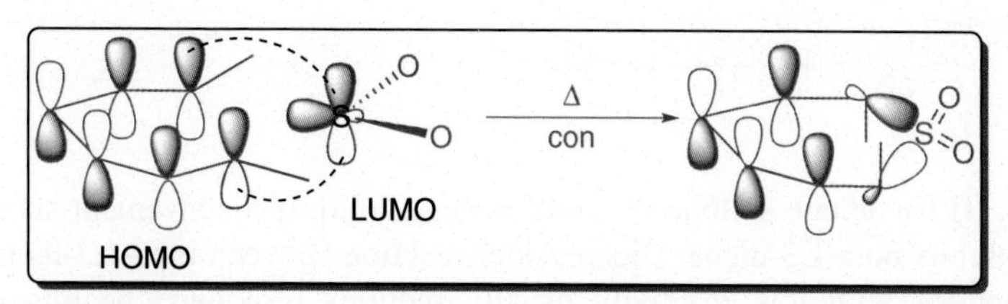

FIGURE 5.6 HOMO–LUMO interactions in the linear approach of SO_2 antarafacial to the (4n + 2) π-system.

The selection rules for the thermal cheletropic reactions are given in Table 5.1, where m is the number of electrons in the π-system and n is an integer including zero.

TABLE 5.1 Selection rules for thermal cheletropic reactions.

m	Allowed reaction	
	Linear	Nonlinear
4n	dis	con
4n + 2	con	dis

5.1.4 Solved Problems

Q 1. In the following reaction, the major product **I** is:

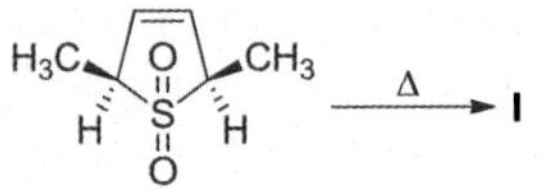

(a) (2*E*,4*E*)-Hexa-2,4-diene (b) (2*Z*,4*E*)-Hexa-2,4-diene
(c) (2*Z*,4*Z*)-Hexa-2,4-diene (d) (2*E*,4*Z*)-Hexa-2,4-diene

Sol 1. (a) On heating, *cis*-sulfone undergoes cheletropic elimination of SO_2 to give stereospecifically the (2*E*,4*E*)-hexa-2,4-diene.

Q 2. Give the products of the following reactions:

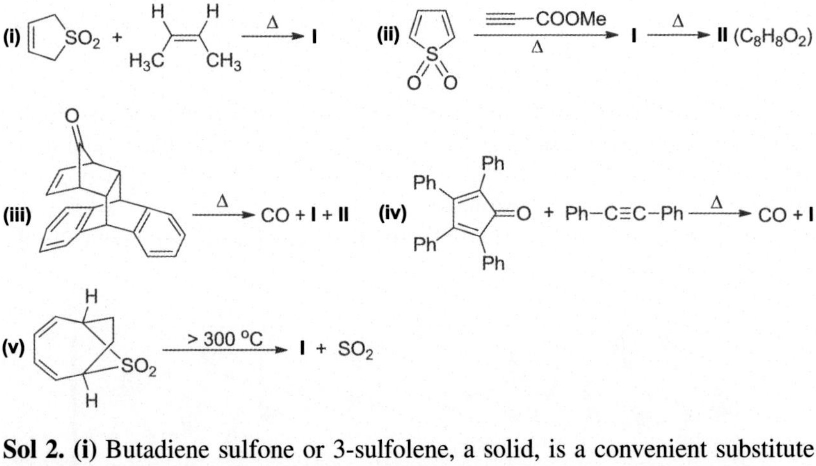

Sol 2. (i) Butadiene sulfone or 3-sulfolene, a solid, is a convenient substitute for gaseous buta-1,3-diene. Diels–Alder reaction between buta-1,3-diene and dienophiles with low reactivity usually requires prolonged heating above 100 °C. This makes a procedure rather dangerous, if neat buta-1,3-diene is used, and requires special equipment for work under elevated pressure. Alternatively, buta-1,3-diene can be generated in situ by thermal sulfur dioxide extrusion from sulfolene, in which case no buildup of buta-1,3-diene pressure could be expected as the liberated diene is consumed in the cycloaddition, and, therefore, the equilibrium of the reversible extrusion reaction acts as an internal "safety valve."

Δ cheletropic elimination of SO_2 Δ [4 + 2] I

cis-4,5-Dimethylcyclohex-1-ene

(ii) The strongly electronegative sulfone unit in thiophene 1,1-dioxide destroys the aromaticity of the thiophene component; hence the compound behaves as a diene in the cycloaddition reactions. Even though thiophene 1,1-dioxide is an electron-deficient diene, it is very reactive and cycloadditions do take place between it and electron-deficient dienophiles. For example, thiophene 1,1-dioxide forms an adduct when it is heated with methyl propiolate. The adduct than undergoes cheletropic elimination of sulfur dioxide to give **II**.

MeOOC Δ [4 + 2] COOMe I Δ cheletropic elimination of SO_2 COOMe + SO_2 II ($C_8H_8O_2$)

However, it should be noted that benzo[b]thiophene 1,1-dioxide acts as a dienophile. In this case also, a reduction in aromaticity takes place, but here the benzene resonance is maintained and the ring fusion blocks a double bond of the thiophene unit. As a result, the compound behaves as an electron-deficient dienophile forming an adduct with, for example, cyclopentadiene.

\+ Δ [4 + 2]

(iii) Δ cheletropic elimination of CO + CO Δ retro D. A. I + II

(iv) Cyclopentadienones readily undergo Diels–Alder reaction because of their *s-cis* conformation and their antiaromaticity. The intermediate 1,2,3,4,5,6-hexaphenylbicyclo[2.2.1]hepta-2,5-dien-7-one undergoes spontaneous cheletropic elimination of carbon monoxide to give hexaphenylbenzene:

Δ [4 + 2] Δ cheletropic elimination of CO I

(v) Due to geometrical constraints, the ring opening in the given bicyclic sulfone cannot take place by the symmetry-allowed linear pathway in a conrotatory fashion. However, the reaction can take place via a symmetry-allowed nonlinear pathway or by a stepwise mechanism. But the experimental evidence suggests the latter mechanism for this reaction.

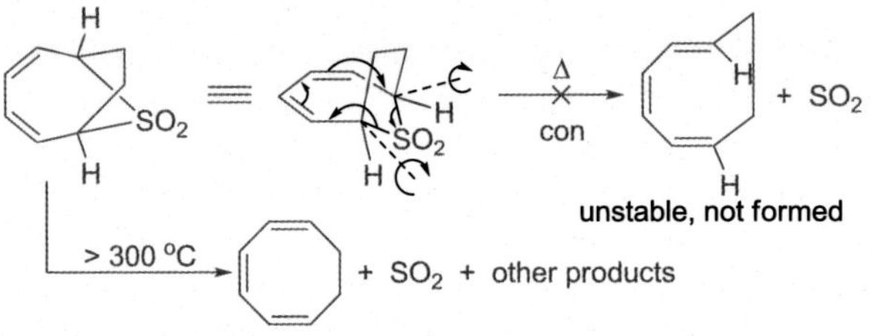

5.2 1,3-DIPOLAR CYCLOADDITIONS

1,3-Dipolar cycloadditions are the reactions that involve atoms with formal charges. Usually four-membered rings are constructed by [2 + 2] cycloaddition, six-membered rings by [4 + 2] cycloaddition, and the *five-membered rings by 1,3-dipolar cycloaddition reactions* (1,3-DPCAs). In general, a 1,3-DPCA may be depicted as shown in Figure 5.7.

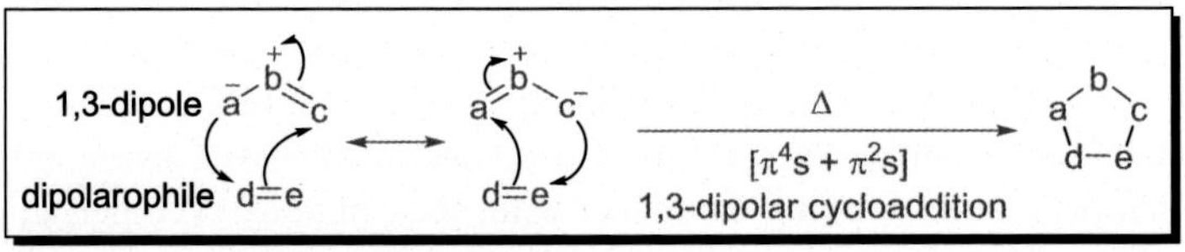

FIGURE 5.7 1,3-Dipolar cycloaddition reaction.

1,3-DPCAs are $[\pi^4s + \pi^2s]$ cycloaddition reactions are thermally allowed according to symmetry rules. Because of structural variety of dipolarophiles (for example, alkenes, alkynes, carbonyls, and nitriles) and dipoles having a three-atom skeleton bearing at least one hetero atom, these reactions provide a very common and versatile route for the synthesis of five-membered heterocyclic compounds.

5.2.1 Dipolarophile and 1,3-Dipole

The 2π-component in a 1,3-DPCA is commonly referred to as the **dipolarophile**. Substituted alkenes and alkynes are typical dipolarophiles, however, other systems containing a π-bond such as carbonyl (C=O), imine (C=N), nitrile (C≡N), and nitroso (N=O) functionalities also act as dipolarophiles. The π-bond may be isolated, conjugated, or part of a cumulene system.

A three-atom π-electron system with four π-electrons delocalized over three atoms is referred to as **1,3-dipole**. Typically, such a structure can be represented by a resonance hybrid structure in which the positive charge is located on the central atom (b) and the negative charge is distributed over the two terminal atoms (a and c). The other way to represent such structures is to have the central atom (b) having two of the four electrons. The resonating hybrid structures thus can be shown as in Figure 5.8.

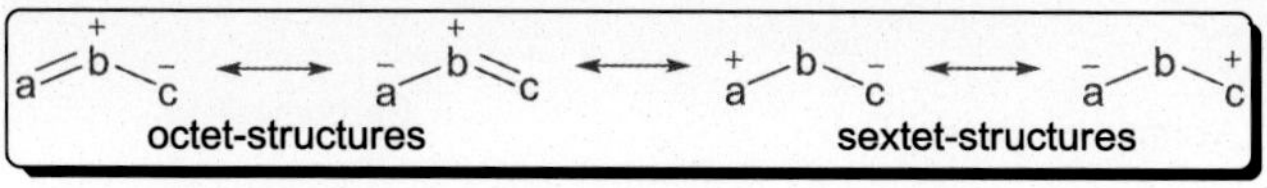

FIGURE 5.8 Octet and sextet structures of 1,3-dipole.

Although sextet structures do not contribute much to the resonance hybrid, they are important to understanding the mechanism, reactivity, and regiochemistry of 1,3-DPCAs.

5.2.1.1 Classification of 1,3-Dipoles

The kind of dipoles that feature in the 1,3-DPCAs are isoelectronic with an **allyl** or **propargyl anion** system. They have a π-electron system consisting of two filled and one empty orbital, and both ends of the dipole have nucleophilic as well as electrophilic properties. 1,3-Dipoles of the allyl anion type are bent as the system has four electrons in three parallel p_z-orbitals perpendicular to the plane of the dipole. On the other hand, the presence of a double bond orthogonal to the delocalized π-system in the propargyl/allenyl anion type causes linearity to the dipole (Figure 5.9).

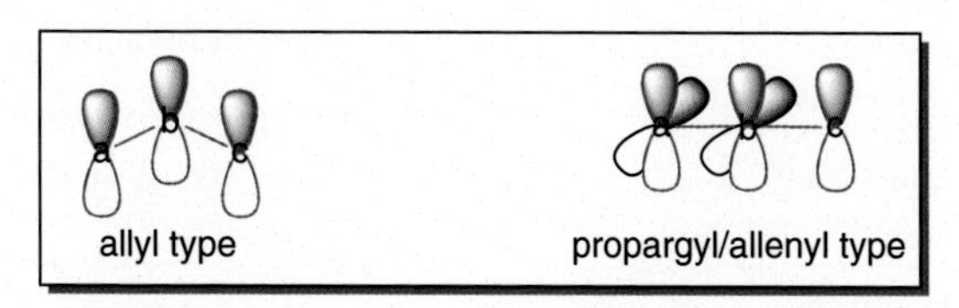

FIGURE 5.9 Arrangement of orbitals in allyl- and propargyl-type dipoles.

The central atom, b, in allyl-type dipoles is commonly a group V element (for example, N or P) or a group VI element (for example, O or S). In propargyl/allenyl type, the central atom, b, is always a group V element as only such an atom can bear a positive charge in the quartervalent state. The terminal atoms (a and b) in 1,3-dipole are second-row elements (C, N, O). A wide variety of 1,3-dipoles, allyl and propargyl/allenyl type, are shown in Table 5.2. The dipoles having higher-row elements such as sulfur and phosphorus are also possible but are much less widely used.

TABLE 5.2 Types and core structures of some important 1,3-dipoles.

Allyl type	
***N*-centered**	***O*-centered**
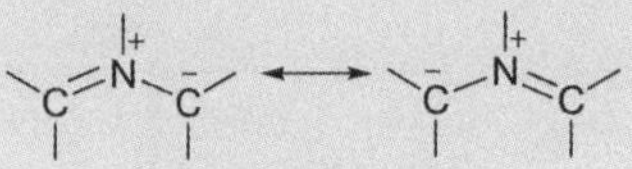 Azomethine ylide	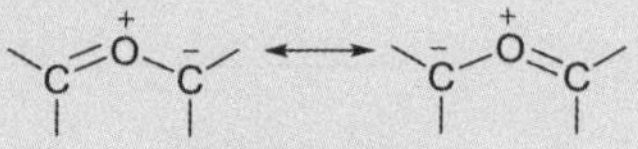 Carbonyl ylide
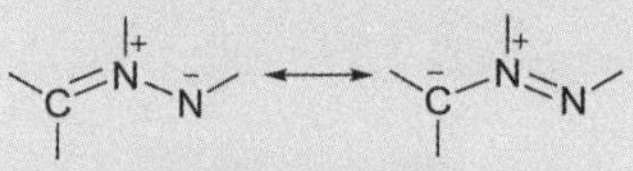 Azomethine imine	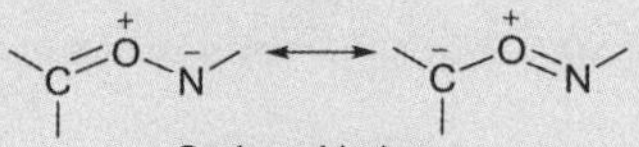 Carbonyl imine
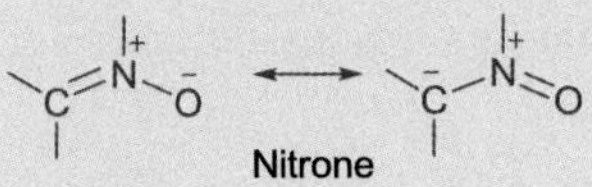 Nitrone	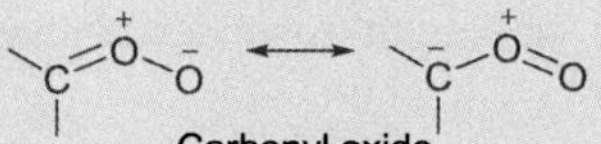 Carbonyl oxide
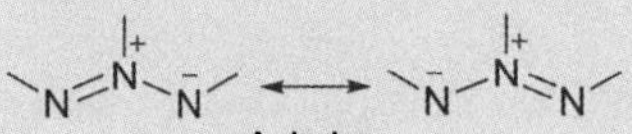 Azimine	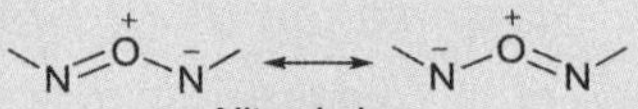 Nitrosimine
 Azoxy compound	 Nitroso oxide
 Nitro compound	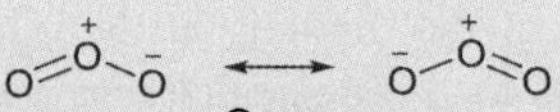 Ozone

Propargyl-allenyl type	
Nitrilium betaines	**Diazonium betaines**
$-C\equiv\overset{+}{N}-\overset{-}{C}<\ \longleftrightarrow\ -\overset{-}{C}=\overset{+}{N}=C<$ Nitrile ylide	$N\equiv\overset{+}{N}-\overset{-}{C}<\ \longleftrightarrow\ \overset{-}{N}=\overset{+}{N}=C<$ Diazoalkane
$-C\equiv\overset{+}{N}-\overset{-}{N}-\ \longleftrightarrow\ -\overset{-}{C}=\overset{+}{N}=N-$ Nitrile imine	$N\equiv\overset{+}{N}-\overset{-}{N}-\ \longleftrightarrow\ \overset{-}{N}=\overset{+}{N}=N-$ Azide
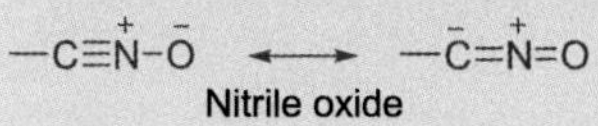 $-C\equiv\overset{+}{N}-\overset{-}{O}\ \longleftrightarrow\ -\overset{-}{C}=\overset{+}{N}=O$ Nitrile oxide	 $N\equiv\overset{+}{N}-\overset{-}{O}\ \longleftrightarrow\ \overset{-}{N}=\overset{+}{N}=O$ Nitrous oxide

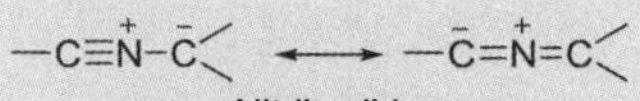

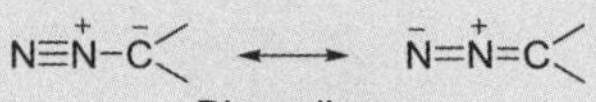

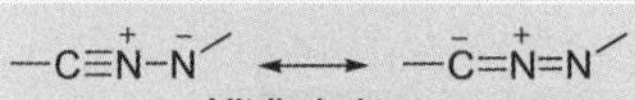

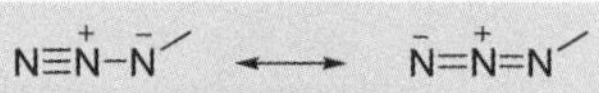

5.2.1.2 Nature of 1,3-Dipoles

Many 1,3-dipolar compounds show very high polarity. In allyl type of π-system, the negative charge is distributed over the terminal atoms, a and c, and the central atom, b, carries positive charge. As expected, the polarity of the system is considerably reduced in case the distribution of the negative charge is more balanced. Assuming a propargyl type of structure, the calculated dipole moment value of diazomethane is about 6.2 D, which is about 5.5 D in the opposite direction if we assume an allenic structure (Figure 5.10). The observed value, $\mu = 1.5$ D, indicates that there is considerable charge cancellation in the resonance hybrid.

$\overset{+}{N}\equiv\overset{-}{N}-CH_2$ ⟷ $\overset{-}{N}=\overset{+}{N}=CH_2$ $N\equiv N=CH_2$

μ = 6.2 D (cal.) μ = 5.5 D (cal.) μ = 1.5 D (exp.)

μ = 3.55 D (exp.)

FIGURE 5.10 Nature of 1,3-dipoles.

It is thus not possible to exactly locate electrophilic and nucleophilic centers within the 1,3-dipolar species and it is safer to say that these compounds display electrophilic or nucleophilic property at either *a* or *c* terminal. Similarly, the dipole moment of *N*-methyl-*C*-phenylnitrone (allyl type), $\mu = 3.55$ D, shows a dominance of azomethine *N*-oxide structure and thus the terminal oxygen carries major fraction of negative charge than the carbon atom.

5.2.2 Analysis and Mechanism of 1,3-Dipolar Cycloadditions

5.2.2.1 FMO Approach

1,3-Dipolar cycloadditions have been well explained using FMO approach (Figure 5.11). In 1,3-dipoles, four π-electrons are distributed over three atoms and they can be considered as the structural variant of the diene component in

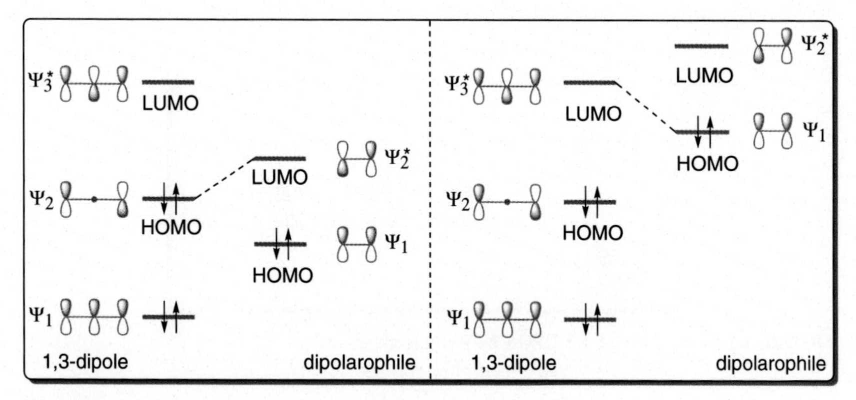

FIGURE 5.11 Molecular orbitals of 1,3-dipole and dipolarophile.

the Diels–Alder reaction. Moreover, the HOMO and LUMO of a 1,3-dipole are of similar symmetry to those of a diene with respect to the two-fold axis and to the mirror plane that bisects the molecule. The HOMO of the 1,3-dipole is Ψ_2 and the LUMO of the dipolarophile is Ψ_2^*. Both components are antisymmetric with respect to mirror symmetry, and, therefore, maximum overlap could be obtained for the reaction to take place in forward direction. Similarly, when HOMO of the dipolarophile Ψ_1 and LUMO of 1,3-dipole Ψ_3^* are symmetric to the mirror plane or antisymmetric to two-fold axis of symmetry, maximum overlap could be obtained for the reaction to take place in forward direction.

Since 1,3-DPCA reactions involve 4π-electrons from the 1,3-dipole and 2π-electrons from the dipolarophiles, it may be considered as symmetry-allowed $[\pi^4s + \pi^2s]$ cycloaddition resembling Diels–Alder reaction. There is HOMO–LUMO interaction in which either reactant can be the electrophilic or nucleophilic component (Figure 5.12).

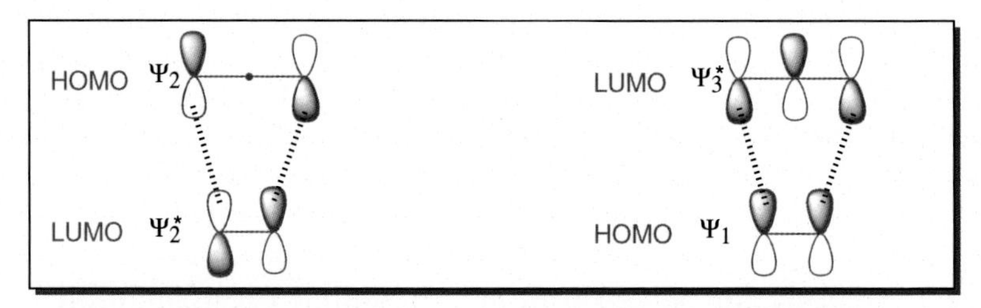

FIGURE 5.12 The specific set of reaction partners that defines the dominant frontier orbitals.

5.2.2.2 PMO Approach

1,3-DPCAs are $[\pi^4s + \pi^2s]$ cycloaddition reactions. The transition state in such reactions has six electrons with zero node and is, therefore, aromatic. Hence, the reactions are thermally allowed (Figure 5.13).

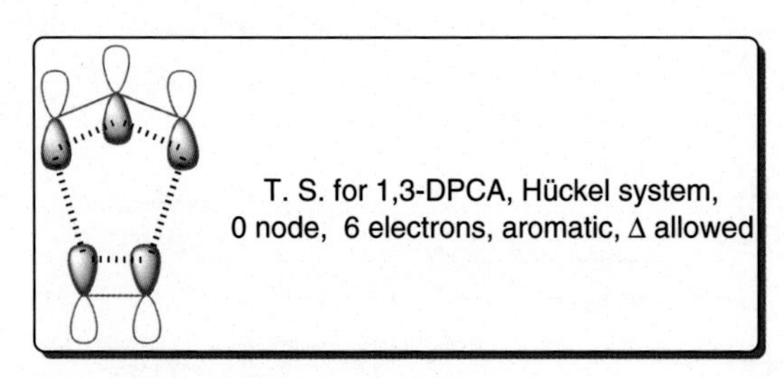

T. S. for 1,3-DPCA, Hückel system,
0 node, 6 electrons, aromatic, Δ allowed

FIGURE 5.13 Analysis of 1,3-DPCA by PMO method.

5.2.3 Regioselectivity and Types of 1,3-Dipolar Cycloadditions

As the transition states of 1,3-DPCAs are controlled by the coefficients of the frontier molecular orbitals, the observed regioselectivity of these reactions can be explained by FMO theory. These reactions have been broadly classified into three types: Type I, Type II, and Type III (Figure 5.14).

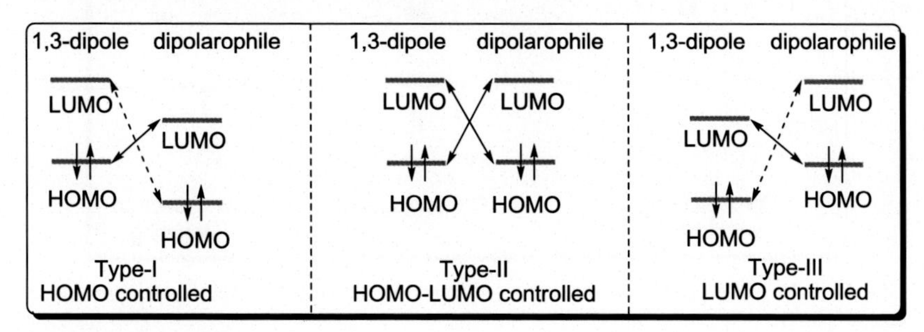

FIGURE 5.14 Types of 1,3-DPCA.

Type I: There is an overlapping of the high-lying HOMO of the dipole with the LUMO of the dipolarophiles. Such a situation is often referred to as a HOMO-controlled dipole or a nucleophilic dipole and includes many commonly used dipoles such as azomethine ylide, carbonyl ylide, nitrile ylide, azomethine imine, carbonyl imine, and diazoalkane. There is a close similarity of these reactions with a normal electron-demand Diels–Alder reaction, which involves the overlapping of the diene HOMO and LUMO of the dienophile.

Type II: Because of the similar energy gap in either direction, HOMO of the dipole can interact with LUMO of the dipolarophiles or HOMO of the dipolarophile can interact with LUMO of the dipole. The situation is referred to as a HOMO–LUMO-controlled dipole or an ambiphilic dipole and includes nitrile imine, nitrone, carbonyl oxide, nitrile oxide, and azide.

Type III: There is an overlapping of the low-lying LUMO of the dipole with the HOMO of the dipolarophiles. Such a situation is often referred to as a LUMO-controlled dipole or an electrophilic dipole and includes nitrous oxide and ozone. There is a close similarity of these reactions with an inverse electron-demand Diels–Alder reaction, which involves the overlapping of the diene LUMO and HOMO of the dienophile.

In 1,3-DPCAs, two regioisomeric products are possible when dipole–dipolarophile pairs are asymmetric. The combination of the largest HOMO and the largest LUMO orbital represents the dominant electronic interaction, therefore, regioselectivity is directed by the atoms that bear the largest orbital

HOMO and LUMO coefficients. The effects of the electron-donating or electron-withdrawing groups on the shapes of the frontier orbitals of dipolarophiles are depicted in Figure 5.15.

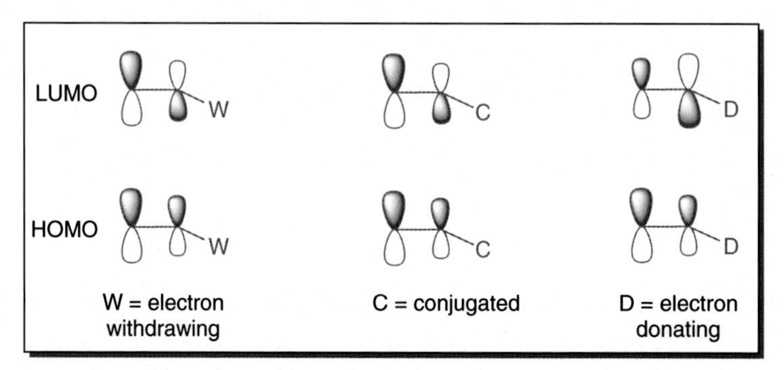

FIGURE 5.15 Effects of the nature of substituents on the shapes of the frontier orbitals of dipolarophiles.

5.2.4 Reactivity of 1,3-Dipoles in Cycloadditions

As the 1,3-DPCA reactions are controlled by the HOMO and the LUMO of the two reactants, the reactivity of 1,3-dipoles greatly varies toward various dipolarophiles. The smaller energy difference between the HOMO and LUMO leads to the stronger interaction and thus makes the reactions fast. The presence of an electron-withdrawing group on either the dipole or dipolarophile lowers the energy level of both the HOMO and LUMO, whereas an electron-donating substituent raises the energy of both. On the other hand, the conjugating substituent increases the energy of the HOMO but decreases the energy of the LUMO. Therefore, the presence of substituents can lead to an increase or decrease of the reaction rate depending on whether the FMO energy gap increases or decreases, respectively.

1,3-DPCA reactions will be favored when one component is highly electrophilic while another is strongly nucleophilic in nature. The reactivity of 1,3-dipoles toward electron-rich and electron-poor dipolarophiles varies significantly as follows:

Type I (nucleophilic dipole): In this case, dipole reacts readily with electrophilic alkene in which a high-lying HOMO of the dipole overlaps with the LUMO of the dipolarophile. In such reactions, the electron-withdrawing groups (EWG) on the dipolarophile accelerate the rate of reaction by lowering the LUMO, whereas the electron-donating groups (EDG) decelerate the reaction by increasing the energy of the HOMO. As an example, the order of reactivity of diazomethane toward various dipolarophiles is as under:

COOEt > Ph > > OBu > NR_2

Type II (ambiphilic dipole): In such cases, the HOMO of the dipole can interact with the LUMO of the dipolarophiles or the HOMO of the dipolarophile can interact with the LUMO of the dipole. Therefore, any substituent on the either dipolarophile or dipole would accelerate the reaction by decreasing the energy gap between the two interacting frontier orbitals, i.e., an electron withdrawing group would decrease the energy of the LUMO while an EDG would increase the energy of the HOMO. For example, azides react with various electron-rich or electron-poor dipolarophiles with about a similar reactivity rate.

Type III (electrophilic dipole): In such cases, the dipole has a low-lying LUMO that overlaps with the HOMO of the dipolarophile. Therefore, the presence of EWGs on the dipolarophile is responsible for decrease in the rate of reaction, while EDGs accelerate the reaction. For example, ozone reacts with 2-methylpropene (electron rich) much faster than the tetrachloroethene (electron poor).

5.2.5 Stereochemistry of 1,3-Dipolar Cycloadditions

Stereospecificity: 1,3-Dipolar cycloadditions usually take place stereospecifically and suprafacially resulting in retention of configuration with respect to the 1,3-dipole as well the dipolarophiles. This high degree of stereospecificity is strong evidence that supports the concerted over the stepwise mechanisms. In 1,3-DPCA reactions, *cis*-substituents on the dipolarophilic alkene remain *cis*, and *trans*-substituents remain *trans* in the resulting five-membered cyclic compound (Figure 5.16).

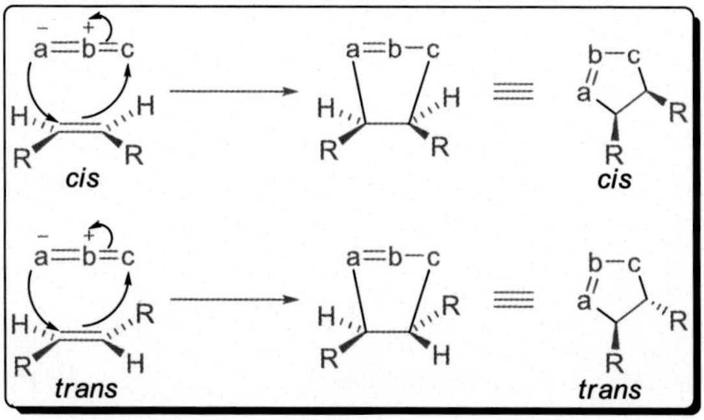

FIGURE 5.16 Stereospecific addition of 1,3-dipoles to *cis*- and *trans*-alkenes.

5.2.6 Cycloaddition with Diazoalkanes

Diazoalkanes can be prepared by the basic hydrolysis of *N*-alkyl-*N*-nitroso compounds (Scheme 5.7). The key step is a base-catalyzed elimination.

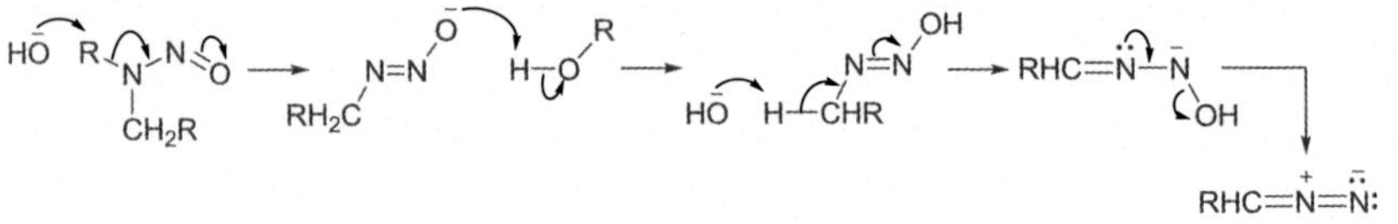

$RHC{=}\overset{+}{N}{=}\ddot{N}{:}$

SCHEME 5.7 The base-catalyzed formation of diazoalkanes.

Some of the reactions used for the preparation of diazomethane are given in Scheme 5.8.

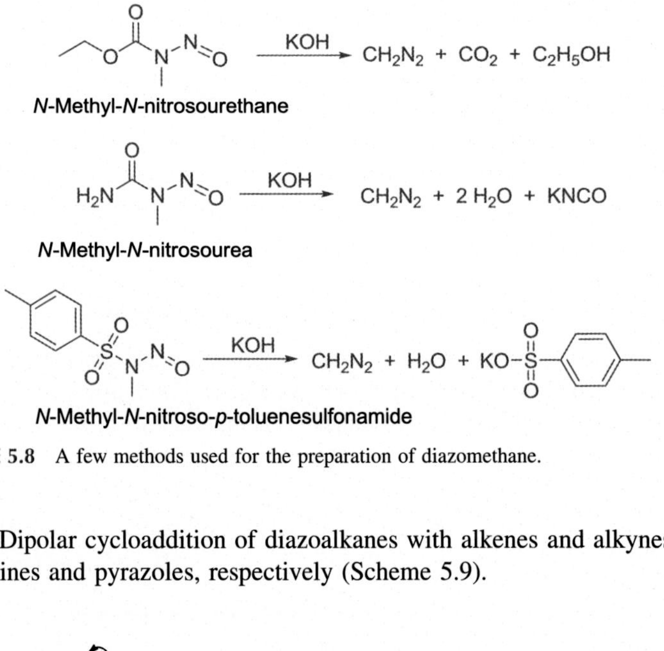

$\xrightarrow{KOH} CH_2N_2 + CO_2 + C_2H_5OH$

N-Methyl-*N*-nitrosourethane

$\xrightarrow{KOH} CH_2N_2 + 2\,H_2O + KNCO$

N-Methyl-*N*-nitrosourea

$\xrightarrow{KOH} CH_2N_2 + H_2O + KO{-}SO_2{-}C_6H_4{-}CH_3$

N-Methyl-*N*-nitroso-*p*-toluenesulfonamide

SCHEME 5.8 A few methods used for the preparation of diazomethane.

1,3-Dipolar cycloaddition of diazoalkanes with alkenes and alkynes gives pyrazolines and pyrazoles, respectively (Scheme 5.9).

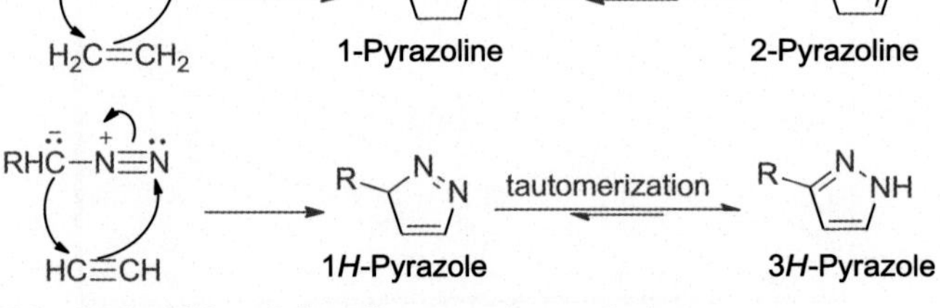

$RHC{-}\overset{+}{N}{\equiv}N$ + $H_2C{=}CH_2$ → 1-Pyrazoline ⇌ (tautomerization) 2-Pyrazoline

$RHC{-}\overset{+}{N}{\equiv}N$ + $HC{\equiv}CH$ → 1*H*-Pyrazole ⇌ (tautomerization) 3*H*-Pyrazole

SCHEME 5.9 Cycloaddition of diazoalkanes to ethene and ethyne.

The cycloaddition of diazoalkanes to alkenes can lead to the formation of two regioisomeric cycloadducts (Scheme 5.10).

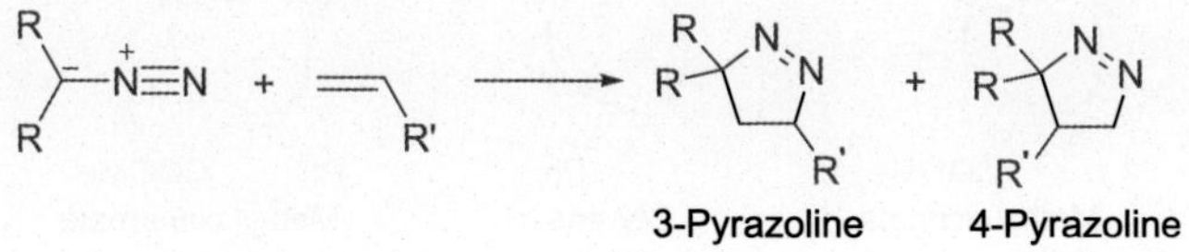

SCHEME 5.10 Regioselective addition of diazoalkanes.

The reaction of simple diazoalkanes with electron-deficient and conjugated alkenes is dipole-HOMO controlled, with the carbon atom of the diazoalkane attacking the terminal carbon of the alkene, resulting in an exclusive formation of the 3-substituted pyrazolines (Figure 5.17).

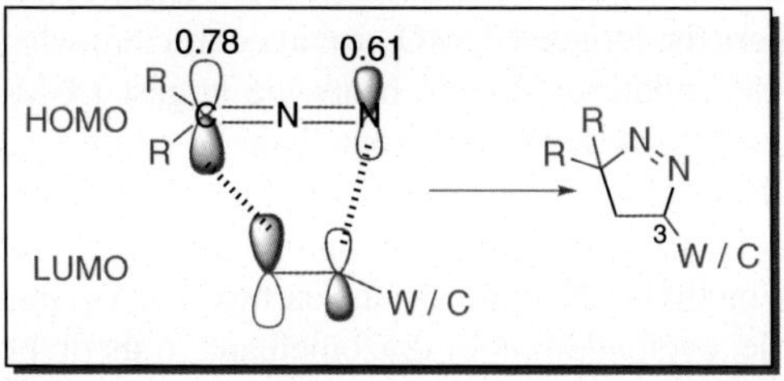

FIGURE 5.17 HOMO–LUMO interaction in cycloaddition of diazoalkanes with electron-deficient and conjugated alkenes.

The $\text{HOMO}_{\text{dipole}}$–$\text{LUMO}_{\text{dipolarophile}}$ and $\text{LUMO}_{\text{dipole}}$–$\text{HOMO}_{\text{dipolarophile}}$ interactions are comparable for electron-rich alkenes. As the coefficients of the $\text{LUMO}_{\text{dipole}}$ are comparable, the regioselectivity is controlled by the $\text{HOMO}_{\text{dipole}}$ leading to the formation of 4-substituted pyrazolines (Figure 5.18).

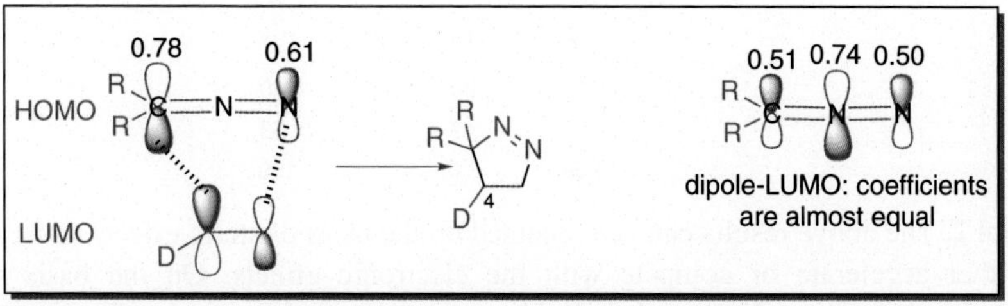

FIGURE 5.18 HOMO–LUMO interaction in cycloaddition of diazoalkanes with electron-rich alkenes.

5.2.6.1 Solved Problems

Q 1. Explain the regioselectivity of the following reactions.

Methyl acrylate

Styrene

Methyl cinnamate

Sol 1. The regioselective formation of the product is governed by the atoms that bear the largest orbital HOMO and LUMO coefficients, which are developed by the electronic/stereoelectronic and steric factors. The carbon of diazomethane bears the largest HOMO orbital coefficient whereas the terminal carbon of methyl acrylate or styrene bears the largest LUMO orbital coefficient. Hence, the cycloaddition of diazomethane to methyl acrylate or styrene gives the C-3 substituted product regioselectively. On the other hand, the two substituents (Ph *vs* COOMe) compete at withdrawing electrons from the alkene in case of methyl cinnamate. As the carboxyl group makes the β-carbon more electrophilic, cycloaddition of diazomethane to methyl cinnamate yields the product bearing the carboxyl group at position-3 selectively.

Q 2. Explain the percentage yields of products obtained when diazomethane is treated with substituted methyl acrylates.

R	Expected Product	Unexpected Product
H	100	0
Me	91	9
Et	80	20
i-Pr	47	53
t-Bu	0	100

Sol 2. The above results can be explained on the basis of steric effects that can either accelerate or compete with the electronic effects. On the basis of electronic effects, diazomethane adds to methyl acrylate to give 3-carboxyl pyrazoline.

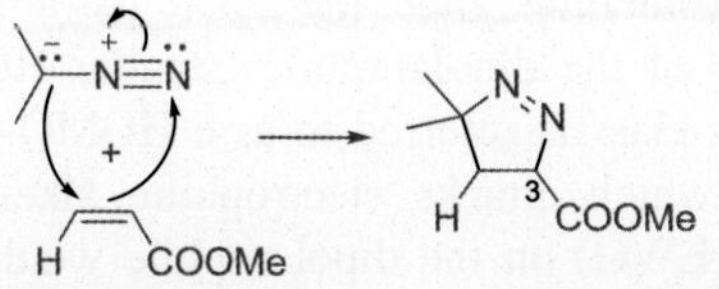

As steric hindrance increases in the system, formation of the opposite regioisomer (4-carboxyl pyrazoline) also begins. The ratio of these two regioisomers depends on the steric factors. At the extreme, increasing the size from hydrogen to *tert*-butyl completely outweighs the electronic preference, giving the opposite regioisomer exclusively.

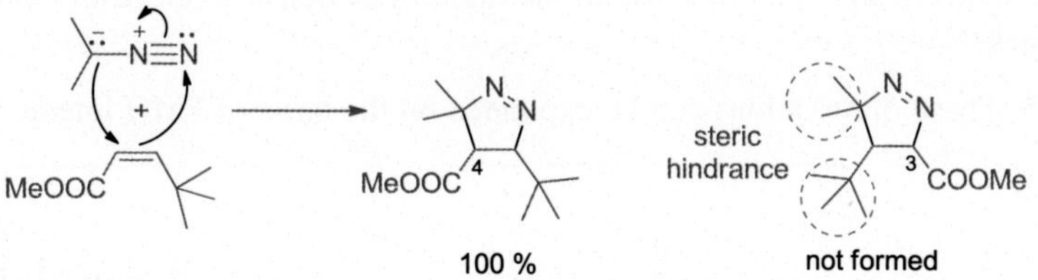

100 %

steric hindrance

not formed

Q 3. Explain the formation of the product in the following reactions.

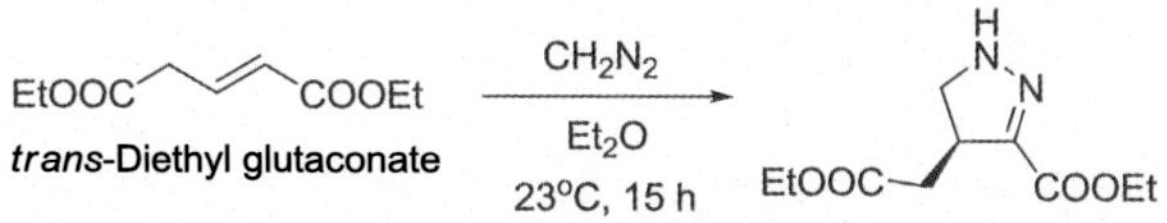

trans-Diethyl glutaconate $\xrightarrow[\text{23°C, 15 h}]{CH_2N_2,\ Et_2O}$

Sol 3. This reaction is regioselective because the diazo terminal carbon atom bonds exclusively to the β-carbon of the ester. The retention of configuration in the product with respect to both the 1,3-dipole and the dipolarophiles is a characteristic feature of 1,3-dipolar cycloadditions. Thus, stereochemistry of the substituents on the resulting five-membered cyclic ring entirely depends upon stereochemistry of the substituents on the dipolarophiles. Such a stereospecificity provides strong support for a concerted mechanism.

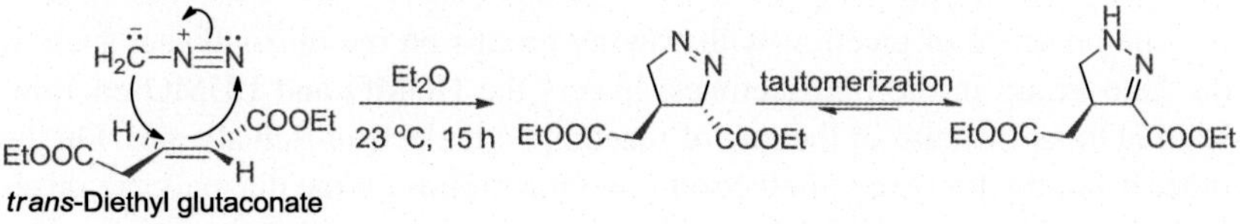

trans-Diethyl glutaconate $\xrightarrow[\text{23 °C, 15 h}]{Et_2O}$ → tautomerization

Q 4. Explain the following order of reactivity of diazomethane with various dipolarophiles.

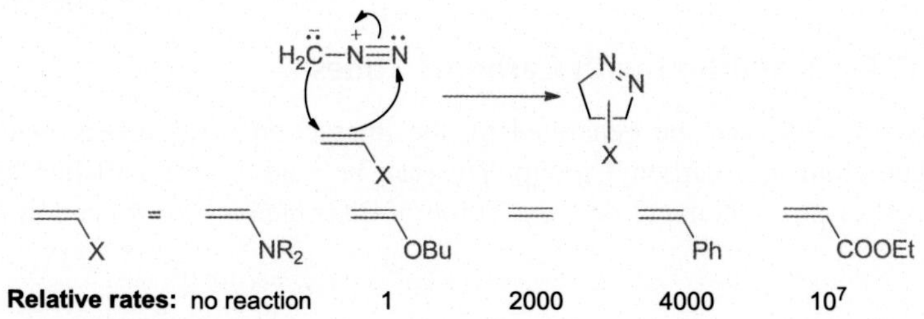

X = NR_2, OBu, H, Ph, COOEt

Relative rates: no reaction, 1, 2000, 4000, 10^7

Sol 4. In case of an azomethine ylide, the dipole has a high-lying HOMO that overlaps with LUMO of the dipolarophile (smaller HOMO–LUMO energy gap). A dipole of this class is referred to as a HOMO-controlled dipole or a nucleophilic dipole, which attacks electrophilic alkenes readily. Electron-withdrawing groups (EWG) on the dipolarophile would accelerate the reaction by lowering the LUMO, while electron-donating groups (EDG) would retard the reaction by raising the HOMO.

Q 5. How do you explain that the introduction of electron-withdrawing groups on the diazoalkane leads to an increase in the rate of reaction with electron-rich alkenes whereas the introduction of electron-donating groups on the diazoalkane leads to an increase in the rate of reaction with electron-deficient alkenes?

Sol 5. These observations can be explained on the basis of FMO interactions.

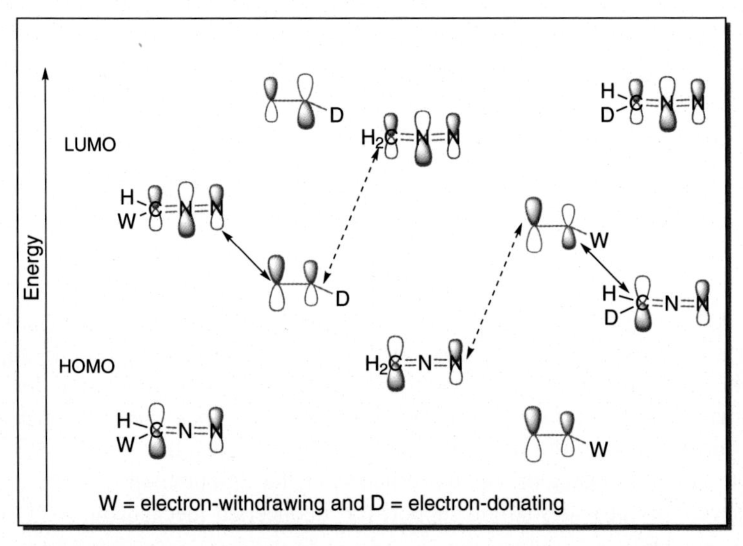

The presence of electron-withdrawing groups on the diazoalkane, such as the keto group in the diazoketones, lowers the HOMO and LUMO energies leading to an increase in the rate of reaction with electron-rich alkenes. On the other hand, the presence of electron-donating groups on the diazoalkane raises both the HOMO and LUMO energies, as a result of which the rate of cycloadditions with electron-deficient alkenes is increased due to decrease in energy separation between the frontier orbitals.

5.2.7 Cycloaddition with Carbonyl Ylides

Carbonyl ylides can be generated by the interaction of a carbene with the oxygen atom of a carbonyl group. This can be readily accomplished by the thermal or photochemical decomposition of diazomethane or transition metal catalyzed decomposition of an α-diazo ketone in the presence of a carbonyl group (Scheme 5.11).

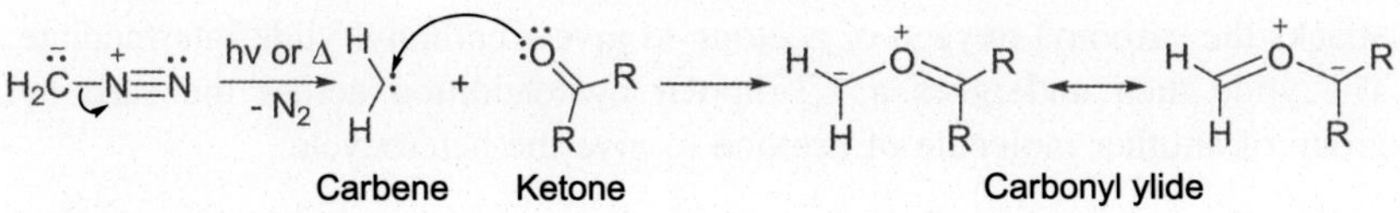

$H_2C-N\equiv N$ hv or Δ, - N_2
Carbene
Ketone
Carbonyl ylide

SCHEME 5.11 Generation of carbonyl ylides by decomposition of diazomethane in the presence of a carbonyl group.

If the reacting carbonyl and the carbene moiety are present in the same molecule, it will result in the formation of a cyclic carbonyl ylide (Scheme 5.12).

$Rh_2(OAc)_4$
Carbonyl ylide

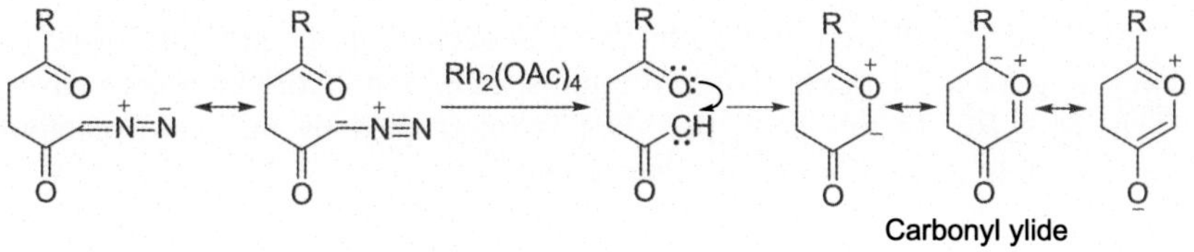

SCHEME 5.12 Intramolecular generation of carbonyl ylides.

Carbonyl ylides can also be prepared by electrocyclic epoxide openings. Under thermal and photochemical conditions, ring opening of an epoxide takes place through con- and disrotatory modes, respectively (Scheme 5.13).

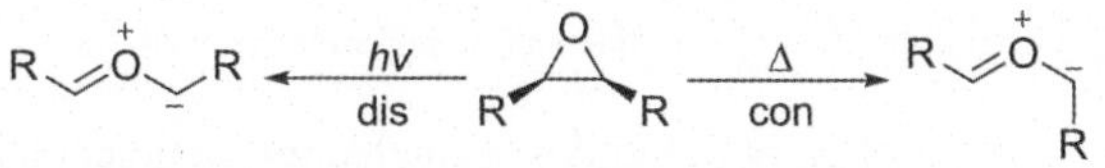

hv, dis
Δ, con

SCHEME 5.13 Thermal and photochemical ring opening of an epoxide.

Carbonyl ylides undergo 1,3-dipolar cycloaddition with alkenes and alkynes to yield tetrahydrofuran and 2,5-dihydrofuran derivatives, respectively (Scheme 5.14).

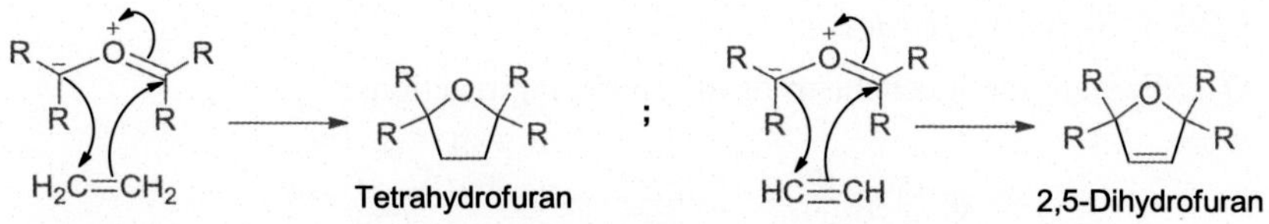

$H_2C{=}CH_2$
Tetrahydrofuran
;
$HC\equiv CH$
2,5-Dihydrofuran

SCHEME 5.14 1,3-Dipolar cycloaddition of carbonyl ylides with ethene and ethyne.

1,3-Dipolar cycloaddition of carbonyl ylides to carbonyl compounds gives dioxalanes. For example, the irradiation of diazomethane in acetone results in the formation of 2,2,4,4-tetramethyl-1,3-dioxolane as the major product (Scheme 5.15). Irradiation of diazomethane generates a singlet carbene, which attacks the carbonyl oxygen of acetone to give a carbonyl ylide intermediate. The ylide then undergoes a 1,3-dipolar cycloaddition across the carbonyl group of another molecule of acetone to give the heterocycle.

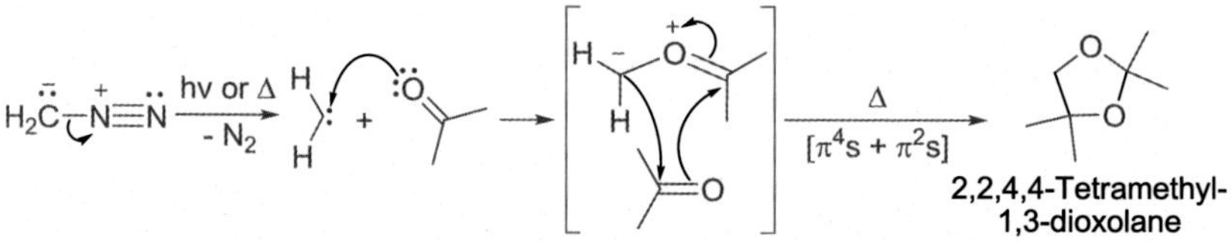

$H_2C-N\equiv N$ → hν or Δ, $- N_2$ → $[\pi^4s + \pi^2s]$, Δ → 2,2,4,4-Tetramethyl-1,3-dioxolane

SCHEME 5.15 Generation of carbonyl ylide and its addition to acetone.

However, irradiation of diazomethane in acetone in the presence of acrylonitrile gave a 2:1 mixture of two regioisomeric tetrahydrofuran derivatives (Scheme 5.16). In this case, the ylide is intercepted by the more reactive dipolarophile present in the reaction mixture.

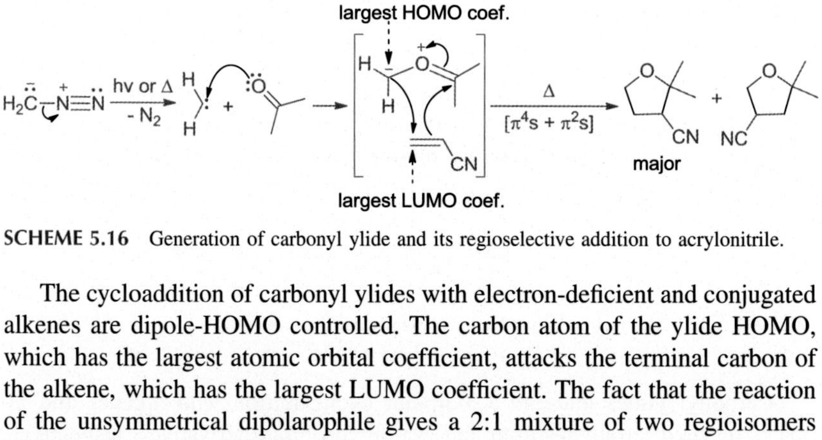

largest HOMO coef.

$H_2C-N\equiv N$ → hν or Δ, $- N_2$ → $[\pi^4s + \pi^2s]$, Δ → major (CN) + (NC)

largest LUMO coef.

SCHEME 5.16 Generation of carbonyl ylide and its regioselective addition to acrylonitrile.

The cycloaddition of carbonyl ylides with electron-deficient and conjugated alkenes are dipole-HOMO controlled. The carbon atom of the ylide HOMO, which has the largest atomic orbital coefficient, attacks the terminal carbon of the alkene, which has the largest LUMO coefficient. The fact that the reaction of the unsymmetrical dipolarophile gives a 2:1 mixture of two regioisomers indicates that in the HOMO of the carbonyl ylide, the electron density at the unsubstituted carbon is greater than that at the substituted carbon atom.

5.2.7.1 Solved Problems

Q 1. Explain the mechanism of the following reactions:

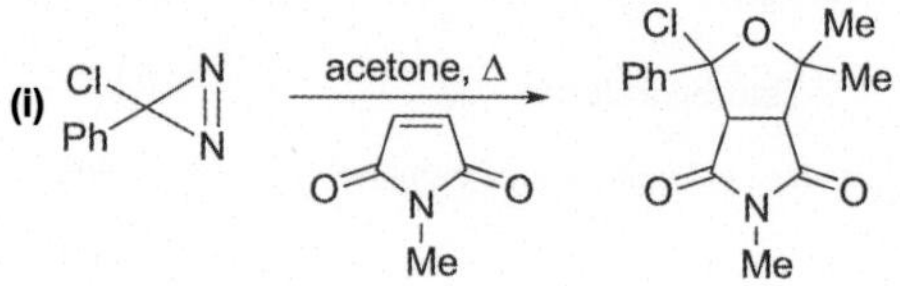

(i) Cl, Ph (diazirine) + N-methylmaleimide (O, N, Me, O) → acetone, Δ → Cl, Ph, O, Me, Me; O, N, Me, O

(ii) $Cu(acac)_2$, PhH, 80 °C

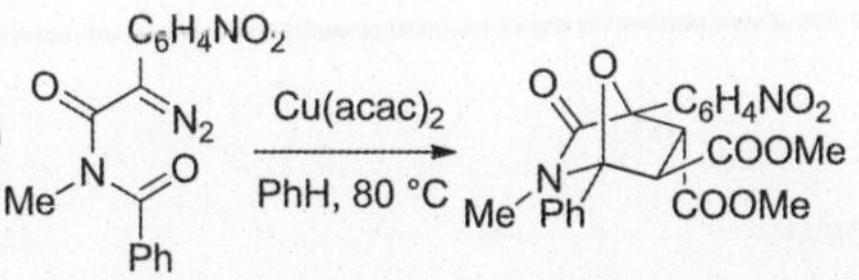

(iii) $Rh_2(OAc)_4$

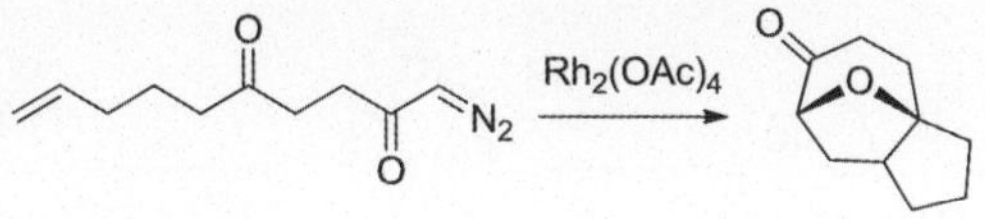

(iv) Δ, DMAD, benzene

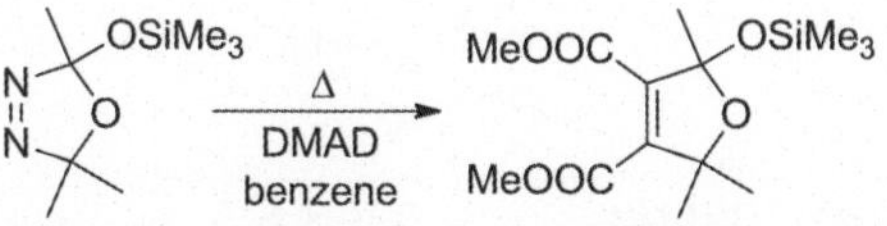

(v) A → B: (i) $Rh_2(OAc)_4$ (ii) H_3O^+

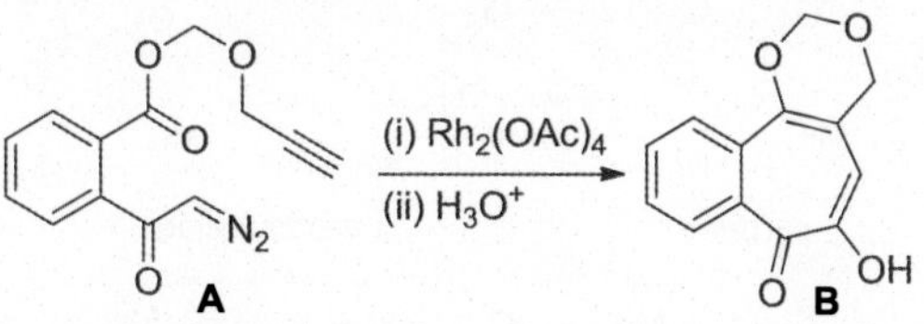

Compound **II** is protected as its methyl ether **III** and then transformed to **IV**. Provide a mechanism for the conversion of **I** to **IV**.

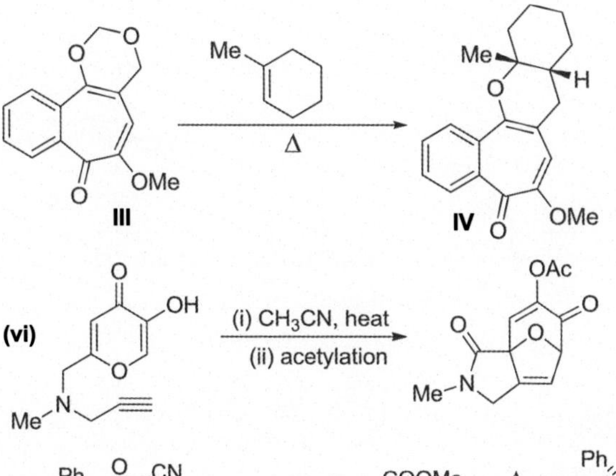

III → **IV**: Δ

(vi) (i) CH_3CN, heat (ii) acetylation

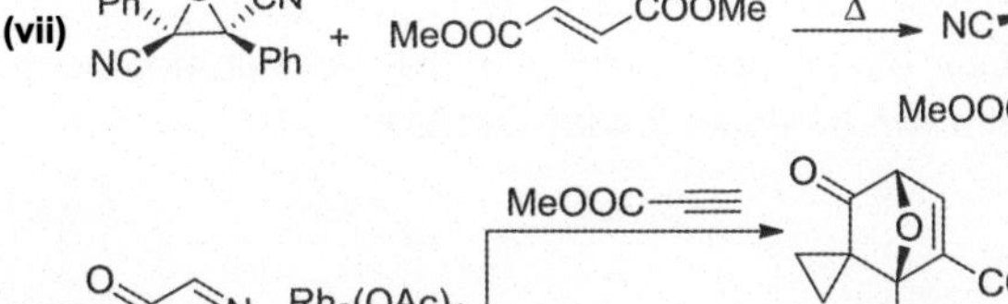

(vii) Δ

(viii) $Rh_2(OAc)_4$

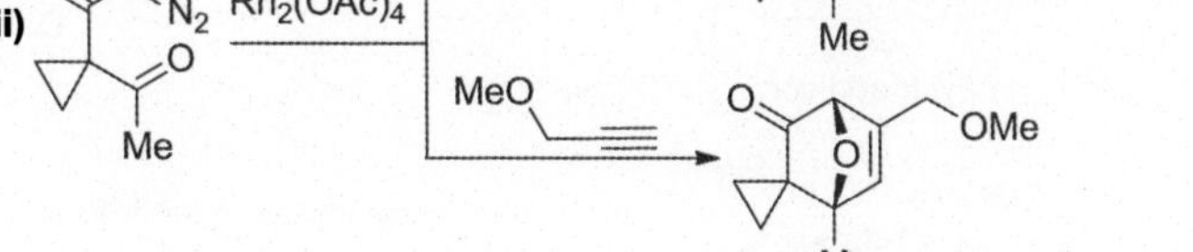

(ix) O, OAc + COOMe, COOMe →Δ→ O, COOMe, COOMe

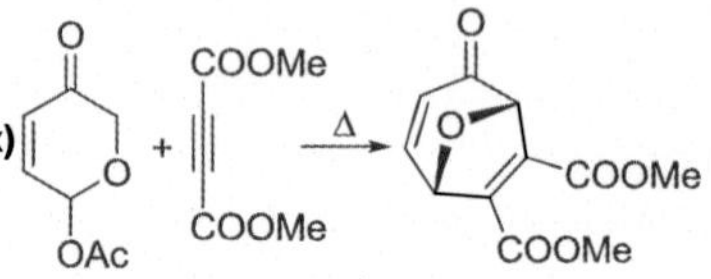

(x) O, Ph, O, Ph + →Δ→ O, Ph, O, Ph

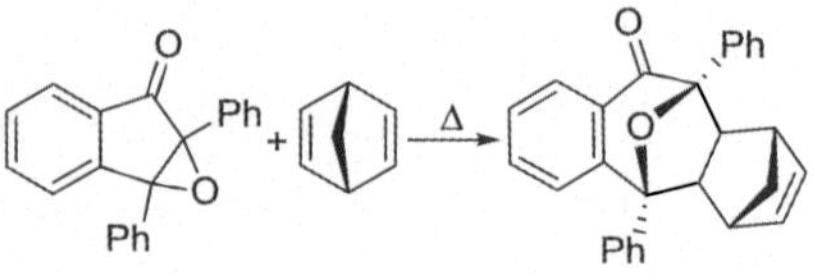

Sol 1. (i) Cl, Ph, N, N →Δ, - N_2→ Cl, Ph + :O: → Cl, Ph, O, + ; O, N, O, Me →Δ [$\pi^4s + \pi^2s$]→ Cl, Ph, O, Me, Me, O, N, O, Me

3-Chloro-3-phenyl
-3*H*-diazirine

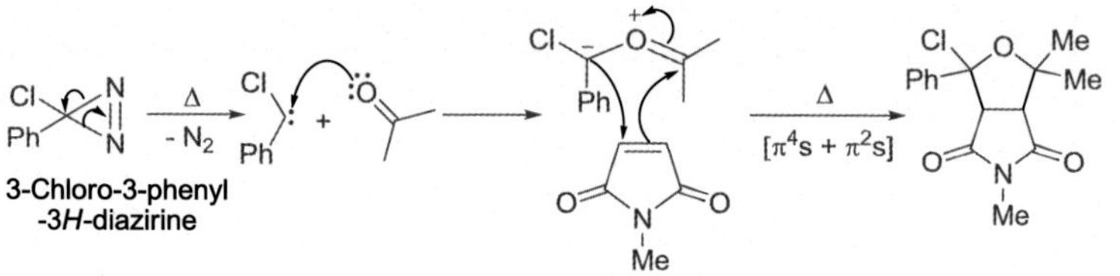

(ii) $C_6H_4NO_2$, O, N_2, Me, N, O, Ph →Cu(acac)$_2$, PhH, 80°C→ $C_6H_4NO_2$, O, O:, Me, N, Ph → $C_6H_4NO_2$, O, O+, Me, N, Ph ⟷ $C_6H_4NO_2$, O, O+, Me, N, Ph ⟷ etc.

carbene

carbonyl ylide

Me, N, O, Ph, O, +, $C_6H_4NO_2$, COOMe, MeOOC →Δ [$\pi^4s + \pi^2s$]→ Me, N, O, Ph, O, $C_6H_4NO_2$, COOMe, MeOOC ≡ Me, N, O, Ph, O, $C_6H_4NO_2$, MeOOC, COOMe ≡ O, O, $C_6H_4NO_2$, COOMe, N, Me, Ph, COOMe

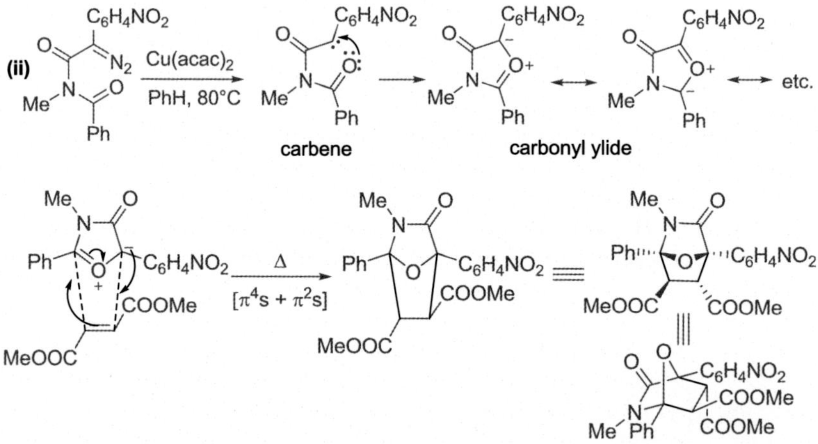

(iii) O, O, N_2 →$Rh_2(OAc)_4$→ O, O: → O, O, + →Δ [$\pi^4s + \pi^2s$]→ [O, O, +] → O, O ≡ O, O

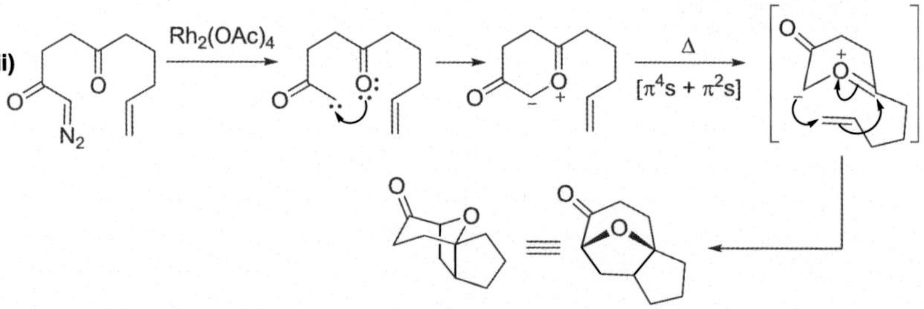

(iv) Oxadiazolines on heating undergo retro-cycloaddition to give the carbonyl ylide, which can be trapped with DMAD.

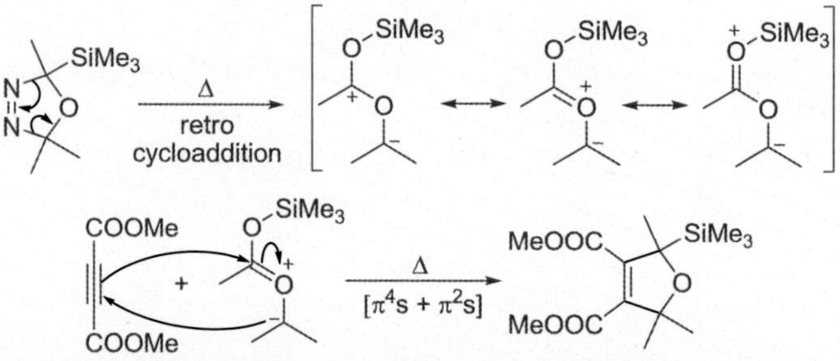

$SiMe_3$, N, N, O →Δ, retro cycloaddition→ [O, $SiMe_3$, +, O ⟷ O, $SiMe_3$, O, + ⟷ +, O, $SiMe_3$, O]

COOMe, COOMe + O, $SiMe_3$, O, + →Δ [$\pi^4s + \pi^2s$]→ MeOOC, MeOOC, $SiMe_3$, O

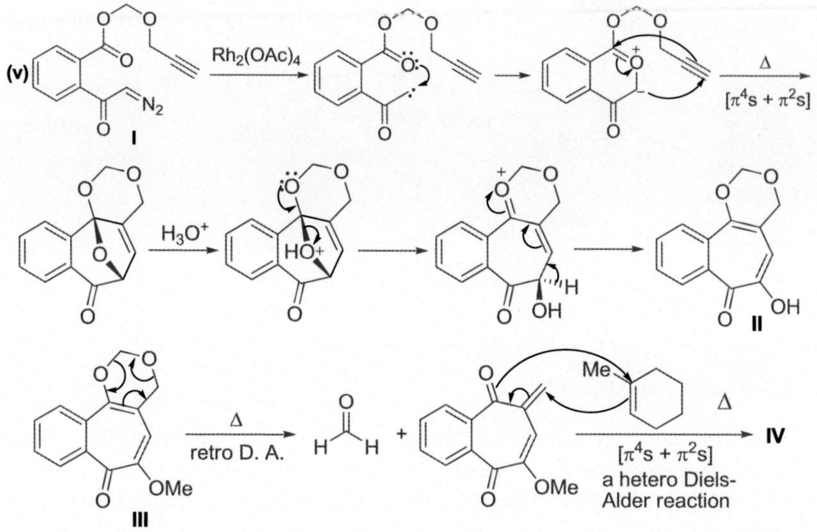

(vi) 5-Hydroxy-4-pyrone can also behave as cyclic carbonyl ylide, which is then trapped intramolecularly by the dipolarophile.

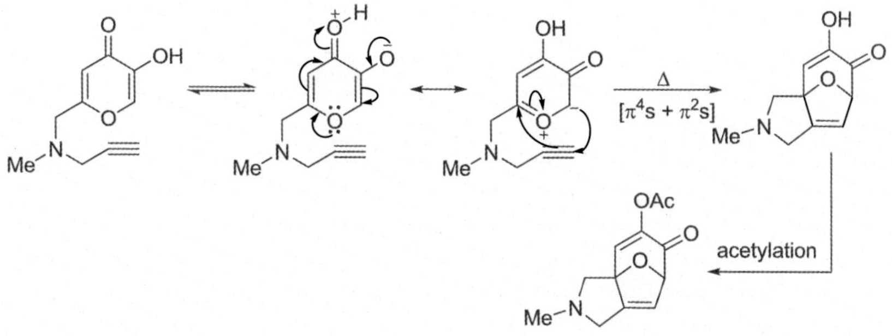

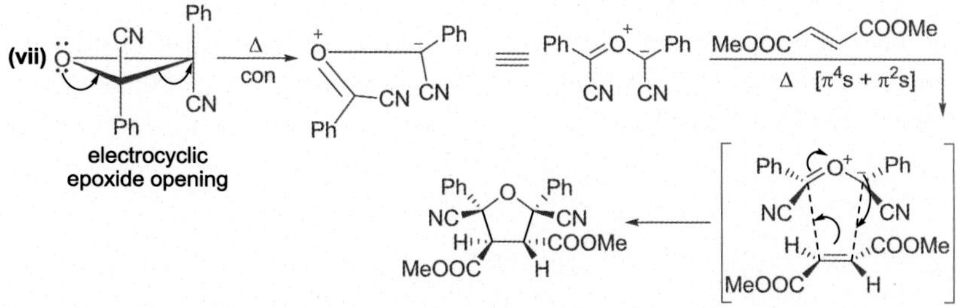

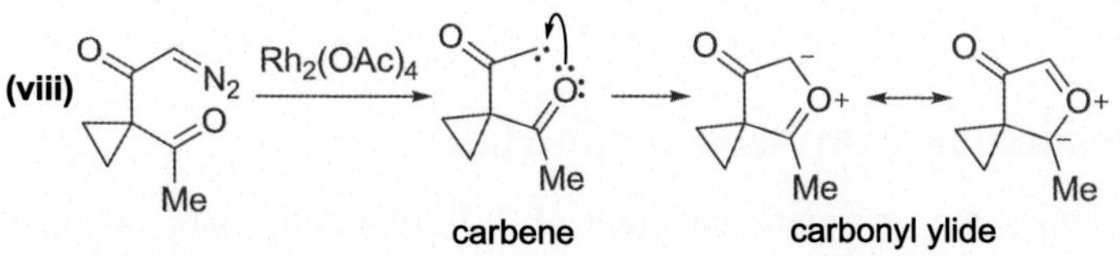

FMO can explain selectivity: $HOMO_{dipole}-LUMO_{dipolarophile}$ for electron-poor dipolarophiles; $LUMO_{dipole}-HOMO_{dipolarophile}$ for electron-rich dipolarophiles.

largest HOMO coef. largest LUMO coef.
O O+ Me COOMe Δ $[\pi^4s + \pi^2s]$ O O Me COOMe ≡ O O Me COOMe

O O+ Me OMe Δ $[\pi^4s + \pi^2s]$ O O Me OMe ≡ O O Me OMe
largest LUMO coef. largest HOMO coef.

(ix) The acetoxypyranone on exposure to either heat or a tertiary base, such as triethylamine, generates 3-oxidopyrylium, which behaves as the carbonyl ylide. The ylide is then trapped with DMAD to furnish the corresponding cycloadduct.

O H O OAc Et_3N or Δ O O: OAc → O O+ ↔ O O+ ↔ O O+
3-Oxidopyrylium

O O+ MeOOC—≡—COOMe Δ $[\pi^4s + \pi^2s]$ O O COOMe COOMe

(x) 2,3-Diphenylindenone oxide, upon thermolysis or photolysis, produces the benzopyrylium oxide. The benzopyrylium oxide behaves as a carbonyl ylide and undergoes cycloadditions with alkenes such as norbornadiene to produce the cycloadducts.

O Ph O: Ph Δ or $h\nu$ O Ph O+ Ph Δ O Ph O Ph

5.2.8 Cycloaddition with Azomethine Ylides

Azomethine ylides are an important class of 1,3-dipoles that are prepared in situ because of their unstable nature. Products are obtained by the subsequent addition of an appropriate dipolarophile. Two common methods for the preparation of azomethine ylides are as follows:

(i) Thermolysis or photolysis of suitably substituted aziridines: Thermolysis or photolysis of suitably substituted aziridines is a very convenient method for the generation of azomethine ylides. Thermal ring opening of aziridines involves a conrotatory pathway like that of a cyclopropyl anion with which it is isoelectronic. As predicted by the orbital symmetry rules, the ring opening of aziridines follows disrotatory mode under the photochemical conditions. The stereochemical consequences of these two modes of ring opening result in the formation of *trans*-azomethine ylide on thermolysis of *cis*-dicarboxylic ester and of *cis*-azomethine ylide on photolysis (Scheme 5.17). The *trans*-dicarboxylic acid ester behaves in the expected manner.

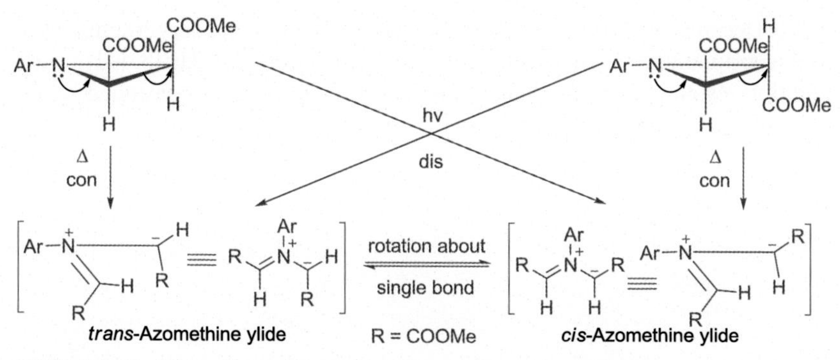

SCHEME 5.17 Thermal and photochemical ring opening of aziridines.

(ii) The fluorine-mediated desilylation of cyanoaminosilanes: Treatment of α-cyanoaminosilanes with silver fluoride results in the desilylation to give the intermediate anion that subsequently loses a cyanide ion to yield the azomethine ylide (Scheme 5.18).

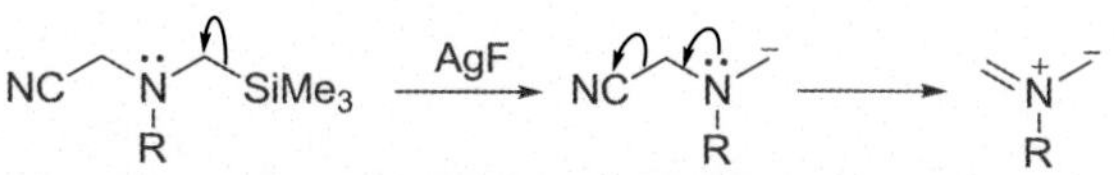

NC N(R) SiMe$_3$ —AgF→ NC N(R) —→ N$^+$(R)

SCHEME 5.18 Generation of azomethine ylide from α-cyanoaminosilanes.

Pyrrolidines and pyrrolines are often obtained by the 1,3-dipolar cycloaddition of azomethine ylides to alkenes and alkynes (Scheme 5.19).

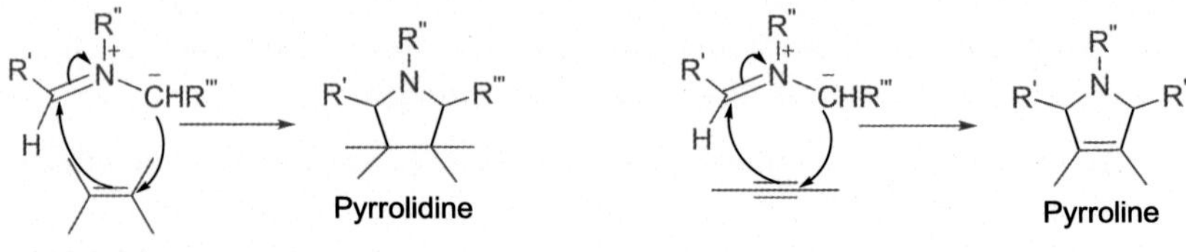

Pyrrolidine

Pyrroline

SCHEME 5.19 1,3-Dipolar cycloaddition of azomethine ylides to alkenes and alkynes.

5.2.8.1 Solved Problems

Q 1. Explain the mechanism of the following reaction:

$\xrightarrow[\text{HNTMS}_2]{\Delta}$

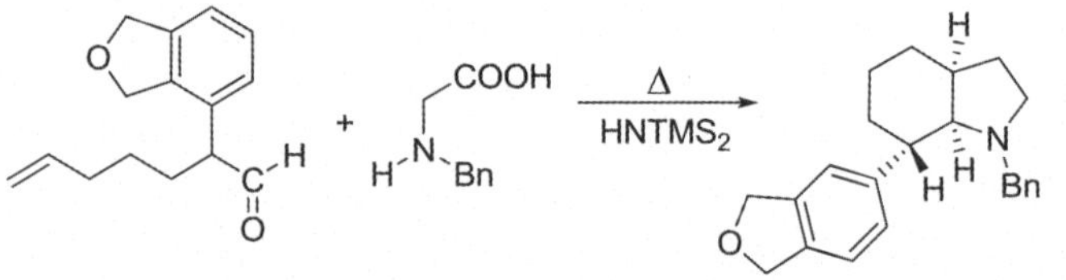

Sol 1. Decarboxylation of imminium ion derived from secondary α-amino acid yields azomethine ylide, which then undergoes intramolecular [3 + 2] cycloaddition reaction with an alkene. The diastereoselectivity is controlled by the formation of a less-strained *cis*-fused ring system.

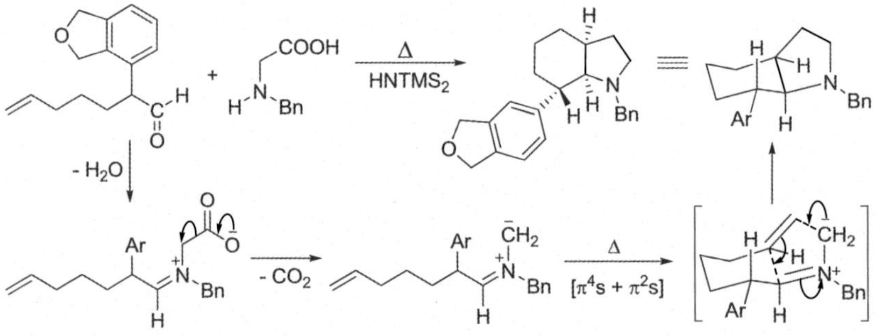

$- H_2O$; $- CO_2$; Δ $[\pi^4s + \pi^2s]$

Q 2. Explain the stereospecificity of the following reactions:

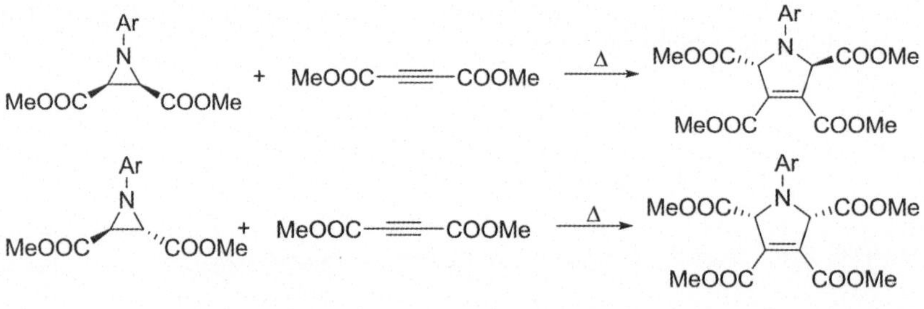

MeOOC—≡—COOMe $\xrightarrow{\Delta}$

Sol 2. 1,3-Dipolar cycloadditions usually take place stereospecifically and suprafacially, resulting in the retention of configuration with respect to both the 1,3-dipole and the dipolarophile.

In most cases, stereochemistry in the dipoles is lost due to the bond rotation as depicted by the resonance structures of such molecules. That the cycloaddition is a stereospecific process with respect to the dipolar species has, however, been demonstrated by a study of cycloaddition reactions of azomethine ylides. *cis*- and *trans*-azomethine ylides, obtained by the ring opening of the corresponding aziridines, are rapidly trapped by alkynes to give stereospecific products.

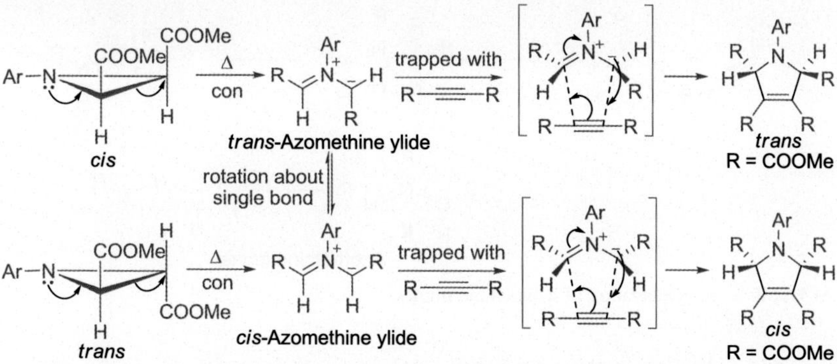

However, it is important for stereospecificity of the cycloaddition that it must take place before the bond rotation in the azomethine ylide. There is a stereospecific addition of both *cis*- and *trans*-azomethines with highly reactive dipolarophiles such as tetracyanoethylene. But in the presence of a weak dipolarophile, conversion of *cis*- to *trans*-isomer (here, azomethine ylides) becomes a competing process.

Q 3. The major product formed in the following reaction is:

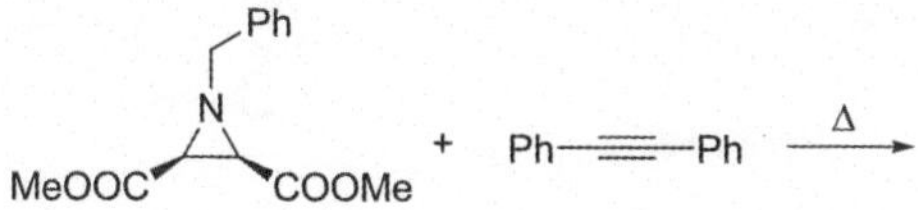

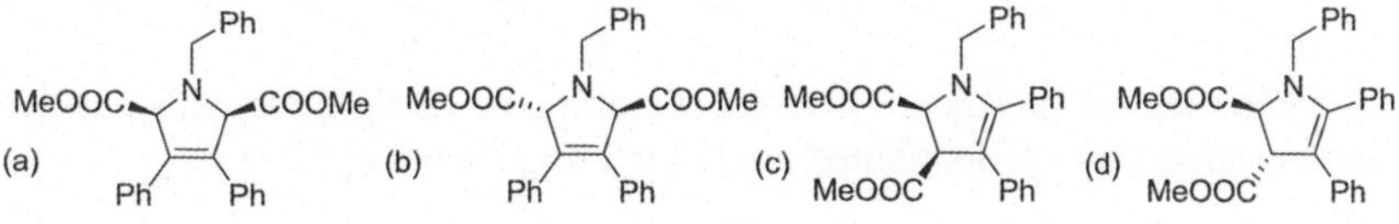

Sol 3. (b)

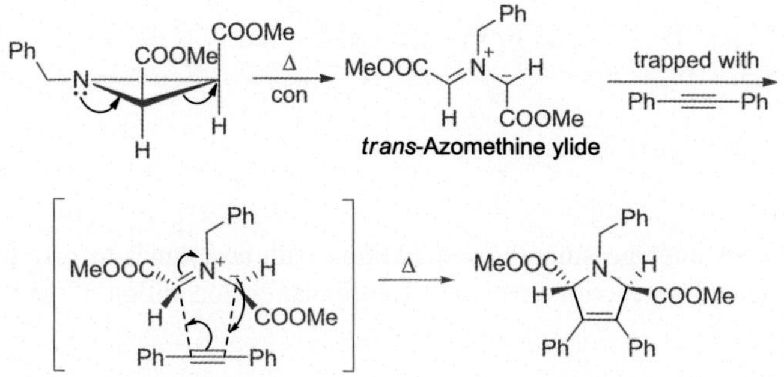

5.2.9 Cycloaddition with Azomethine Imines

High reactivity of these dipoles prevents their isolation and they are normally generated in situ, most commonly by the reaction of *N,N'*-disubstituted hydrazines with an aldehyde (Scheme 5.20).

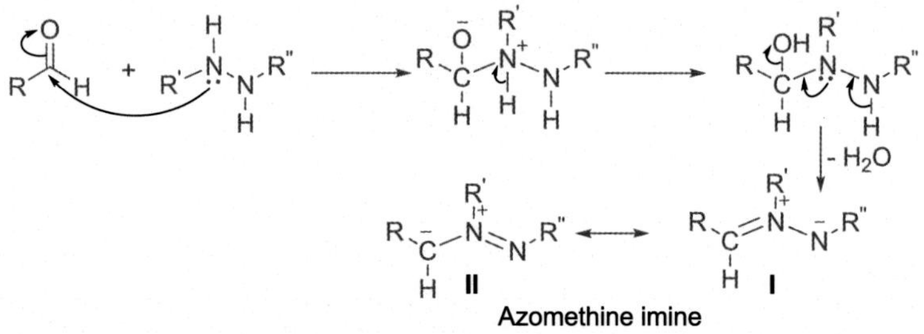

Azomethine imine

SCHEME 5.20 Generation of azomethine imines.

The resonance form **I** contributes more to the hybrid as a result of the higher electronegativity of nitrogen relative to carbon.

Azomethine imines readily undergo 1,3-dipolar cycloaddition reactions with alkenes and alkynes to furnish pyrazolidines and pyrazolines, respectively (Scheme 5.21).

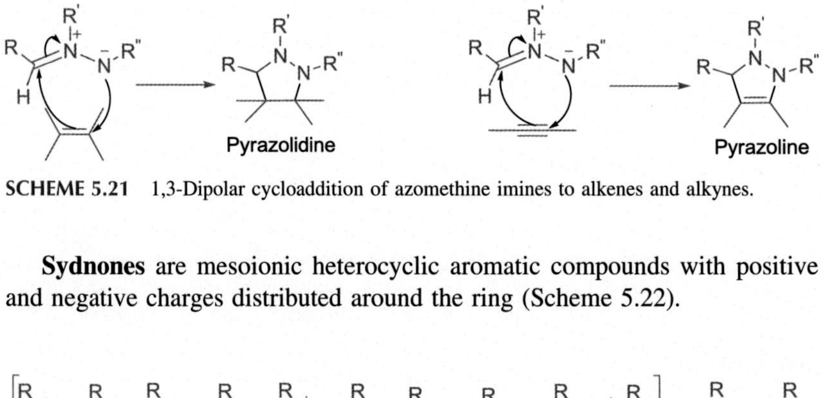

Pyrazolidine

Pyrazoline

SCHEME 5.21 1,3-Dipolar cycloaddition of azomethine imines to alkenes and alkynes.

Sydnones are mesoionic heterocyclic aromatic compounds with positive and negative charges distributed around the ring (Scheme 5.22).

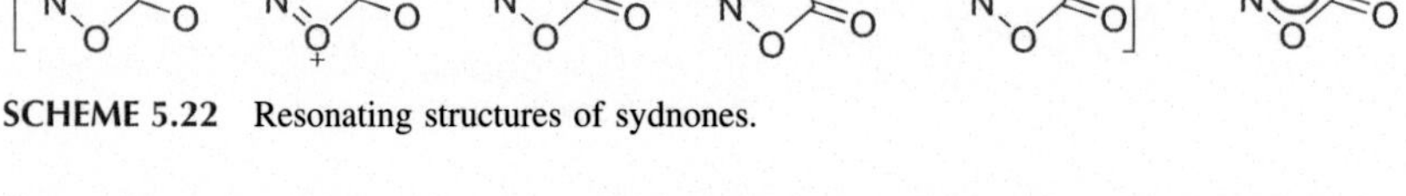

SCHEME 5.22 Resonating structures of sydnones.

Sydnones undergo smooth cycloaddition with acetylenes to give pyrazoles in high yield. The reaction involves a 1,3-dipolar cycloaddition of the sydnones, behaving like a cyclic azomethine imine to the corresponding acetylene, followed by carbon dioxide evolution and aromatization (Scheme 5.23).

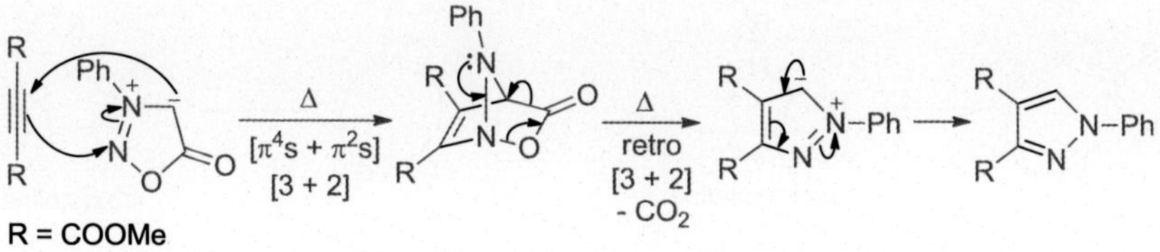

SCHEME 5.23 Cycloaddition of sydnones to acetylene derivative.

The 1,3-dipolar cycloaddition between a sydnone and 1,5-cyclooctadiene provides 9,10-diazatetracycloundecanes (Scheme 5.24).

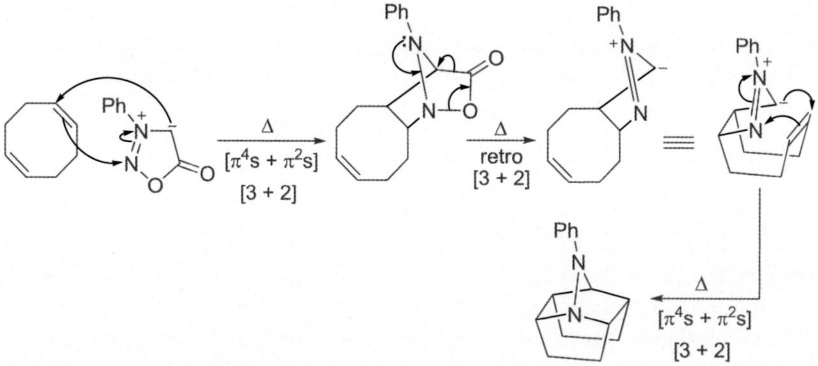

SCHEME 5.24 1,3-Dipolar cycloaddition reaction between sydnones and 1,5-cyclooctadiene.

5.2.10 Cycloaddition with Nitrones

Nitrones are stable compounds and thus do not require in situ generation. The two commonly employed methods for the preparation of nitrones are: (1) reaction of an aldehyde or ketone with a monosubstituted hydroxylamine and (2) oxidation of a hydroxylamine with yellow mercuric oxide (Scheme 5.25).

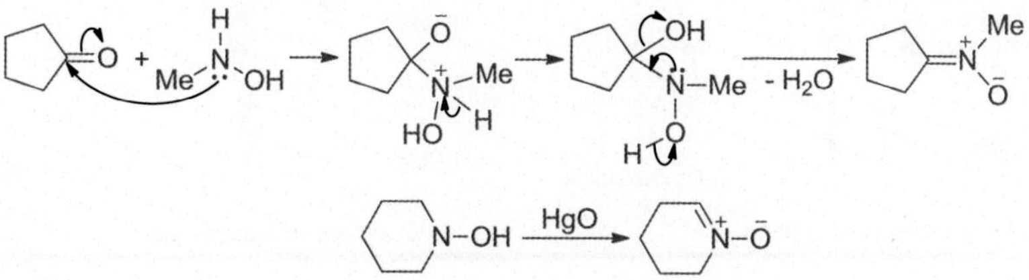

SCHEME 5.25 Two commonly employed methods to generate nitrones.

Nitrones undergo cycloadditions with alkenes and alkynes to generate isoxazolidines and isoxazolines, respectively (Scheme 5.26).

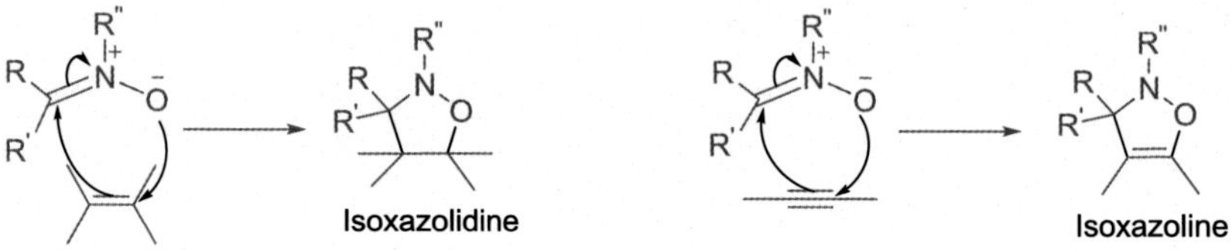

Isoxazolidine

Isoxazoline

SCHEME 5.26 Cycloaddition of nitrones with alkenes and alkynes.

Two regioisomeric cycloadducts are obtained in the cycloaddition reaction of nitrones and monosubstituted alkenes (Scheme 5.27).

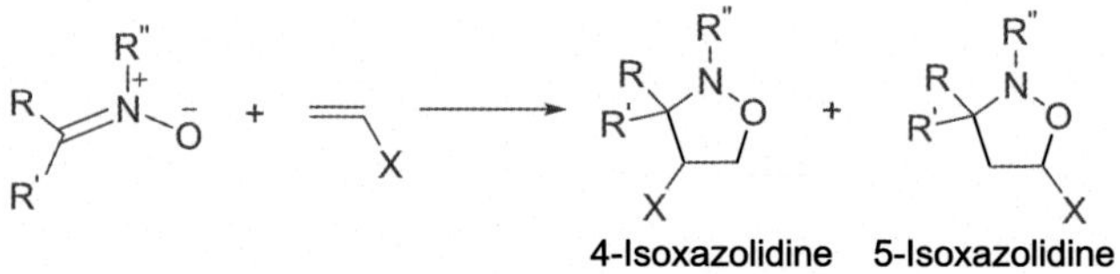

4-Isoxazolidine 5-Isoxazolidine

SCHEME 5.27 Cycloaddition of nitrone to an unsymmetrical alkene.

The $LUMO_{dipole}$–$HOMO_{dipolarophile}$ interaction controls the stereochemistry of the products in the cycloaddition reaction between nitrone and an electron-rich terminal alkene. The regioselective formation of the 5-isoxazolidine in this reaction can be explained on the basis that $LUMO_{dipole}$ has a larger coefficient at the carbon atom and the $HOMO_{dipolarophile}$ has a larger coefficient at the terminal carbon atom (Figure 5.19). On the other hand, cycloaddition reaction between nitrone and terminal alkenes with electron-withdrawing group favors the formation of the other isomer, 4-isoxazolidine. The reaction is thus controlled by the $HOMO_{dipole}$–$LUMO_{dipolarophile}$

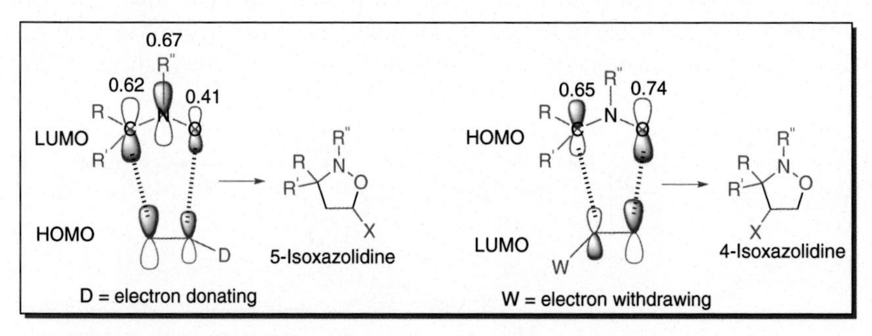

FIGURE 5.19 HOMO–LUMO interactions in cycloaddition reactions involving nitrones.

interaction. In this situation, the $HOMO_{dipole}$ has the largest coefficient at the oxygen atom and the $LUMO_{dipolarophile}$ has the largest coefficient at the terminal carbon atom. Formation of the 4-isomer is thus a preferred process.

As compared to the Diels–Alder reaction, the 1,3-dipolar cycloaddition of nitrones with olefins generally exhibits lower levels of regio- and stereo-control (*exo*/*endo* selectivity), which is a consequence of significant contribution by both $LUMO_{dipole}$–$HOMO_{dipolarophile}$ and $HOMO_{dipole}$–$LUMO_{dipolarophile}$ interactions (Scheme 5.28). The process is further complicated by the possibility of interconversion of the nitrone geometry in the case of acyclic nitrones. For example, the cycloaddition between methyl acrylate and C,N-diphenylnitrone is only marginally in favor of the *exo* mode.

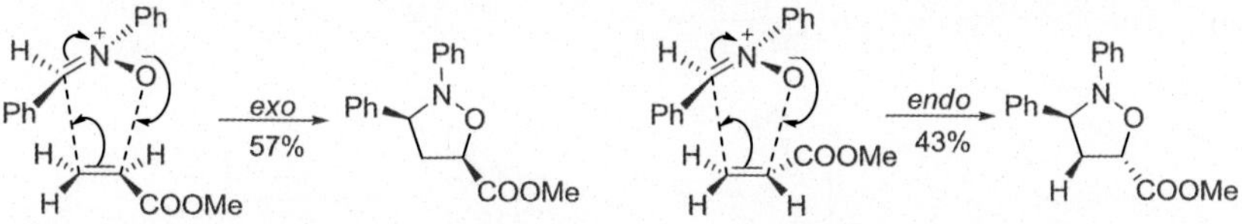

SCHEME 5.28 The *exo*/*endo* selectivity in 1,3-dipolar cycloaddition of nitrones with olefins.

In the absence of *E*/*Z* isomerization, cyclic nitrones show greater stereoselectivity. Further, the *exo* transition state is usually sterically favored with cyclic nitrones (Scheme 5.29).

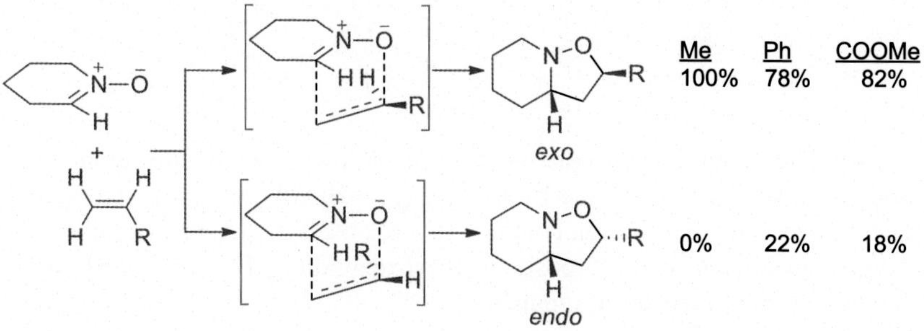

SCHEME 5.29 The *exo*/*endo* selectivity in 1,3-dipolar cycloaddition of cyclic nitrones with olefins.

5.2.10.1 Solved Problems

Q 1. Explain the mechanism of the following reactions:

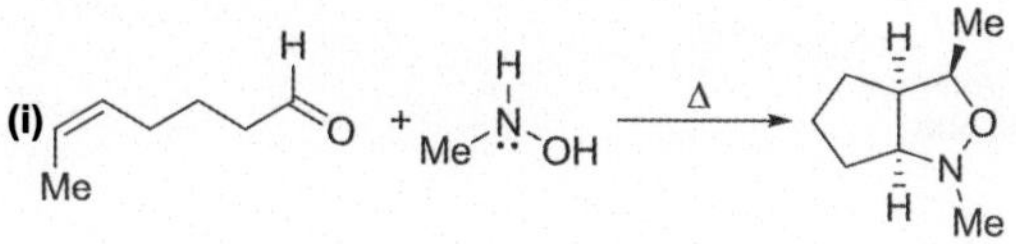

(i)

(ii) N OH H + O H H Δ O N + O N

(iii) N O + O Δ N O H O H H major + N O H O H H minor

Sol 1.

(i) H O Me + H N Me OH Δ - H_2O Me N Me O ≡ Me N Me O

Δ $[\pi^4s + \pi^2s]$

Me H O H N+ Me H

H Me O N H Me

orientation probably driven by ring strain

(ii) N OH H + O H H Δ - H_2O

O N+ Δ $[\pi^4s + \pi^2s]$ O N ≡ N O ≡ O N

O N+ Δ $[\pi^4s + \pi^2s]$ O N ≡ O N ≡ O N

(iii) The cycloaddition reaction between nitrone and dihydrofuran (electron-rich alkene) is controlled by $LUMO_{dipole}$–$HOMO_{dipolarophile}$ interaction. The nitrone and dihydrofuran combine in a regioselective manner to place the oxygen substituent at the 5-position of isoxazolidine. The *exo*-selectivity is obtained to minimize steric repulsion.

largest LUMO coeff.

largest HOMO coeff.

N O O

electron-rich alkene

N O O → N O O H H H ≡ N O H O H H

endo (minor)

N O O → N O H O H H ≡ N O H O H H

exo (major)

5.2.11 Cycloaddition with Nitrile Ylides

Nitrile ylides can be prepared by the following methods:

(i) By the dehydrohalogenation of an imidoyl chloride on treatment with a base such as triethylamine (Scheme 5.30).

$$R{-}C(Cl){=}N{-}CHR'_2 \xrightarrow[-\ Et_3N.HCl]{Et_3N} \underset{\text{Nitrile ylide}}{R{-}C{\equiv}\overset{+}{N}{-}\overset{-}{C}R'_2}$$

SCHEME 5.30 Nitrile ylide from imidoyl chloride.

(ii) By the desilylation of silylthiomidate with silver fluoride (Scheme 5.31).

$$Me{-}C(SPh){=}N{-}CH_2SiMe_3 \xrightarrow[-\ Me_3SiF,\ -\ ^{-}SPh]{AgF} \underset{\text{Nitrile ylide}}{Me{-}C{\equiv}\overset{+}{N}{-}\overset{-}{C}H_2}$$

SCHEME 5.31 Nitrile ylide from desilylation of silylthiomidate.

(iii) By the reaction of carbene with nitrile (Scheme 5.32).

$$Me{-}C{\equiv}N + :CH_2 \longrightarrow \underset{\text{Nitrile ylide}}{Me{-}C{\equiv}\overset{+}{N}{-}\overset{-}{C}H_2}$$

$$\underset{\text{Diazomethane}}{CH_2N_2} \text{ or } \underset{\text{Diazirine}}{H_2C{<}(N{=}N)} \xrightarrow[-\ N_2]{hv} :CH_2$$

SCHEME 5.32 Nitrile ylide from the reaction of carbene with nitrile.

(iv) By the photolytic ring opening of azirines (Scheme 5.33).

$$\underset{\text{Azirine}}{\text{Azirine (R, R')}} \xrightarrow{hv} \underset{\text{Nitrile ylide}}{R{-}C{\equiv}\overset{+}{N}{-}\overset{-}{C}HR'}$$

SCHEME 5.33 Nitrile ylide from photolytic ring opening of azirines.

Nitrile ylides undergo cycloadditions with alkenes and alkynes to yield 1-pyrroline and 2*H*-pyrroles, respectively (Scheme 5.34).

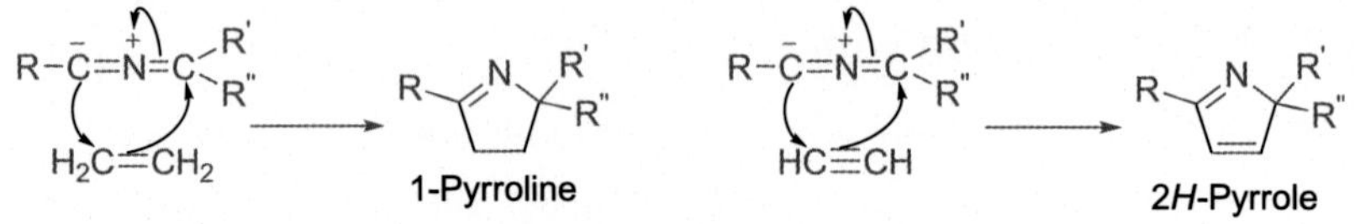

SCHEME 5.34 Cycloaddition involving nitrile ylide.

When a solution of 3-phenyl-2*H*-azirine in excess methyl acrylate or acrylonitrile is photolyzed, 2-phenyl-4-substituted-1-pyrroline is obtained (Scheme 5.35).

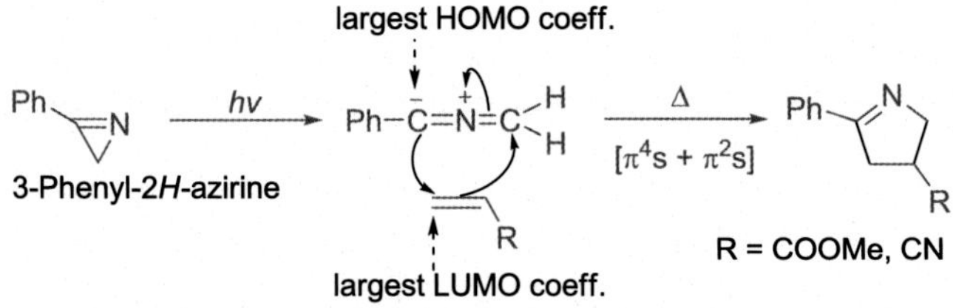

SCHEME 5.35 Regioselective cycloaddition involving nitrile ylide.

With nitrile ylides, the favored cycloadduct is the one formed by the bonding of atoms with the largest coefficients in the $HOMO_{dipole}$ and $LUMO_{dipolarophile}$. In the HOMO of nitrile ylide under consideration, the electron density at the disubstituted carbon is somewhat greater than at the trisubstituted carbon. In electron-deficient olefins, the largest coefficient in the LUMO is on the unsubstituted carbon. This treatment satisfactorily explains the reaction of nitrile ylide obtained from diphenyl azirine with methyl acrylate to yield only the 4-substituted regioisomer.

5.2.11.1 Solved Problem

Q 1. The following conversion involves:

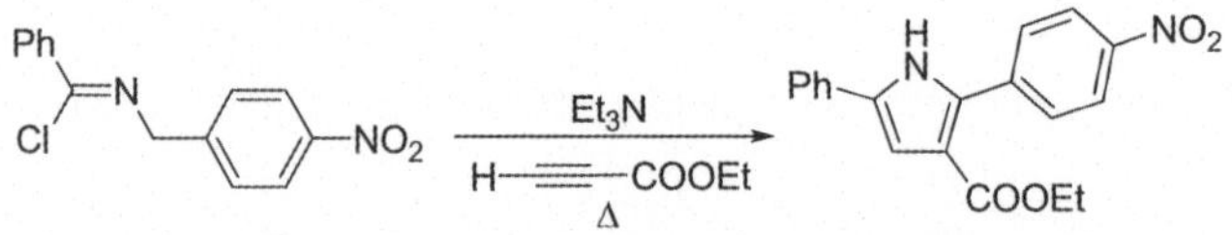

(a) A 1,3-dipolar species as reactive intermediate, and a cycloaddition
(b) A carbocation as reactive intermediate, and a cycloaddition
(c) A 1,3-dipolar species as reactive intermediate, and an aza-Wittig reaction
(d) A carbanion as reactive intermediate, and an aza-Cope rearrangement

Sol 1. (a) The treatment of *N*-*p*-nitrobenzylbenzimidoyl chloride with triethylamine gives the nitrile ylide (1,3-dipolar species). The dipole is then trapped with ethyl propiolate by [3 + 2] cycloaddition reaction.

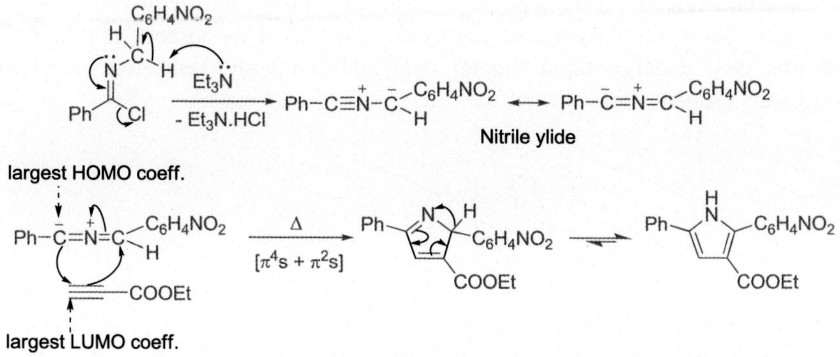

5.2.12 Cycloaddition with Nitrile Oxides

A convenient method to prepare nitrile oxides involves dehydrohalogenation of hydroximic acid halides, which are readily obtained from the corresponding aldoximes by treating them with *N*-chlorosuccinimide (NCS) or *N*-bromosuccinimide (NBS) or nitrosyl chloride or X_2 (Scheme 5.36). A tertiary amine base, such as triethylamine, is commonly used for the dehydrohalogenation of hydroximic acid halides.

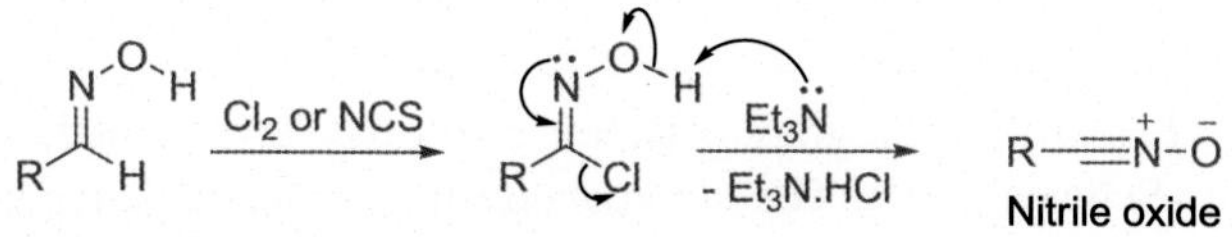

SCHEME 5.36 Generation of nitrile oxides from hydroximic acid halides.

Aliphatic nitrile oxides can also be prepared by the dehydration of primary nitro compounds with phenylisocyanate in the presence of a catalytic amount of a tertiary base such as triethylamine. Phenylcarbamate generated in situ on reaction with phenylisocyanate is subsequently converted into urea derivative (Scheme 5.37).

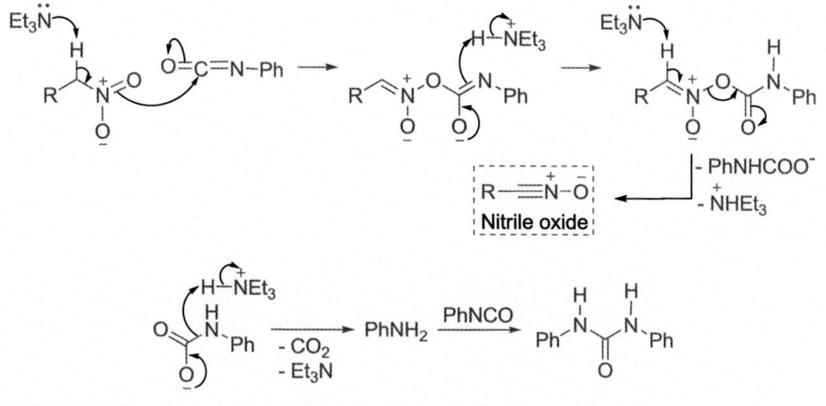

SCHEME 5.37 Generation of nitrile oxides from primary nitro compounds.

Nitrile oxides are usually prepared in the presence of the olefin or acetylene acceptor. Most nitrile oxides are highly reactive and in the absence of trapping agents they undergo rapid dipolar cycloaddition with themselves to give furoxans (Scheme 5.38).

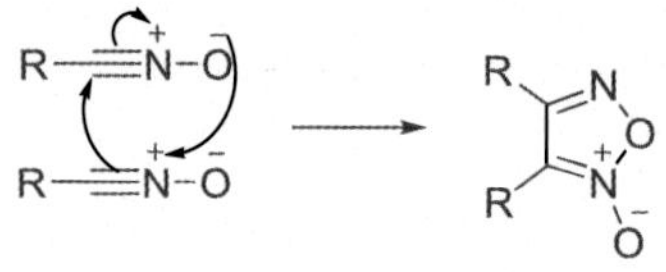

SCHEME 5.38 Cycloaddition of nitrile oxides to give furoxans.

Cycloaddition between a nitrile oxide and an alkene or alkyne yields cyclic products, isoxazoline or isoxazole, respectively (Scheme 5.39).

SCHEME 5.39 Cycloaddition reaction between a nitrile oxide and an alkene or alkyne.

The cycloaddition reactions between nitrile oxide and olefins are stereospecific (Scheme 5.40). The stereochemistry of the substituents present in alkene remains preserved in the final product. The treatment of nitrile oxide with *trans*-alkene gives *trans*-product whereas with *cis*-alkene gives *cis*-product.

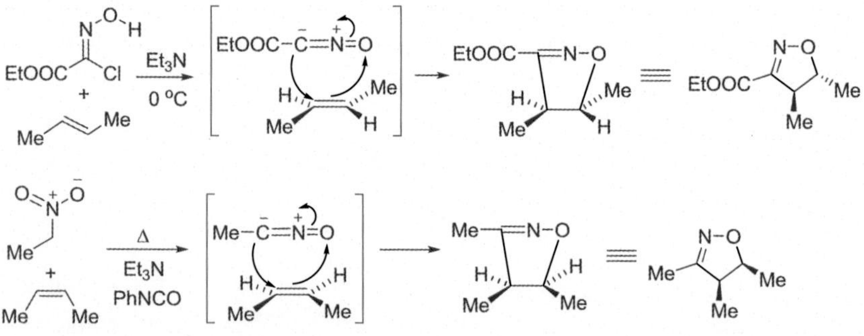

SCHEME 5.40 Stereospecificity in the cycloaddition reaction between nitrile oxide and olefins.

With the regioselectivity depending upon the electronic and steric factors, 1,3-dipolar cycloaddition reaction of a nitrile oxide and a monosubstituted alkene yields two regioisomeric cyclic structures, either the 4-substituted or 5-substituted isoxazolines (Scheme 5.41).

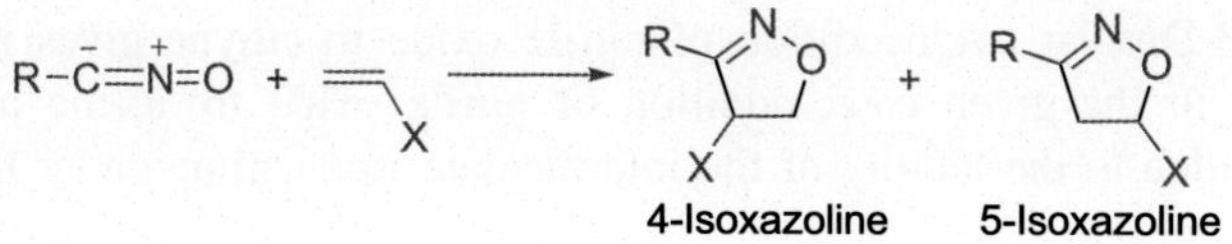

SCHEME 5.41 Regioselectivity in the cycloaddition reactions between nitrile oxide and a monosubstituted alkene.

However, there is an exclusive formation of 5-substituted isoxazolines in the cycloaddition reaction of nitrile oxides with electron-rich and conjugated alkenes. These reactions are $LUMO_{dipole}$ controlled, and the carbon atom of the nitrile oxide attacks the terminal carbon of the alkene (Figure 5.20).

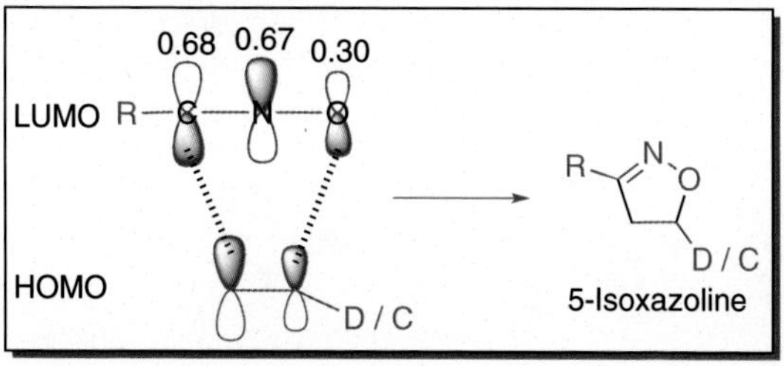

FIGURE 5.20 $LUMO_{dipole}$-controlled cycloaddition of nitrile oxide with an electron-rich alkene.

For electron-deficient dipolarophiles, both the dipole HOMO and LUMO interactions are significant and a mixture of regioisomers results (Figure 5.21). The 4-substituted isoxazoline is favored when strongly electron-withdrawing substituents (such as the sulfono group) are present on the dipolarophile.

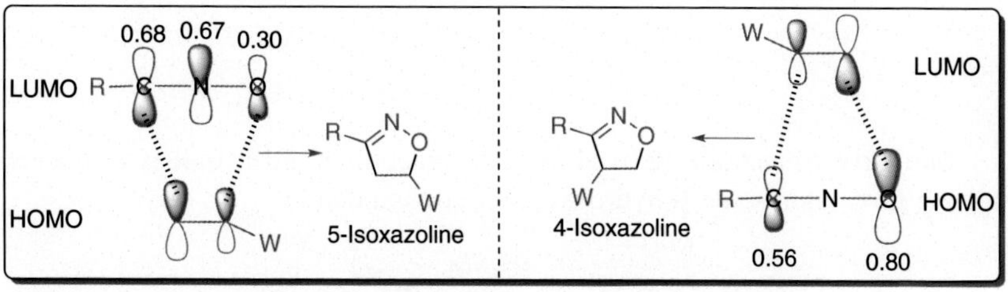

FIGURE 5.21 Cycloaddition of nitrile oxide with an electron-deficient alkene.

5.2.12.1 Solved Problems

Q 1. The major product **I** observed in the following reaction sequence is:

$$\mathrm{H{-}C{\equiv}\overset{+}{N}{-}\overset{-}{O}} + \text{CH}_2{=}\text{CH}{-}\text{OAc} \xrightarrow{\Delta} \mathbf{I} \xleftarrow{\Delta} \mathrm{HC{\equiv}CH} + \mathrm{H{-}C{\equiv}\overset{+}{N}{-}\overset{-}{O}}$$

Sol 1. 1,3-Dipolar cycloaddition of nitrile oxide to ethyne gives isoxazole. However, in the given case, addition of nitrile oxide to alkene also gives isoxazole due to the lability of the intermediate isoxazoline under the experimental conditions.

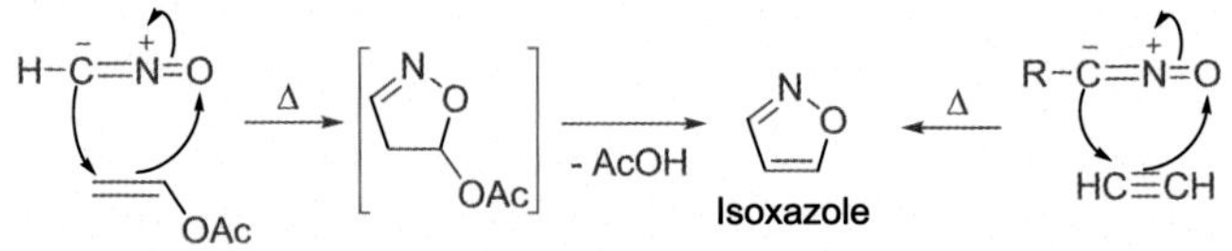

Isoxazole

Q 2. The major products **I** and **II**, respectively, in the following reaction sequence are:

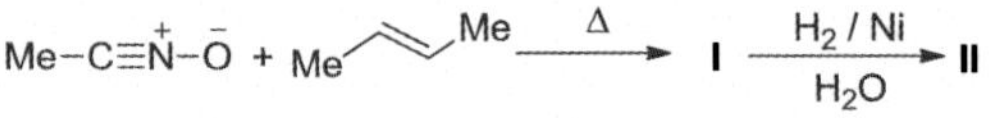

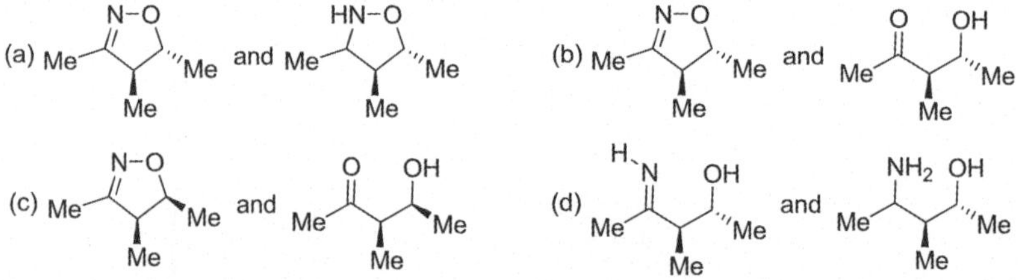

$$\mathrm{Me{-}C{\equiv}\overset{+}{N}{-}\overset{-}{O}} + \text{Me–CH=CH–Me} \xrightarrow{\Delta} \mathbf{I} \xrightarrow[H_2O]{H_2/Ni} \mathbf{II}$$

(a) ... and ...
(b) ... and ...
(c) ... and ...
(d) ... and ...

Sol 2. (b) It is an example of a widely used masked-aldol reaction. 1,3-Dipolar cycloaddition of nitrile oxides with an alkene (or alkyne) gives a cyclic product, isoxazoline (or isoxazole). These cyclic compounds are readily cleaved by reduction of the N—O bond and subsequent hydrolysis of the resulting imine to give aldol-type β-hydroxycarbonyl (or Claisen-type β-dicarbonyl) products.

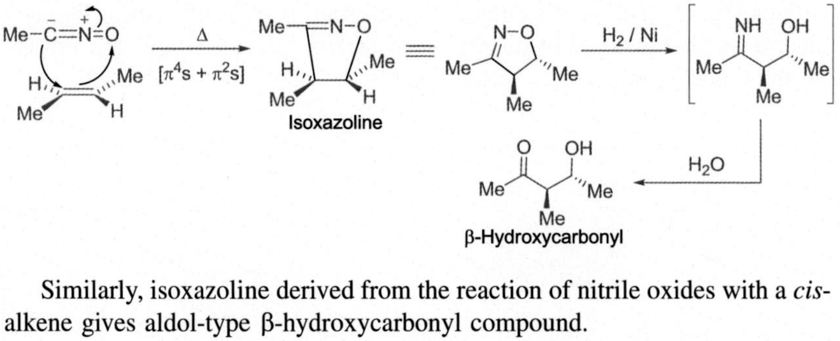

Δ $[\pi^4s + \pi^2s]$ — Isoxazoline — H_2 / Ni — H_2O — β-Hydroxycarbonyl

Similarly, isoxazoline derived from the reaction of nitrile oxides with a *cis*-alkene gives aldol-type β-hydroxycarbonyl compound.

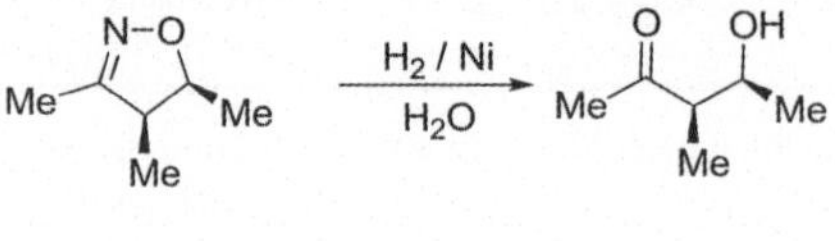

H_2 / Ni, H_2O

However, isoxazole gives β-dicarbonyl compound.

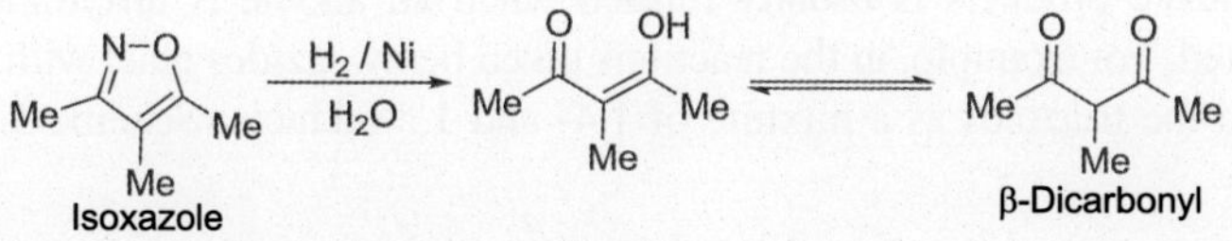

Isoxazole H_2 / Ni, H_2O β-Dicarbonyl

5.2.13 Cycloaddition with Azides

Organic azides (RN_3) can be prepared by several methods, for example, by nucleophilic substitution using ionic azides, addition reactions, insertion of a N_2 group (diazo transfer), insertion of a nitrogen atom (diazotization), cleavage of triazines and analogous compounds, and rearrangement of azides (Figure 5.22).

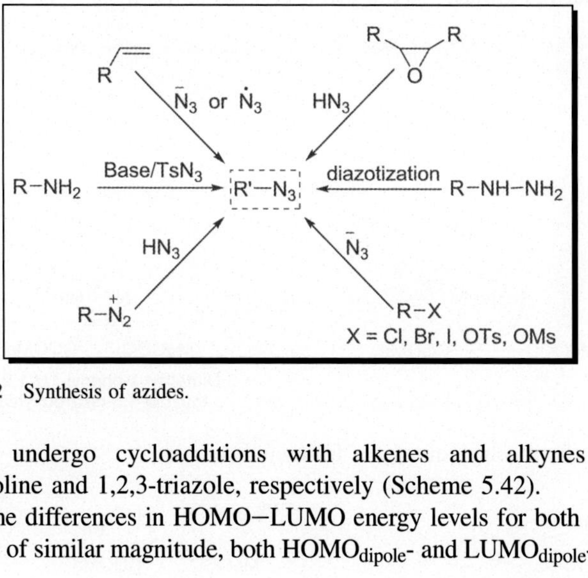

FIGURE 5.22 Synthesis of azides.

Azides undergo cycloadditions with alkenes and alkynes to yield 1,2,3-triazoline and 1,2,3-triazole, respectively (Scheme 5.42).

Since the differences in HOMO–LUMO energy levels for both azides and alkynes are of similar magnitude, both $HOMO_{dipole}$- and $LUMO_{dipole}$-controlled

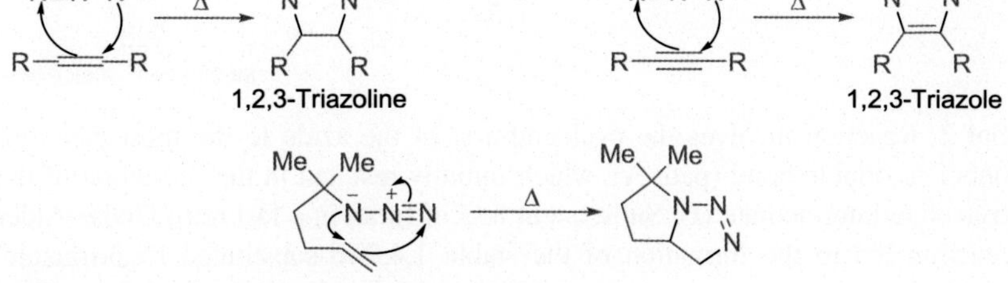

1,2,3-Triazoline 1,2,3-Triazole

SCHEME 5.42 Cycloadditions of azides to alkenes or alkynes.

pathways operate in these cycloadditions. As a result, a mixture of regioisomeric 1,2,3-triazole products is usually formed when an alkyne is unsymmetrically substituted. For example, in the reactions given below, azides react with alkynes to afford the triazoles as a mixture of 1,4- and 1,5-adducts (Scheme 5.43).

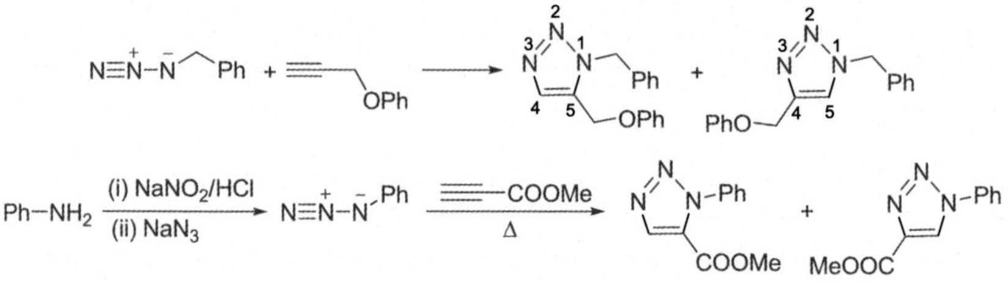

SCHEME 5.43 Cycloadditions of azides to unsymmetrical alkenes or alkynes.

5.2.13.1 Solved Problems

Q 1. Predict the product of the given dipolar addition.

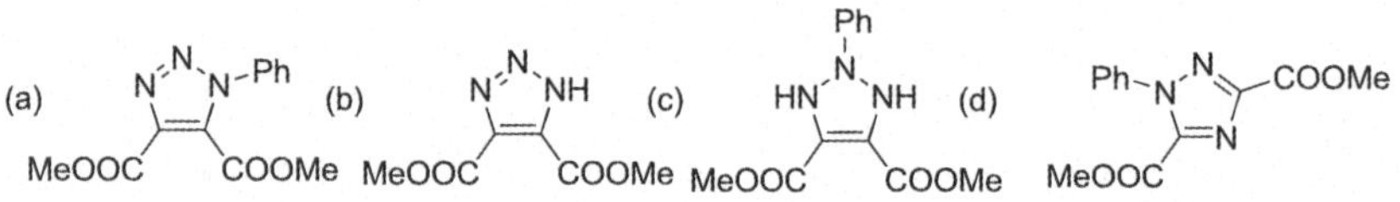

Sol 1. (a)

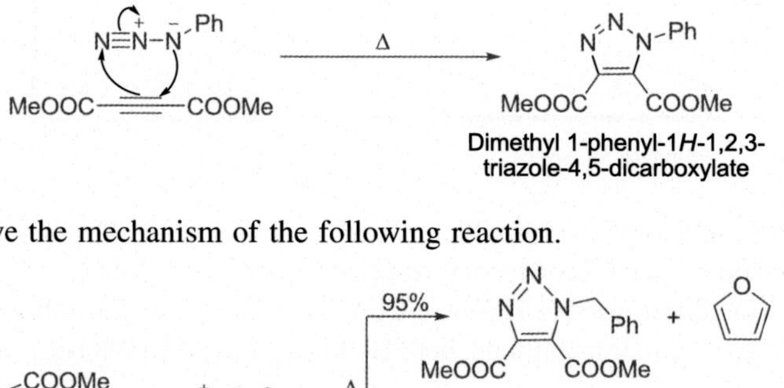

Dimethyl 1-phenyl-1*H*-1,2,3-triazole-4,5-dicarboxylate

Q 2. Give the mechanism of the following reaction.

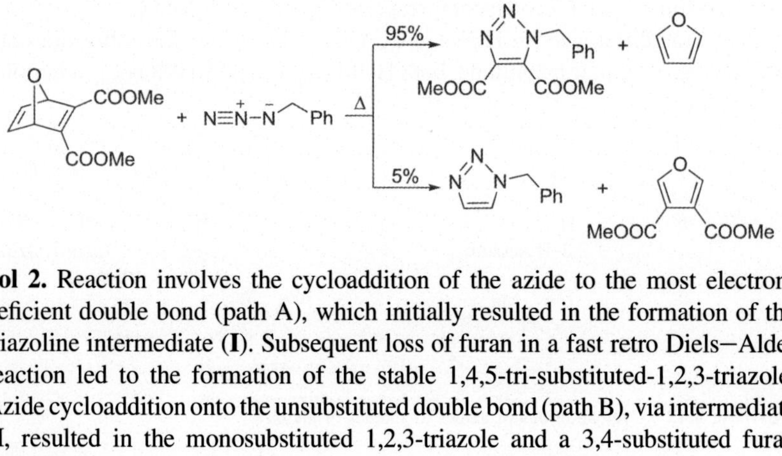

Sol 2. Reaction involves the cycloaddition of the azide to the most electron-deficient double bond (path A), which initially resulted in the formation of the triazoline intermediate (**I**). Subsequent loss of furan in a fast retro Diels–Alder reaction led to the formation of the stable 1,4,5-tri-substituted-1,2,3-triazole. Azide cycloaddition onto the unsubstituted double bond (path B), via intermediate **II**, resulted in the monosubstituted 1,2,3-triazole and a 3,4-substituted furan derivative.

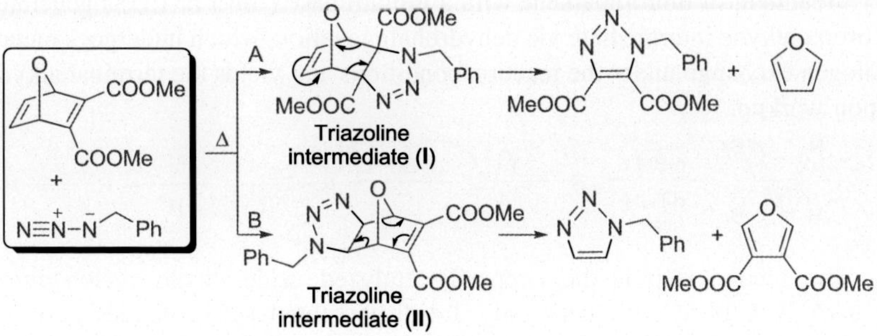

Although both double bonds can react in the [3 + 2] cycloaddition, the retro Diels–Alder reaction of the intermediates occurs rapidly to give 1,4,5-tri-substituted-1,2,3-triazole predominantly.

Q 3. The major products **I** and **II** in the following reaction sequence are -

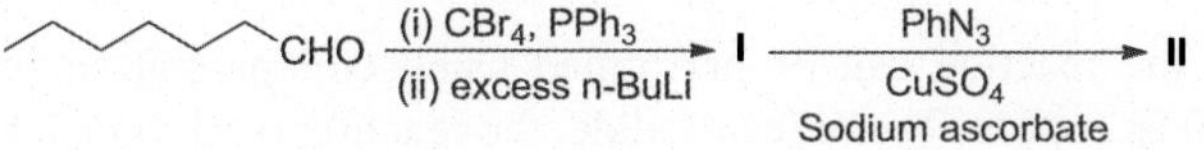

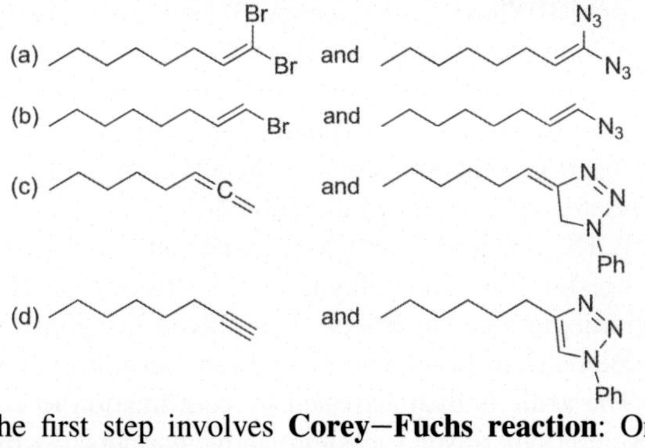

(a) and

(b) and

(c) and

(d) and

Sol 3. (d) The first step involves **Corey–Fuchs reaction**: One-carbon homologation of an aldehyde to dibromoolefin, which is then treated with n-BuLi to produce a terminal alkyne.

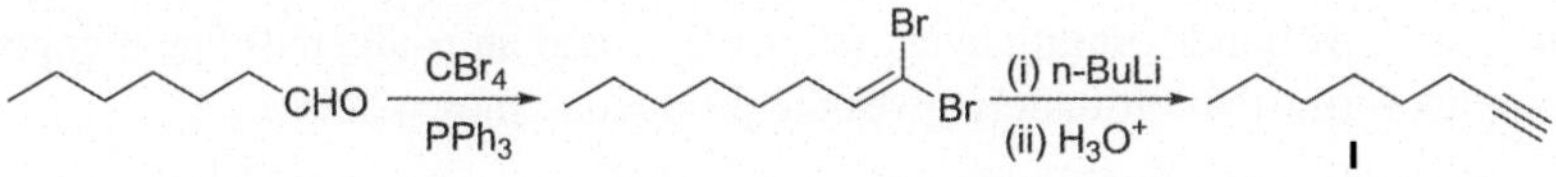

Mechanism: In the formation of the ylide from CBr_4, two equivalents of triphenylphosphine are used. One equivalent forms the ylide while the other acts as reducing agent and bromine scavenger. The addition of the ylide to the aldehyde is comparable to a Wittig reaction and leads to a dibromoalkene.

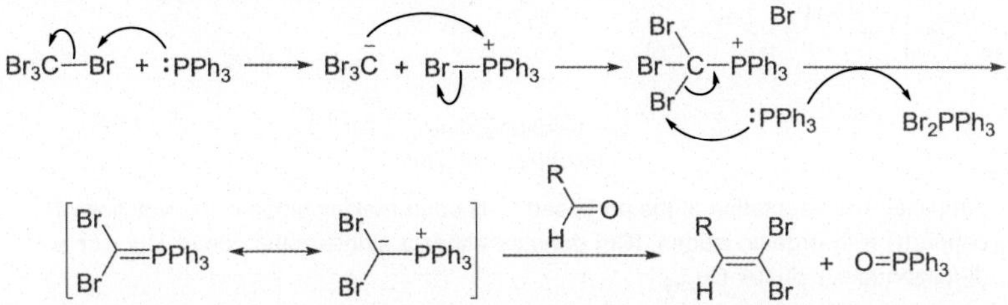

Treatment of dibromoalkene with a lithium base (BuLi or LDA) generates a bromoalkyne intermediate via dehydrohalogenation, which undergoes metal-halogen exchange under the reaction conditions and yields the terminal alkyne upon workup.

R Br n-BuLi - n-BuH - LiBr H Br Bu R—≡—Br n-Bu—Li R—≡—Li H_2O H^+ R—≡—H

The second step is the copper(I)-catalyzed azide-alkyne cycloaddition (CuAAC). CuAAC produces only 1,4-disubstituted-1,2,3-triazoles at room temperature in excellent yields. Formally, it is not a 1,3-dipolar cycloaddition and thus should not be termed a Huisgen cycloaddition.

I + N≡N⁺–N⁻–Ph Δ Cu^+ II 1 2 3 4 5 Ph

While the reaction can be performed using commercial sources of copper(I) such as cuprous bromide or iodide, the reaction works much better using a mixture of copper(II) (for example, copper(II) sulfate) and a reducing agent (for example, sodium ascorbate) to produce Cu(I) in situ. The advantage of generating the Cu(I) species in this manner is to eliminate the need of a base in the reaction. Also, the presence of reducing agent makes up for any oxygen that may have been present into the system. Oxygen oxidizes the Cu(I) to Cu(II), which retards the reaction and results in low yields.

Mechanism: The copper(I) species generated in situ forms a π-complex with the triple bond of a terminal alkyne. π-coordination of alkyne to copper significantly acidifies terminal hydrogen of the alkyne, bringing it into the proper range to be deprotonated in an aqueous medium and resulting in the formation of a σ-acetylide. The azide is then activated by coordination to copper, and the coordination event is synergistic for both reactive partners: coordination of the azide reveals the β-nucleophilic, vinylidene-like properties of the acetylide, whereas the azide's terminus becomes even more electrophilic. In the next step, the first C—N bond-forming event takes place, and an α-alkenylidine dicopper complex forms that ultimately gives the product as shown below:

Cu^{2+} $+ e^-$ Cu^+ R—≡—H R—≡—H [Cu] $+ Cu^+$ $- H^+$ R—≡—[Cu] [Cu] N≡N⁺–N⁻–Ph N≡N⁺–N–Ph R—≡—[Cu] [Cu] α-Alkenylidine dicopper complex $- Cu^+$ $+ H^+$ [Cu] R H N–Ph 1 2 3 4 5

Simplified representation of the proposed C–N bond-making steps in the reaction of copper(I) with organic azides. [Cu] denotes either a single-metal center CuL_x or a di-/oligonuclear cluster Cu_xL_y.

5.2.14 Cycloaddition with Nitrile Imines

Nitrile imines are prepared in situ, generally by base-induced dehydrohalogenation of hydrazonoyl halides (Scheme 5.44).

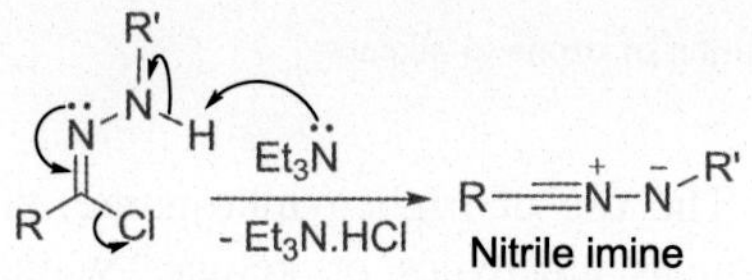

SCHEME 5.44 Generation of nitrile imines of hydrazonoyl halides.

Nitrile imines can also be prepared conveniently by the oxidation of easily accessible hydrazones (Scheme 5.45). The oxidant generally used is chloramine-T (*N*-chloro-*N*-sodio-*p*-toluenesulfonamide, CAT). Other oxidants such as iodobenzene diacetate (IBD) and 2,3-dichloro-5,6-dicyano-1,4-benzoquinone have also been used.

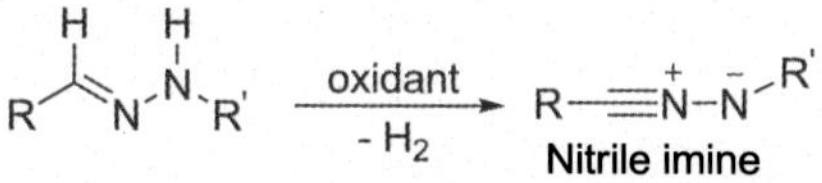

SCHEME 5.45 Generation of nitrile imines by oxidation of hydrazones.

Nitrile imines smoothly undergo 1,3-dipolar cycloadditions with alkenes and alkynes to generate 2-pyrazolines and pyrazoles, respectively (Scheme 5.46).

SCHEME 5.46 1,3-dipolar cycloaddition of nitrile imines with symmetrical alkenes or alkynes.

5.2.15 Cycloaddition with Ozone

Mechanism of Ozonolysis (Criegee mechanism): The initial step of the reaction involves a 1,3-dipolar cycloaddition of ozone to the alkene leading to the formation of the primary ozonide (molozonide or 1,2,3-trioxolane), which decomposes to give a carbonyl oxide and a carbonyl compound. The carbonyl oxides are similar to ozone in being 1,3-dipolar compounds and undergo 1,3-dipolar cycloaddition to the carbonyl compound with the reverse regiochemistry, leading to a relatively stable secondary ozonide (1,2,4-trioxolane) (Scheme 5.47).

A reagent is then added to convert the intermediate ozonide to a carbonyl derivative. Reductive workup conditions are far more commonly used than

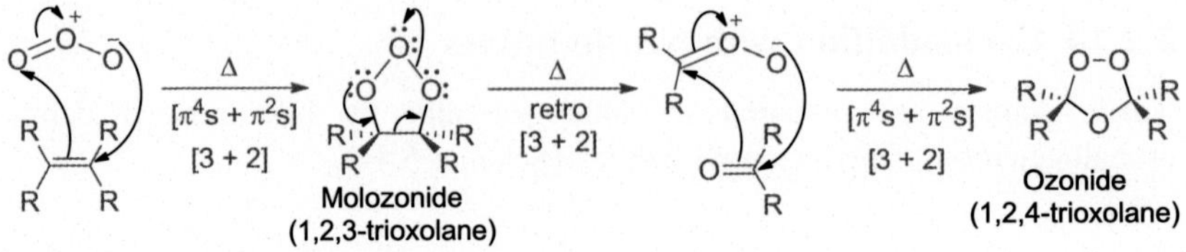

SCHEME 5.47 Cycloaddition of ozone to alkenes.

oxidative conditions. The use of triphenylphosphine, thiourea, zinc dust, or dimethyl sulfide produces aldehydes or ketones while the use of sodium borohydride produces alcohols. The use of hydrogen peroxide produces carboxylic acids.

Chapter 6

Group Transfer, Elimination, and Related Reactions

Chapter Outline

6.1 GROUP TRANSFER REACTIONS

The concerted transfer of a group from one molecule to another due to concomitant movement of a σ-bond (from one molecule to another) and formation of a new σ-bond at the expense of a π-bond is generally referred to as **group transfer** *pericyclic reaction.* These reactions resemble sigmatropic rearrangements since a σ-bond moves. However, sigmatropic reactions are unimolecular reactions whereas the group transfer reactions are bimolecular. They also resemble cycloadditions, since a new σ-bond is formed at the expense of a π-bond. However, in group transfer reactions, no ring is formed. Let us consider the general case of the synchronous double group transfer reaction that can be represented as an interaction between a

Pericyclic Reactions. http://dx.doi.org/10.1016/B978-0-12-803640-2.00006-3
Copyright © 2016 Elsevier Inc. All rights reserved.

component with p π-electrons and a component with q π-electrons as shown in Figure 6.1.

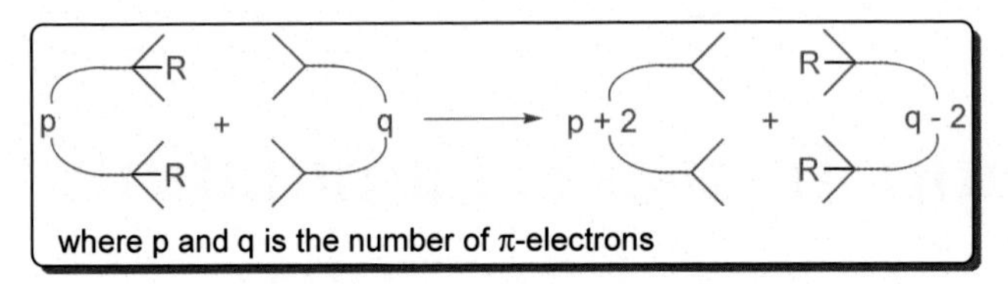

FIGURE 6.1 A general representation of synchronous double group transfer reaction.

The Woodward–Hofmann selection rules for the generalized case are the same as for cycloaddition reactions: (1) For supra–supra or an antara–antara double group transfer, the reaction is symmetry allowed under thermal conditions when $p + q = 4n + 2$, and it is photochemically allowed when $p + q = 4n$. (2) For supra–antara double group transfer, the reaction is symmetry allowed under thermal conditions when $p + q = 4n$, and it is photochemically allowed when $p + q = 4n + 2$.

The simplest example of group transfer reaction is the transfer of two hydrogen atoms from ethane to ethene ($p + q = 0 + 2 = 2$). This reaction is symmetry allowed under thermal conditions. Obviously, the reaction in which two hydrogen atoms are transferred from ethane to the termini of butadiene ($p + q = 0 + 4 = 4$) is symmetry forbidden under thermal conditions (Scheme 6.1).

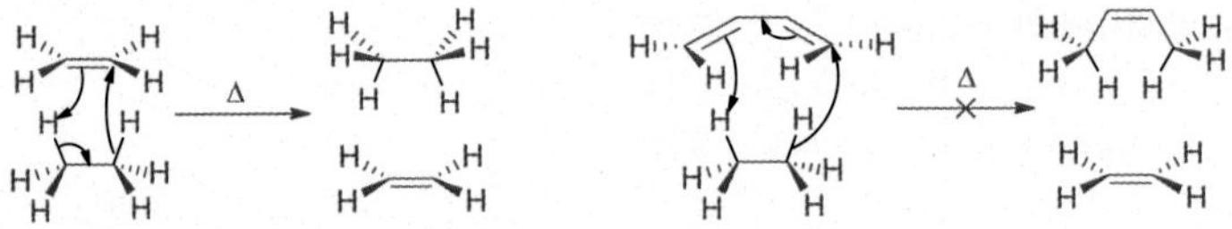

SCHEME 6.1 Group transfer reaction in ethane–ethene and ethane–butadiene systems.

A well-known example of group transfer reaction with $p = 0$, $q = 2$ is represented by the concerted reduction of alkenes and alkynes with the reactive intermediate diimide stereoselectively in the *cis* fashion (Scheme 6.2). The driving force of the reaction is the formation of the stable nitrogen molecule.

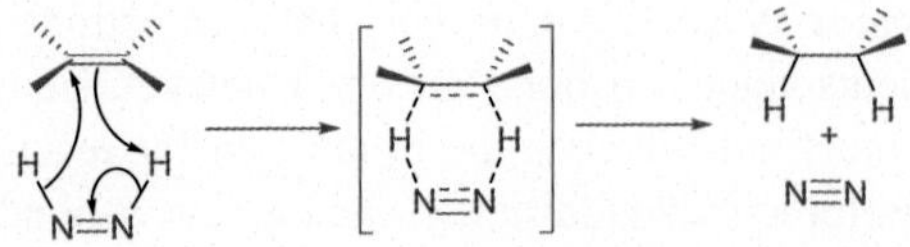

SCHEME 6.2 Stereoselective concerted reduction of an alkene derivative using diimide.

Hydrocarbon analogues with p = 0, q = 2 provide another interesting example of a group transfer reaction in which the driving force of the reaction is aromatization of cyclohexadiene to benzene system (Scheme 6.3).

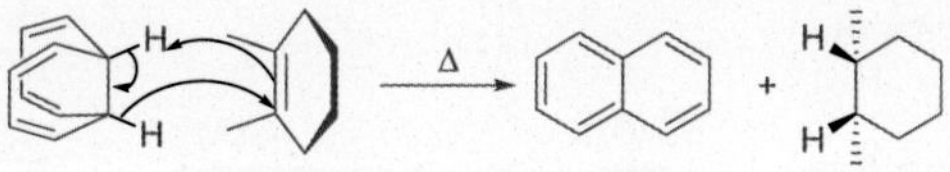

SCHEME 6.3 Concerted hydrogen transfer and aromatization of the cyclohexadiene to benzene system.

A case with p = 2, q = 4 has also been observed. Loss of aromaticity in the middle ring of anthracene is less important because of the smaller stabilization energy of this ring as compared to benzene (Scheme 6.4).

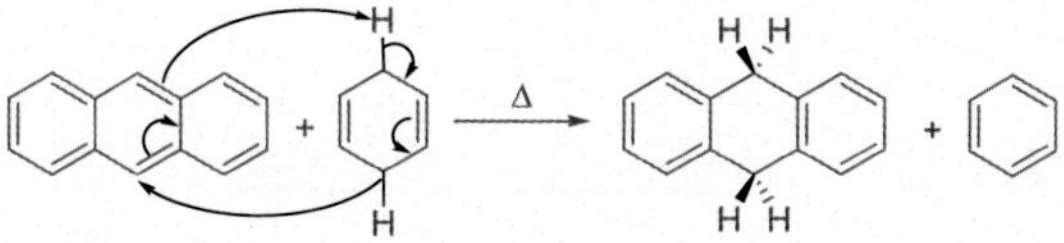

SCHEME 6.4 Concerted hydrogen transfer reaction and aromatization in anthracene-cyclohexadiene system.

6.1.1 Analysis of Group Transfer Reactions by Orbital Symmetry Correlation-Diagram Method

Let us consider the transfer of two hydrogen atoms from ethane to ethene. In this reaction, a mirror plane symmetry (*m*) is conserved throughout the transformation. The molecular orbitals of the system can be obtained by considering the symmetric and antisymmetric combinations of σ and σ^* C−H bonds in ethane, and π and π^* bonds in ethene. The molecular orbitals (MOs) that are lowest in energy are formed from linear combination of the σ C−H bonds, lowest is the symmetric sum of the two C−H σ-bond (σ_1), followed by the antisymmetric sum (σ_2). Similarly, the two highest energy MOs are formed from linear combination of the two C−H σ^*-bonds, highest is the antisymmetric σ_4^*, preceded by the symmetric σ_3^* at a slightly lower energy. MOs of ethene, i.e., π and π^*, are in the center of the energy scale. Since there is a correlation between the ground state orbitals of the reactants and products, the reaction is symmetry allowed under thermal conditions (Figure 6.2).

6.1.2 Analysis of Group Transfer Reactions by Perturbation Molecular Orbital Method

The group transfer reactions can be easily explained by perturbation molecular orbital (PMO) approach. If both the R groups migrate without inversion, in a

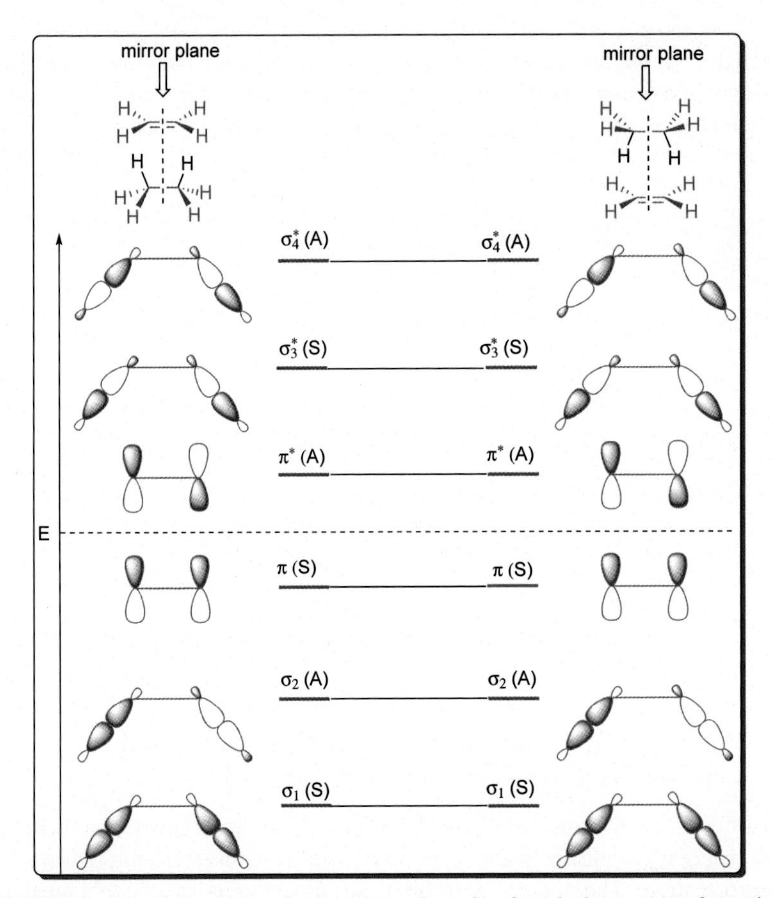

FIGURE 6.2 Correlation diagram for the concerted transfer of two hydrogen atoms from ethane to ethene.

suprafacial manner, then transition state becomes aromatic and hence the reaction is thermally allowed (Figure 6.3). If any one of R group undergoes inversion, there will be one node and the transition state will be antiaromatic and the reaction is thermally forbidden. The same situation will be obtained if it is antara on the ethene moiety and supra on both C—R bonds.

6.1.3 Solved Problems

Q 1. Predict the products of the following reactions under thermal conditions.

(i) HOOC, H / H, COOH **I** $\xrightarrow{N_2D_2}$

(ii) H, HOOC / H, COOH **II** $\xrightarrow{N_2D_2}$

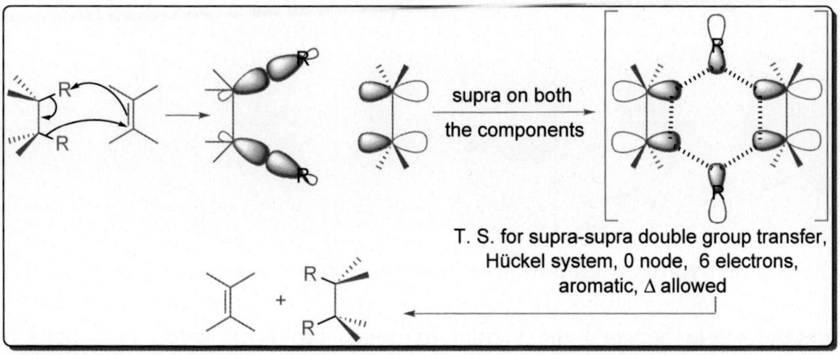

FIGURE 6.3 PMO approach for a group transfer reaction when both the R groups migrate suprafacially.

Sol 1. The transfer of deuterium from diimide to C=C bond occurs in a completely *syn* manner. Therefore, the reduction of fumaric acid (**I**) and maleic acid (**II**) with dideuteriodiimide resulted in the formation of racemic and meso products, respectively.

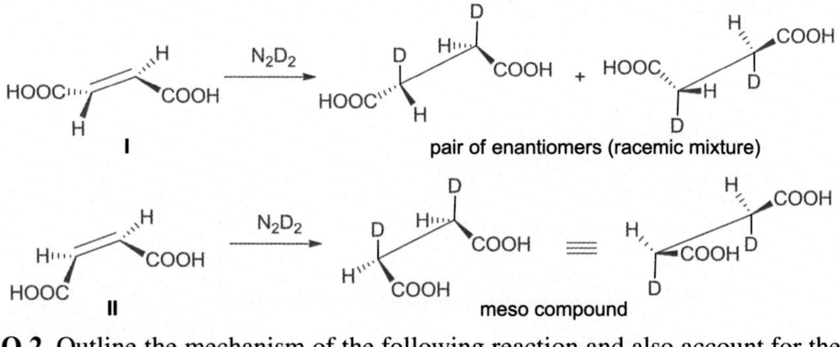

Q 2. Outline the mechanism of the following reaction and also account for the fact that when cyclohexane is used as the solvent in this reaction the product yield is only 20%.

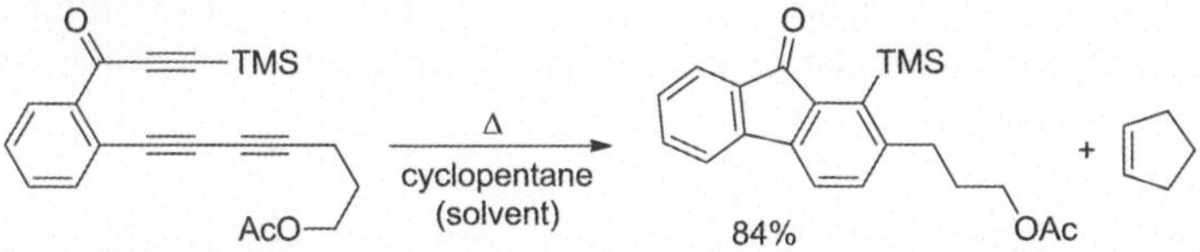

Sol 2. Initial step of the reaction involves an intramolecular hexadehydro-Diels–Alder (HDDA) reaction to give a highly reactive benzyne intermediate. In the absence of external trapping reagents, the benzyne intermediate can simultaneously accept two vicinal hydrogen atoms from a suitable alkane 2H-donor, often the reaction solvent. This desaturates the donor alkane, forming an alkene, and traps the benzyne to a dihydrobenzenoid product.

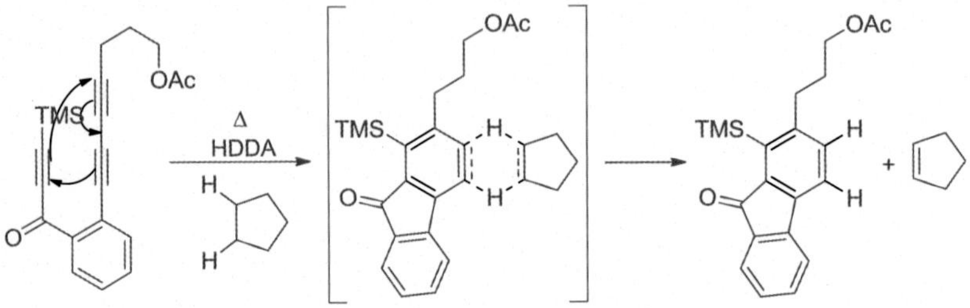

The double hydrogen atom transfer process involves a nearly planar geometry of the six reacting atoms. In the low-energy conformer of cyclopentane, the dihedral angle between the vicinal hydrogen atoms is small, i.e., there is greater degree of eclipsing, so it will be more reactive as a hydrogen donor as compared to cyclohexane.

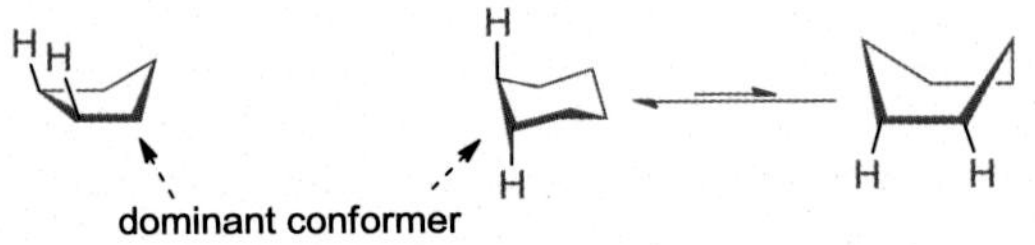

6.2 ELIMINATION REACTIONS

In the generalized case when q = 0, we get concerted elimination as shown in Figure 6.4.

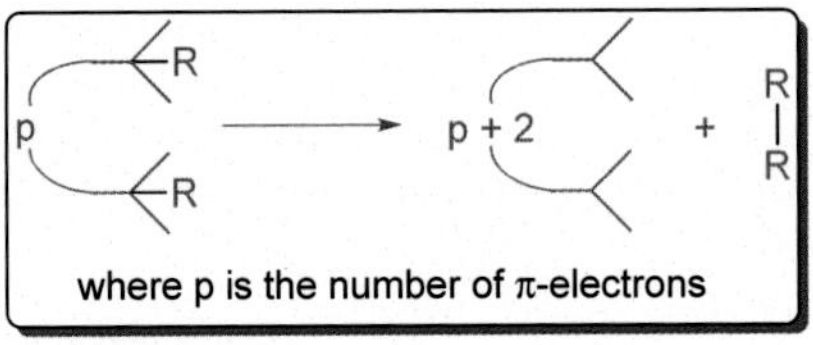

where p is the number of π-electrons

FIGURE 6.4 Concerted elimination as a group transfer reaction when q = 0.

Thus, according to the selection rules, the concerted noncatalytic elimination should be 1,4 rather than 1,2. For example, 2,5-dihydrofuran loses hydrogen in a unimolecular process, while 2,3-dihydrofuran does not react in this way (Scheme 6.5).

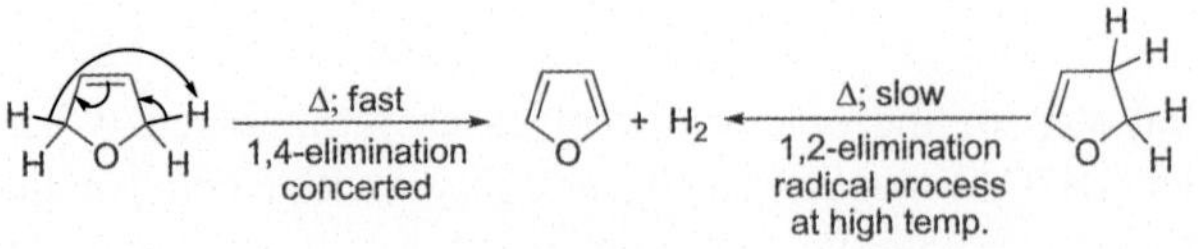

SCHEME 6.5 Concerted elimination in dihydrofurans.

Similarly, cyclohexa-1,4-diene decomposes unimolecularly to benzene and hydrogen; while at the same temperature the cyclohexa-1,3-diene is thermally stable, and at higher temperatures it decomposes mainly by radical-chain processes (Scheme 6.6).

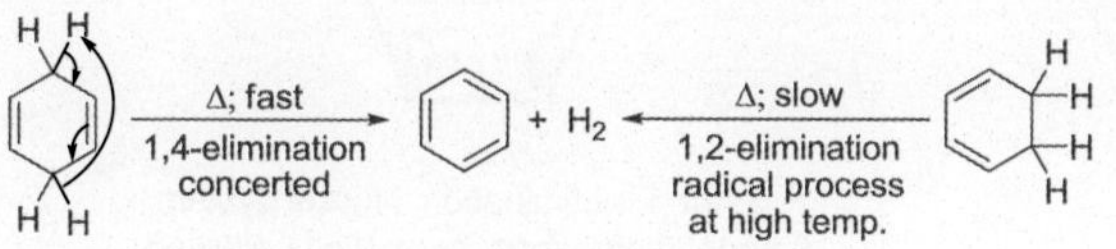

SCHEME 6.6 Concerted elimination in cyclohexadiene systems.

Elimination of hydrogen in dihydrofurans can be explained by the application of PMO method as shown in Figure 6.5.

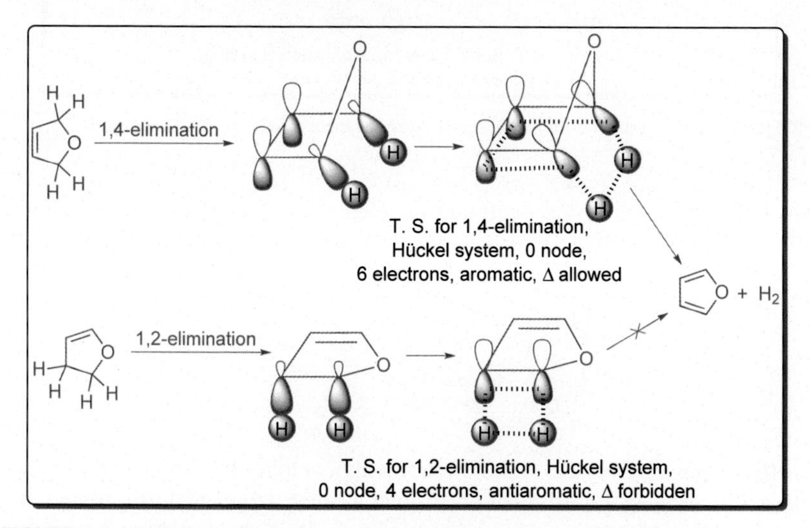

FIGURE 6.5 Analysis of elimination reactions in dihydrofurans by PMO approach.

PMO method also explains eliminations in cyclohexa-1,3- or 1,4-diene (Figure 6.6). However, in the case of the 1,4-isomer, the six-electron transition state involving two hydrogens must be derived from the boat form of the cyclohexadiene because of the conformational requirements.

6.2.1 Solved Problems

Q 1. Rationalize the following reactions:

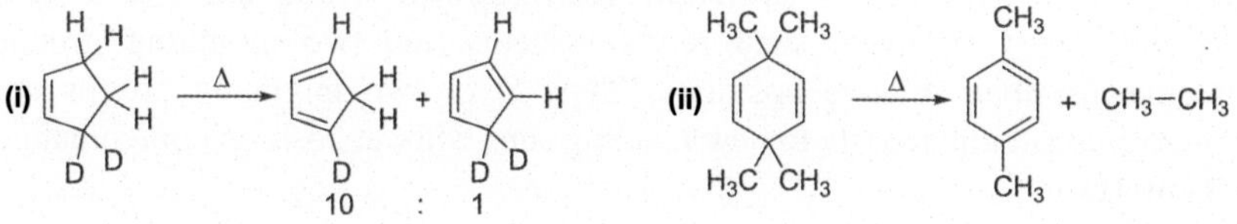

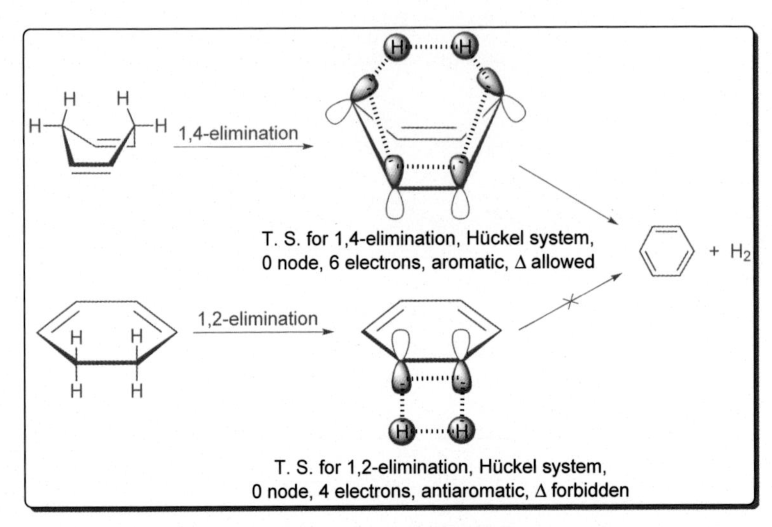

FIGURE 6.6 Analysis of elimination reactions in cyclohexadienes by PMO approach.

Sol 1. (i) Pyrolysis of the deuterium-labeled cyclopentene has clearly shown that the major product is formed via concerted 1,4-elimination rather than 1,2-elimination.

(ii) Upon pyrolysis, 3,3,6,6-tetramethylcyclohexa-1,4-diene undergoes thermally allowed 1,4-elimination to give a much larger yield of ethane and *p*-xylene than could have arisen from a free radical chain reaction.

6.3 DYOTROPIC REARRANGEMENTS

An intramolecular pericyclic reaction involving simultaneous migration of two sigma bonds is called a **dyotropic** rearrangement (from the Greek *dyo*, meaning two). In many such rearrangements, the two migrating groups interchange their relative positions (**Type I**) or two migrating groups are transferred simultaneously to new bonding sites without positional interchange (**Type II**).

In a typical Type I dyotropic rearrangement, the migrating groups having an assumed *anti* conformation migrate suprafacially to opposite sides of the molecular framework resulting in the inversion of configuration at both the migration origin and migration terminus (Figure 6.7). Obviously, the transition state for such a system involves four electrons.

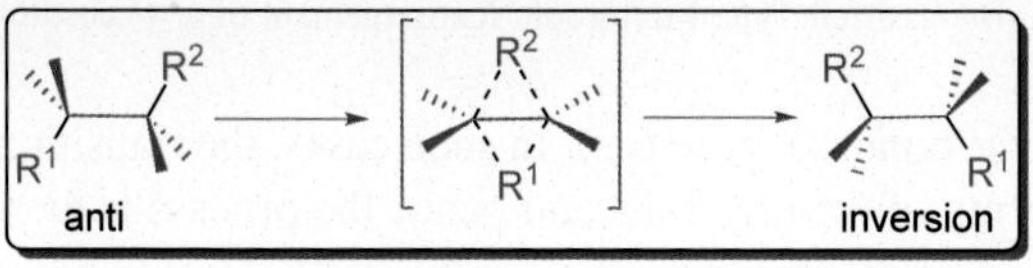

FIGURE 6.7 Type I dyotropic rearrangement.

If there is retention of configuration in both the groups, the process will be thermally forbidden. However, it will be thermally allowed if one of the groups migrates with inversion of configuration. Also, where the migrating group possesses an additional set of nonbonding electron pair(s), which help in completing the cyclic array of electrons, the process is again thermally allowed.

Good examples of Type I dyotropic rearrangements involving C–C bond as the stationary surface are provided by the interconversion of vicinal dibromides in cyclohexane and cyclohexanone ring systems (Scheme 6.7).

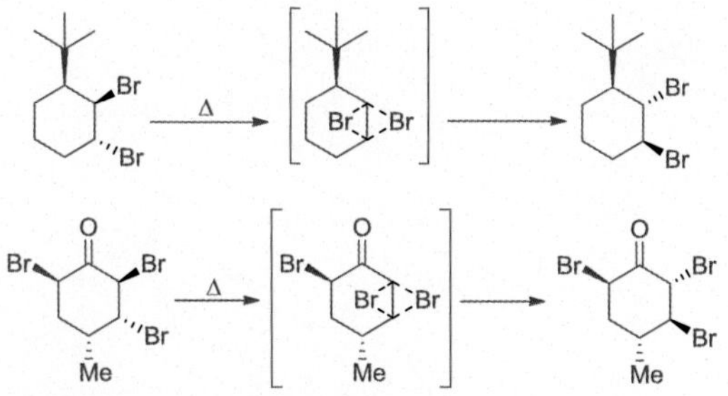

SCHEME 6.7 Interconversion of vicinal dibromides.

Another example of Type I dyotropic rearrangement is provided by the rearrangement of a 4-substituted-β-lactone (**1**) to a γ-lactone (**2**) under the influence of ethereal $MgBr_2$. The stereospecificity of the reaction is maintained during the ring expansion process where the carbon–oxygen and carbon-migrating methyl bonds are aligned in an anti-coplanar manner in **1** (Scheme 6.8). The reaction closely resembles Wagner–Meerwein-type carbocation rearrangements.

In **Type II** rearrangement, two migrating groups are transferred simultaneously to the new bonding sites without positional interchange, resulting in

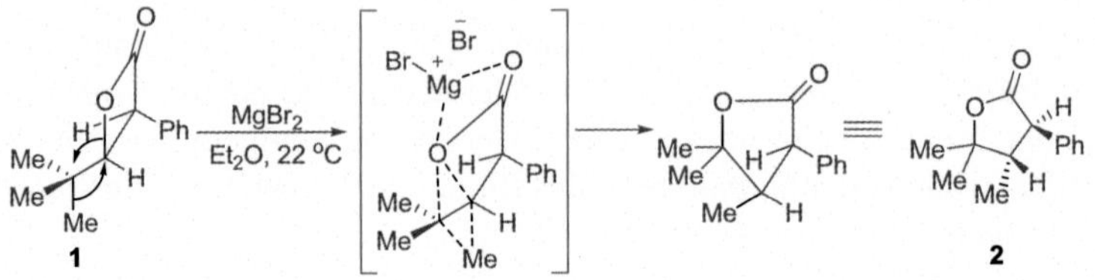

SCHEME 6.8 $MgBr_2$-assisted Type I dyotropic rearrangement of a 4-substituted-β-lactone.

transposition of π-bonds (Figure 6.8). In such cases, the transition state involves a monocyclic array of 6π-electrons and hence the process is thermally allowed.

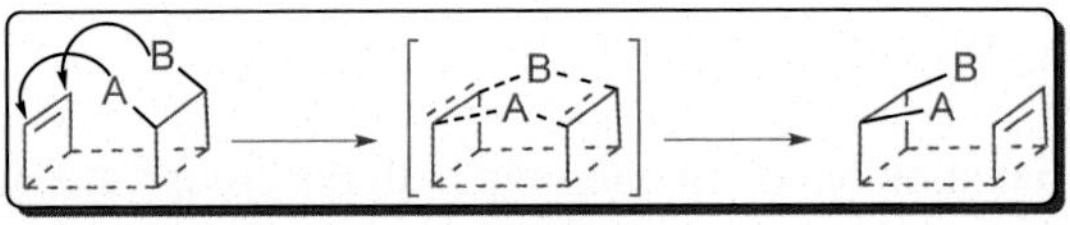

FIGURE 6.8 Type II dyotropic rearrangement.

Some examples of Type II dyotropic rearrangements by double C–H transfer are given in Scheme 6.9. The driving force for the reaction is provided by the proximity of the interacting orbitals and the rigid carbon framework. Furthermore, the reaction is accelerated by the release of the ring strain at the receptor π-bond and aromatization of the diene ring.

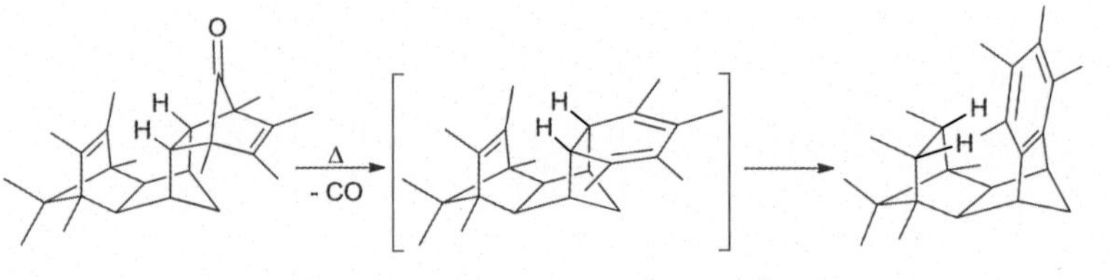

SCHEME 6.9 Examples of Type II dyotropic rearrangements.

6.3.1 Solved Problems

Q 1. i-Pr substituted *trans* (**I**) and *cis* (**V**) β-lactones undergo Type I dyotropic shifts to give different products as shown below. Explain the mechanism of their formation.

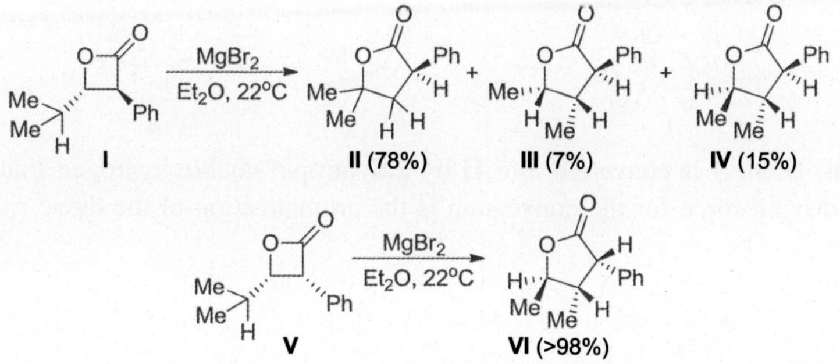

Sol 1. In case of *trans* β-lactone (**I**), the 1,2 hydride shift to provide **II** was the predominant reaction pathway due to unrestricted bond rotation in isopropyl group, which aligns proton anti periplanar to the carbon−oxygen bond, and also due to the higher migratory aptitude of hydrogen atom as compared to alkyl groups. In the case of *cis* β-lactone (**V**), steric hindrance between the 3-phenyl group and the Me group allows the more favored 1,2-methyl shift to result in almost quantitative yield of **VI**.

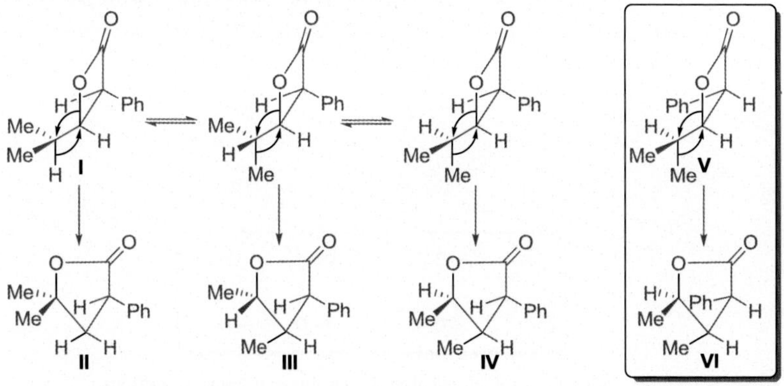

Q 2. Give the mechanism of the following reactions:

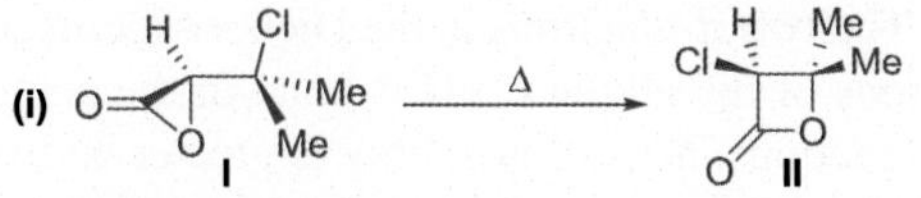

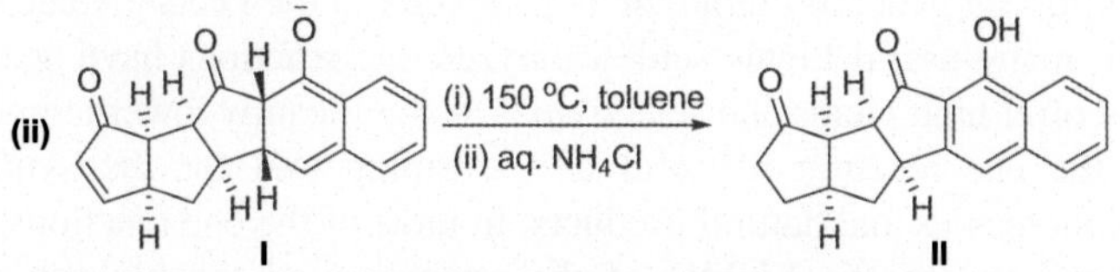

Sol 2. **(i)** The conversion of β-dimethyl substituted α-lactone (**I**) to the ring expanded β-lactone (**II**) proceeds through a dyotropic transition state.

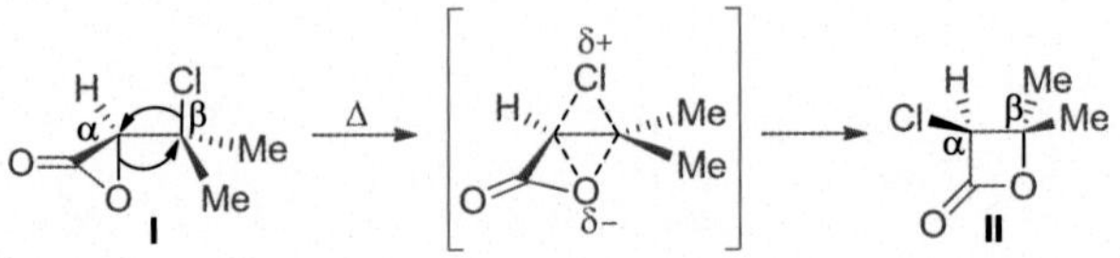

(ii) Dione **I** is converted into **II** by a dyotropic double hydrogen transfer. The driving force for the conversion is the aromatization of the diene ring.

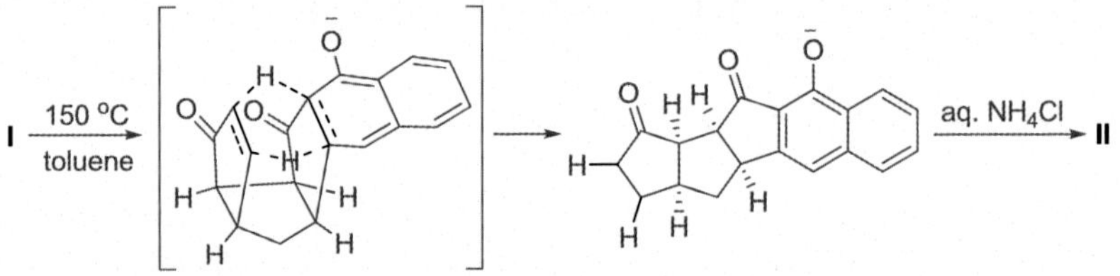

6.4 ENE REACTIONS

Ene reaction is a classic example of group transfer reactions. *It involves the addition of a compound with a double bond having an allylic hydrogen (the "ene") to a compound with a multiple bond (the "enophile") with transfer of the allylic hydrogen and a concomitant reorganization of the bonding*, as illustrated in Scheme 6.10.

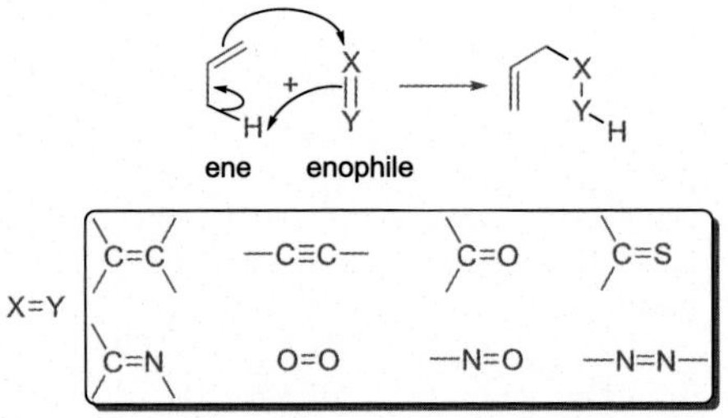

ene enophile

X=Y: C=C, —C≡C—, C=O, C=S, C=N, O=O, —N=O, —N=N—

SCHEME 6.10 The Ene reaction: a classical example of group transfer reactions.

Although the ene reaction is not a cycloaddition reaction, it does resemble the Diels–Alder reaction. The two π-electrons in the Diels–Alder reaction are replaced by the two electrons of the allylic C–H σ-bond in the ene reaction. Therefore, in case of ene reactions, activation energy is greater and so higher temperatures are generally required as compared to the Diels–Alder reaction. However, many useful Lewis acid–catalyzed ene reactions have been developed that offer high yields and selectivities at significantly lower temperatures, making the ene reaction a useful C–C-forming tool for the synthesis of complex molecules and natural products. In most of the ene reactions, there is interaction between the LUMO of the enophile and the highest occupied molecular orbital (HOMO) of the ene component in a suprafacial–suprafacial manner. Mechanistically, the concerted ene reaction can be classified as a $[\pi^2s + \pi^2s + \sigma^2s]$ cycloaddition reaction and is symmetry allowed under thermal conditions. As in the Diels–Alder reaction, the catalysts exert their effect by lowering energy of the lowest unoccupied molecular orbital (LUMO) of the enophile (Figure 6.9).

The ene reaction can also be analyzed by using PMO method. As shown in Figure 6.10, in the transition state of ene reaction, addition to the double bond takes place from the same side (suprafacial), therefore, a supra–supra mode of addition leads to a Hückel array that is aromatic with (4n + 2) π-electrons. Hence, the reaction is thermally allowed and photochemically forbidden.

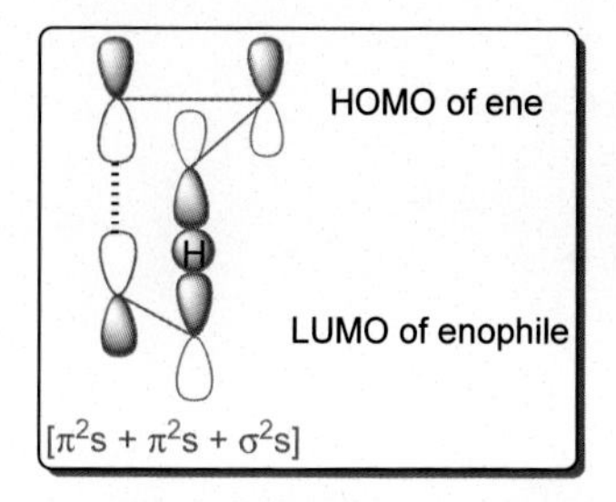

FIGURE 6.9 HOMO–LUMO interactions in ene reaction.

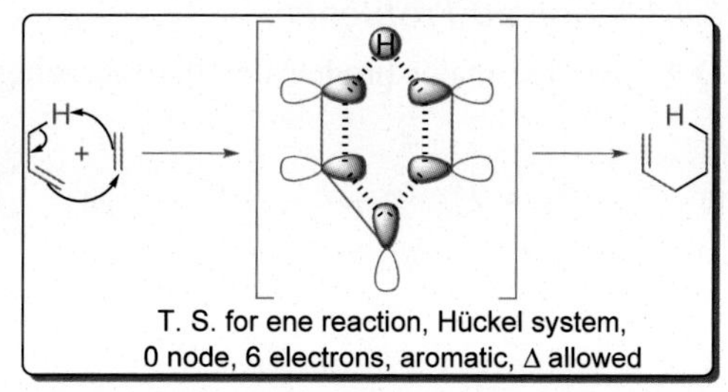

FIGURE 6.10 PMO analysis of ene reaction.

6.4.1 Intermolecular Ene Reactions

β-Pinene is a reactive alkene and undergoes ene reaction with a variety of enophiles such as acroyl chloride and maleic anhydride (Scheme 6.11). In the first case, the major interaction is between the nucleophilic end of the exocyclic double bond of the β-pinene and electophilic end of the acroyl chloride, as these atoms have the largest coefficients in the HOMO and LUMO, respectively.

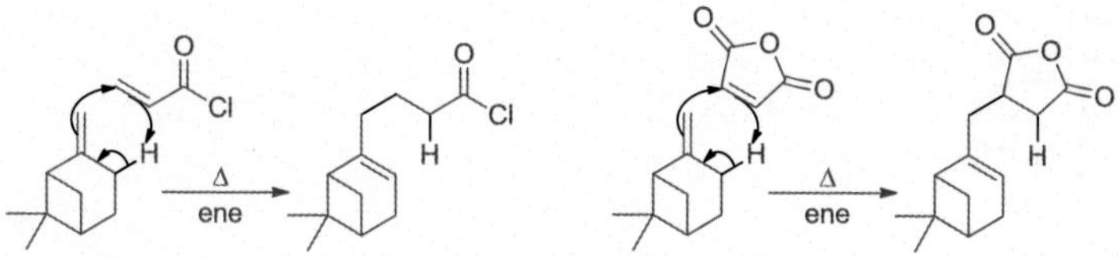

SCHEME 6.11 Ene reactions of β-pinene.

Ene reaction is concerted and involves *syn* addition. For example, in the reaction of hept-1-ene with dimethyl acetylenedicarboxylate, the adduct formed has two ester groups on the same side of alkene. It is formed by the addition of the hydrogen and allyl groups on the same side of the triple bond of the enophile. The corresponding fumaric ester is not obtained (Scheme 6.12).

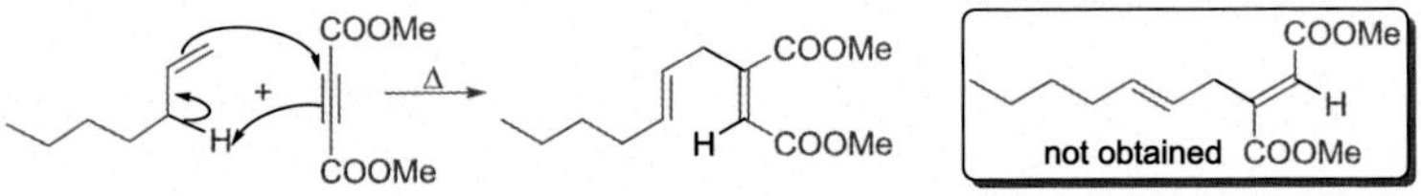

COOMe Δ COOMe H COOMe COOMe COOMe H not obtained COOMe

SCHEME 6.12 Ene reaction of hept-1-ene with dimethyl acetylenedicarboxylate.

6.4.1.1 Solved Problems

Q 1. Give the major products of the following reactions.

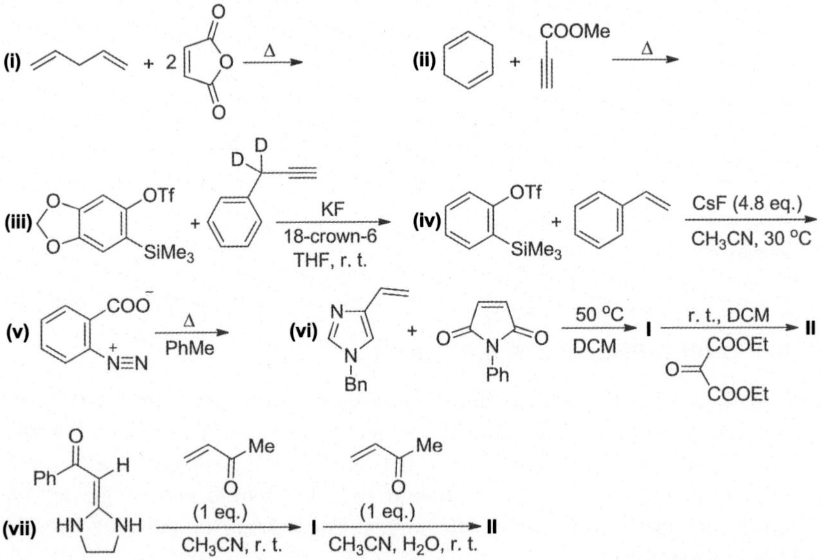

(i) + 2 (maleic anhydride) Δ → (ii) + COOMe Δ →

(iii) OTf SiMe$_3$ + D D KF 18-crown-6 THF, r. t. → (iv) OTf SiMe$_3$ + CsF (4.8 eq.) CH$_3$CN, 30 °C →

(v) COO$^-$ $\overset{+}{N}$≡N Δ PhMe → (vi) N Bn + O N Ph O 50 °C DCM → I r. t., DCM COOEt O= COOEt → II

(vii) Ph O H HN NH Me O (1 eq.) CH$_3$CN, r. t. → I Me O (1 eq.) CH$_3$CN, H$_2$O, r. t. → II

Sol 1. **(i)** Tandem ene/Diels–Alder reaction of penta-1,4-diene and maleic anhydride results in the formation of 1:2 adduct.

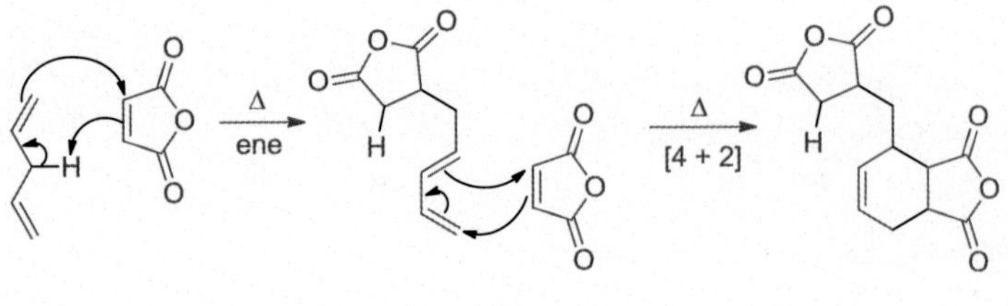

H O O O Δ ene → O O H O O O Δ [4 + 2] → O O H O O O

(ii) Tandem ene/intramolecular Diels–Alder reaction of cyclohexa-1,4-diene and methyl propiolate gives a tricylic adduct.

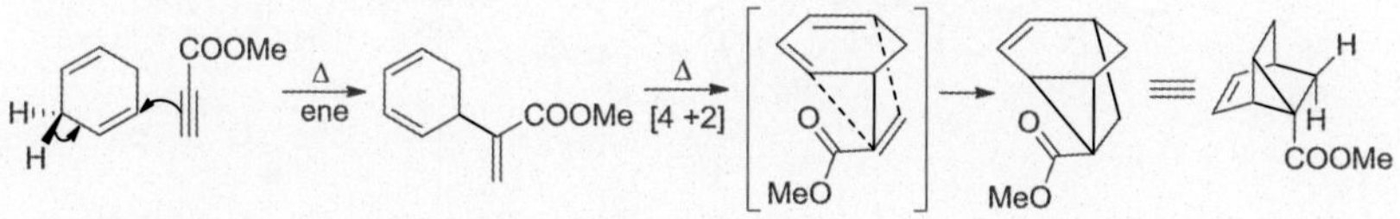

(iii) The aryne generated by the fluoride-induced 1,2-elimination of 2-(trimethylsilyl)aryl triflates undergoes intermolecular ene reaction with dideuteriobenzylethyne possessing two propargylic deuteriums to give an allene. As predicted for an ene reaction, one of the deuteriums is on the allenyl moiety and the other is on the benzodioxole ring *ortho* to the allenyl moiety.

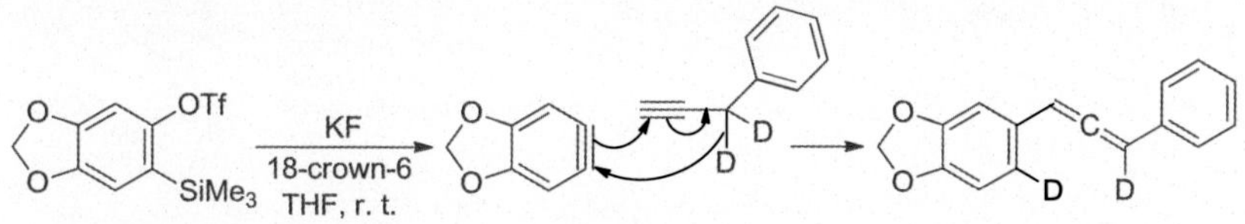

(iv) Styrene undergoes Diels–Alder reaction with benzyne to give an intermediate adduct, which then undergoes a concerted ene reaction with another molecule of the benzyne to give 9-phenyl-9,10-dihydrophenanthrene.

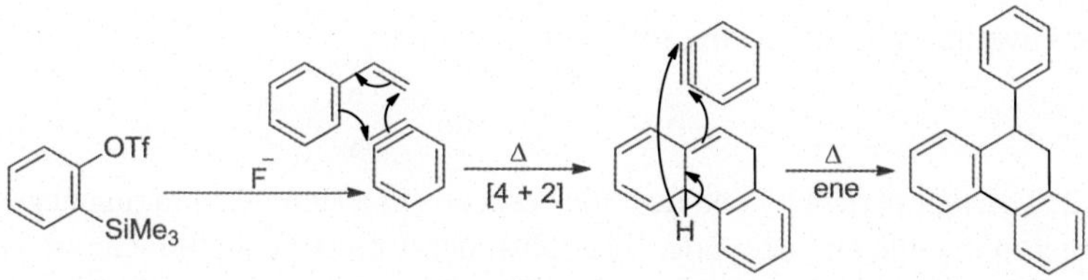

(v) The reaction involves an aromatic ene reaction between the benzyne (enophile) and toluene bearing a benzylic C–H bond (ene donor) to form an isotoluene intermediate, which is rearomatized by reaction with another molecule of the benzyne to give the final product.

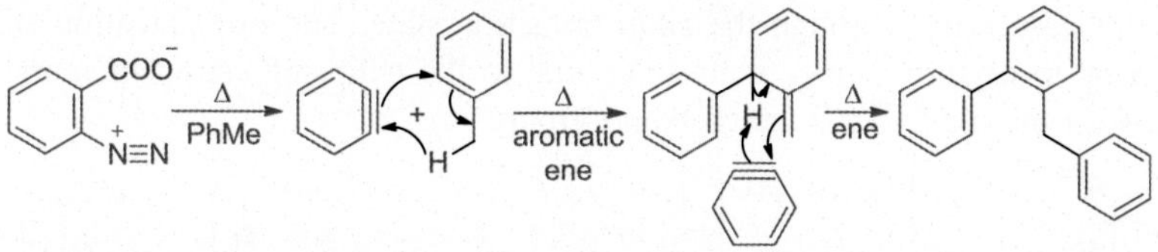

(vi) The Diels–Alder reaction of 1-benzyl-4-vinyl-1*H*-imidazole and *N*-phenylmaleimide gives the *endo* cycloadduct (**I**), which then undergoes diastereoselective intermolecular ene reaction to give 4,5,6,7-tetrahydro-1*H*-benzo [d]imidazole derivative (**II**). Rearomatization of the imidazole ring provides highly favorable thermodynamics for the second step.

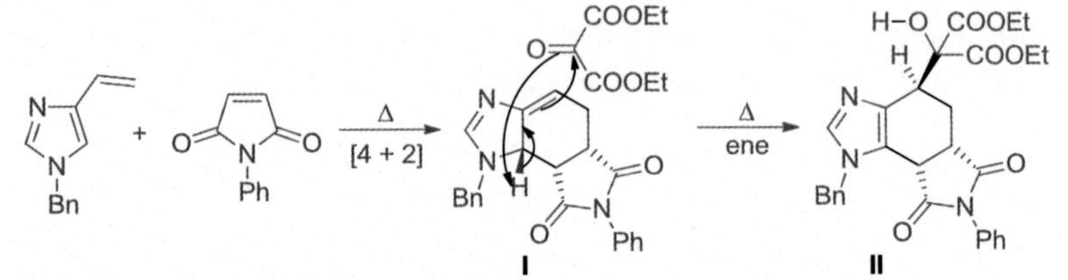

(vii) The reaction of heterocyclic ketene aminal with methyl vinyl ketone proceeded via an aza-ene addition followed by intramolecular cyclization to afford imidazo[1,2-a]pyridine derivative (**I**). When excess of methyl vinyl ketone is used, the initially formed **I** underwent another aza-ene addition to afford imidazo[1,2,3-ij][1,8]naphthyridine derivative (**II**).

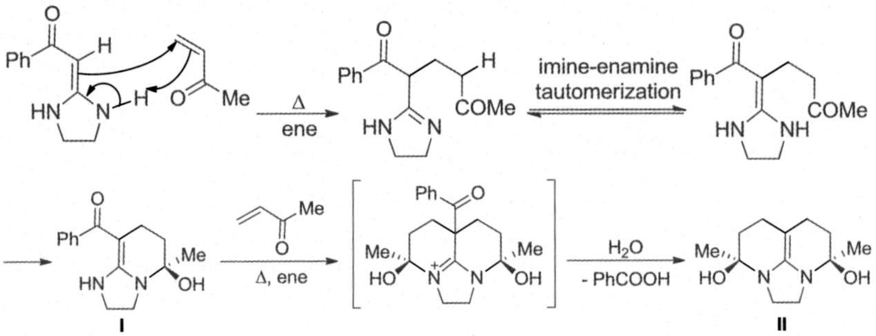

Q 2. Give the products of the following reactions:

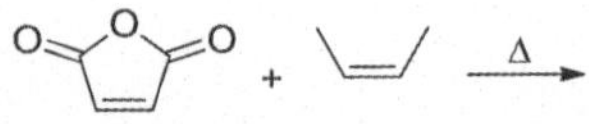

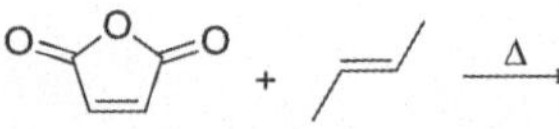

Sol 2. Thermal pericyclic ene reactions proceed in a supra–supra manner with respect to the ene and enophile. The substituents on the enophile can be *endo* or *exo*. In terms of the diastereoselection with respect to the newly created chiral centers, an *endo* preference has been qualitatively observed, but steric effects can easily modify this preference. For example, the reaction between (*Z*)-but-2-ene and maleic anhydride gave erythro- and threo-(1-methylalkyl)-succinic anhydride in a kinetically controlled ratio of 1:4. The major product in this case arises through the *endo* transition state. The *exo* transition state leading to erythro isomer is unstable due to the eclipsing steric interactions between the methyl and the carbonyl group.

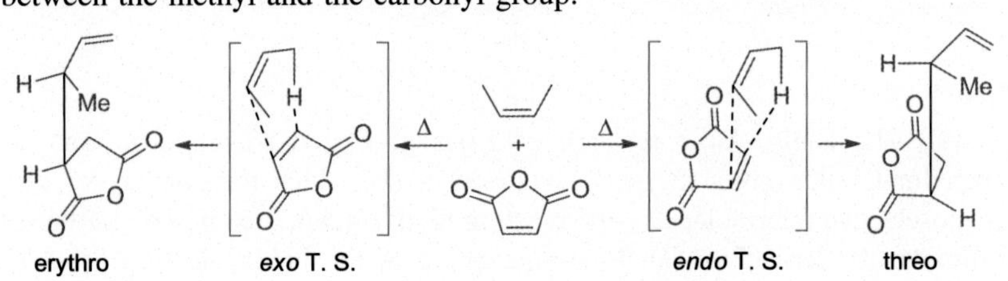

In the reaction between (*E*)-but-2-ene and maleic anhydride, the major product was the erythro adduct but the selectivity was poor due to the eclipsing steric interactions of the *trans* methyl group in the *endo* transition state.

threo *exo* T. S. *endo* T. S. erythro

Q 3. Explain the yield of the products obtained in the following ene reactions under thermal and Lewis acid-catalyzed conditions.

(i)

Conditions	Product 1	Product 2
180 °C; 48 h	92%	8%
$SnCl_4$ (0.2 eq.); 0 °C; 5 min	2.5%	97.5%

(ii)

Conditions	Product 1	Product 2
220 °C	93%	7%
$AlCl_3$; 25 °C	100%	0%

Sol 3. **(i)** In case of carbonyl ene reactions, addition of Lewis acids makes the enophiles more electrophilic, so the reaction can be performed at lower temperature. This also improves the yields, regio- and stereoselectivities of the carbonyl ene reactions. For example, in the Lewis acid–catalyzed carbonyl ene reaction between 2-methylhepta-2,6-diene and oxamalonate, a positive charge develops at the ene component due to which the trisubstituted alkene becomes more reactive than the monosubstituted alkene.

However, in the case of thermal ene reaction, steric accessibility of the double bond and allylic hydrogen are of primary concern.

(ii) Thermal ene reaction of 2-methylprop-1-ene with methyl propiolate gives a regioisomeric mixture. However, Lewis acid–mediated ene reaction gives only one regioisomer. Complexation of a Lewis acid to the ester makes the double bond electron-deficient and polarizes it so that reaction occurs regiospecifically at the electron-deficient β-carbon.

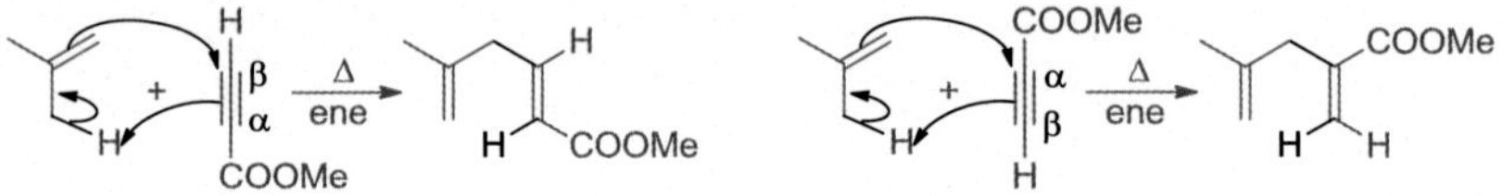

Q 4. Give the major products of the following reactions.

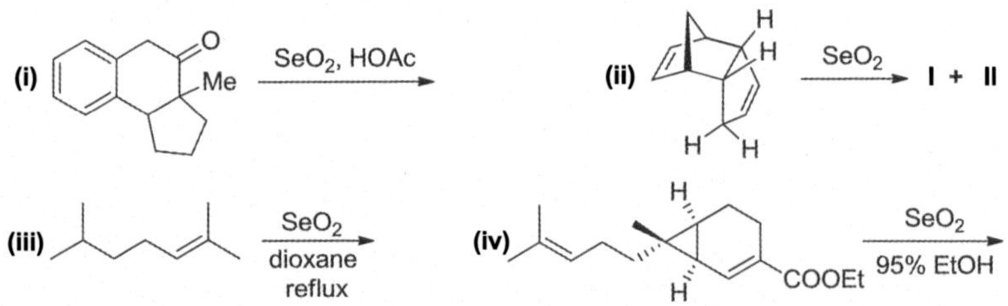

Sol 4. (i) SeO_2 oxidizes active methylene or methyl group present adjacent to the carbonyl group to give 1,2-dicarbonyl compounds. This reaction is called **Riley oxidation**.

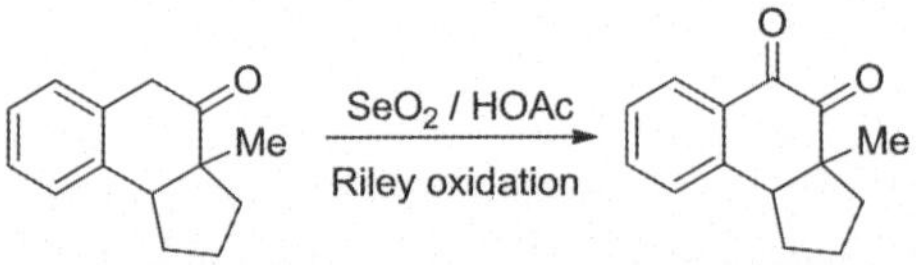

The mechanism involves ene reaction followed by [2,3]-sigmatropic rearrangement and then elimination to give the desired 1,2-dicarbonyl compounds.

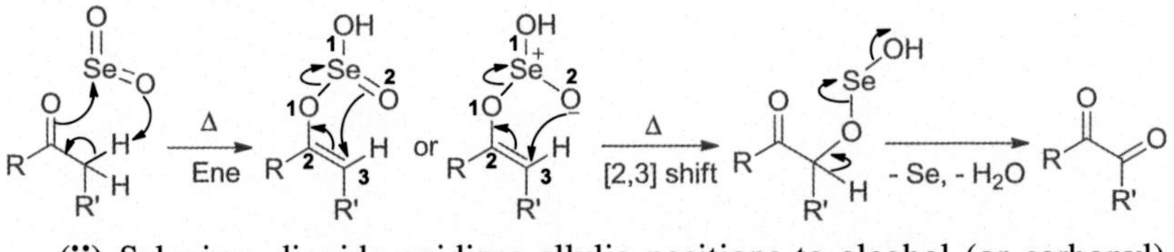

(ii) Selenium dioxide oxidizes allylic positions to alcohol (or carbonyl) group.

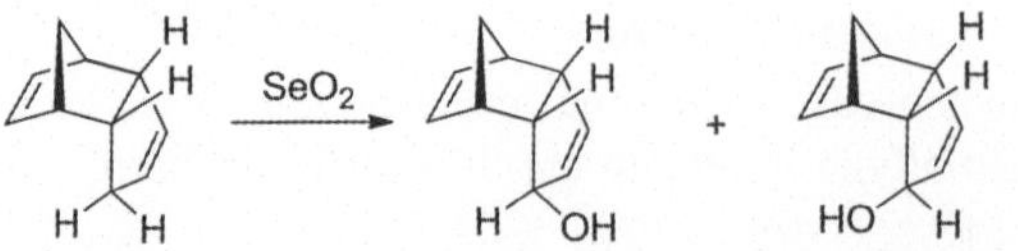

major product (I) minor product (II)

Mechanism: The first step is a cycloaddition between SeO_2 and the allylic substrate similar to the carbonyl ene reaction. The allylic seleninic acid produced in the first step undergoes a [2,3]-sigmatropic rearrangement to reinstate the double bond position. Rapid decomposition of the selenium(II) intermediate leads to an allylic alcohol or an allylic carbonyl compound as shown below:

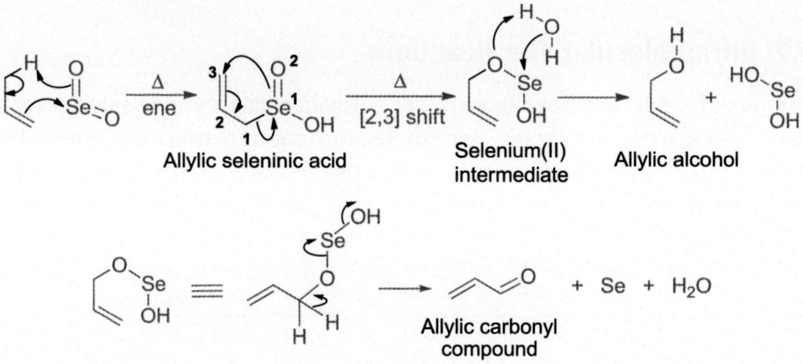

Allylic seleninic acid

Selenium(II) intermediate

Allylic alcohol

Allylic carbonyl compound

(iii) Trisubstituted alkenes are oxidized selectively at the more substituted end of the carbon–carbon double bond, indicating that the ene reaction step is electrophilic in character. Selenium dioxide reveals a useful stereoselectivity when applied to trisubstituted *gem*-dimethyl alkenes. The products are predominantly the *E*-allylic alcohols or unsaturated aldehydes. This stereoselectivity can be explained by a five-membered transition state for the sigmatropic rearrangement step. The observed *E*-stereochemistry results if the larger alkyl substituent adopts a pseudoequatorial conformation.

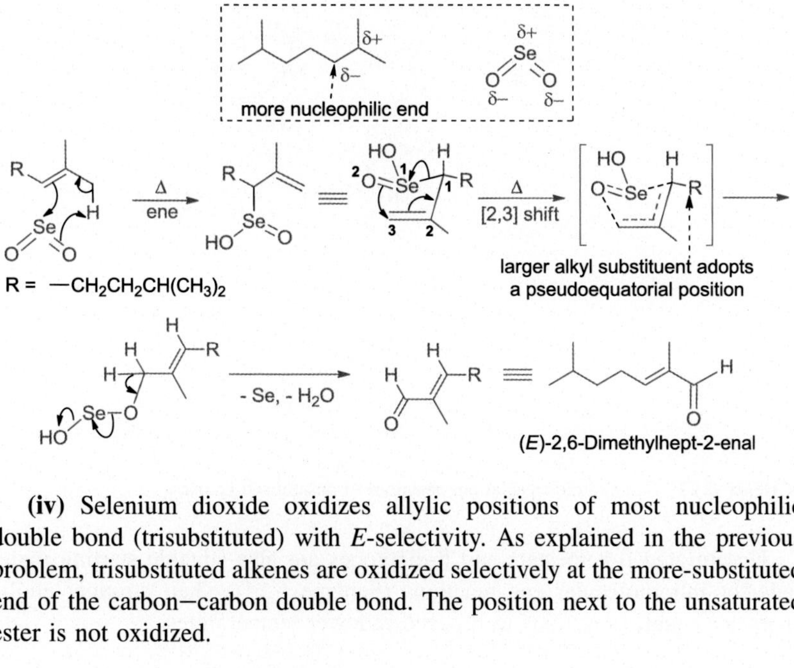

more nucleophilic end

$R = -CH_2CH_2CH(CH_3)_2$

larger alkyl substituent adopts a pseudoequatorial position

(*E*)-2,6-Dimethylhept-2-enal

(iv) Selenium dioxide oxidizes allylic positions of most nucleophilic double bond (trisubstituted) with *E*-selectivity. As explained in the previous problem, trisubstituted alkenes are oxidized selectively at the more-substituted end of the carbon–carbon double bond. The position next to the unsaturated ester is not oxidized.

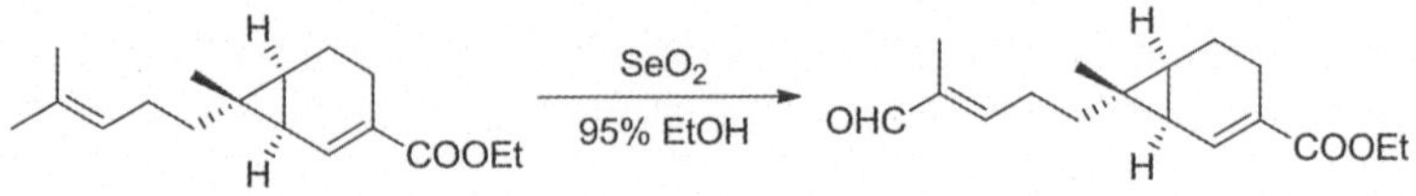

6.4.2 Intramolecular Ene Reactions

Intramolecular ene reactions have lower activation energy than intermolecular because of entropic advantage. Intramolecular ene reactions are often classified into three main types. These types differ only in the position of the attachment of the tether that connects the ene and enophile components (Figure 6.11).

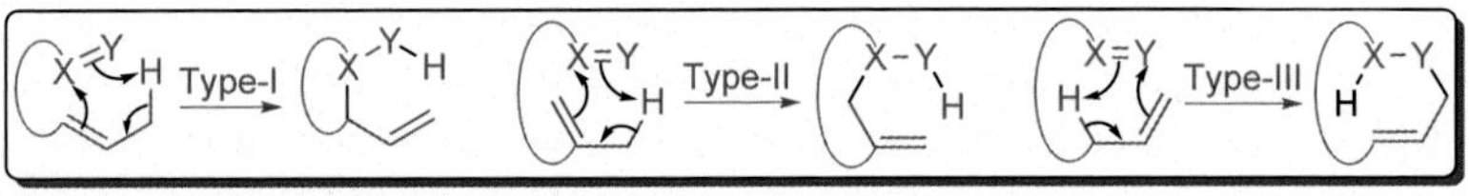

FIGURE 6.11 Classification of intramolecular ene reactions.

The **Conia-Ene reaction** is an intramolecular ene reaction of unsaturated ketones, in which the carbonyl functionality serves as the ene component, via its tautomer, and olefinic moiety serves as the enophile. Some of the important examples are summarized in Scheme 6.13.

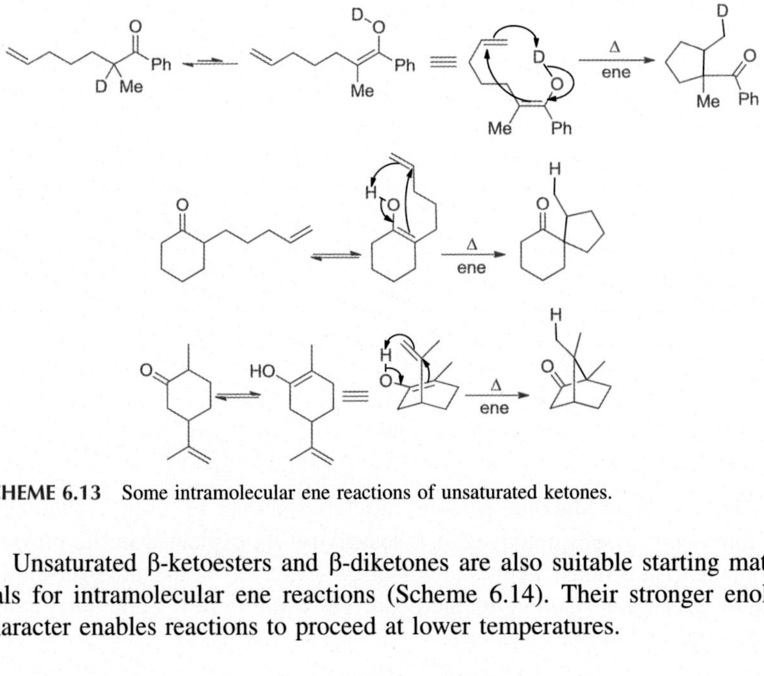

SCHEME 6.13 Some intramolecular ene reactions of unsaturated ketones.

Unsaturated β-ketoesters and β-diketones are also suitable starting materials for intramolecular ene reactions (Scheme 6.14). Their stronger enolic character enables reactions to proceed at lower temperatures.

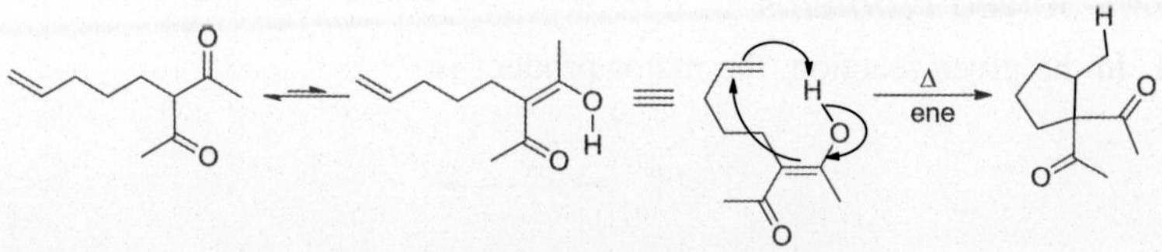

SCHEME 6.14 Intramolecular ene reaction of a β-diketone derivative.

The arynes generated by various methods can act as enophiles and undergo intramolecular reactions with various types of ene donors as shown in Scheme 6.15.

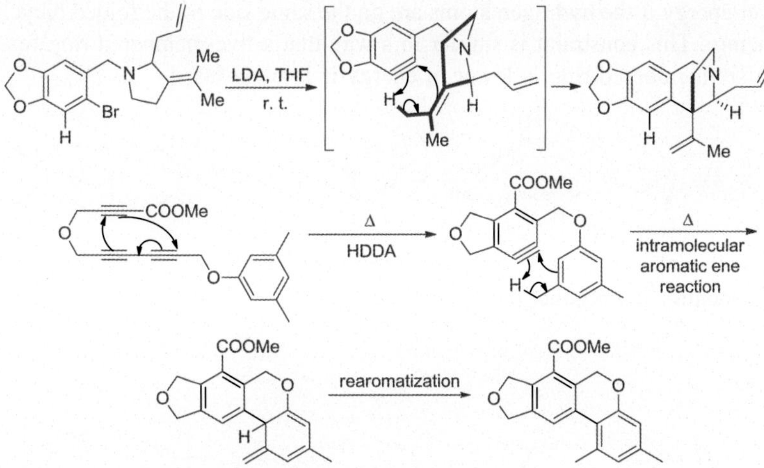

SCHEME 6.15 Intramolecular ene reactions of arynes.

Some examples of Type-II and Type-III intramolecular ene reactions are given in Scheme 6.16.

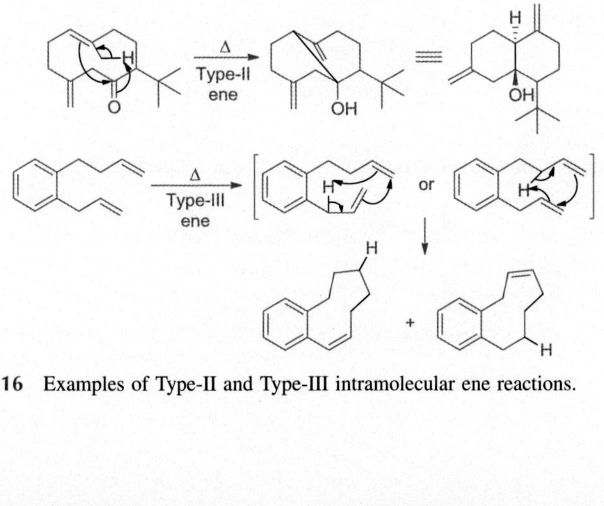

SCHEME 6.16 Examples of Type-II and Type-III intramolecular ene reactions.

6.4.2.1 Solved Problems

Q 1. In the given reaction, the major product is:

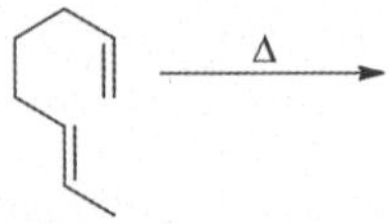

Δ

(a) 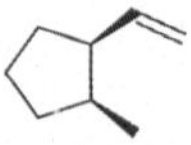 (b) 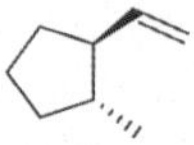 (c) 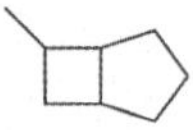 (d)

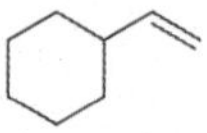

Sol 1. (a) Selective formation of the *cis*-disubstituted cyclopentane via intramolecular ene reaction is due to the fact that the transition state will be of lower energy if the hydrogen atoms are on the same side of the folded bicyclic structure. This constraint is similar in a way that a five-membered ring fused to a six-membered ring is lower in energy if it is *cis*-fused.

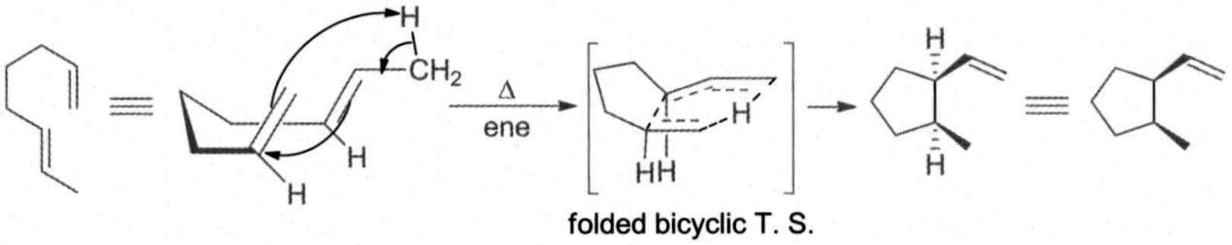

H CH_2 H H Δ ene H HH H H

folded bicyclic T. S.

Q 2. Identify the product of the reaction:

Δ

(a) 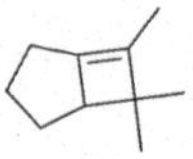 (b) 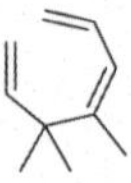 (c) 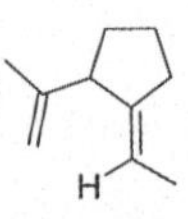 H (d)

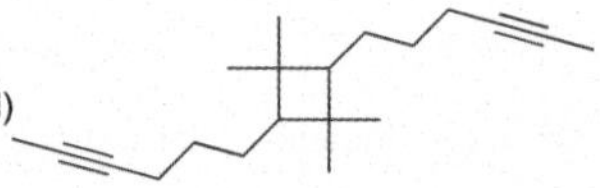

Sol 2. (c) The reactant undergoes intramolecular ene reaction under thermal conditions. In this reaction, allyl hydrogen is transferred to alkyne system instead of the transfer of propargylic hydrogen to alkene part, which may give allene derivative.

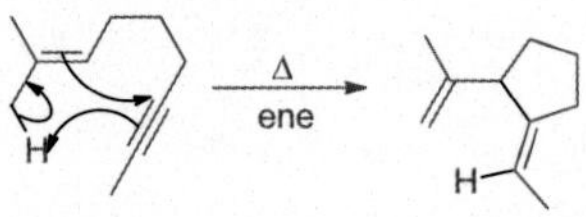

H Δ ene H

Q 3. The major product formed in the reaction sequence is:

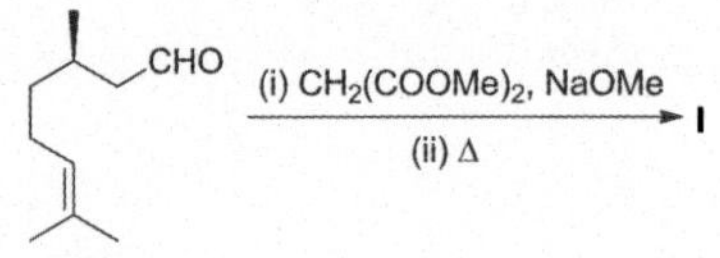

CHO (i) $CH_2(COOMe)_2$, NaOMe (ii) Δ I

Citronellal

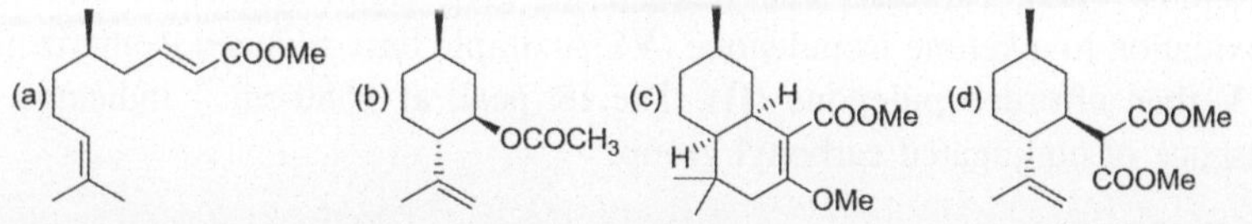

(a) COOMe (b) OCOCH3 (c) H COOMe H OMe (d) COOMe COOMe

Sol 3. (d) Initially dimethyl malonate (active methylene compound) undergoes Knoevenagel condensation with aldehyde group of citronellal to give the diene, which on heating undergoes intramolecular ene reaction via the formation of a stable six-membered chair-like transition state to give **I**.

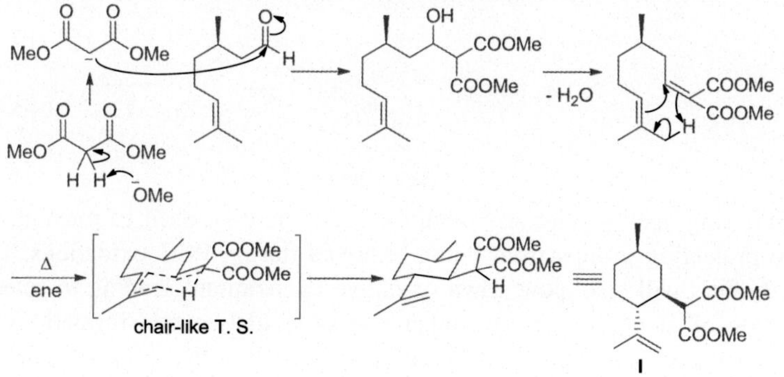

MeO OMe H COOMe COOMe - H_2O COOMe COOMe H MeO OMe H H OMe

Δ ene COOMe COOMe H chair-like T. S. COOMe COOMe H COOMe COOMe I

Decarboxylation does not occur since the Knoevenagel product is heated without adding acid.

Q 4. Citronellol **I** on oxidation with pyridinium chlorochromate (PCC) followed by treatment with aq. sodium hydroxide gives the product **II** (IR: 1680 cm^{-1}); whereas oxidation with PCC in the presence of sodium acetate gives product **III** (IR: 1720 cm^{-1}). Compounds **II** and **III** are:

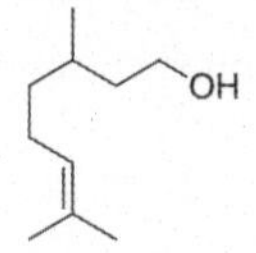

OH
Citronellol (I)

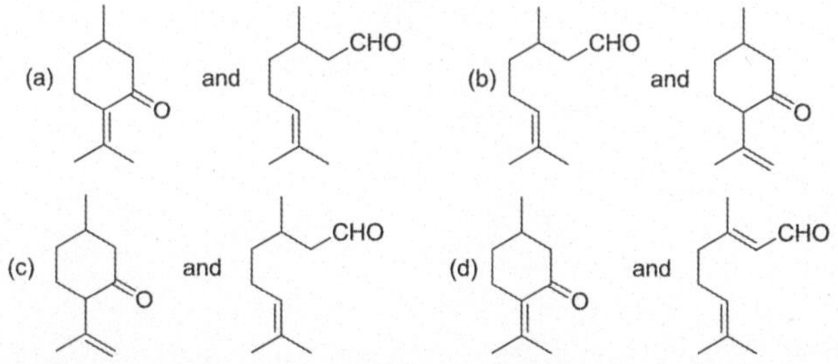

(a) O and CHO (b) CHO and O

(c) O and CHO (d) O and CHO

Sol 4. (a) Oxidative annulation of alkenols to form six-membered rings may be accomplished with mildly acidic oxidant pyridinium chlorochromate (PCC). This process is postulated to occur via initial oxidation of the citronellol (**I**) to citronellal (**III**), carbonyl ene reaction to give isopulegol (**IV**), and then reoxidation to a ketone isopulegone (**V**). A simple base-induced isomerization of **V** then afforded pulegone (**II**). The IR peak at 1680 cm^{-1} indicates the presence of conjugated carbonyl group.

Citronellol (**I**) —PCC, oxdn.→ [Citronellal (**III**) —Δ, ene→ Isopulegol (**IV**)] —oxdn.→ Isopulegone (**V**) —aq. NaOH→ ... —H_2O→ Pulegone (**II**) (IR: C=O str. 1680 cm^{-1})

Buffering agents (such as sodium acetate) may be used to prevent acid-labile protecting groups from being removed during PCC oxidations. However, buffers will also slow down oxidative cyclizations, leading to selective oxidation of alcohols over any other sort of oxidative transformations. For example, citronellol, which cyclizes in the presence of PCC, does not undergo cyclization when buffers are used. The IR peak at 1720 cm^{-1} indicates the presence of a carbonyl group without conjugation.

Citronellol (**I**) —PCC, CH_3COONa→ Citronellal (**III**) (IR: C=O str. 1720 cm^{-1})

Q 5. Rationalize the following transformations:

(i) —$ZnBr_2$, benzene, 5-10 °C→

(ii) —Δ→ ... OHC ... + ... OHC

(iii) Ph, HO —210 °C, benzene→ Ph, O

(iv) + —Δ→ + OH

(v) Br, MeO, OMe —LDA, THF, r. t.→ MeO, OMe

(vi) + —Me_2AlCl, CH_2Cl_2, 0 °C→ HO

(vii)

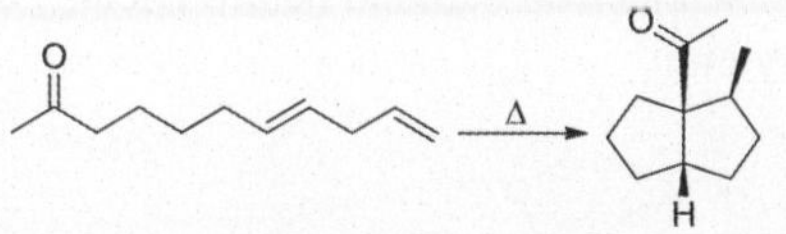

(viii)

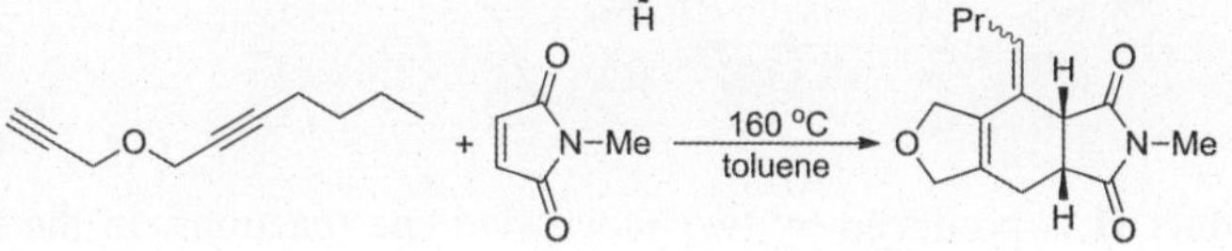

(ix)

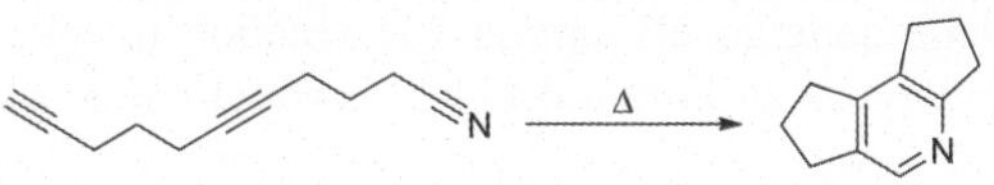

Sol 5. **(i)** The initial step involves the formation of an oxenium ion by the coordination of the Lewis acid to the carbonyl group of (R)-citronellal. This then undergoes an intramolecular carbonyl ene reaction via a chair-like transition state having all the substituents in the equatorial positions to give pure l-isopulegol.

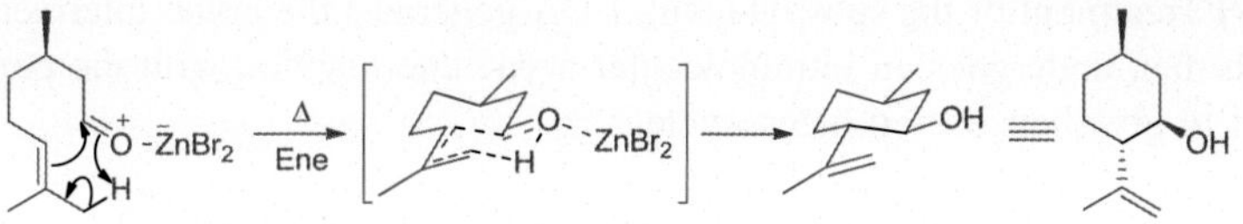

(ii)

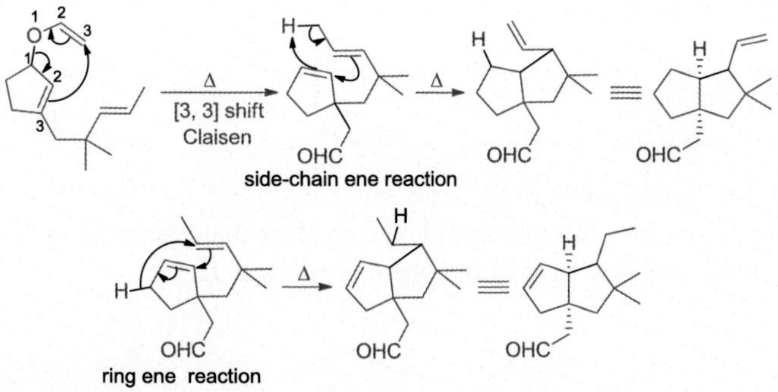

side-chain ene reaction

ring ene reaction

(iii) The starting compound on heating undergoes diastereoselective oxy ene reaction to give the *trans*-substituted product. When an allyl alcohol is used as the ene component, the reaction is known as an oxy ene reaction.

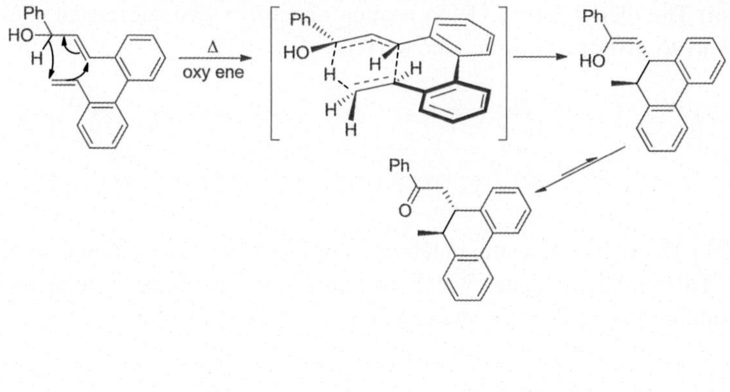

(iv) The product **I** arises by an inverse electron demand hetero Diels–Alder reaction between acrolein and isobutylene.

Δ
$[\pi^4s + \pi^2s]$

I

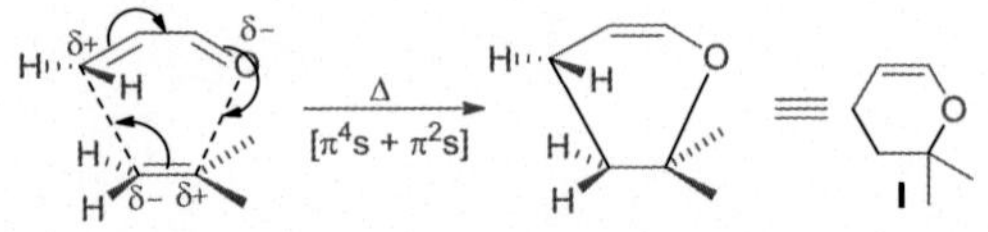

The product **II** is produced by two successive ene reactions. In the first step, acrolein and isobutylene undergo all carbon ene reaction to give an intermediate, which then undergoes an intramolecular carbonyl ene reaction to give **II**.

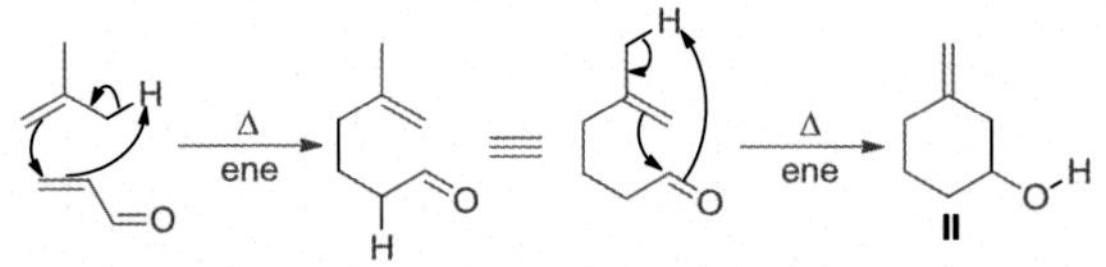

Δ ene
Δ ene
II

(v) Treatment of the substrate with LDA generates the aryne intermediate, which then undergoes an intramolecular aryne ene reaction with the pendant olefin to give benzofused heterocycle.

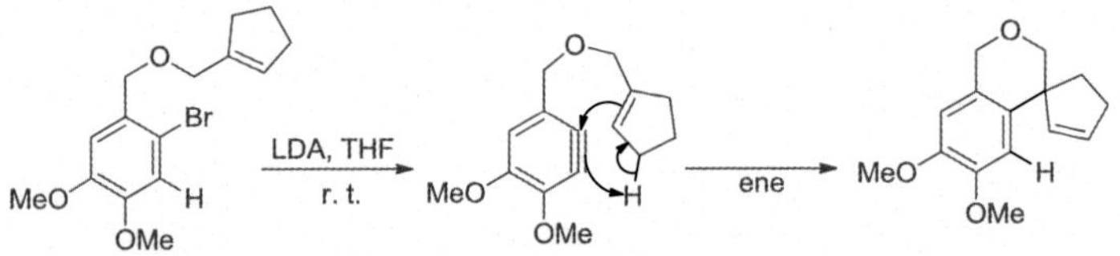

LDA, THF
r. t.
ene

(vi) Ethylidenecyclohexane and but-3-en-2-one undergo all carbon ene reaction to give the intermediate **I**, which then undergoes an intramolecular carbonyl ene reaction to give the bicyclic product **II**.

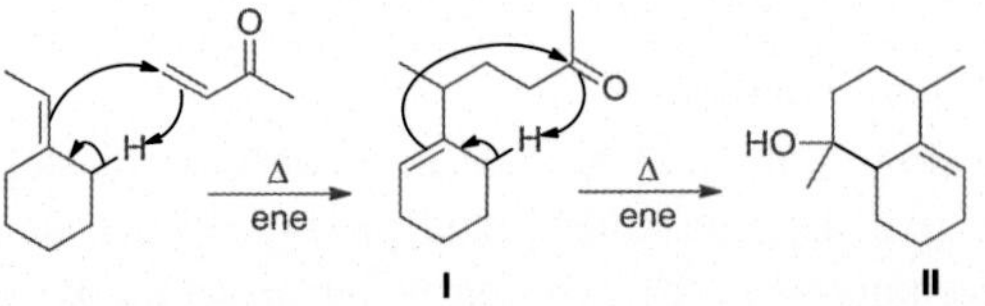

Δ ene
I
Δ ene
II

(vii) The dienyl ketone (**I**) on heating undergoes two successive enol ene reactions to give the bicyclic ketone (**II**).

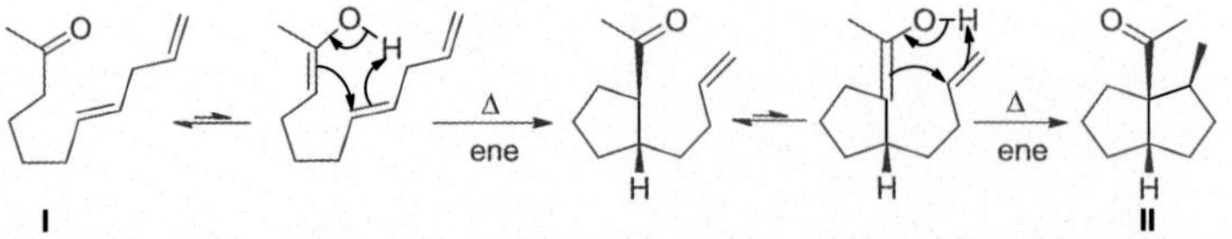

I
Δ ene
Δ ene
II

(viii) The diyne on heating undergoes propargylic ene reaction to generate a vinylallene intermediate, which is trapped in an intermolecular [4 + 2] cycloaddition with *N*-methylmaleimide to give an *endo* adduct.

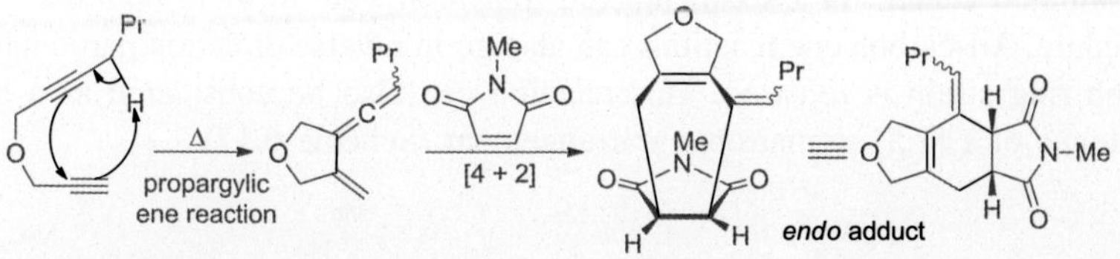

(ix) Thermolysis of cyanodiyne (I) in toluene produces substituted pyridine (II). The formation of II can proceed through three possible mechanistic pathways. The compound II can be directly generated by a thermal [2 + 2 + 2] cycloaddition in a single step. The second pathway involves a cascade mechanism involving a propargylic ene reaction, followed by an intramolecular Diels–Alder reaction involving the vinylallene and the cyano group. After a [1,5] hydrogen shift, the substituted pyridine (II) is generated. The third pathway begins with an intramolecular propargylic ene reaction in which the cyano group serves as the enophile. The resulting allenylimine then undergoes hetero Diels–Alder reaction followed by tautomerization to give the same pyridine product (II).

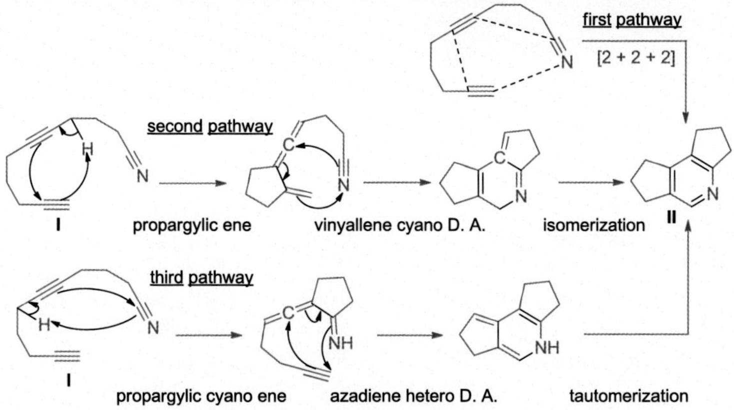

Experimental evidences and computational studies have shown that the second pathway is the most favorable pathway in this formal [2 + 2 + 2] pyridine synthesis. These results have been justified on the basis of the high stability and difficulty of distortion of the cyano group, and the high distortion energy of the [2 + 2 + 2] transition state, which makes the ene reaction of the cyano group or the direct [2 + 2 + 2] cycloaddition more difficult than the second pathway. However, it has been shown that if the hydrogen atom at the propargylic carbon in I is substituted by some other groups, then the initial propargylic ene step cannot take place. In that situation the reaction follows the third pathway and leads to the formation of similar substituted pyridine products.

6.4.3 Retro Ene Reactions

Retro ene reaction or retro ene fragmentation involves the intramolecular thermolysis of unsaturated compounds by the transfer of a γ-hydrogen to the unsaturated center via a six-membered transition state to yield both ene and enophile. All-carbon ene reactions can also go in reverse direction particularly when ring strain is released. This reaction can also be considered as a homologue of a [1,5] sigmatropic rearrangement (Scheme 6.17).

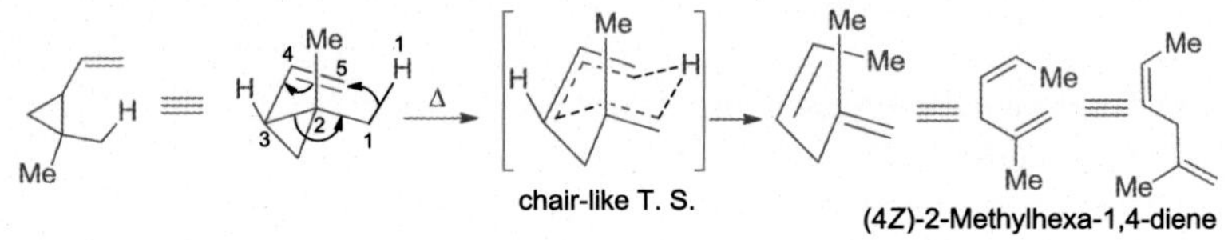

Me H Me 4 5 1 H H 3 2 1 Δ Me H H chair-like T. S. Me Me Me Me Me Me (4Z)-2-Methylhexa-1,4-diene

SCHEME 6.17 Retro ene reaction: a homologue of a [1,5] sigmatropic rearrangement.

6.4.3.1 Solved Problems

Q 1. Explain the mechanism of the following reactions:

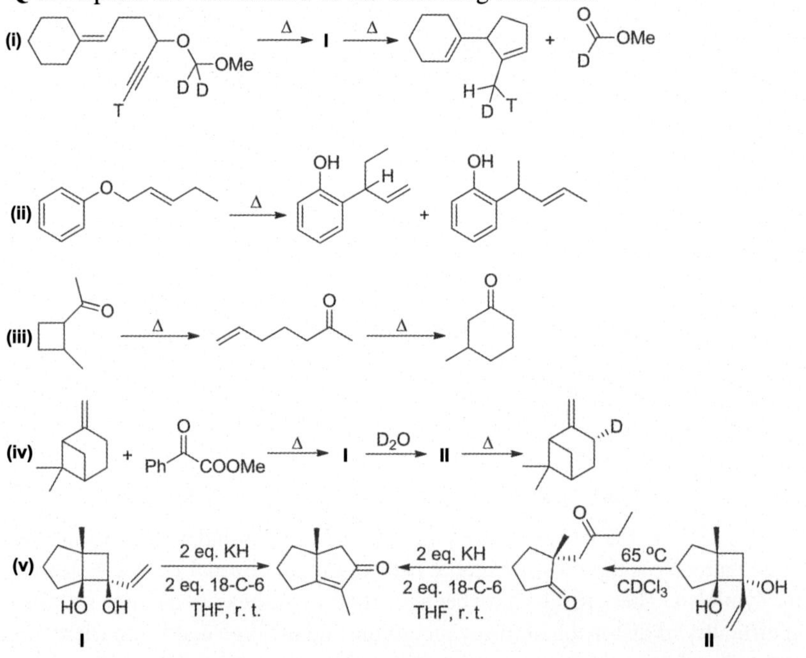

(i) O OMe D D T Δ I Δ H D T + O OMe D

(ii) O Δ OH H + OH

(iii) O Δ O Δ O

(iv) + O Ph COOMe Δ I D_2O II Δ D

(v) HO OH I 2 eq. KH 2 eq. 18-C-6 THF, r. t. O 2 eq. KH 2 eq. 18-C-6 THF, r. t. O O 65 °C $CDCl_3$ HO OH II

Also explain the fact that the thermal retro ene did not occur when the *cis*-diol **I** was refluxed in toluene for a week. However, the reaction occurs at a reasonable temperature in case of the *trans*-diol **II**.

Sol 1. (i) The synthesis involves two consecutive reactions; an ene reaction followed by retro ene reaction. The driving force for the former is formation of a new σ-bond at the expense of a carbon–carbon triple bond and for the latter is the formation of a C=O π-bond.

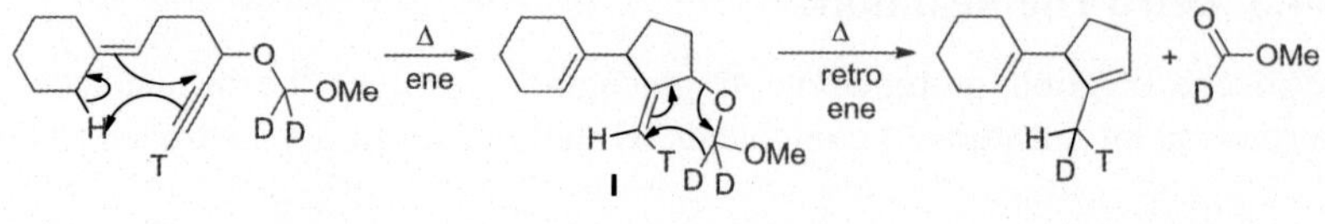

O H OMe D D T Δ ene H O T OMe D D I Δ retro ene H D T + O OMe D

(ii) Further rearrangement of the first *o*-allyl phenol (normal Claisen rearrangement product) by an ene reaction followed by retro ene reaction gives the second *o*-allyl phenol (abnormal Claisen rearrangement product).

Δ
[3,3] shift
tautomerization

Δ
ene
Δ
retro ene

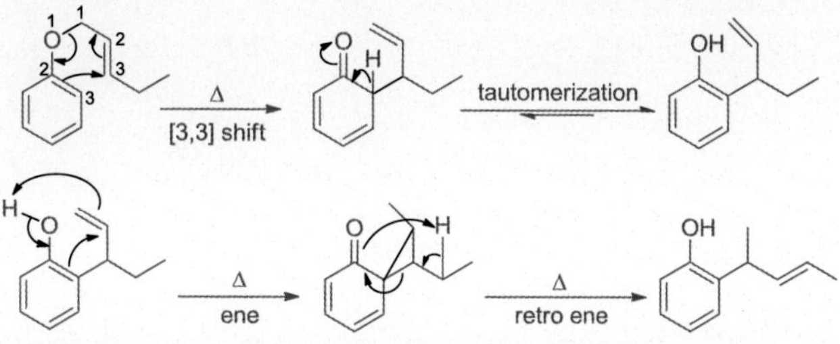

(iii) The reactant on heating undergoes retro ene reaction to release the ring strain. The unsaturated ketone thus obtained undergoes an intramolecular ene reaction in which the carbonyl functionality serves as the ene component via its tautomer and olefinic moiety serves as the enophile.

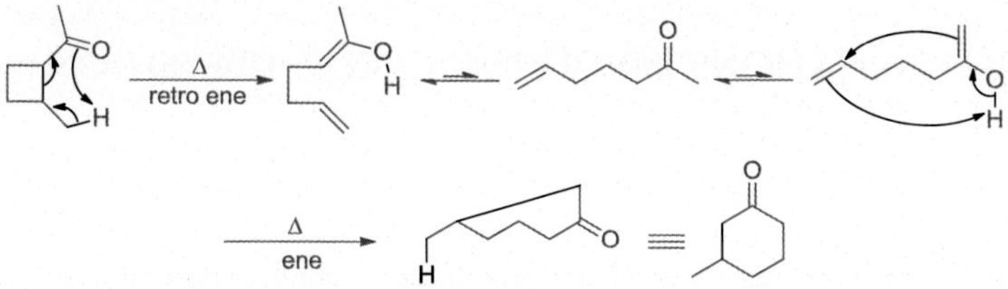

Δ
retro ene

Δ
ene

(iv) β-Pinene on heating with methyl 2-oxo-2-phenylacetate undergoes carbonyl ene reaction to give **I**, which undergoes reversible deuterium exchange to give **II**. On heating, **II** is transformed to *trans*-labeled isomer of β-pinene by a stereospecific retro ene reaction.

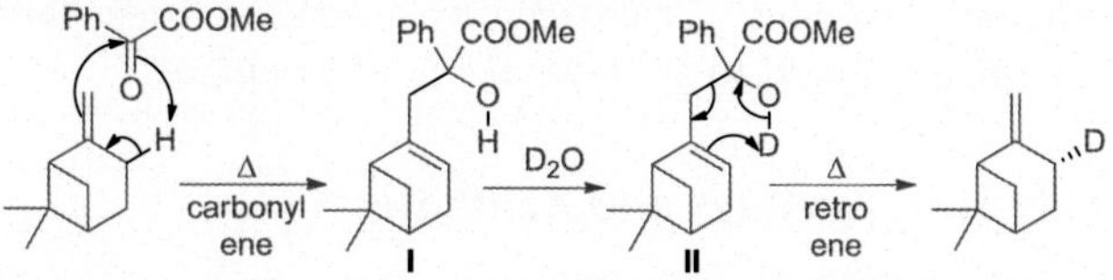

Δ
carbonyl ene
D_2O
Δ
retro ene

(v) Treatment of the *cis*-diol **I** with KH and 18-Crown-6 at room temperature gives an alkoxide, which undergoes a facile **anionic oxy retro ene reaction** to yield the enol enolate, which is perfectly set up for the tandem intramolecular aldol condensation.

It should be noted that the *cis*-diol **I** undergoes retro ene reaction at room temperature because of the alkoxide accelerating effect. However, the thermal retro ene does not occur when the *cis*-diol **I** was refluxed in toluene. This is due to the fact that thermal retro ene reactions tend to occur between 180 and 400 °C. However, if the migrating hydrogen comes from an oxygen atom rather than a carbon, then the transition state is significantly lowered, and the rearrangement can occur within the 130 °C range. Combination of this effect with the strain release of a cyclobutane ring in the case of the *trans*-diol **II** lowers the temperature for the reaction to occur within the 40–60 °C range.

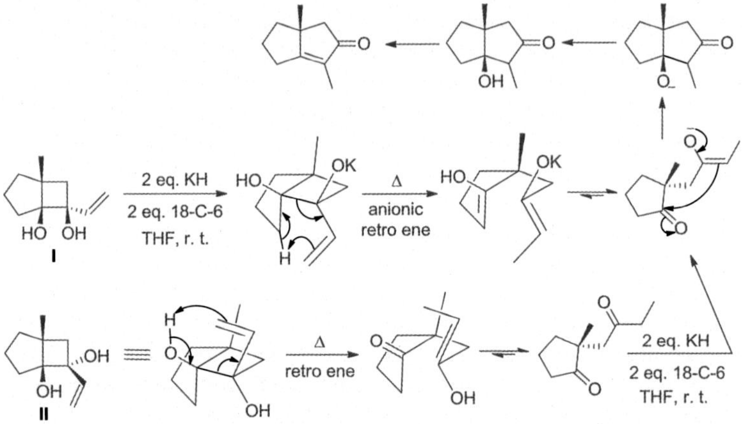

6.5 β-ELIMINATIONS INVOLVING CYCLIC TRANSITION STRUCTURES

Another important family of elimination reactions having cyclic transition states involves an intramolecular hydrogen transfer that accompanies elimination to form a new carbon–carbon double bond. These are thermally activated unimolecular reactions that normally do not involve acidic or basic catalysts. There is, however, a wide variation in the temperature at which elimination proceeds at a convenient rate. The cyclic transition state makes it mandatory that elimination occurs with *syn* stereochemistry. At least in principle, all such reactions can proceed by a concerted mechanism. Such reactions are often referred to as *thermal syn eliminations*.

6.5.1 Pyrolysis of Acetates and Xanthates

Sometimes retro ene reactions are of great synthetic value in stereoselective eliminations as in the pyrolysis of esters (Scheme 6.18). However, high temperature (300–500 °C) required for such reactions restricts their utility.

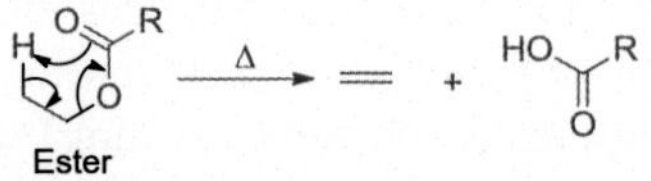

SCHEME 6.18 Pyrolysis of ester derivatives.

The ready decarboxylation of β-ketocarboxylic acids is shown to occur in a similar manner (Scheme 6.19).

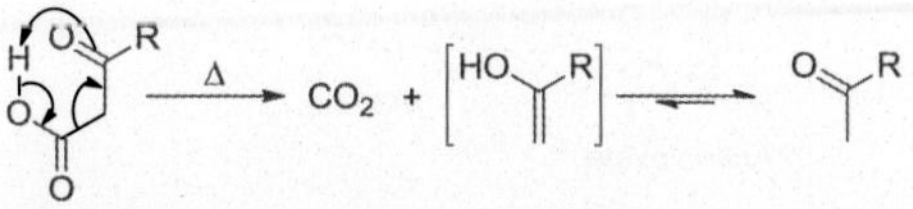

SCHEME 6.19 Decarboxylation of β-ketocarboxylic acids.

For the related **Chugaev reaction**, which involves the pyrolysis of xanthate esters (*O*-alkyl-*S*-methyl xanthates), conditions are much milder (100–200 °C) (Scheme 6.20).

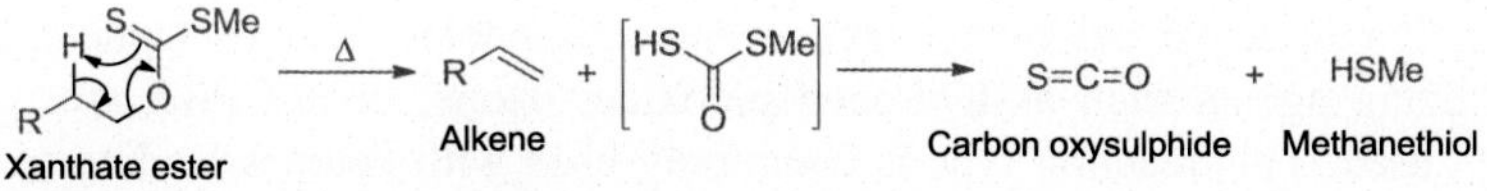

SCHEME 6.20 Pyrolysis of xanthate esters.

These pyrolytic *syn*-eliminations in acetates and xanthates proceed via a six-membered transition state in which three electron pairs are shifted at the same time. They can be easily explained by PMO approach (Figure 6.12).

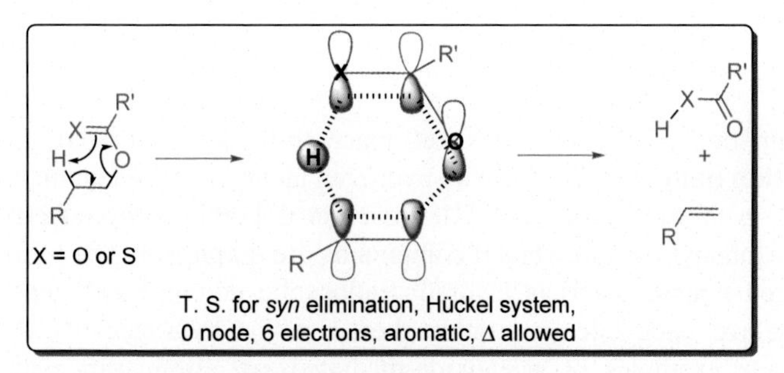

FIGURE 6.12 PMO analysis of pyrolytic *syn* eliminations in acetates and xanthates.

Some cycloeliminations use a ring of five atoms instead of six but still involve six electrons. This is no longer a retro ene reaction but it is still a retro group transfer and is allowed in the all-suprafacial mode. Some common examples of such reactions are pyrolysis of amine oxides, sulphoxides, and selenoxides. All these reactions are *syn* stereospecific.

6.5.2 Pyrolysis of Amine Oxides

The pyrolysis of amine oxides is called **Cope elimination** and typically takes place at 120 °C (Scheme 6.21). The reaction is a *syn* periplanar elimination in which six electrons move in a five-membered ring according to a concerted, thermally induced mechanism to yield an alkene and a hydroxylamine.

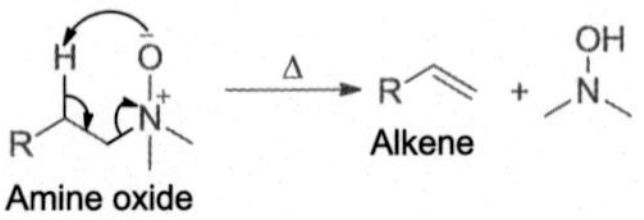

SCHEME 6.21 Pyrolytic *syn*-eliminations in amine oxides: Cope elimination.

6.5.3 Pyrolysis of Selenoxides

The pyrolysis of selenoxides takes place at room temperature or below. In the presence of a β-hydrogen, a selenite will give an elimination reaction after oxidation to leave behind an alkene and a selenenic acid (Scheme 6.22). Oxidizing agents such as hydrogen peroxide, ozone, or *m*-CPBA are quite often used. This reaction type is commonly used with ketones leading to the formation of enones.

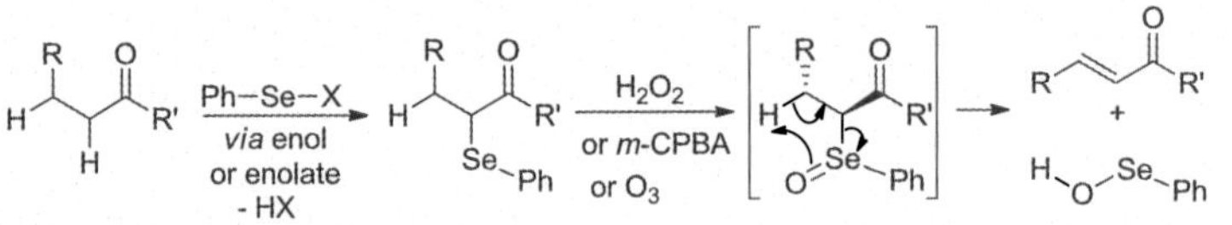

SCHEME 6.22 Pyrolysis of selenoxides.

Elimination of selenoxides takes place through an intramolecular, ***syn* elimination pathway**. The carbon—hydrogen and carbon—selenium bonds are coplanar in the transition state. The reaction is highly ***trans*-selective** when acyclic α-phenylseleno carbonyl compounds are employed. The formation of conjugated double bonds is favored. Endocyclic double bonds tend to predominate over exocyclic ones, unless there is no *syn* hydrogen available in the ring. Some examples of selenoxide-mediated *syn* elimination reaction are given in Scheme 6.23.

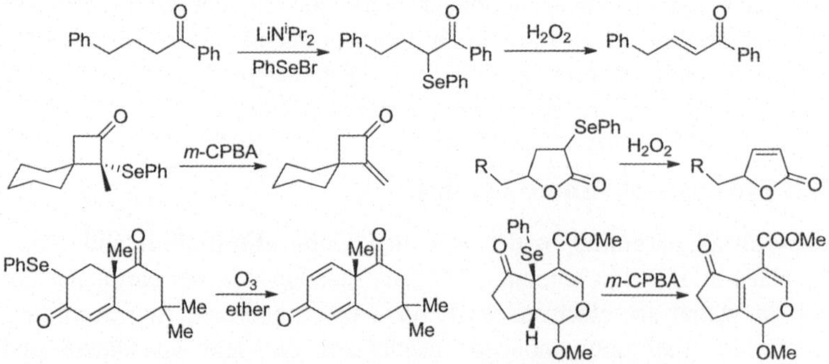

SCHEME 6.23 Some important *syn* elimination reactions through selenoxides.

6.5.4 Pyrolysis of Sulphoxides

On pyrolysis, sulphoxides with a β-hydrogen undergo *syn* elimination to give alkenes (Scheme 6.24). The pyrolysis of sulphoxides takes place at about 80 °C. Like other concerted pericyclic reactions, this reaction is also highly stereoselective. For example, the *anti*-sulphoxide (**1**) on pyrolysis gives mainly *trans*-methylstilbine (**2**), whereas the corresponding *syn*-sulphoxide (**3**) gives predominantly *cis*-methylstilbine (**4**).

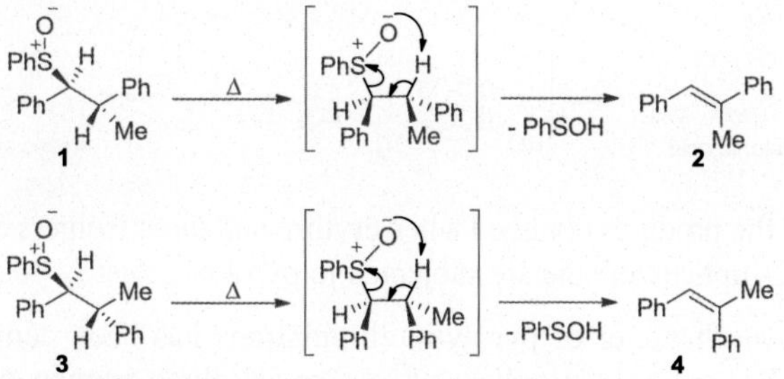

SCHEME 6.24 Pyrolysis of sulphoxides.

6.5.5 Solved Problems

Q 1. Diols (**I–IV**), which react with CrO_3 in aqueous H_2SO_4 and yield products that readily undergo decarboxylation on heating, are:

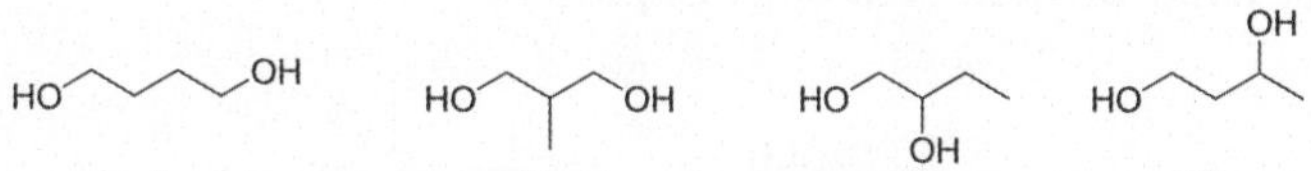

(a) I and II (b) II and III (c) II and IV (d) I and IV

Sol 1. (c) Geminal dicarboxylic acid (i.e., compounds in which the two carboxyl groups are on the same carbon atom) and β-ketocarboxylic acid obtained by the oxidation of diols **II** and **IV**, respectively, readily undergo decarboxylation on heating.

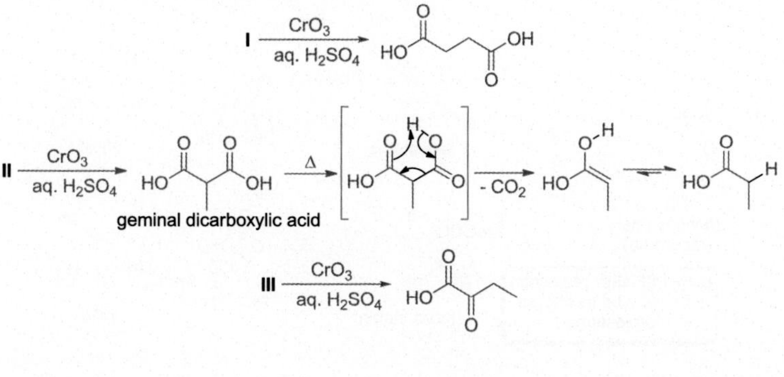

IV $\xrightarrow[\text{aq. } H_2SO_4]{CrO_3}$ β-Ketocarboxylic acid $\xrightarrow{\Delta}$ [ ] $\xrightarrow{- CO_2}$

Q 2. Name the dicarboxylic acid that on strong heating produces acetic acid.

(a) Succinic acid (b) Malonic acid (c) Oxalic acid (d) Glutaric acid

Sol 2. (b) Malonic acid being a geminal dicarboxylic acid, readily undergoes decarboxylation on heating to produce acetic acid.

Malonic acid $\xrightarrow{\Delta}$ [ ] $\xrightarrow{- CO_2}$ Acetic acid

Q 3. Identify the products obtained when erythro and threo isomers of 1-acetoxy-2-deutero-1,2-diphenylethane are subjected to pyrolysis. Justify your answer.

Sol 3. The *syn* character of pyrolytic eliminations has been demonstrated in many ways. For example, pyrolysis of erythro and threo isomers of 1-acetoxy-2-deutero-1,2-diphenylethane gave in each case *trans*-stilbene as a major product, but the stilbene from the erythro compound retained nearly all its deuterium, whereas the stilbene from the threo compound had lost most of its deuterium. Either the hydrogen or deuterium could be *syn* to the acetoxy group, but the preferred conformations are those in which the phenyl groups are as far removed from each other as possible.

erythro isomer (racemate)

convert Fisher projection into eclipsed sawhorse projections

less stable — 5%

more stable — 95%

threo isomer (racemate)

convert Fisher projection into eclipsed sawhorse projections

less stable — 26%

more stable — 74%

Q 4. Explain the mechanism of the following reaction:

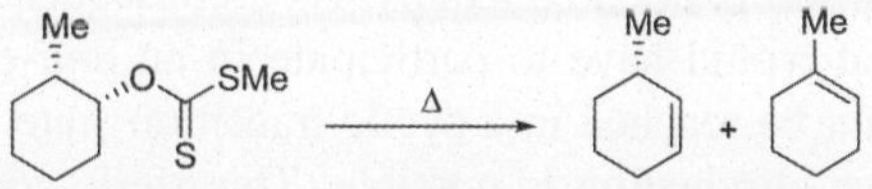

Sol 4. In the Chugaev reaction, the favored transition state is the one in which the leaving group and the *syn*-hydrogen atom in the β-position leave the substrate i.e., a *syn*-transition state. But the reaction is not necessarily *syn* selective. The six-membered transition state in the Chugaev reaction provides a stereochemical flexibility that the leaving group and an *anti*-hydrogen atom in the β-position can leave the substrate jointly as a thiocarboxylic acid half ester. For example, in the Chugaev reaction shown below, we get a mixture of two olefins. Fifty percent of the resulting olefin mixture represents the thermodynamically more stable Saytzeff's product. It stems from an *anti*-elimination in a substrate conformer with an equatorially disposed xanthate group because the equally equatorially disposed *anti*-hydrogen at β-carbon is within reach in the corresponding cyclic transition state.

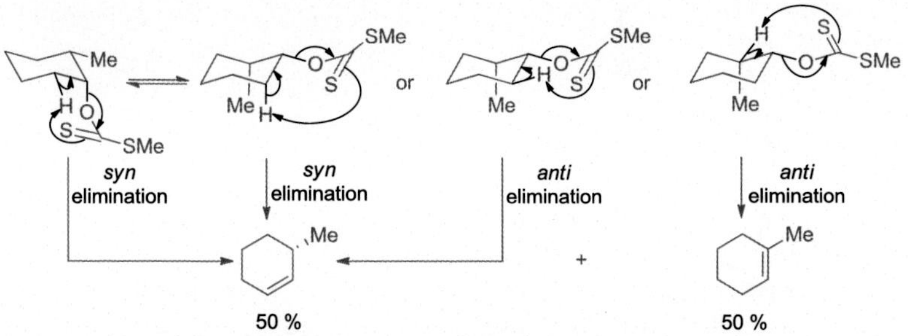

Q 5. The major product formed in the following reaction is:

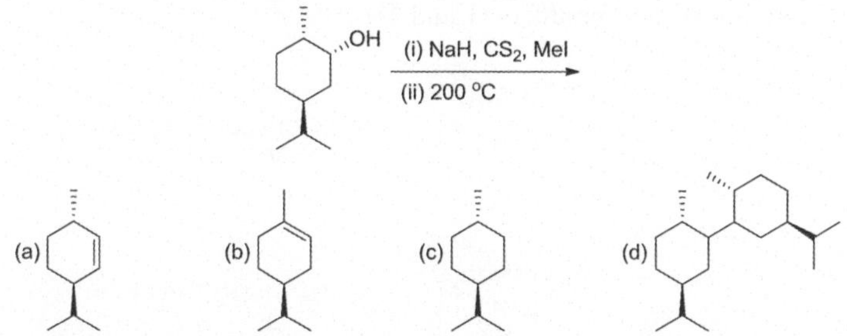

(a) (b) (c) (d)

Sol 5. (a) The reaction involves the conversion of alcohol into olefins by the pyrolysis of the corresponding xanthate esters (Chugaev reaction). In the first step, xanthate ester is synthesized from the alcohol as shown below:

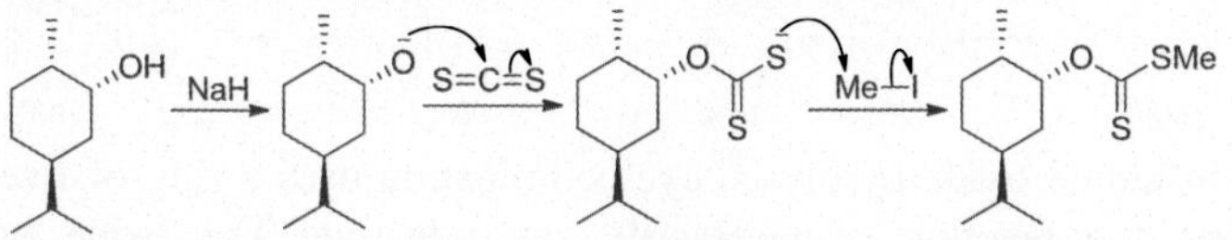

In the presence of the isopropyl group, the cyclohexane ring exists preferentially in the chair conformation in which this group is oriented equatorially and the xanthate group is oriented axially. As a consequence, the H atoms in the β-positions that would have to participate in an *anti*-elimination are far removed and cannot be reached in a cyclic transition state. Therefore, only a highly *syn*-selective elimination is possible. This steric course explains why the thermodynamically less stable Hofmann product is produced regioselectively in this case.

Unstable conformations due to the presence of the bulky isopropyl group in axial position

syn elimination | *syn* elimination | *anti* elimination | *anti* elimination

100 % 0 %

Q 6. Pyrolysis of $Me-\overset{\bar{O}}{\underset{Me}{N^+}}-CH(Me)(CD_3)$ would give

(a) A mixture of CH_2=CH-CD_3 and CH_3-CH=CD_2

(b) CH_3-CH=CD_2

(c) Me_2N^+=C(Me)(CD_3)

(d) CH_2=CH-CD_3

Sol 6. (a) *Syn* elimination of hydrogen or deuterium present at β-position leads to the formation of two products (**I** and **II**).

CH_3-CH=CD_2 + Me_2N–OD **I**

CD_3-CH=CH_2 + Me_2N–OH **II**

Q 7. Rationalize the following transformations:

trans → 85% + 15%; *cis* → 2% + 98%

Sol 7. In amine oxide pyrolysis, cycloelimination uses a ring of five atoms, therefore, the reaction is necessarily *syn* selective. The oxide of *trans*-1-phenyl-2-dimethylaminocyclohexane undergoes *syn* elimination to give

1-phenylcyclohexene as the major product. This may be due to increased acidity of the proton α to the phenyl ring and the stabilizing effect of the developing conjugation in the transition state. In the *cis* isomer, there is no *syn* hydrogen at the phenyl-substituted carbon and the nonconjugated regioisomer is formed. The elimination does not involve the *anti*-hydrogen even though it is activated by the phenyl group but it involves one of the adjacent methylene hydrogens to give 3-phenylcyclohex-1-ene.

Q 8. The major product formed in the following reaction is:

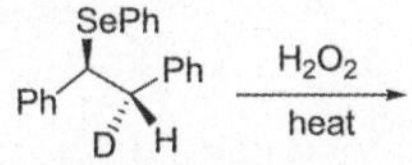

SePh, Ph, Ph, D, H — H_2O_2, heat →

(a) (b) (c) (d)

Sol 8. (c) In the intermediate selenoxide, either β-hydrogen or β-deuterium, can attain the eclipsed conformation of the cyclic transition state; therefore, a mixture of alkenes is formed. The product ratio parallels the relative stability of the competing transition states. Generally, more of the *E*-alkene is formed because of the larger steric interactions present in the transition state leading to the formation of *Z*-alkene, but the selectivity is generally not high.

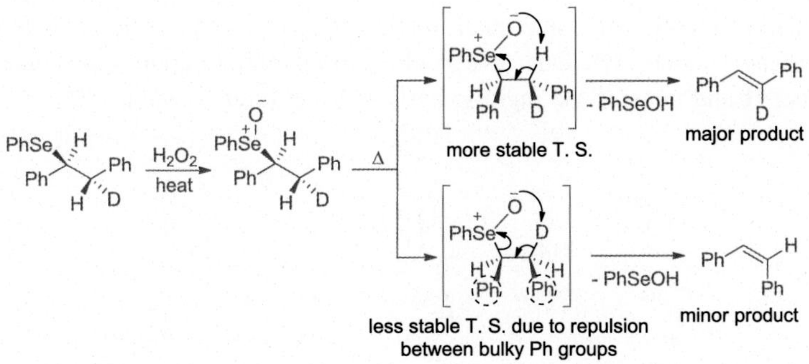

H_2O_2, heat; Δ; − PhSeOH

more stable T. S. → major product

less stable T. S. due to repulsion between bulky Ph groups → minor product

Q 9. In the reaction sequence given below, the major products **I** and **II**, respectively, are:

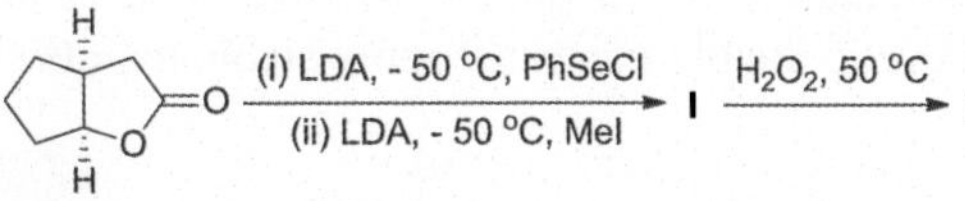

(i) LDA, - 50 °C, PhSeCl; (ii) LDA, - 50 °C, MeI → **I**; H_2O_2, 50 °C → **II**

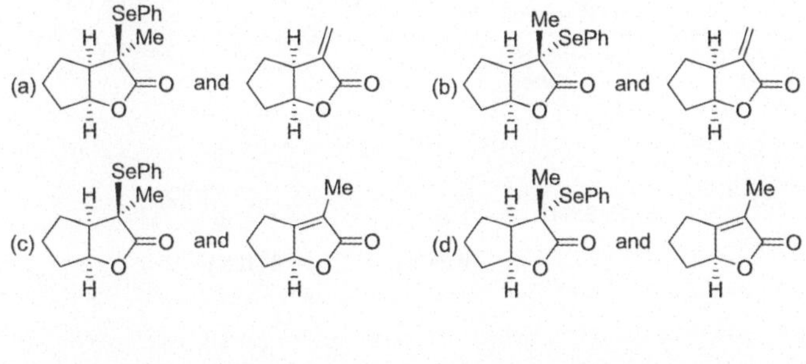

(a) and (b) and (c) and (d) and

Sol 9. (a) Carbonyl compounds can be converted to α,β-unsaturated carbonyl compounds by selenoxide eliminations. Initial deprotonation with LDA gives an enolate anion, which acts as a nucleophile and attacks the phenylselenyl chloride. The nucleophile approaches from the less-hindered side, i.e., from the side of hydrogen to avoid steric interaction with the ring carbon. However, addition of LDA to the above product gives a planar enolate anion, which is attacked by the methyl group, again from the less-hindered side to push the already present PhSe group to the side opposite to the hydrogen.

endo
exo
LDA
- 50 °C
Ph–Se–Cl
planar enolate attacks the electrophile from the less hindered *exo* face
SePh
LDA
- 50 °C
Me–I
Me
I

The final step involves oxidation of selenide (**I**) to selenoxide using hydrogen peroxide followed by pyrolytic ***syn*** **elimination** to give a double bond. In this case, the oxygen cannot approach the hydrogen at the ring junction, so elimination involves the side chain methyl group to give the less stable *exo*-methylene lactone (**II**). Due to *syn* elimination, the reaction gives only the product **II** and none of the regioisomeric α,β-unsaturated lactone (**IV**).

H_2O_2
50 °C
Ph–Se
CH_2
- PhSeOH
I
II

However, if we change the sequence of the first two steps, we get different products **III** and **IV**. In this case, the diastereomer **III** has two *syn* hydrogens for elimination, but it prefers the hydrogen at the ring junction because the endocyclic double bonds are more stable than the exocyclic ones.

(i) LDA, - 50 °C, MeI
(ii) LDA, - 50 °C, PhSeCl
III
H_2O_2
50 °C
- PhSeOH
IV (90%) + **II (10%)**

Q 10. The major product of the following reaction is:

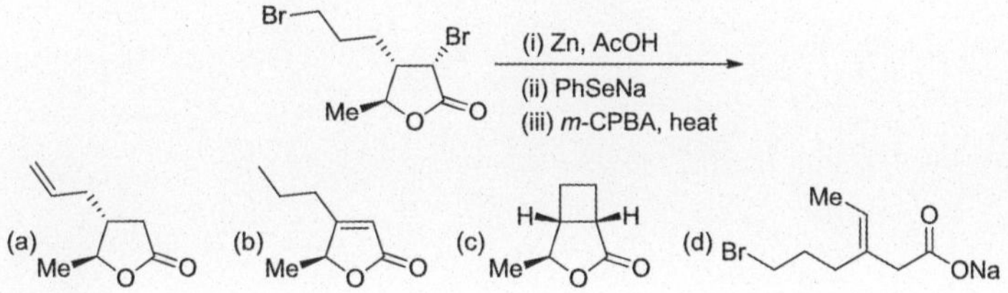

(i) Zn, AcOH
(ii) PhSeNa
(iii) *m*-CPBA, heat

(a) (b) (c) (d)

Sol 10. (a)

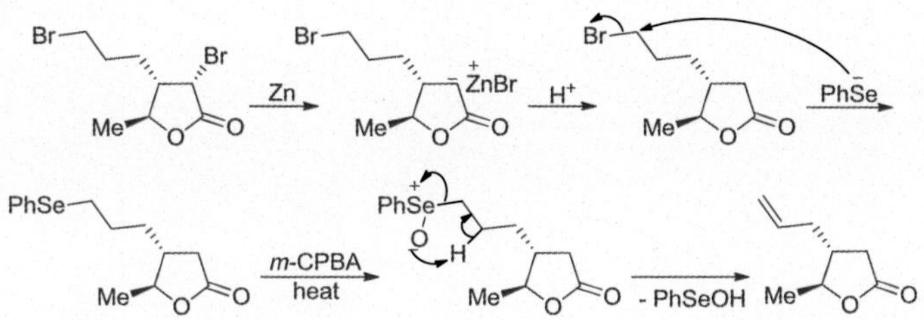

Zn → H^+ → $\overset{-}{PhSe}$ → *m*-CPBA heat → - PhSeOH

Chapter 7

Unsolved Problems

Q 1. Rationalize the following transformations:

(i) O, N, H; $\xrightarrow[\text{PhNO}_2]{\Delta}$ O, N

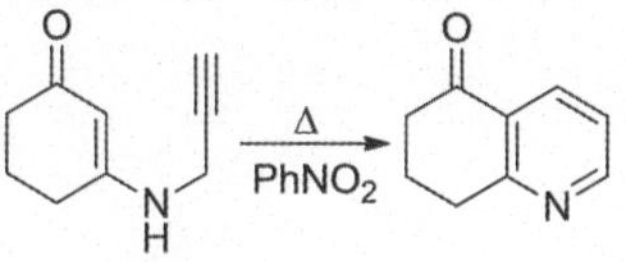

(ii) Ph, O, OH; $\xrightarrow{\Delta}$ HO, O, Ph

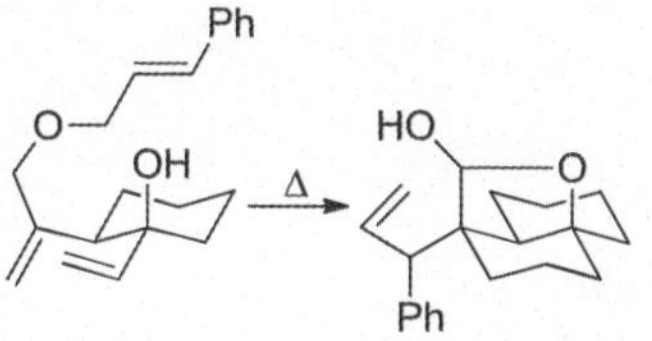

(iii) O, O; $\xrightarrow[\text{PhNEt}_2]{\text{reflux}}$ O, O

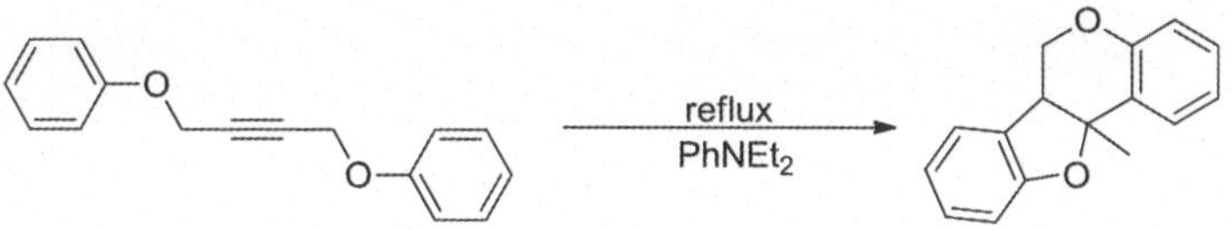

(iv) NH_2, N, H + O, OMe, OMe $\xrightarrow[\text{H}^+]{\Delta}$ N, H, OMe, OMe

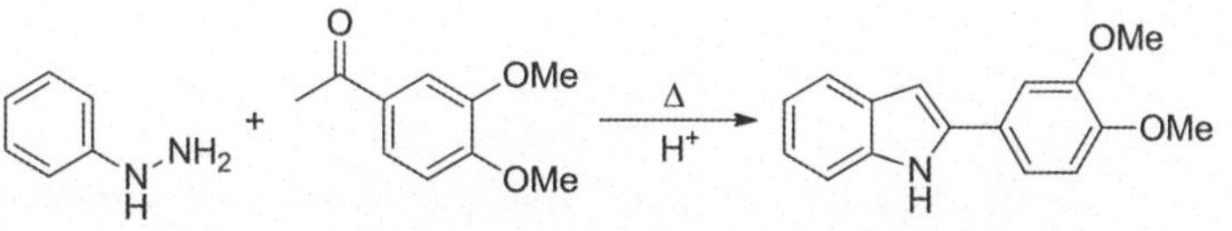

Q 2. Give plausible mechanisms for the given transformations:

(i) Me, $^{-}$O, $^{+}$SePh $\xrightarrow{\Delta}$ Me

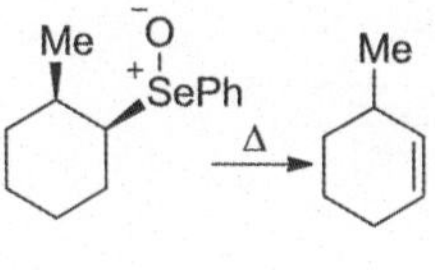

(ii) O, Ph, Ph, O $\xrightarrow[\text{acetone, N}_2]{h\nu}$ O, Ph, OH

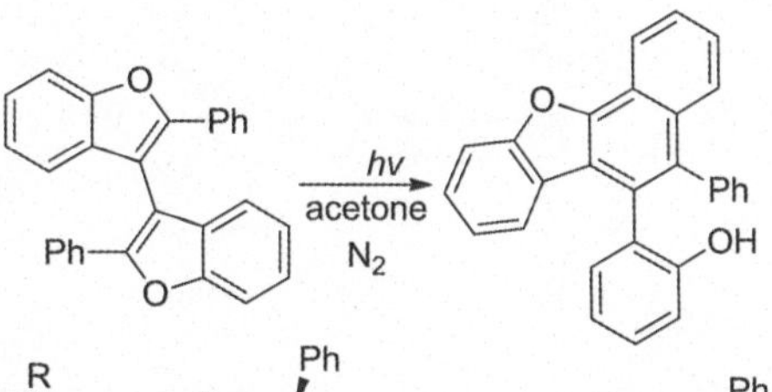

(iii) O, N, R, Me $\xrightarrow[\text{DMF, 180 °C}]{\text{CuI, Cs}_2\text{CO}_3}$ O, N, H, R, H, Me

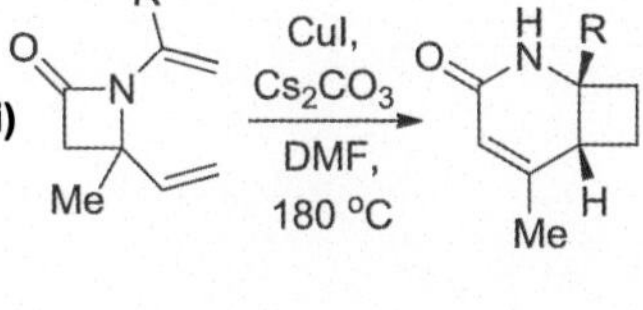

(iv) Ph, N–NH_2, Ph $\xrightarrow[\text{MnO}_2 \text{ or HgO}]{\Delta}$ Ph, Ph

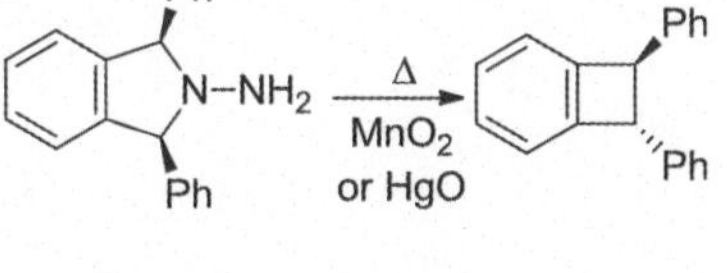

Pericyclic Reactions. http://dx.doi.org/10.1016/B978-0-12-803640-2.00007-5
Copyright © 2016 Elsevier Inc. All rights reserved.

Q 3. How do you explain the formation of the products in the following reaction sequences?

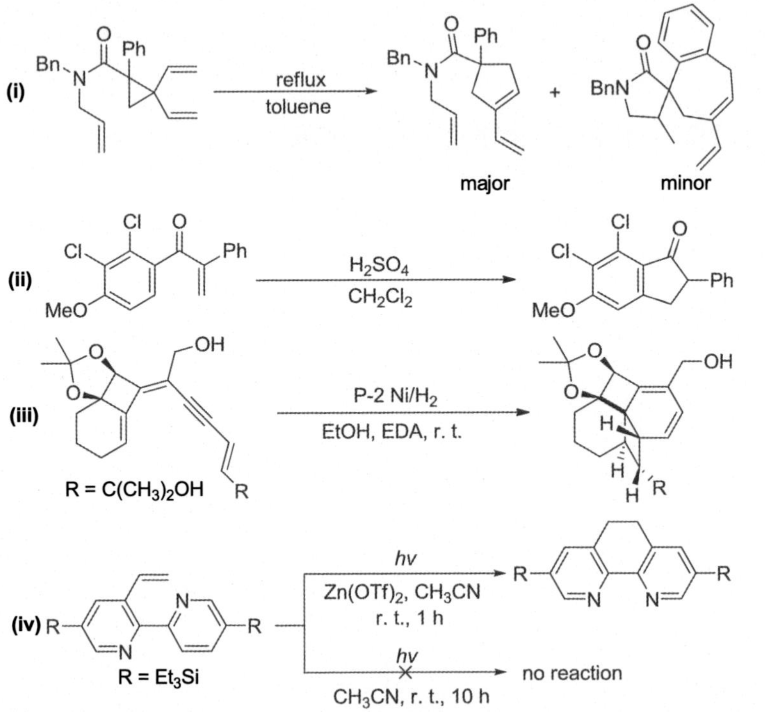

(i) reflux, toluene → major + minor

(ii) H_2SO_4, CH_2Cl_2

(iii) P-2 Ni/H_2, EtOH, EDA, r. t.; R = $C(CH_3)_2OH$

(iv) R = Et_3Si; hv, $Zn(OTf)_2$, CH_3CN, r. t., 1 h; hv, CH_3CN, r. t., 10 h → no reaction

Q 4. Explain the following chemical transformations with suitable mechanisms:

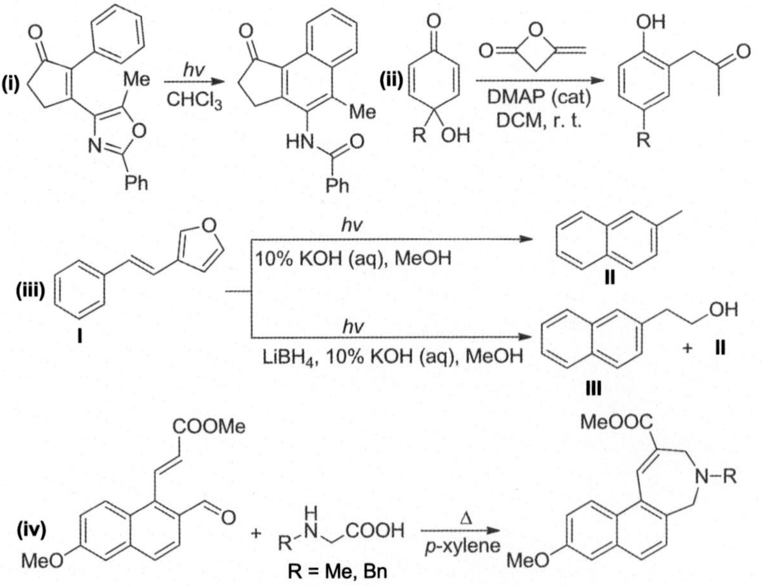

(i) hv, $CHCl_3$

(ii) DMAP (cat), DCM, r. t.

(iii) I; hv, 10% KOH (aq), MeOH → II; hv, $LiBH_4$, 10% KOH (aq), MeOH → III + II

(iv) R = Me, Bn; Δ, p-xylene

Q 5. Provide suitable mechanisms for the following transformations:

(i) 130 °C, xylene

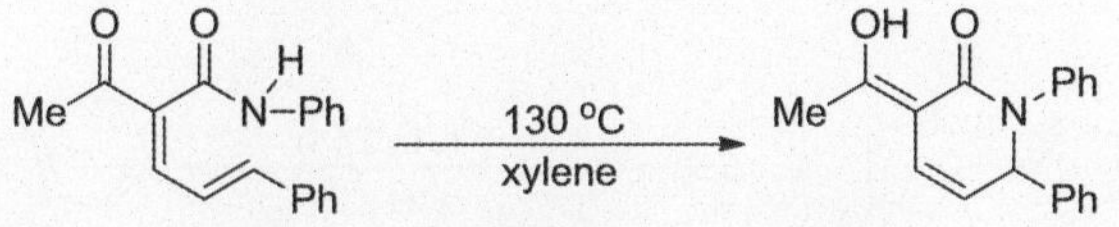

(ii) 170 °C, toluene

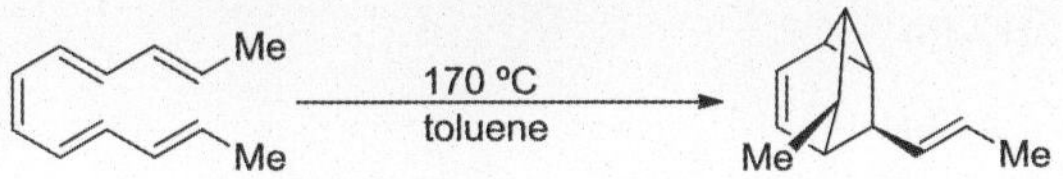

(iii) 18 °C, $AgClO_4$, TFE

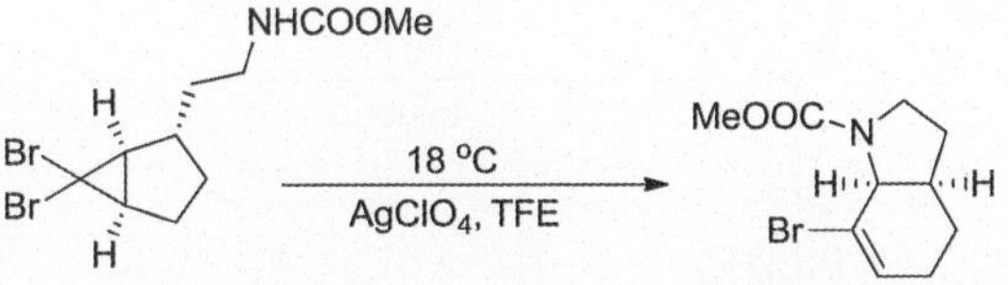

(iv) 80 °C, toluene

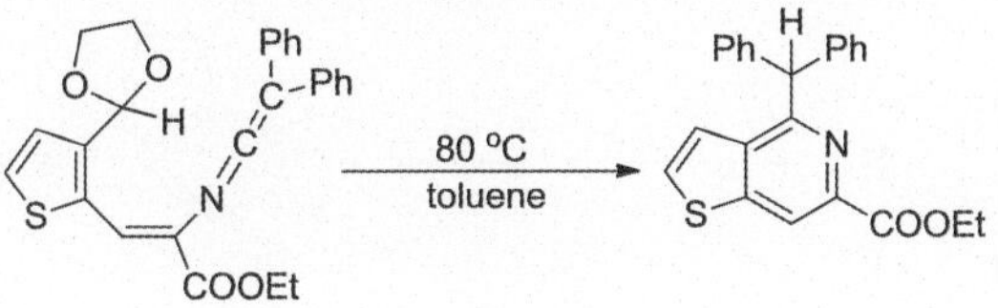

Q 6. Provide the mechanistic rationalization for each of the following reactions:

(i) (i) Me, Li; THF, – 78 °C (ii) – 78 °C to 25 °C, overnight (iii) H_2O, 0 °C — I → II + III

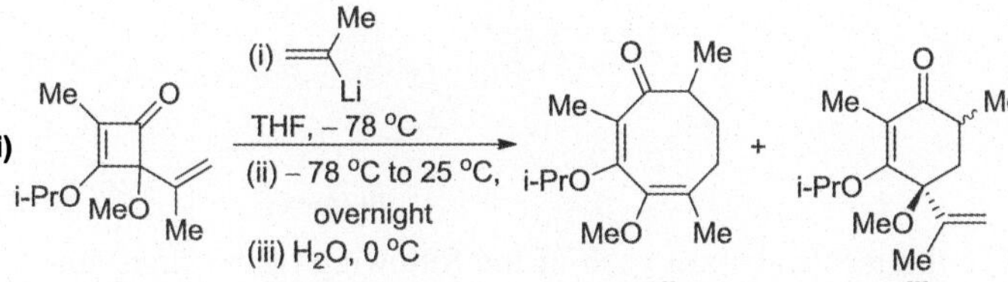

(ii) *hv*, THF, DBU

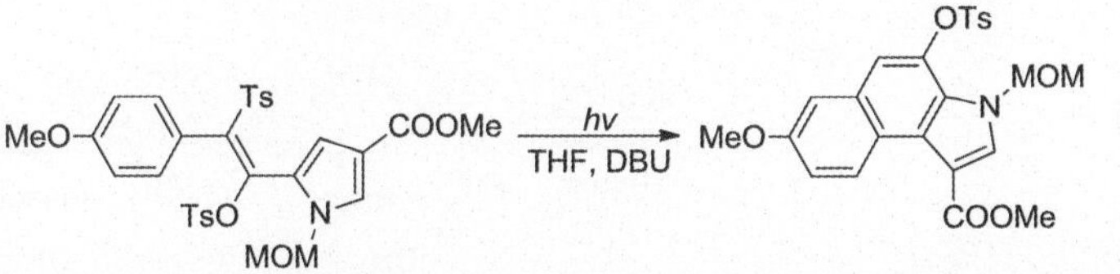

MOM = methoxymethyl ether and DBU = 1,8-diazabicyclo[5.4.0]undec-7-ene, a non-nucleophilic base.

(iii) 2 (OTf, TMS) + (Ph, Ph) CsF, MeCN, 30 °C, 24 h

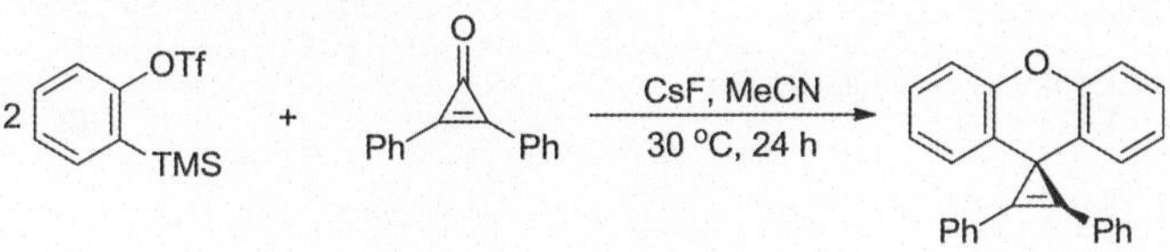

(iv) EtOOC, =N₂ (2 eq.); LDA (2 eq.), THF, – 78 °C r. t., 1h — major + minor

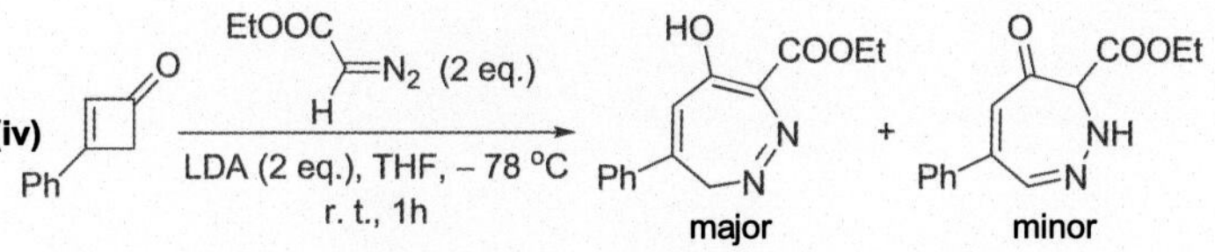

Q 7. Provide the mechanistic rationalization for each of the following reactions:

(i) Bn N SO$_2$Ph + ≡—Ts → r. t., 24 h; CH_3CN → H Ts H Bn N H SO$_2$Ph

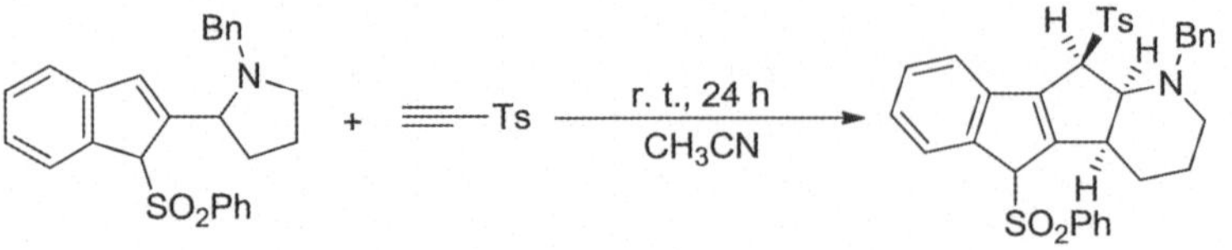

(ii) RN NR NR → 110 °C; toluene → RN H NR H H NR

$R = SO_2(2,4,6\text{-}^iPrC_6H_2)$

(iii) S → Δ → S + S

I II 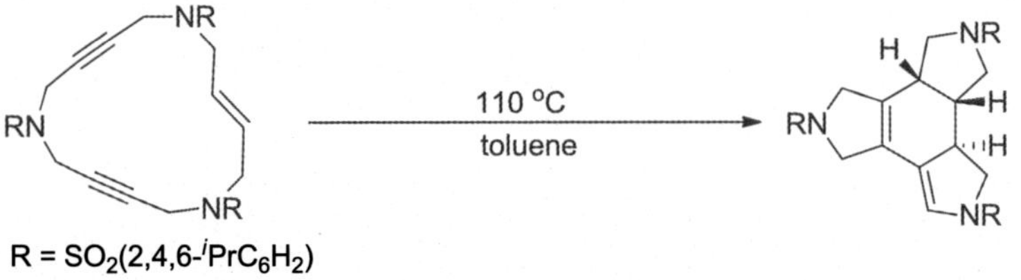 III

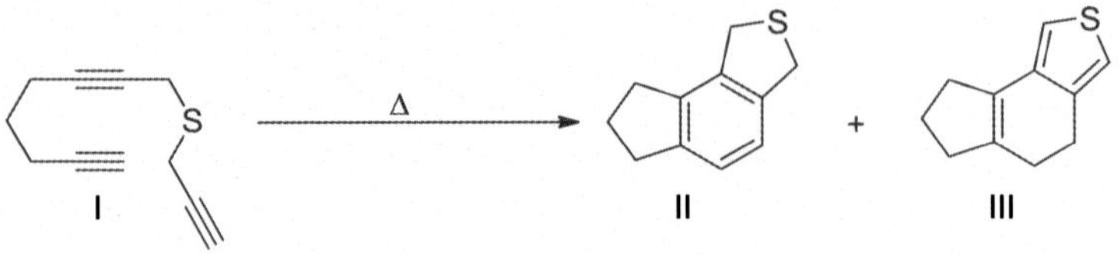

(iv) Br Br Me → $AgBF_4$ (2 eq.); DCE, 65 °C → Br Me + Br Me

I III (45%) IV (55%)

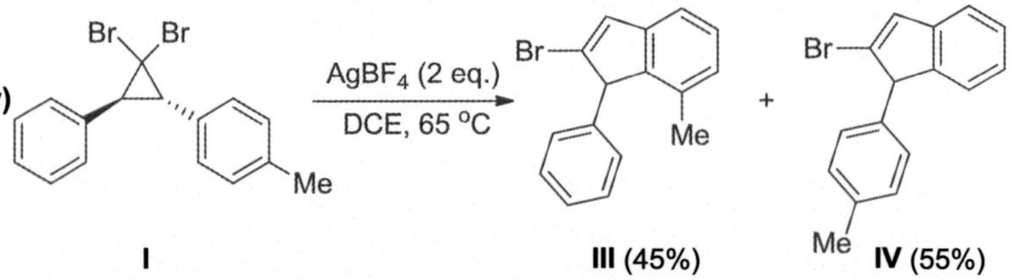

Q 8. Give the mechanism of each of the following conversions and comment on the reactions:

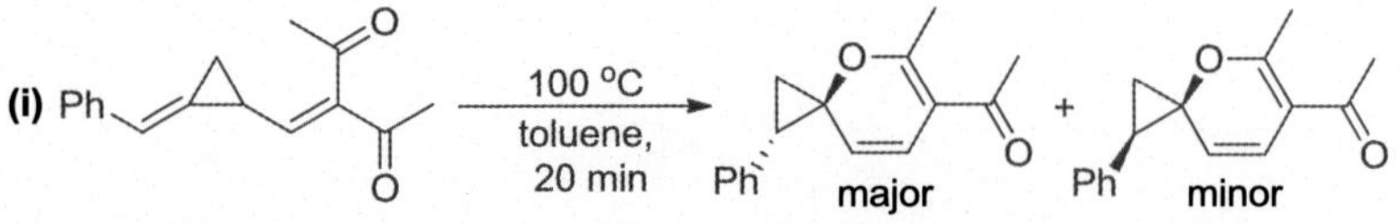

(i) Ph O O → 100 °C; toluene, 20 min → O O Ph major + O O Ph minor

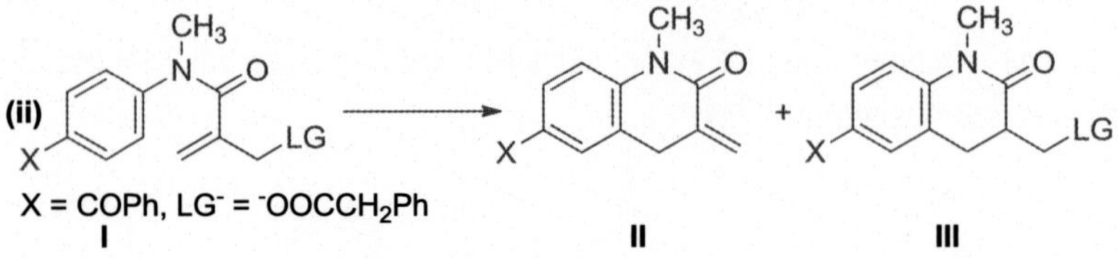

(ii) CH_3 N O LG X → CH_3 N O X + CH_3 N O LG X

X = COPh, $LG^- = {}^-OOCCH_2Ph$

I II III

(iii) I → II: $h\nu$, 17% aq. phosphate buffer in CH_3CN

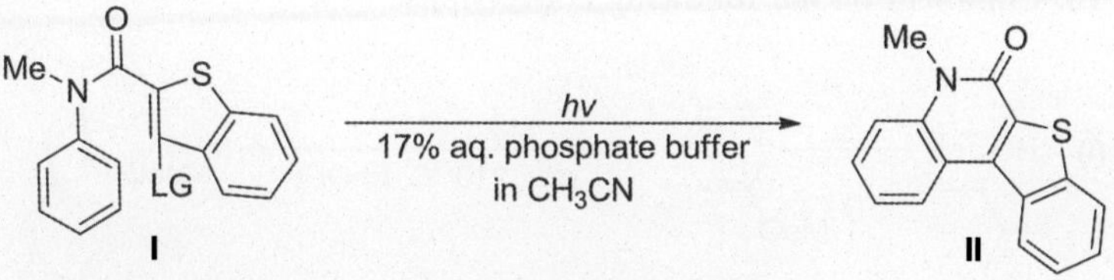

$LG^- = Cl^-, PhCH_2COO^-, PhS^-, PhCH_2S^-, PhO^-, HO^-$

(iv) I → II (44%) + III (34%): 220 °C, toluene

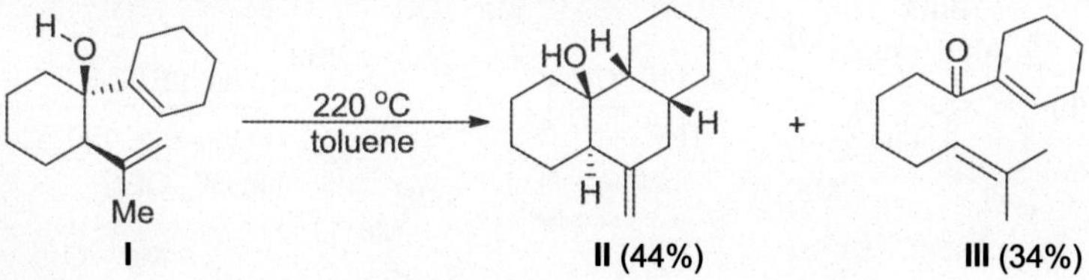

Q 9. Explain the formation of the products of each of the following chemical conversions:

(i) Δ

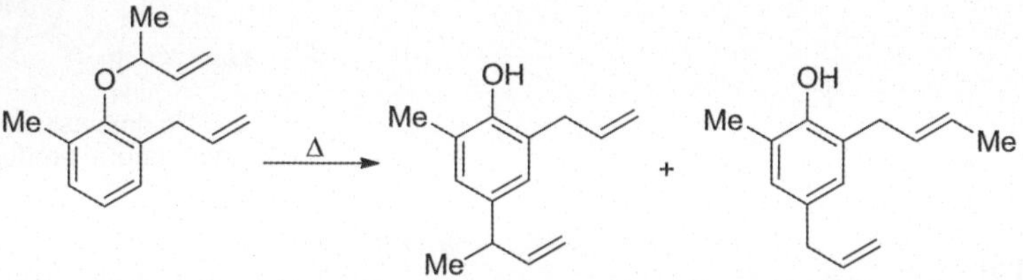

(ii) (i) *t*-BuLi, DMPU, THF, – 78 °C; (ii) NH_4Cl

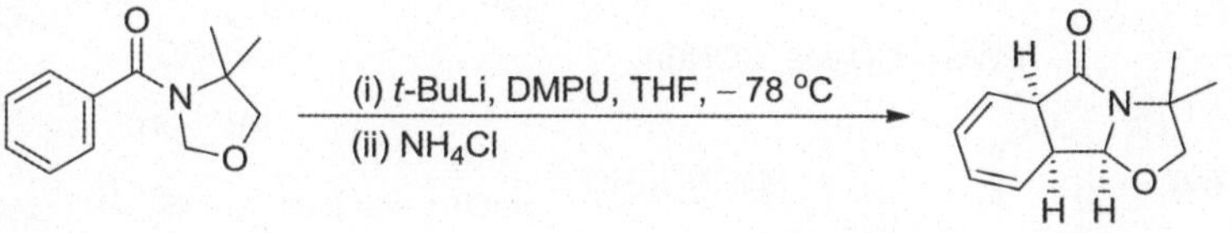

(iii) Δ, toluene, reflux → major + minor

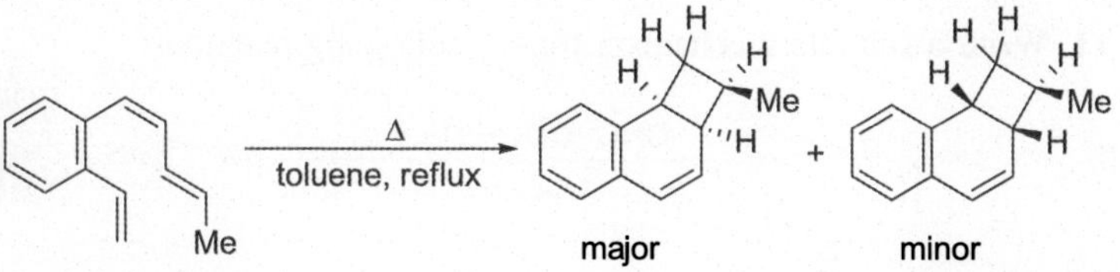

(iv) Δ

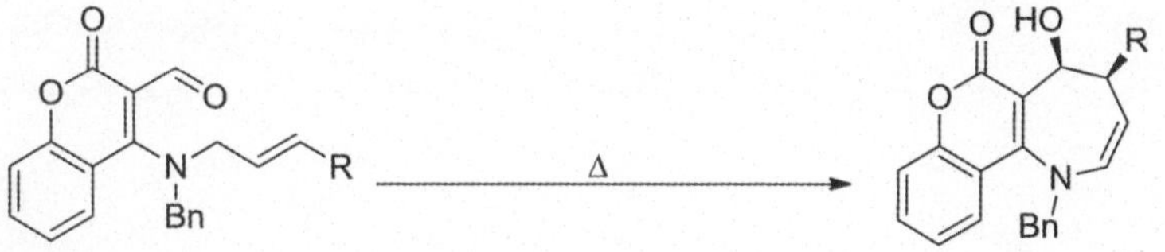

Q 10. Comment on the following conversions:

(i) MeLi; mw, 210 °C, PhOEt

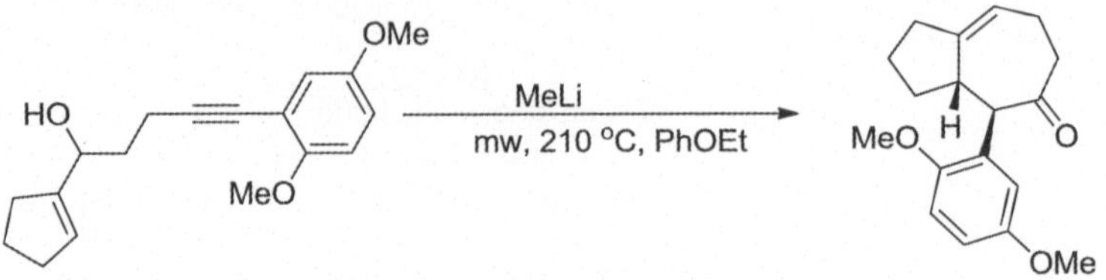

(ii) mw, 110 °C; THF → (i) air, 25 °C (ii) toluene, 150 °C (iii) $BF_3.OEt_2$

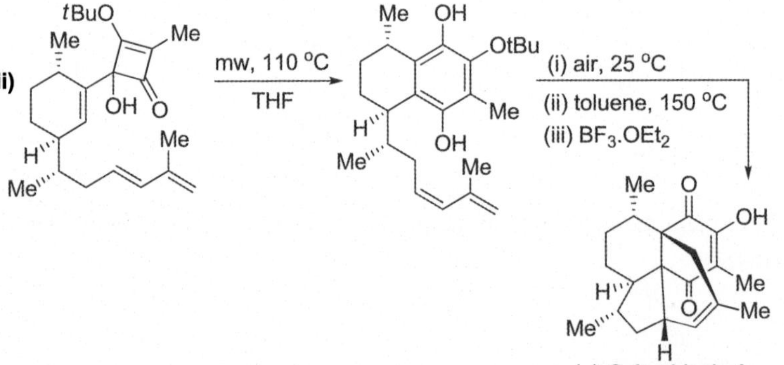

(−)-Colombiasin A (marine natural product)

(iii) mw, 150 °C; DMF, 1 h

EWG = CO_2Me, $CONBn_2$

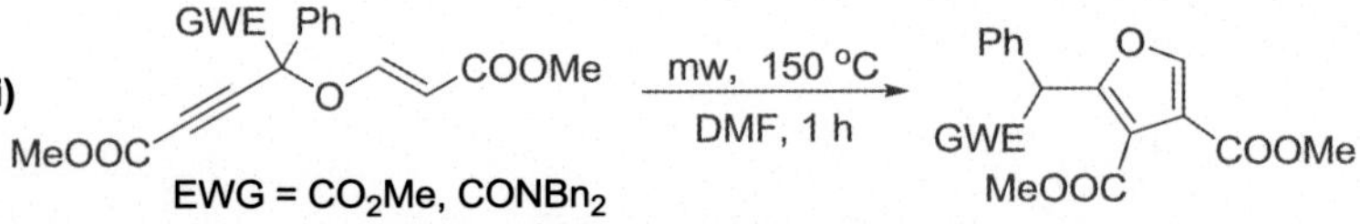

(iv) + $MeONH_2.HCl$; mw, 120 °C; NaOAc, ethanol

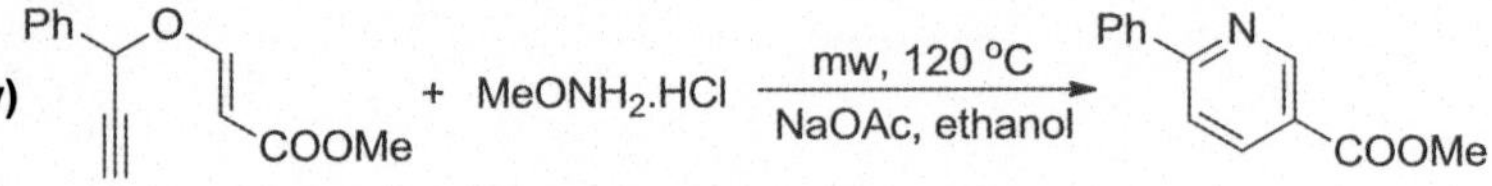

Q 11. Write a suitable mechanism for the following reactions:

(i) $OHC-COOH.H_2O$; MeOH

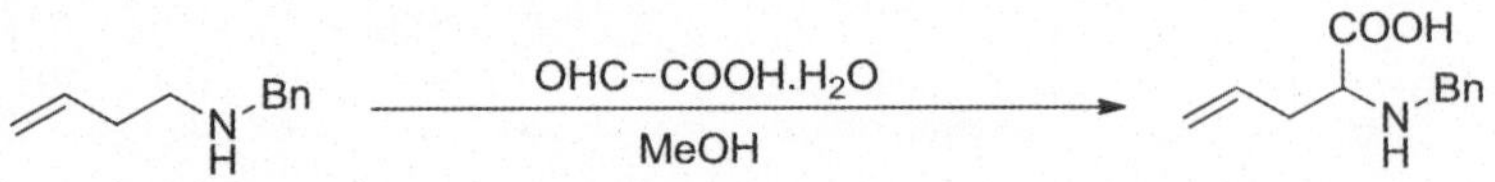

(ii) 120 °C, DMF; 20 min

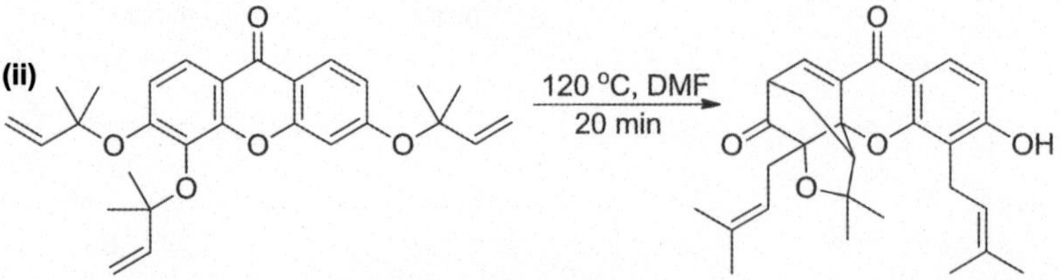

(iii) (i) LDA, TMSCl, THF, − 78 °C (ii) 65 °C (iii) H_3O^+

(i) LDA, TMSCl, THF/HMPA, − 78 °C (ii) 65 °C (iii) H_3O^+

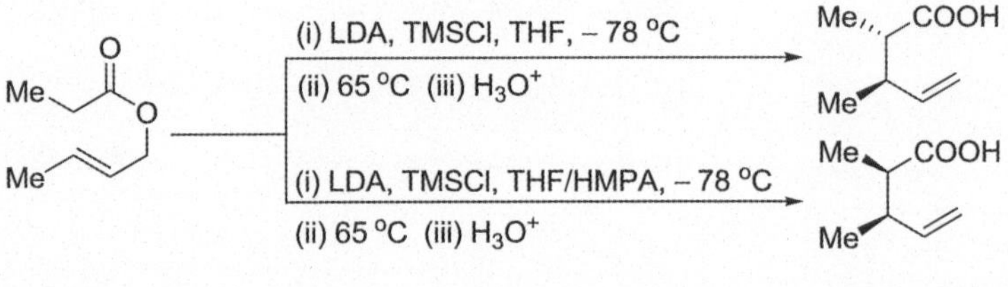

Q 12. Explain the mechanisms of the following reactions:

(i) 2 eq. LDA, Δ

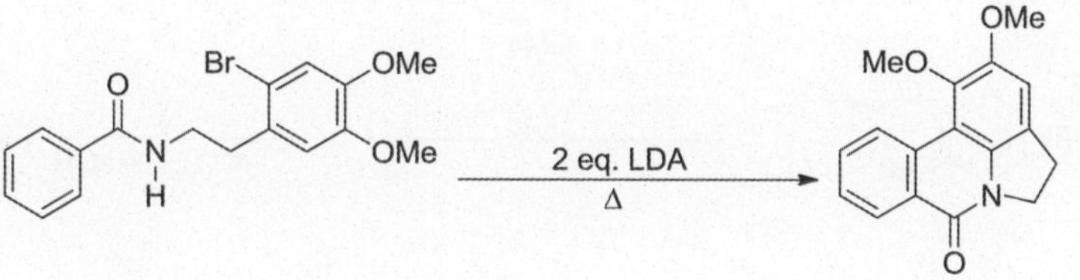

(ii) (i) LDA, THF, – 78 °C; (ii) Me–CH=CH–COOMe, THF, r. t.

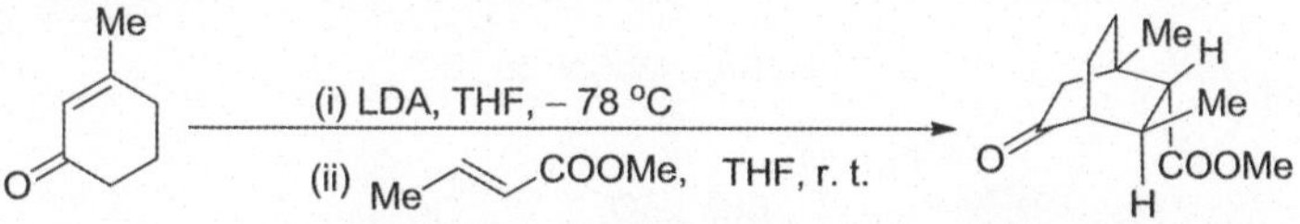

(iii) (i) $CH_3S(O)CH_2Li$, 10 min, 0 °C; (ii) MeO-furyl–CH=CH–C(O)OMe, 1.25 h, 0 °C; (iii) H_3O^+

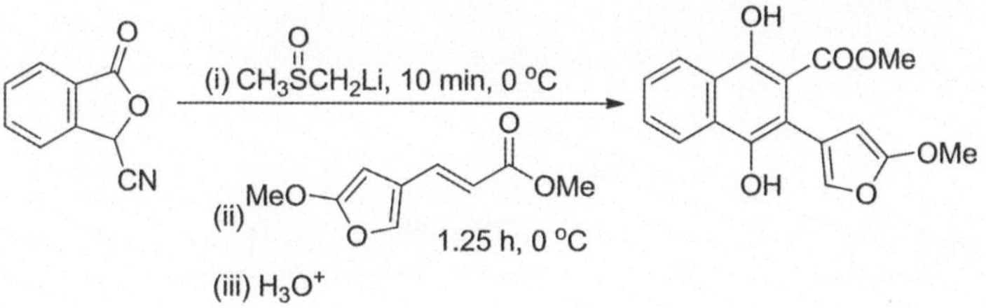

(iv) (i) NaH, 0 °C; (ii) EtOOC–≡–COOEt, 25 °C

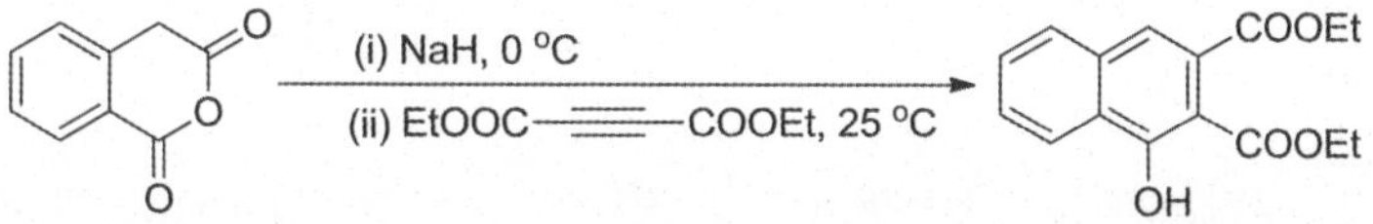

Q 13. Consider the following reaction and account for the control of the regiochemistry of **I** in the cycloaddition and also predict the relative configurations of the three chiral centers in **I**:

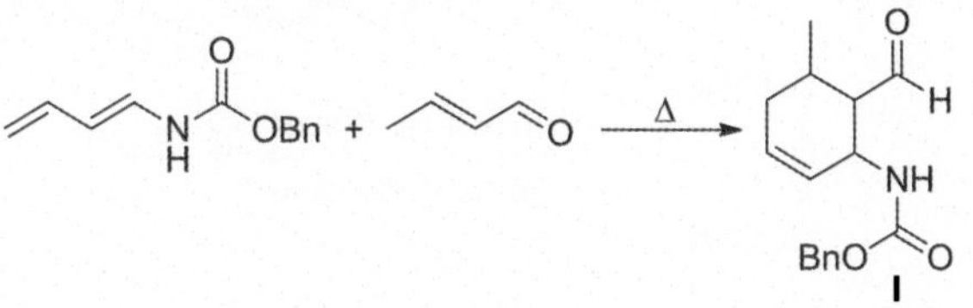

Q 14. Deduce the sequence of symmetry-allowed changes under thermal conditions, which takes place in the following transformations:

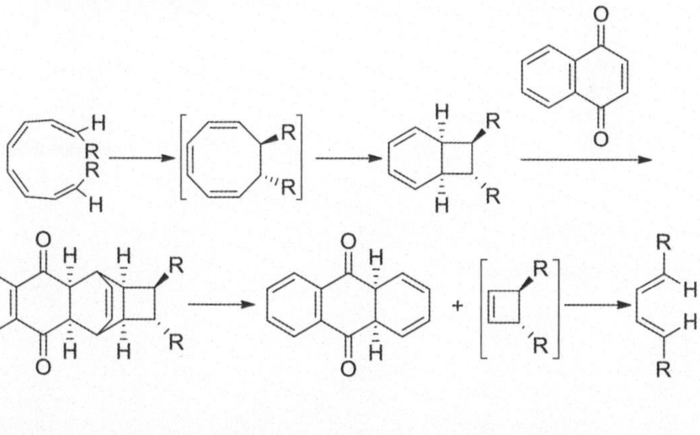

Q 15. Draw the structures for **I**, **II**, and **III**, and outline the mechanisms of the following reactions:

(i) OMe, OMe, Δ → I; Δ, $-CO_2$, $-$MeOH → II; COOMe

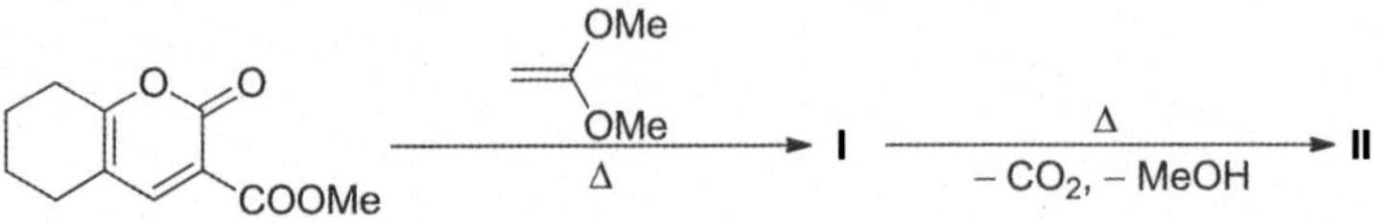

(ii) SPh, N, N, $SiMe_3$, CO_2Me; (i) TBAF, MeCN, 50 °C (ii) EtO, O, Et → I; Δ → II

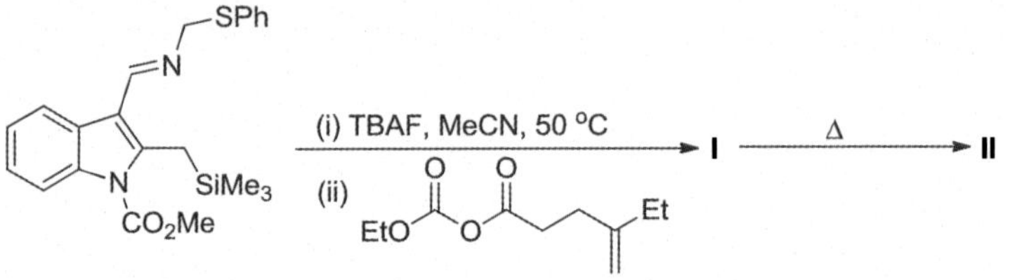

(iii) N, Me, N, Bn, O; $BF_3.OEt_2$, toluene, 100 °C → [+ I] → II

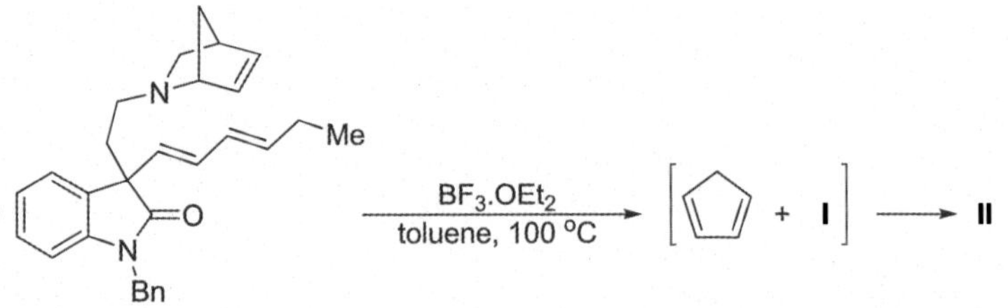

(iv) Bn, N, O, N, Cl, S; 110 °C, toluene → I; *m*-CPBA, DCM → II; Br, BuLi (2.1 eq.), -78 °C → III; 150 °C, *o*-dichlorobenzene → BnN, O

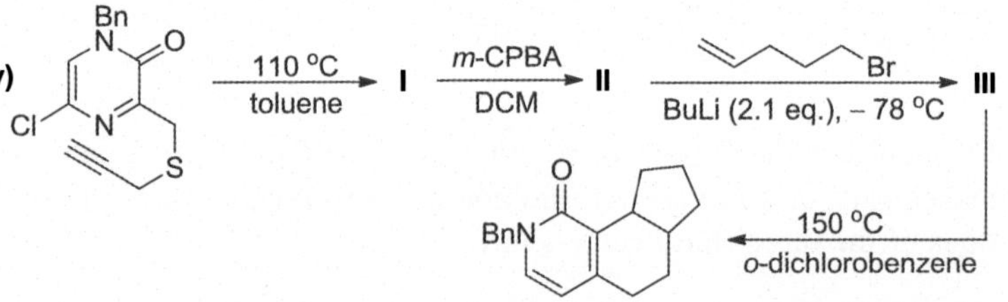

(v) O + COOMe, COOMe; Δ → I; (i) $NaBH_4$, MeOH (ii) Ac_2O, pyridine → II; Δ → COOMe, COOMe + OAc

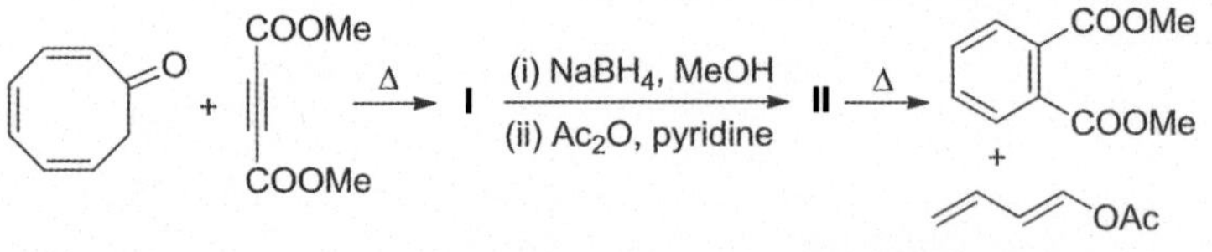

(vi) O, O, O; LDA (2 eq.), THF, -78 °C → I; 23 °C, 3.5 h → II; (i) O_3, Me_2S, CH_2Cl_2, -78 °C (ii) Ac_2O, pyridine → Me, Me, MeOC, H, O, O, OAc

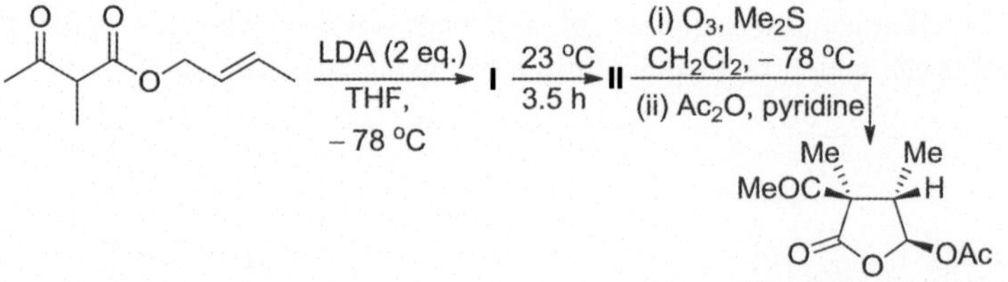

(±)-4-epi-acetomycin

Q 16. Suggest the mechanisms for the following reactions, which involve a retro Diels–Alder reaction as one of the steps:

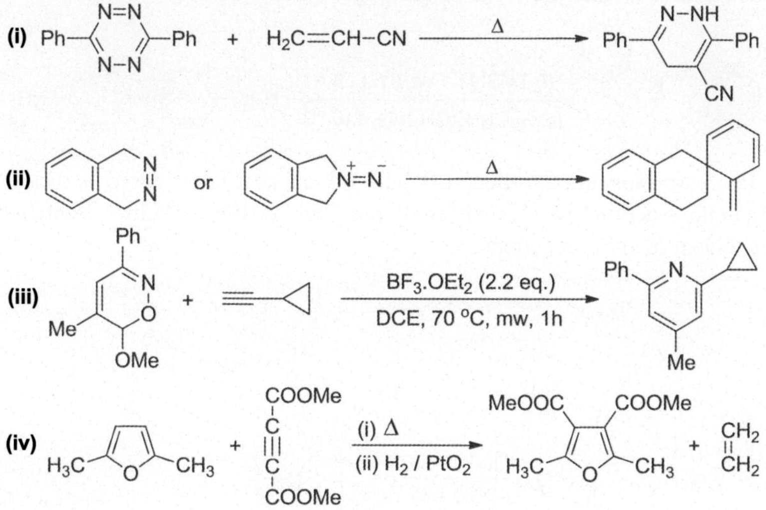

(i) Ph-tetrazine-Ph + $H_2C{=}CH{-}CN$ →(Δ) dihydropyridazine (N–NH) bearing Ph, Ph and CN

(ii) benzo-fused cyclic azo compound (N=N) or isoindoline N-diazene ($\overset{+}{N}{=}\overset{-}{N}$) →(Δ) spiro product

(iii) oxazine (Ph, Me, OMe) + cyclopropylacetylene →($BF_3.OEt_2$ (2.2 eq.), DCE, 70 °C, mw, 1h) 2-phenyl-6-cyclopropyl-4-methylpyridine

(iv) 2,5-dimethylfuran (H_3C, CH_3) + MeOOC–C≡C–COOMe →((i) Δ; (ii) H_2 / PtO_2) dimethyl 2,5-dimethylfuran-3,4-dicarboxylate (MeOOC, COOMe, H_3C, CH_3) + $CH_2{=}CH_2$

Q 17. The diamine **I** on heating in ethanol at ambient temperature gives **II** in good yield by the Diaza–Cope rearrangement. However, on heating diamine **III** or **IV** in $CDCl_3$ at 25 °C, an equilibrium mixture is obtained that contains **III** and **IV** in a ration of 14:1. Explain the observations.

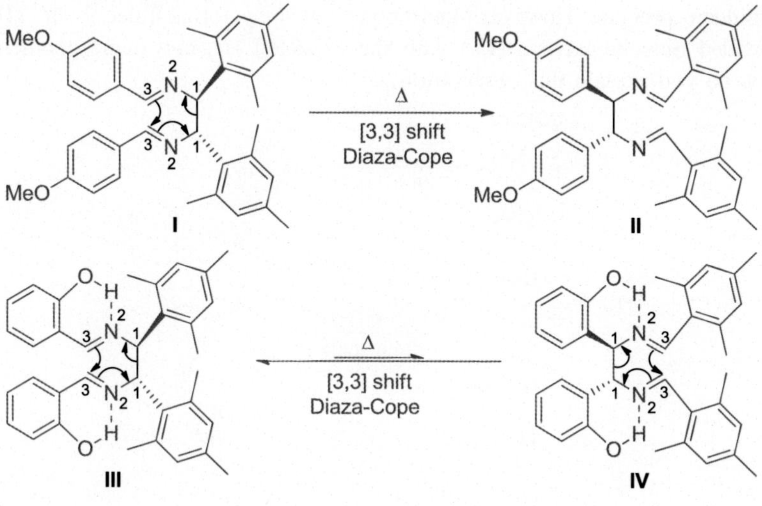

I →(Δ, [3,3] shift, Diaza-Cope) **II**

III ⇌(Δ, [3,3] shift, Diaza-Cope) **IV**

Q 18. Give the mechanism of the following reaction and also explain the fact that when this reaction is carried out without forming a pyridinium salt with trimethylsilyl iodide, the yield of the target compound **II** is very poor:

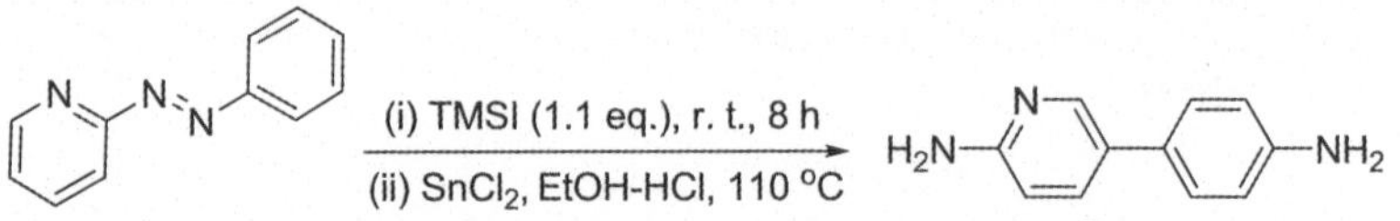

(i) TMSI (1.1 eq.), r. t., 8 h
(ii) $SnCl_2$, EtOH-HCl, 110 °C

H_2N NH_2

I II

Q 19. 5-Aminopenta-2,4-dienals **I** and **III** on heating undergo a series of pericyclic reactions to give different products as shown below. Sketch the mechanism of these reactions.

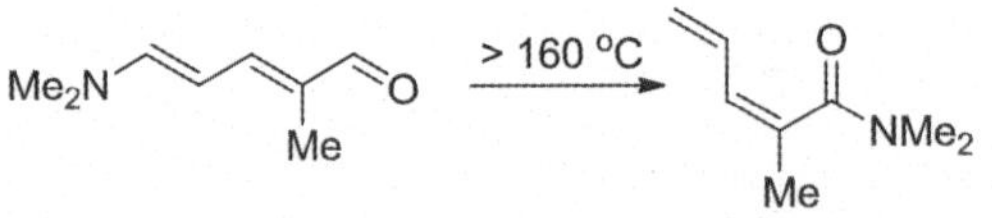

Me_2N Me O > 160 °C O NMe_2 Me

I II

Me 200-220 °C o-DCB, mw Me O H

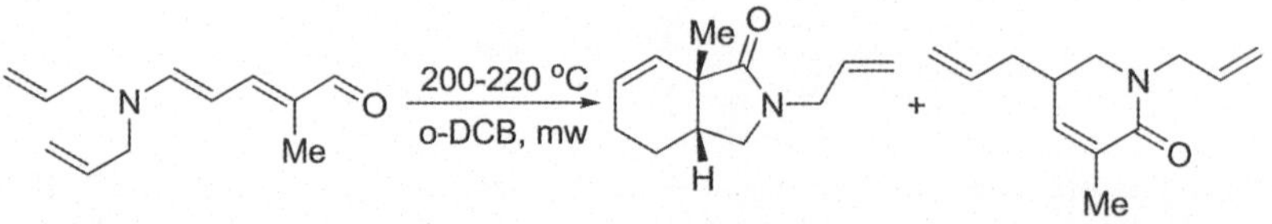

III IV major V minor

Q 20. Thermolysis of bicyclo[3.2.0]hept-6-ene (**I**) provided bicyclic lactone (**II**) with the *cis,cis*-diene geometry that formally represents a thermally forbidden process. However, thermolysis of bicyclo[6.2.0]dec-9-ene (**III**) provided bicyclic lactone (**IV**) with the expected *cis,trans*-diene geometry. How do you explain this observation?

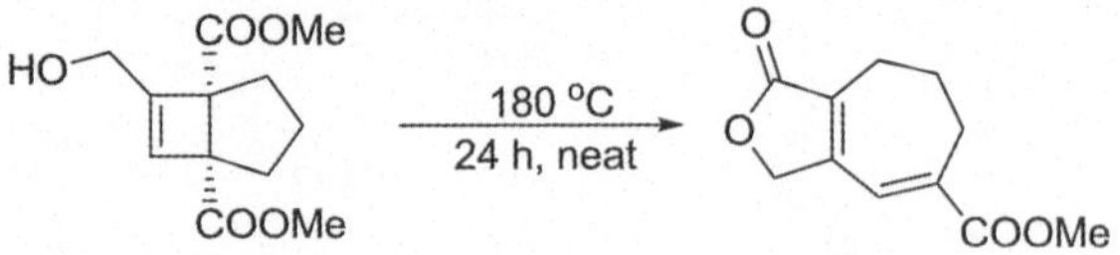

COOMe HO COOMe 180 °C 24 h, neat O COOMe

I II

COOMe HO COOMe 90 °C 24 h, toluene O MeOOC

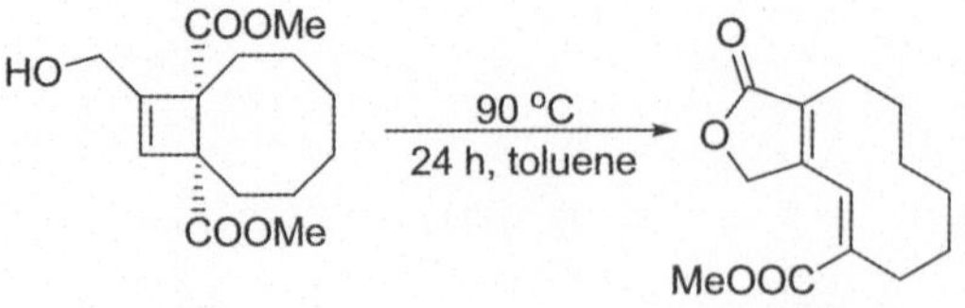

III IV

Appendix

Solution Manual

Sol 1. (i) On heating, the starting compound **I** undergoes a [3,3] sigmatropic shift to give an allene (**II**). In the second step, **II** forms an extended enol (**III**), and movement of the double bond in conjugation follows to give **IV**. This is followed by a thermally-allowed [1,5] hydrogen shift from nitrogen to the middle of the allene in **IV**. The compound **V** then undergoes six-electron disrotatory electrocyclization to give **VI**, which is finally oxidized by nitrobenzene to give a pyridine derivative (**VII**).

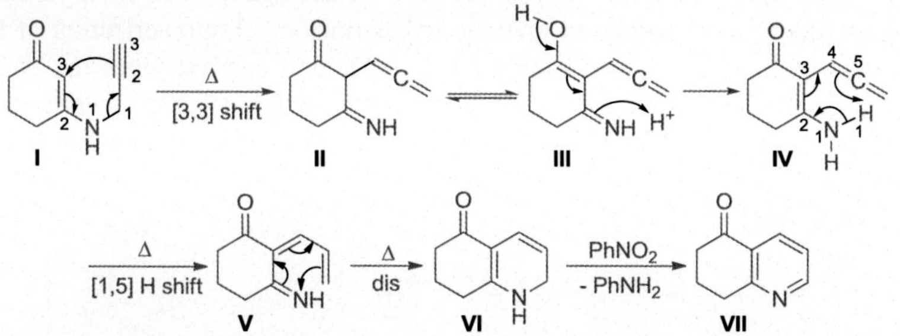

(ii) The reaction involves cascade sequence of three successive pericyclic reactions. The thermal Oxy-Cope rearrangement of 1,2-divinyl cyclohexanol allyl ether (**I**) leads to enol **II**, which tautomerizes to produce the ketone **III**. Transannular ene reaction of **III** generates an intermediate **IV**, which is properly set up for the subsequent Claisen rearrangement to afford a bicyclic lactol **VI**.

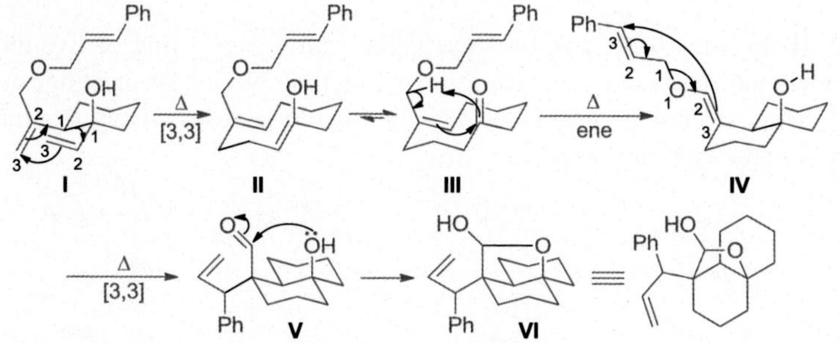

Copyright © 2016 Elsevier Inc. All rights reserved.

(iii) On refluxing in *N,N*-diethylaniline, 1,4-diphenoxybut-2-yne gives a benzofurobenzopyran ring system through a series of pericyclic reactions as shown below.

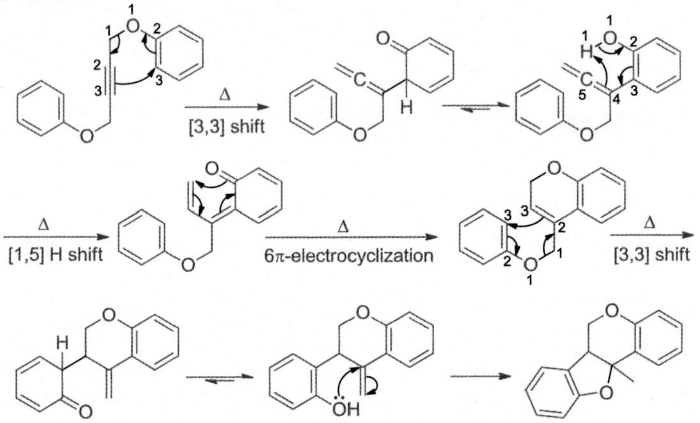

Δ [3,3] shift

Δ [1,5] H shift

Δ 6π-electrocyclization

Δ [3,3] shift

(iv) The Fischer indole synthesis involves condensation of an arylhydrazine with an aldehyde or ketone to give an aryl hydrazone. Tautomerization of the obtained hydrazone in the presence of an acid as a catalyst gives the corresponding ene-hydrazinium species, which then undergoes [3,3] sigmatropic rearrangement followed by loss of ammonia to give the corresponding indole derivative (**I**).

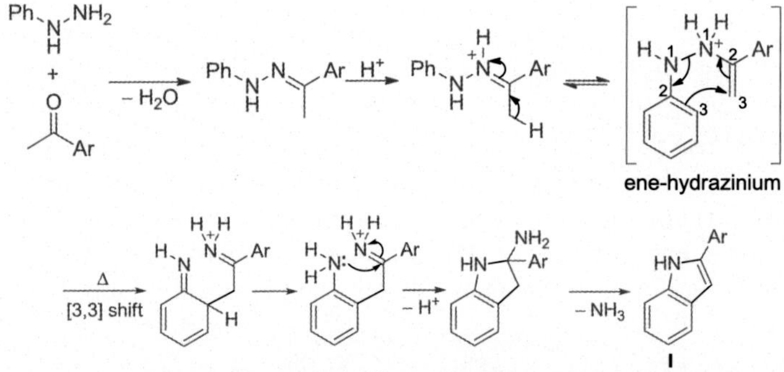

$-H_2O$; H^+ ; ene-hydrazinium

Δ [3,3] shift ; $-H^+$; $-NH_3$; I

Sol 2. (i) In selenoxide, pyrolysis cycloelimination uses a ring of five atoms; therefore, the reaction is necessarily *syn*-selective. Hence, on heating, cyclohexylphenyl selenoxide undergoes *syn*-elimination regioselectively to produce the less stable Hofmann product.

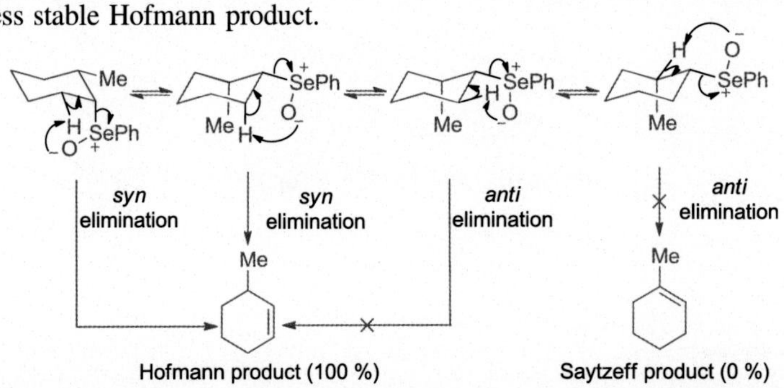

syn elimination ; *syn* elimination ; *anti* elimination ; *anti* elimination

Hofmann product (100 %) ; Saytzeff product (0 %)

(ii) On irradiation, 2,2′-diphenyl-3,3′-bibenzofuran (**I**) undergoes conrotatory 6π-electron cyclization to give an intermediate **II**, which then undergoes cleavage of one furan ring via a *syn*-elimination to yield a phenol derivative (**III**).

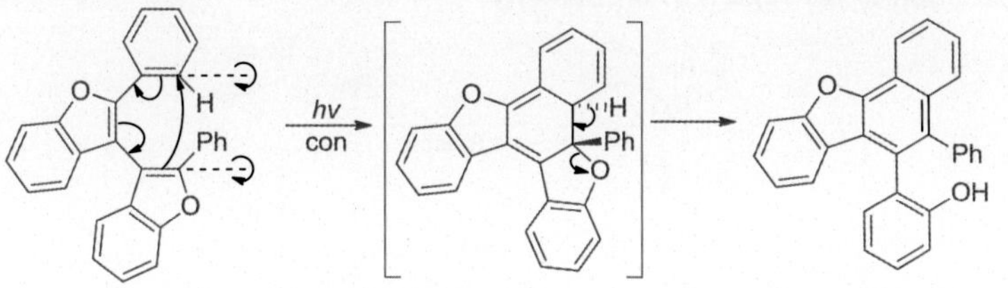

I II III

(iii) *N*-vinyl β-lactam (**I**) on heating undergoes a domino one-pot [3,3]-sigmatropic/6π-electrocyclization process to give a bicyclo product (**IV**). When the reaction mixture is heated in the absence of base and CuI, the reaction stops at the formation of the azacylocta-1,5-diene (**II**). Addition of CuI and Cs_2CO_3 enables the 6π-electrocyclization to occur in a one-pot sequence. CuI undergoes chelation with the nitrogen and/or oxygen atoms, thus catalyzing the proton abstraction, which generates the required triene (**III**) for the electrocyclization.

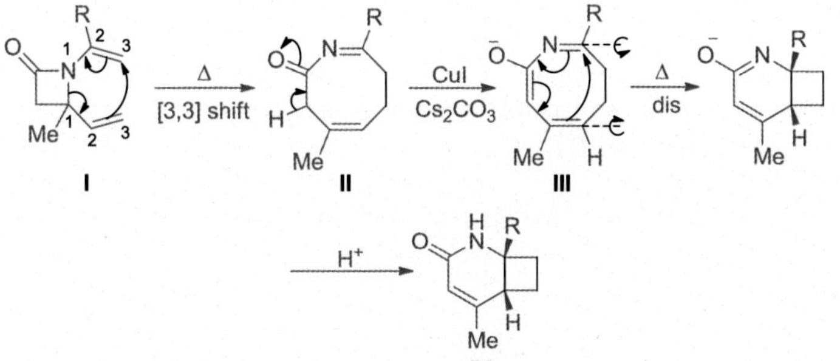

I II III IV

(iv) Oxidation of the *cis* isomer of the cyclic hydrazine (**I**) with MnO_2 or HgO gives an intermediate diazene (**II**), which undergoes symmetry-allowed cheletropic elimination of nitrogen in a disrotatory manner to give the *o*-quinodimethane intermediate (**III**), which then undergoes a symmetry-allowed conrotatory ring closure to give *trans*-diphenylbenzocyclobutane (**IV**).

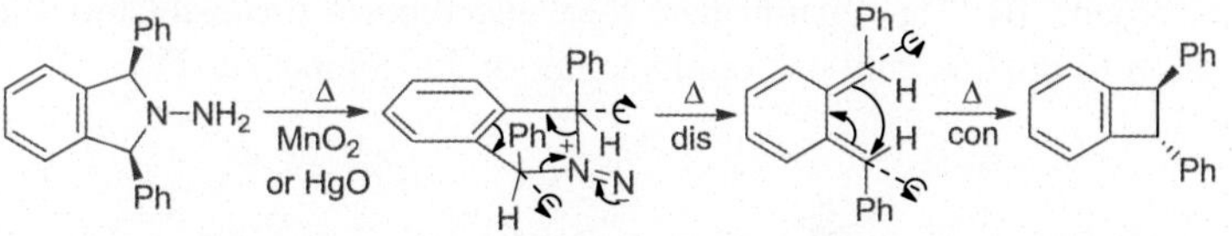

I II III IV

Sol 3. (i) 1,1-Divinyl-2-phenylcyclopropane derivative (**I**) on heating undergoes a vinylcyclopropane rearrangement to give vinylcyclopentene derivative (**II**). In this reaction one of the vinyl groups of **I** participates in the reaction and the other is a substituent. The major spirolactam (**III**) is formed by a reversible aromatic Cope rearrangement followed by an irreversible ene reaction. In this case, the driving force for the Cope rearrangement is the release of strain energy of the cyclopropane ring, which partially compensates for the loss of aromaticity of the phenyl ring. Similarly, the driving force for the ensuing intramolecular ene reaction is the regain of aromaticity.

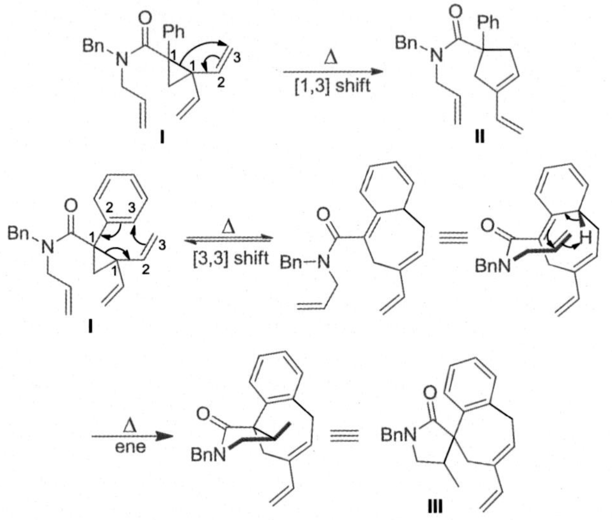

(ii) Treatment of compound **I** with sulfuric acid generates a cation (**II**), which then undergoes conrotatory ring closure (Nazarov cyclization) to afford a fused cyclopentenone (**III**).

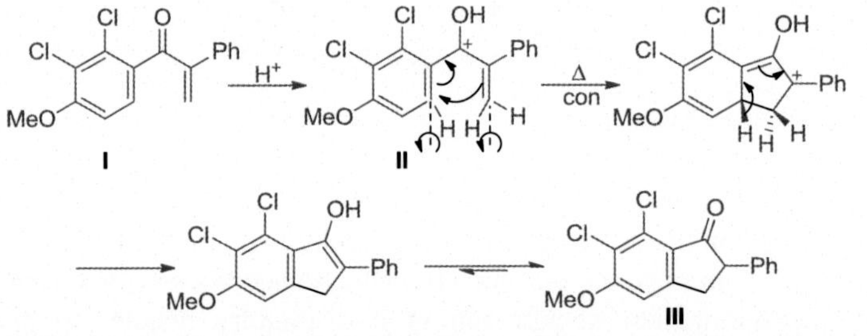

(iii) Partial reduction of the alkyne functionality in en-yn-en (**I**) generates tetraene **II**, which can undergo an 8π-conrotatory electrocyclization to form cyclooctatriene **III**. This intermediate then undergoes a 6π-disrotatory electrocyclization to provide only one diastereomer of the compound (**IV**).

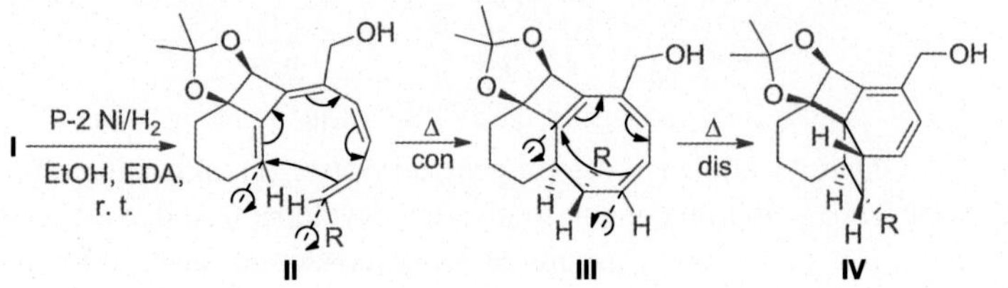

(iv) The *s-trans* conformation of the bipyridine, which is not suitable for electrocyclic reactions, is more stable than its *s-cis* conformation. Therefore, irradiation of a CH_3CN solution of 5,5′-bis(triethylsilyl)-3-vinyl-2,2′-bipyridine (**I**) does not give the desired product (**II**). However, the addition of a chelating agent fixes the bipyridine in its *s-cis* conformation suitable for electrocyclization. Moreover, formation of a metal chelate would induce a significant bathochromic shift of the absorption band, which would enable the efficient photoexcitation of the substrate.

favored
I (R = Et_3Si)
disfavored
[Zn]
hv
6π-electrocyclization
[1,5] H shift
II

Sol 4. (i) The first step involves a conrotatory 6π-electron cyclization of **I** to give an intermediate **II**, which undergoes [1,9] hydrogen shift to give a stable benzenoid system **III**. This is followed by a [1,3] hydrogen shift and opening of an oxazoline ring to form a stable naphthalene system **V**.

hv
con
[1,9]
H shift
[1,3]
H shift
I
II
III
IV
V

(ii) Reaction between *p*-quinol and diketene in the presence of a catalytic DMAP afforded acetoacetate, which underwent **Carroll–Claisen rearrangement** to afford substituted arylacetone. This rearrangement involves the transformation of a β-ketoallylester into α-allyl-β-ketocarboxylic acid accompanied by decarboxylation. The final product is an allyl ketone.

DMAP (cat)
DCM, r. t.
Δ
[3,3] shift

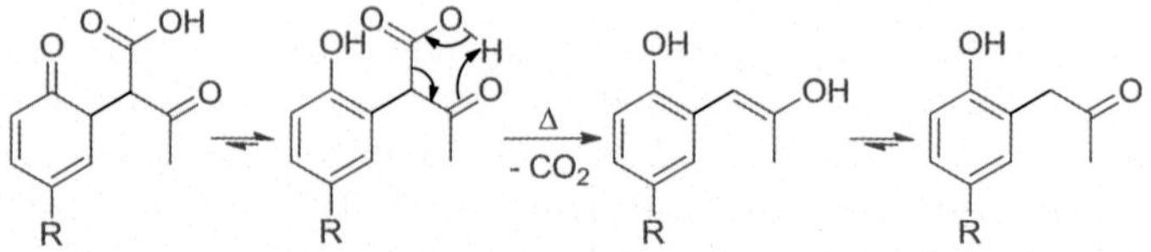

(iii) On irradiation, 3-styrylfuran (**I**) undergoes an initial equilibration to a photostationary mixture of *E* and *Z* isomers, which is followed by a 6π-electron conrotatory ring closure of the *Z*-isomer to give *trans*-9a,9b-dihydronaphtho [2,1-b]furan intermediate (**IV**). As the leaving groups are in a *cis* configuration, **IV** cannot undergo the E2 reaction directly. Therefore, **IV** undergoes E1cB (unimolecular elimination of conjugate base) reaction to produce the intermediate **V**, which undergoes protonation and tautomerization to give **VI**. Finally, a Norrish Type I photoreaction occurs to give 2-methylnaphthalene (**II**). However, in the presence of a reducing agent the reductive product, 2-(naphthalen-2-yl) ethanol (**III**), is also obtained by the reduction of the intermediate **VI**.

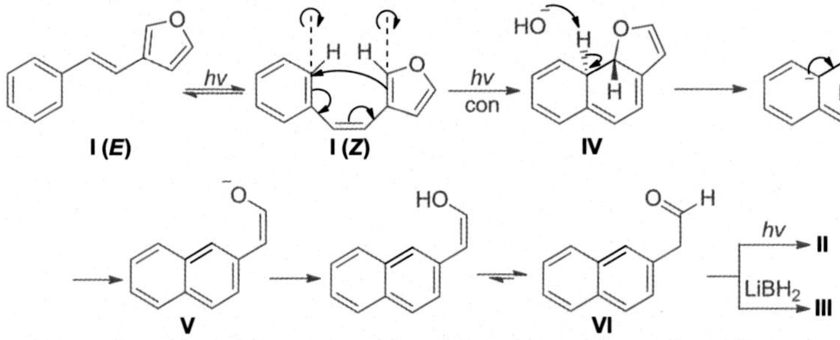

(iv) Condensation of 1-alkenylnaphthalene-2-carbaldehyde (**I**) with *N*-monosubstituted α-amino acid (**II**) generates unstabilized α,β:γ,δ-unsaturated azomethine ylide (**III**), which then undergoes 8π-electron cyclization followed by [1,5] hydrogen shift to give polycondensed dihydroazepine (**IV**).

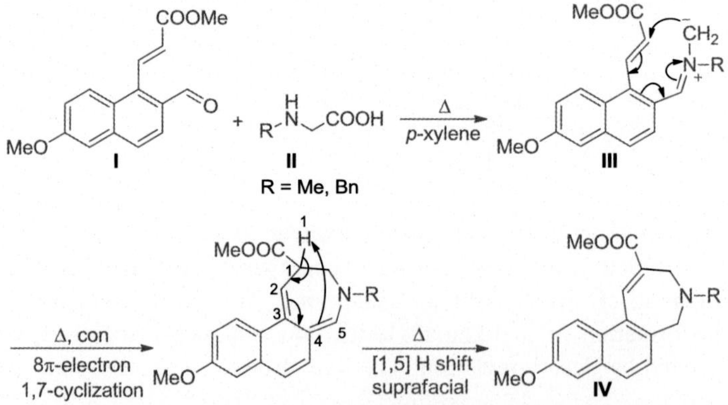

Sol 5. **(i)** (2*Z*,4*E*)-2-Acetyl-*N*,5-diphenylpenta-2,4-dienamide (**I**) is in equilibrium with its tautomer imidic acid (**II**), which is stabilized by the

intramolecular hydrogen bond. Moreover, this intramolecular hydrogen bonding also keeps the azadiene N=C−C=C in a *cis* conformation. As a result, **II** readily undergoes 6π-azaelectrocyclization to give the intermediate **III**, which then undergoes a [1,5] hydrogen shift to give the product **IV**, which is also stabilized by a hydrogen bond.

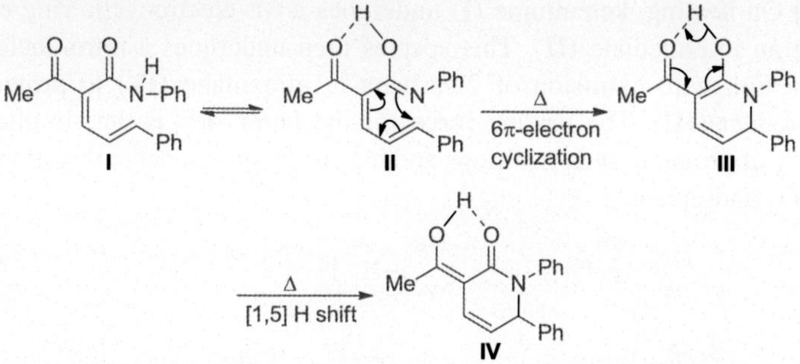

(ii) Under thermal conditions, (2*E*,4*E*,6*Z*,8*E*,10*E*)-dodeca-2,4, 6,8,10-pentaene (**I**) undergoes symmetry-allowed 6π-electrocyclization of its central (*E*,*Z*,*E*)-triene moiety to give the corresponding *cis*-5,6-disubstituted 1,3-cyclohexadiene (**II**), followed by an IMDA reaction involving one of its (*E*)-prop-1-enyl side chains to give the *endo*,*exo*-6-methyl-8-((*E*)-prop-1-en-1-yl)tricyclo[3.2.1.0^{2,7}]oct-3-ene (**III**). It should be noted that *E* configuration of the inner olefinic bonds precludes adoption of a conformation suitable for 8π-electrocyclization, forcing the alternative 6π process to prevail.

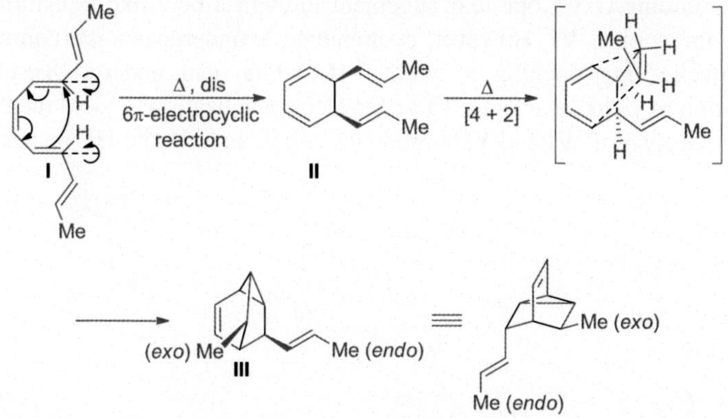

(iii) The *endo* bromine atom is preferentially expelled by the dis-in mode of opening of cyclopropane σ-bond to give the more stable allyl cation, which is trapped by the pendant nitrogen nucleophile to yield the hexahydroindole derivative.

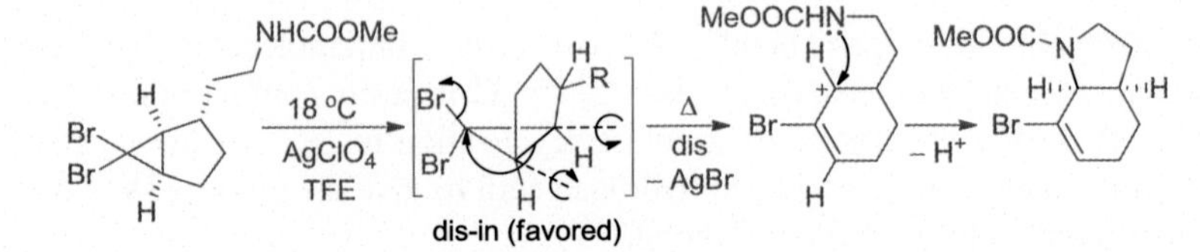

(iv) On heating, ketenimine (I) undergoes a 6π-electrocyclic ring closure to give an intermediate (II). This species then undergoes a retro-cheletropic ene reaction with extrusion of 2-carbena-1,3-dioxolane (IV) to provide the final product (III). The driving force for the latter step is the simultaneous recovery of aromaticity at two rings and highly exergonic decomposition of **IV** into CO_2 and ethene.

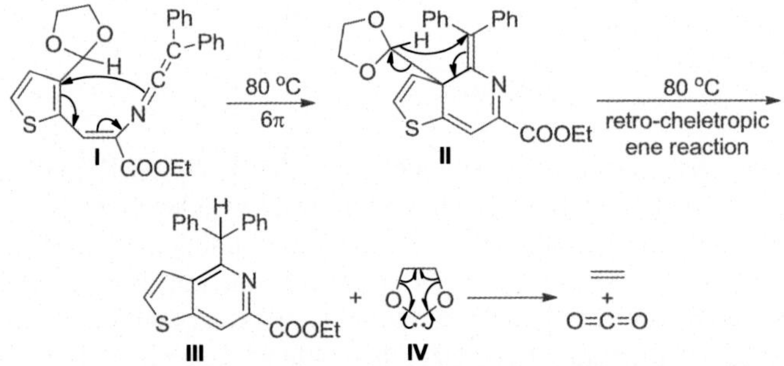

Sol 6. (i) Addition of vinyllithium to cyclobutenone **I**, gives a mixture of *cis*- and *trans*-divinyl substituted cyclobutenes **IV** and **V**. Cyclobutene **IV** undergoes anionic Oxy-Cope rearrangement through a boat-like transition state to give the enolate **VI**. However, cyclobutene **V** undergoes a 4π-conrotatory electrocyclic ring opening to yield **VII**, which can undergo 8π-electrocyclization to yield the enolate **VI** or 6π-electrocyclization to yield the enolate **VIII**. Hydrolysis of **VI** and **VIII** produces compounds **II** and **III**, respectively.

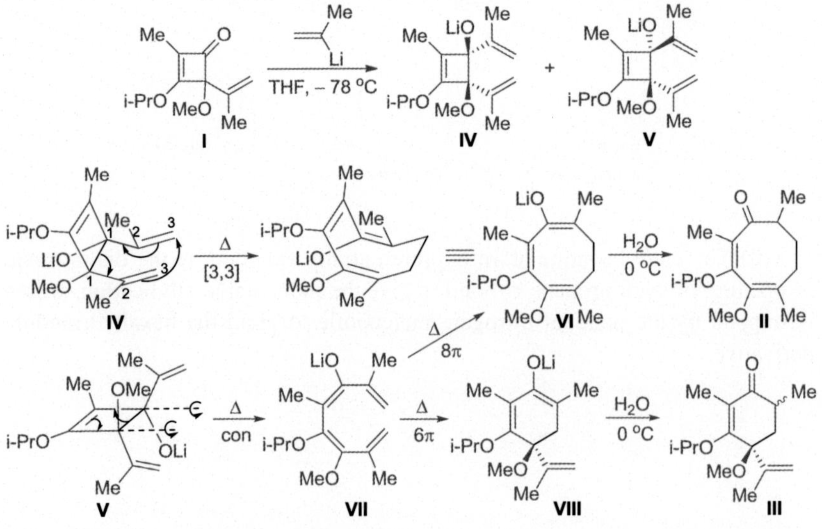

(ii) On irradiation, stilbenoid compound **I** undergoes an initial equilibration to a photostationary mixture of *E* and *Z* isomers, which is followed by a photochemically-allowed ring closure of the *Z* isomer to an unstable compound **II**. Interestingly, **II** loses its tosyl group (*p*-Me-Ph-SO_2-) rather than the tosyloxy (*p*-Me-Ph-SO_2O-) substituent, which is a much better leaving group. This can be explained by the selective deprotonation of **II** to the sulfone-stabilized anion **III**. Protonation of this anion leads to intermediate **IV**, which yields phenanthrenoid **V** by losing tosyl group. An alternate mechanism involving the expulsion of the tosyloxy group is not viable, because no sulfone-stabilized anion is possible.

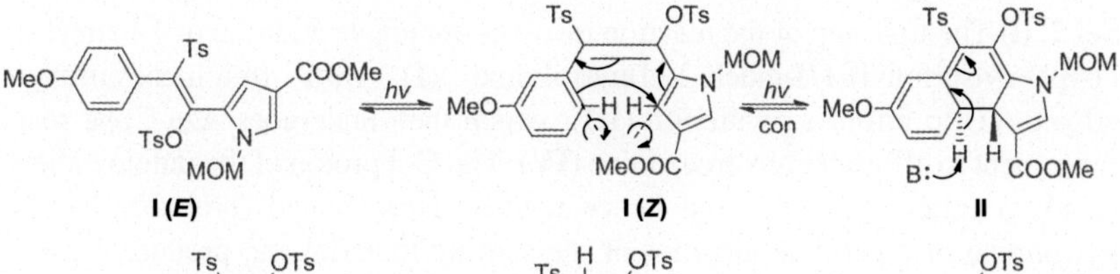

(iii) The benzyne generated by the fluoride-induced 1,2-elimination of **I** is attacked by the strongly nucleophilic oxygen of the cyclopropenone to give a zwitterionic intermediate **II**. This species then undergoes a ring closure to give the spirocyclic benzoxete derivative **III**. Electrocyclic ring opening of **III** affords reactive *o*-quinone methide **IV**, which undergoes a [4 + 2] cycloaddition with a second benzyne to generate the spirocyclic compound **V**.

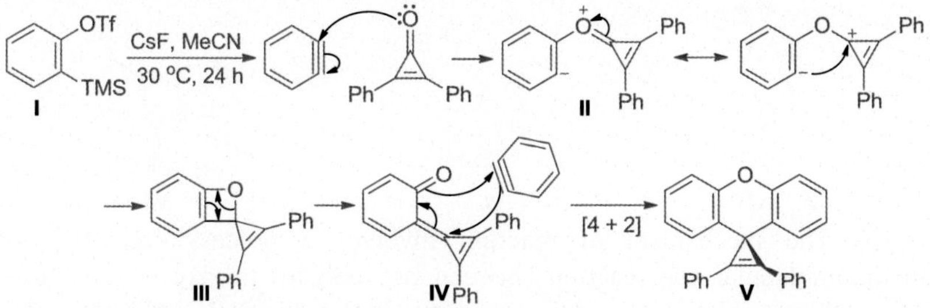

(iv) Reaction of 3-phenylcyclobut-2-enone (**I**) with lithiodiazoacetate (generated in situ) gives an intermediate oxy anion (**II**), which under thermal conditions undergoes conrotatory ring opening. In the case of **II**, the ring can open by two conrotatory paths. However, in practice the oxy anion substituent undergoes exclusive outward rotation (torquoselectivity) to give a diazo-diene intermediate (**III**), which then undergoes an 8π-electron cyclization to give the 1,2-diazepine product as a mixture of **V** and its tautomer **VI**. It should be noted that both electrocyclic reactions take place under extremely mild reaction conditions due to oxy anion acceleration.

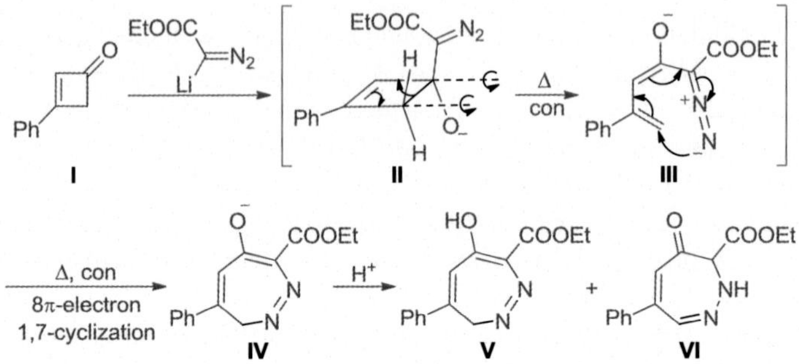

Sol 7. **(i)** The first step of the reaction involves conjugate addition of 1-benzyl-2-(1-(phenylsulfonyl)-1*H*-inden-2-yl)pyrrolidine (**I**) with (*p*-toluenesulfonyl) ethyne (**II**) to afford a zwitterion (**III**), which then undergoes Aza-Cope rearrangement to afford tetrahydroazonine (**IV**). The C-3 proton of the indole moiety of **IV** is relatively acidic and hence readily deprotonated (probably by the pyrrolidine or piperidine moieties of the starting material and product, respectively) to give an anion **V**. This is followed by an anionic 6π-electron electrocyclic ring closure of anion **V** to give **VI**. The *cis*-fusion of cyclopentene and piperidine rings in **VI** is due to the disrotatory nature of the 6π-electrocyclic reaction under thermal conditions. Protonation of **VI** gives the final product **VII**.

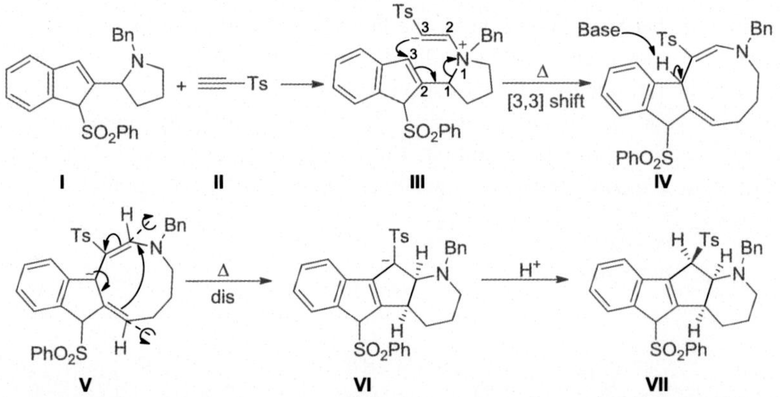

(ii) The mechanism of reaction involves a process consisting of an intramolecular ene reaction between two alkynes to give a vinylallene intermediate, which then undergoes an intramolecular Diels–Alder reaction with the double bond to give the fused tetracyclic product.

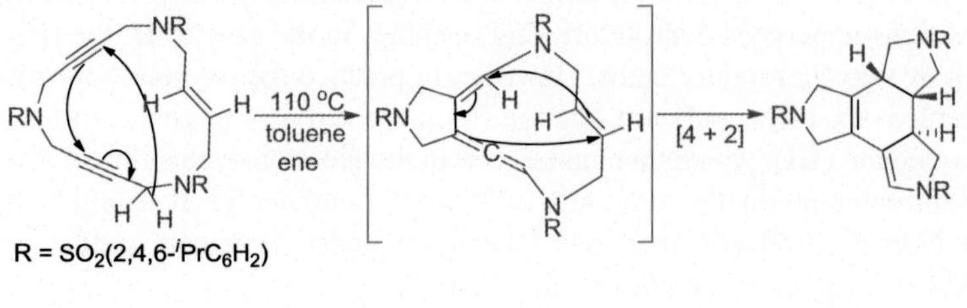

R = $SO_2(2,4,6\text{-}^iPrC_6H_2)$

(iii) As the triyne **I** is an asymmetric triyne, the reaction can proceed via two pathways. In both cases the first step of the cascade reaction corresponds to an intramolecular propargylic ene-type reaction of a 1,6-diyne to generate a vinylallene followed by an intramolecular Diels–Alder reaction and tautomerization. Tautomerization by a thermal suprafacial [1,3] hydrogen transfer is symmetry forbidden. A thermal suprafacial [1,5] hydrogen transfer is symmetry allowed, but in this case a six-membered ring transition state cannot be generated due to the rigidity of the planar conjugated diene moiety. It has been shown that tautomerization in such systems actually involves water supported [1,3] or [1,5] hydrogen transfer.

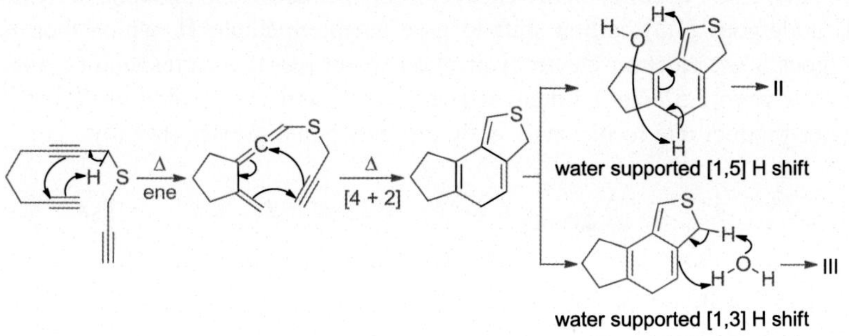

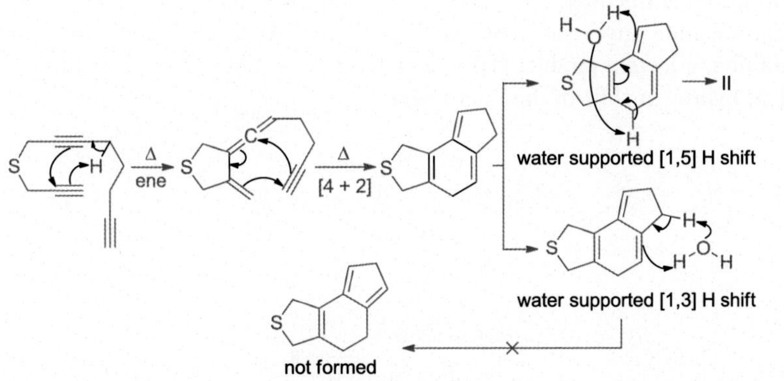

Δ ene

Δ [4 + 2]

water supported [1,5] H shift → II

water supported [1,3] H shift → III

Δ ene

Δ [4 + 2]

water supported [1,5] H shift → II

water supported [1,3] H shift

not formed

(iv) 1,2-Diaryl substituted gem-dibromocyclopropane (**I**) on treatment with silver tetrafluoroborate ($AgBF_4$) in dichloroethane at 65 °C led to the formation of 2-bromo-1-aryl substituted indenes via a domino reaction sequence. Initial silver(I) promoted 2π-disrotatory electrocyclic ring opening of **I** led to the formation of allylic carbocation intermediates **IIa** and **IIb**, which under the reaction conditions could equilibrate with each other. 4π-Conrotatory electrocyclization of **IIa** and **IIb** followed by deprotonation led to the formation of isomeric indenes **III** and **IV**, respectively.

Br Br
I
Me
Δ, Ag (I)
2π-disrotatory
electrocyclic
ring-opening
Br
+
Me
H H
IIa
Δ
con
Br
+
Me
H H
- H+
III
Δ, Ag (I)
2π-disrotatory
electrocyclic
ring-opening
Br
+
Me
H H
IIb
Δ
con
Br
+
H H
Me
- H+
IV

Sol 8. (i) Upon heating, methylenecyclopropane methylene diketone derivative (**I**) undergoes [1,3] carbon shift to give an intermediate **II**, which then undergoes a 6π-electron electrocyclization to furnish the corresponding *trans*- and *cis*-spiro[2.5]octa-3,5-diene derivatives **III** and **IV**. In addition, **III** is the major product due to its small steric hindrance and greater stability.

Ph
1
2
3
1
O
O
I
Δ
[1,3] shift
Ph
O
O
II
Δ
O
O
Ph
III (major)
+
O
O
Ph
IV

(ii) *N*-Methyl-*N*-arylacrylamide (**I**), with allylic leaving groups (LG), undergoes a photoinduced conrotatory 6π-electrocyclization to produce a zwitterionic intermediate. Expulsion of the leaving group followed by deprotonation gives an α-methylene lactam (**II**) as the major product. In addition, a lactam product (**III**) that retains the leaving group is formed via a [1,5] hydrogen shift in the intermediate.

I
CH3
+N
O−
X
LG
hv
con
CH3
+N
O−
X
H
LG
[1,5]
H shift
CH3
+N
O−
X
H
LG
III
− LG−
CH3
+N
O
X
H
− H+
II

(iii) Benzothiophene carboxanilide (**I**) undergoes a photoinduced conrotatory 6π-electrocyclization to produce a zwitterionic intermediate. Expulsion of the leaving group followed by deprotonation gives the cyclized product (**II**).

I
Me
+N
O−
S
H LG
hv
con
Me
+N
O−
S
H LG
− LG−
Me
+N
O
S
H
− H+
II

(iv) In the first step a thermal Oxy-Cope rearrangement of divinylcyclohexanol (**I**) affords enol intermediate **IV**. Subsequent tautomerization of **IV** generates ketone intermediate **V**, which reacts via transannular ene reaction to furnish tricyclic product **II**. The unsaturated ketone **III** is formed by a retro ene reaction of **I**.

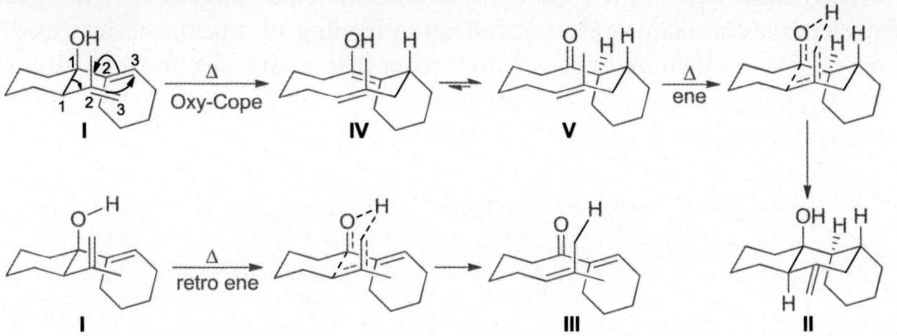

Sol 9. **(i)** When compound **I** is subjected to Claisen rearrangement, the initial [3,3] shift may give rise to two diallyl cyclohexadiene intermediates, i.e., **II** or **III**. In the case of **II**, migration of either allyl group occurs with equal ease to the *para* position to give the rearranged products **IV** and **V**. However, migration of the allyl group from **III** gives only **V**. This experiment proves the intermediacy of cyclohexadiene intermediates.

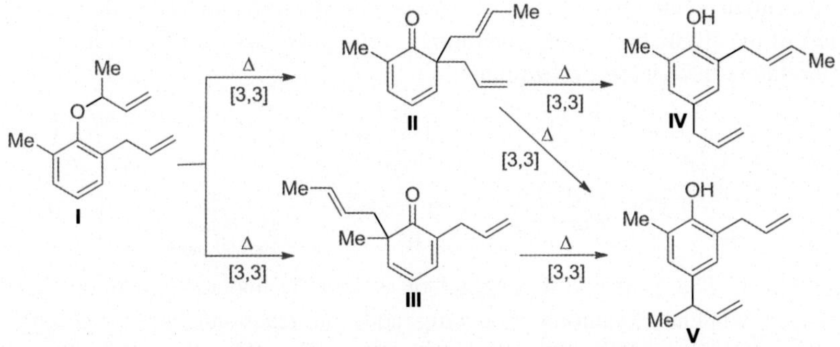

(ii) *N*-Benzoyl oxazolidine derivative (**I**) on treatment with *t*-BuLi in the presence of 1,3-dimethylhexahydro-2-pyrimidinone (DMPU) at −78 °C firstly led to the formation of an *ortho*-lithiated compound **II**, which slowly undergoes anion translocation to give **III**. A 6π-electron disrotatory thermal electrocyclic ring closure of **III** provides enolate **IV** with a *cis* tricyclic structure. Finally, protonation with ammonium chloride gives **V**.

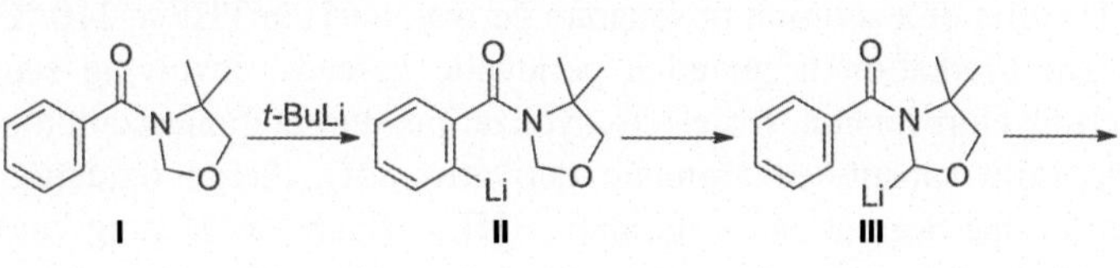

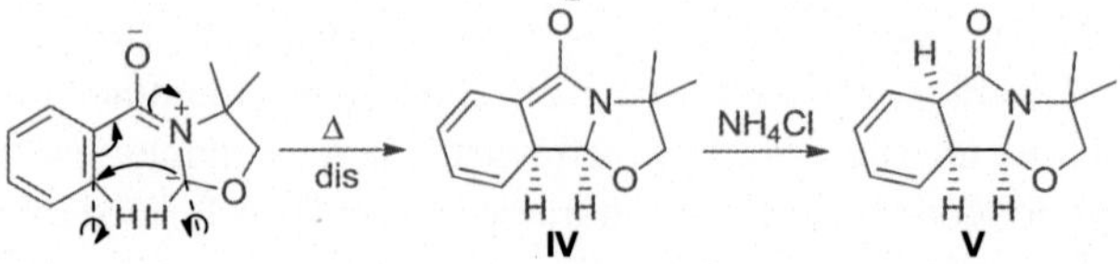

(iii) Under thermal conditions, *cis,trans*-benzooctatetraene (**I**) undergoes 8π-electron conrotatory electrocyclization leading to a pentaene intermediate (**II**), which then undergoes 6π-electron disrotatory electrocyclization to give a mixture of two isomeric products, *endo*- and *exo*-2-methyl-1,2,2a,8b-tetrahydrocyclobuta[a]naphthalene (**III**) in a 3:1 ratio.

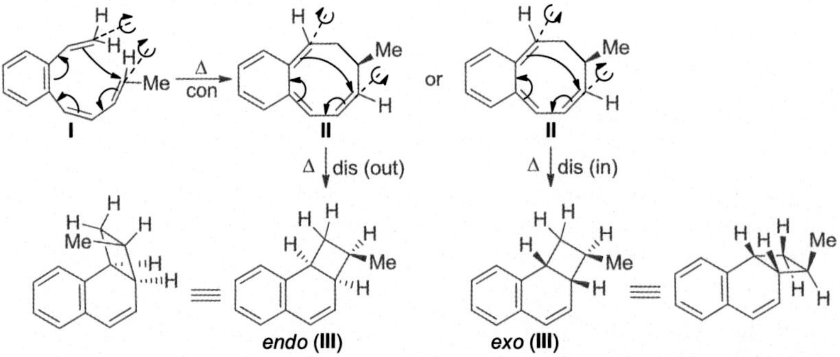

(iv) The synthesis of heterocycle-annulated azepine derivative involves two consecutive symmetry-allowed reactions: the antarafacial [1,6] sigmatropic shift of the allylic hydrogen, generating a conjugated azomethine ylide, and its conrotatory 1,7-electrocyclization.

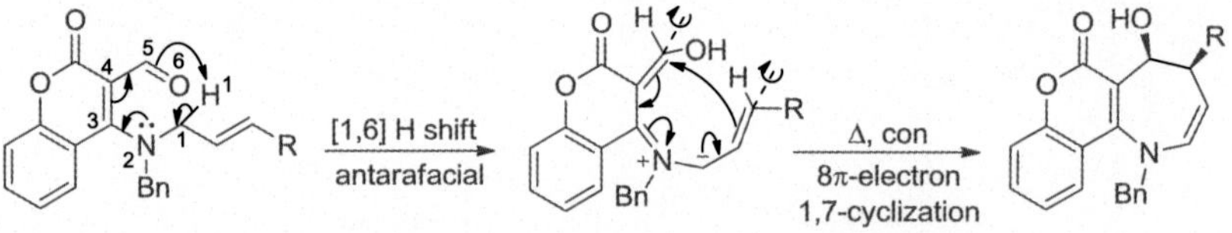

Sol 10. (i) The 1-alkenyl-4-pentyn-1-ol system (**I**) undergoes a microwave-assisted tandem oxyanionic 5-*exo*-dig addition reaction/Claisen rearrangement sequence to give cyclohept-4-enone derivative (**II**).

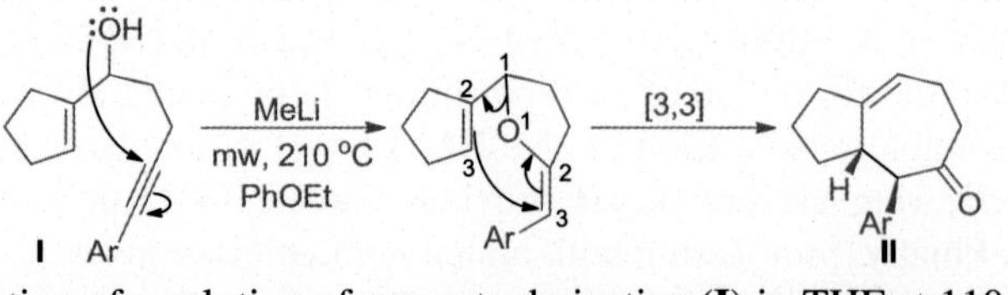

(ii) Heating of a solution of squarate derivative (**I**) in THF at 110 °C under microwave irradiation triggered a pericyclic cascade, involving sequential 4π-electrocycloreversion, 6π-electrocyclization, and tautomerization steps to give the fully substituted aromatic derivative (**II**). Aerial oxidation of **II** resulted in the formation of quinone (**III**), which on heating undergoes intramolecular Diels–Alder reaction to give **IV**. Cleavage of the *tert*-butyl protecting group completed the total synthesis of marine natural product (**V**).

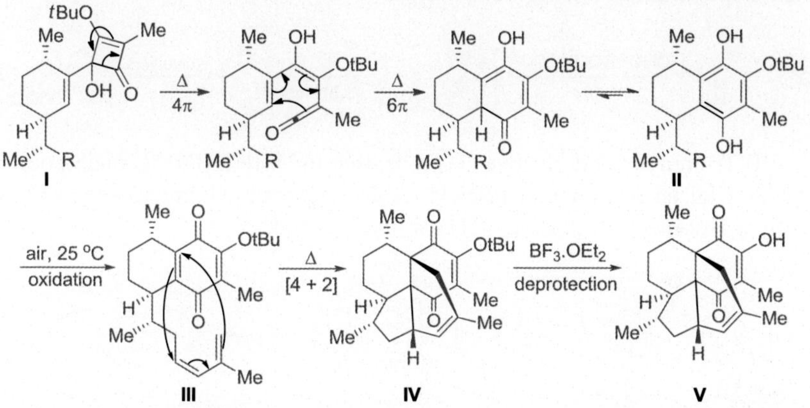

(iii) Tertiary propargyl vinyl ether (**I**) having an electron withdrawing group (amide or ester) on heating undergoes [3,3] propargyl Claisen rearrangement to afford the mixture of the corresponding β-allenal (**II**) and its enolic tautomer (**III**), which possess a highly electrophilic central allenic carbon and a reactive hydroxyl enol functionality conveniently placed to perform a 5-exo-dig O-cyclization reaction to afford trisubstituted furan (**IV**).

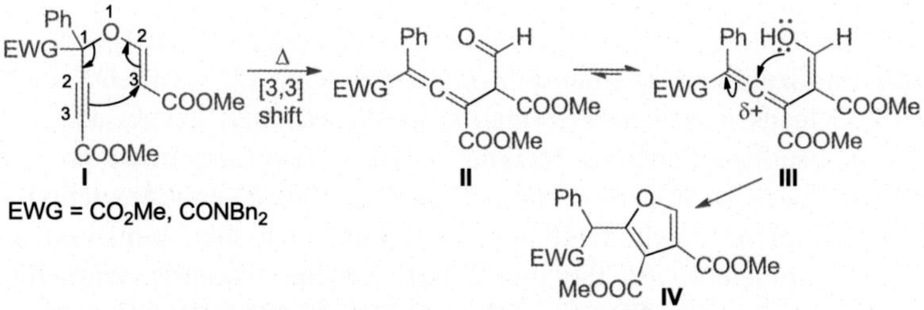

EWG = CO_2Me, $CONBn_2$

(iv) Under microwave irradiation propargyl vinyl ether (**I**) undergoes propargyl Claisen rearrangement to allene intermediate (**II**), which by 1,3-protropic isomerization gives the corresponding diene (**III**). Condensation of **III** with methoxyamine hydrochloride gives imine intermediate (**IV**), which undergoes 6π-aza-electrocyclization followed by an elimination/aromatization step leading to functionalized nicotinic acid (**V**).

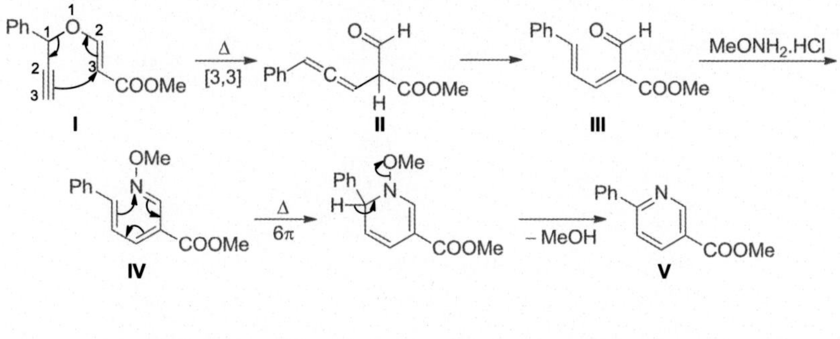

Sol 11. **(i)** On heating *N*-benzylbut-3-en-1-amine with glyoxylic acid monohydrate in methanol, *N*-benzylallylglycine is formed by involving a tandem Aza-Cope/iminium ion solvolysis reaction.

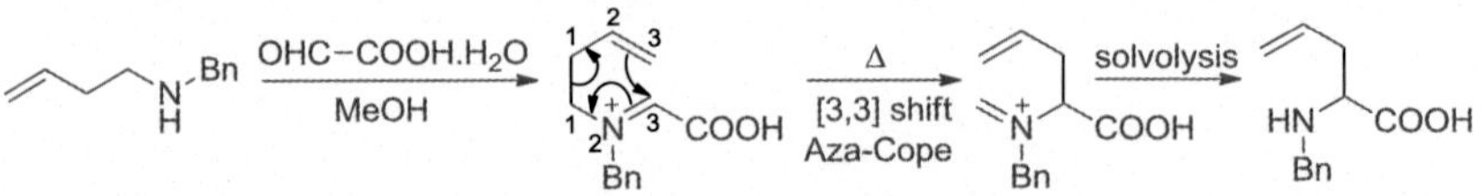

(ii) Triprenylated derivative (**I**) on heating in DMF at 120 °C undergoes a double Claisen rearrangement/Diels–Alder sequence, via intermediate **II** to furnish 1-*O*-methylforbesione (**III**), a natural product.

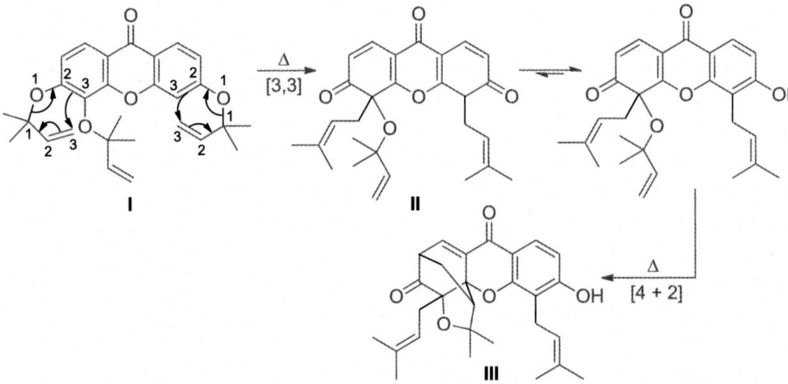

(iii) Enolization of (*E*)-but-2-en-1-yl propionate (**I**) with LDA in THF at −78 °C leads to selective formation of the (*E*)-silyl ketene acetal (**II**), which on heating undergoes **Ireland–Claisen rearrangement** preferably through a chair-like transition state. Therefore, the (*E*)-enolate methyl substituents adopt a pseudo-axial position in the chair-like transition state, leading predominantly to the *anti*-2,3-dimethylpent-4-enoic acid (**III**) in 87:13 dr. However, enolization of **I** in a 23% HMPA/THF solvent system leads to selective formation of the (*Z*)-silyl ketene acetal (**IV**), which on rearrangement gives *syn*-2,3-dimethylpent-4-enoic acid (**V**) in 81:19 dr. Thus, Ireland–Claisen rearrangement can direct the diastereoselectivity of the process in a highly predictable manner by controlling the geometry of enolization.

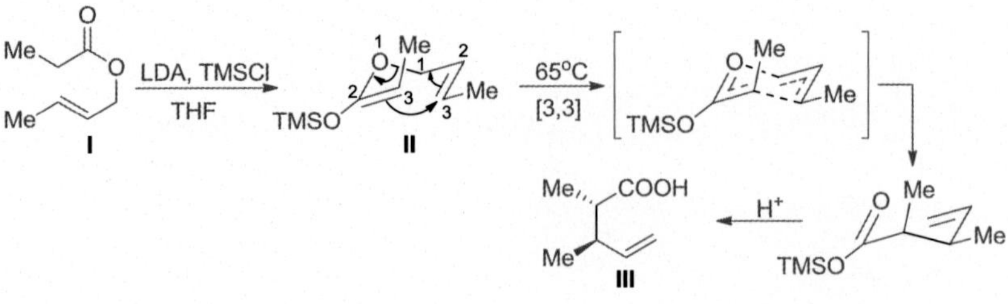

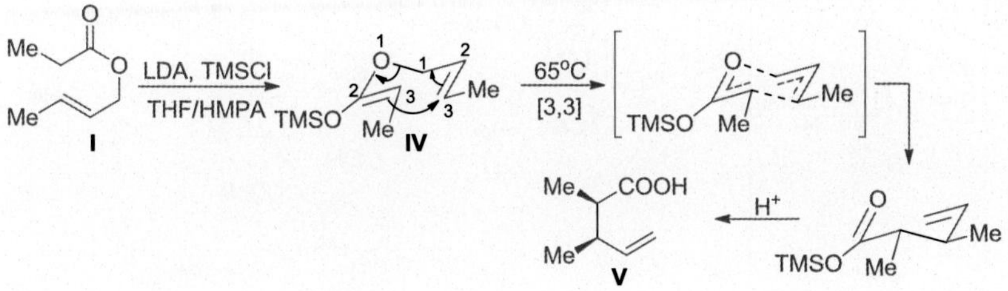

LDA, TMSCl
THF/HMPA
65°C
[3,3]
H+
I
IV
V

Sol 12. **(i)** The first equivalent of LDA creates an intermediate as a diene component by deprotonation of nitrogen and the second equivalent induces an E2 elimination to generate a benzyne. The intermediate benzyne is then trapped by an intramolecular Diels–Alder reaction to give the enolate, which after protonation and oxidation by air gives the observed aromatic product.

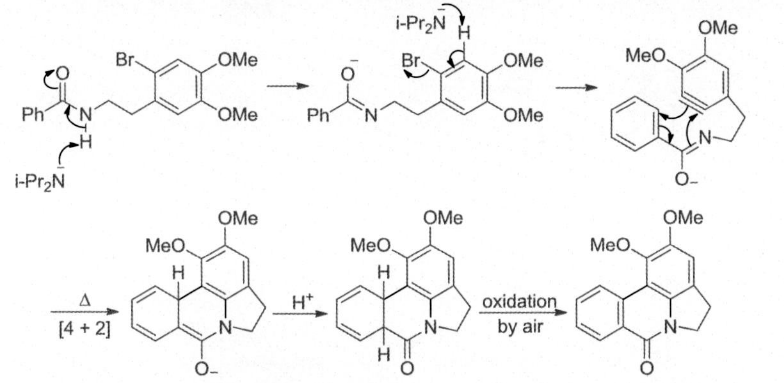

Δ
[4 + 2]
H+
oxidation
by air

(ii)

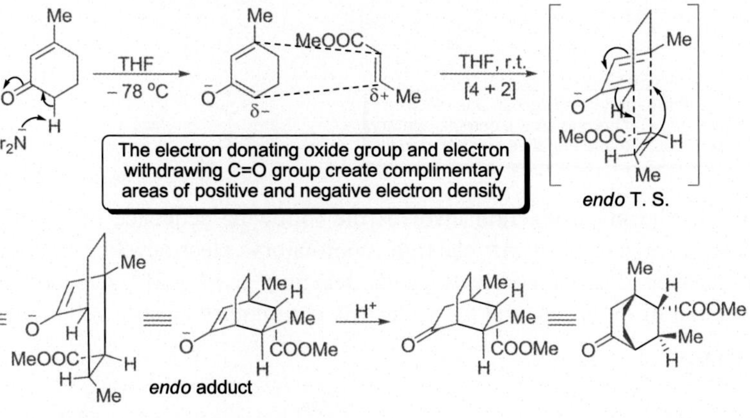

THF
− 78 °C
THF, r.t.
[4 + 2]
The electron donating oxide group and electron
withdrawing C=O group create complimentary
areas of positive and negative electron density
endo T. S.
endo adduct
H+

(iii)

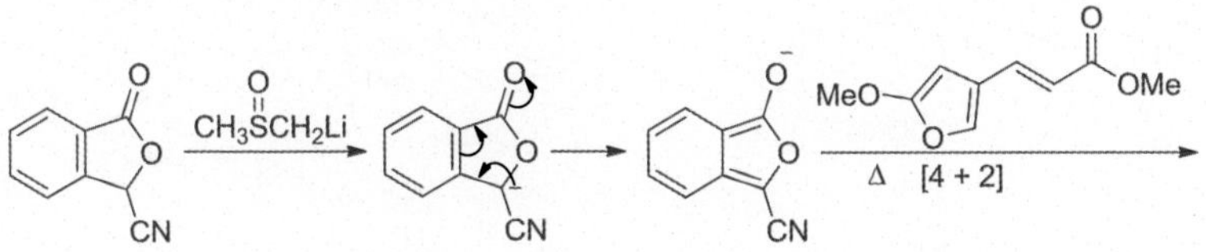

Δ [4 + 2]

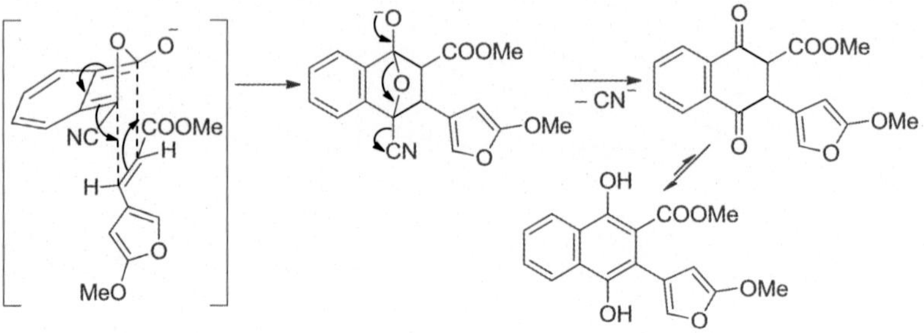

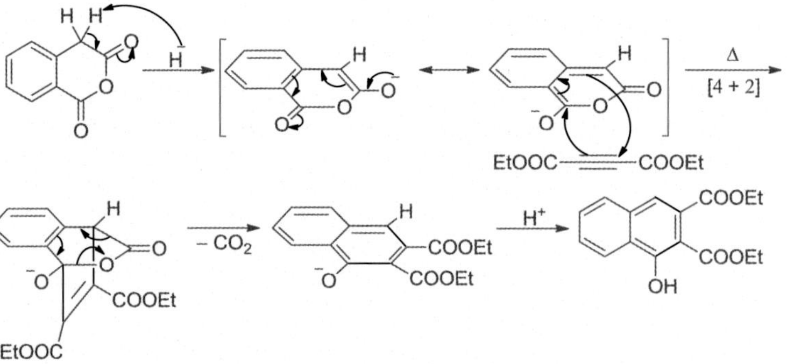

Sol 13.

The electron donating N group and electron withdrawing C=O group create complementary areas of positive and negative electron density

endo T. S.

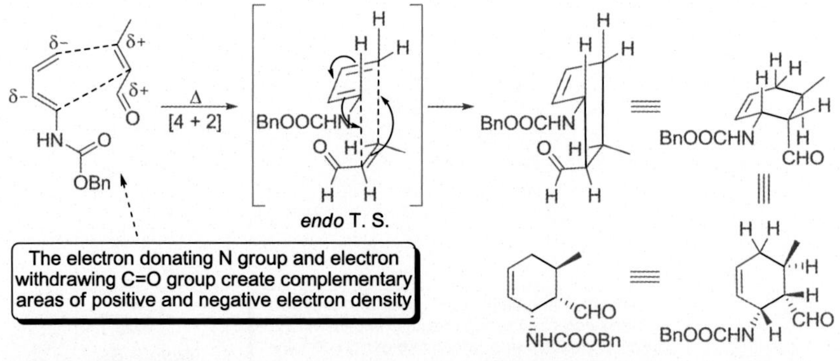

Sol 14. The given reaction involves the following sequence of symmetry-allowed changes: an 8π-electron conrotatory electrocyclic reaction, a 6π-electron disrotatory electrocyclic reaction, a [4 + 2] cycloaddition, a reversion of the same, and a 4π-electron conrotatory electrocyclic reaction.

Sol 15.

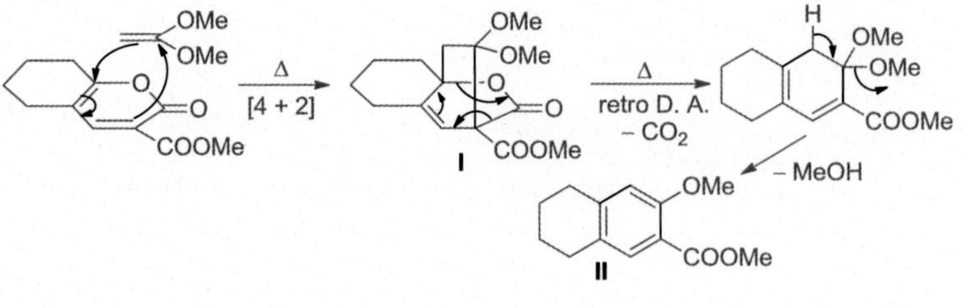

(iii) A $BF_3\backslash OEt_2$-induced retro Diels–Alder reaction, which leads to expulsion of cyclopentadiene and results in the formation of an imino dienophile (**I**). With a diene system appended from the same carbon atom of the indole skeleton in proximity, a facile intramolecular aza Diels–Alder reaction generates **II** as a 1.5:1 mixture of diastereomers.

(iv) 5-Chloropyrazine-2(1*H*)-one acts as the aza-diene in the intramolecular hetero Diels–Alder reaction to give an intermediate, which then loses chloronitrile group in the retro Diels–Alder step to give **I**. Oxidation of **I** with *m*-CPBA gives **II** and regioselective deprotonation with *n*-butyl lithium followed by nucleophilic substitution places the tether group at an appropriate position to the sulfone group in **III**. On heating, **III** undergoes cheletropic loss of SO_2, producing heteroaromatic *o*-quinodimethane intermediate, which readily undergoes an intramolecular Diels–Alder reaction resulting in the formation of the final product.

Bn N O Cl N S Δ [4 + 2] BnN O N S Cl Δ retro D. A. BnN O S I [O] BnN O S O O II

Br BuLi (2.1 eq.) − 78 °C BnN O S O O III Δ − SO_2 BnN O Δ [4 + 2] BnN O

(v) H O H Δ dis H O H MeO_2C CO_2Me Δ [4 + 2] O H H COOMe MeOOC

O H H COOMe COOMe ≡ O COOMe COOMe I (i) $NaBH_4$, MeOH (ii) Ac_2O, pyridine AcO COOMe COOMe II Δ retro D. A.

COOMe COOMe + OAc Δ con OAc OAc

(vi) The reaction involves the conversion of (*E*)-but-2-en-1-yl 2-methyl-3-oxobutanoate to a dienolate (**I**) that smoothly undergoes a **Carroll–Claisen rearrangement** to give a β-ketocarboxylic acid (**II**). The stereochemistry of the major isomer **II** results from the rearrangement of **I** through a chair-like transition state. The ozonolysis reaction enables spontaneous cyclization to occur concurrently. This is followed by in situ acetylation to give the desired (±)-4-epi-acetomycin (an antibiotic). It should be noted that in this case the decarboxylation of the initially formed β-ketocarboxylic acid (**II**) is avoided by carrying out its intramolecular lactonization at low temperature.

O O O LDA (2 eq.) THF, − 78 °C OLi OLi O I ≡ LiO LiO H 2 1 O 1 3 3 2 23 °C [3,3] shift

LiO LiO H O → HO O H O ≡ Me Me MeOC H HOOC II O_3, Me_2S Me Me MeOC H O OH O

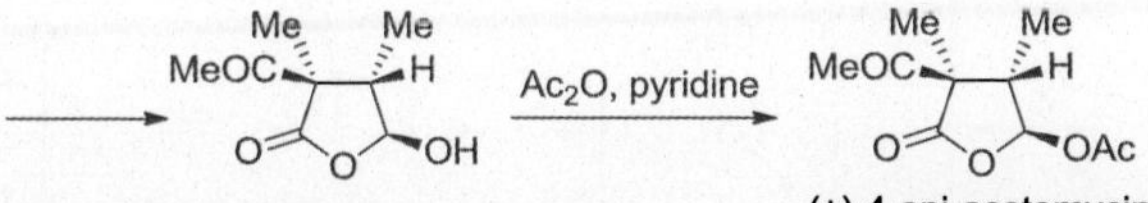

(±)-4-epi-acetomycin

Sol 16. (i) 3,6-Diphenyl-1,2,4,5-tetrazine (**I**) undergoes Diels−Alder reaction with acrylonitrile to give a bicyclic product (**II**), which undergoes retro Diels−Alder reaction generating a 4,5-dihydropyridazine (**III**), which on tautomerization gives a 1,4-dihydropyridazine (**IV**).

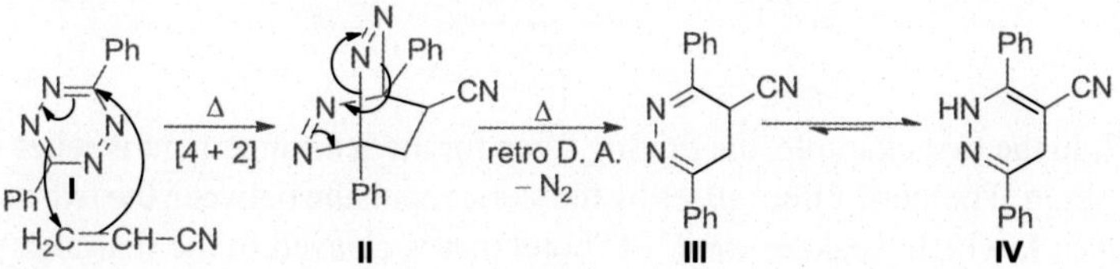

(ii) Both the compounds are unstable toward loss of nitrogen at room temperature and readily undergo retro Diels−Alder reaction to give off N_2 and an *o*-quinodimethane. In the absence of any other trapping agent, this highly reactive intermediate dimerizes in another Diels−Alder reaction to give the product. A strong driving force for the [4 + 2] cycloaddition of such species is a result of the establishment (or re-establishment) of aromaticity.

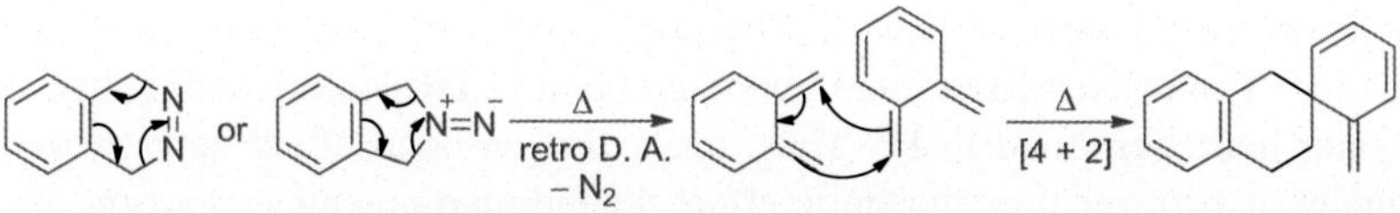

(iii) Lewis acid-mediated ionization of 6-methoxy-5-methyl-3-phenyl-6*H*-1,2-oxazine (**I**) gives azapyrylium intermediate (**II**). The considerable electron deficiency of a cationic heteroaromatic system presumably imparts a very low energy to the 1-azadiene LUMO. Therefore, **II** readily undergoes [4 + 2] cycloaddition with less activated dienophiles such as ethynylcyclopropane to give **III**, which undergoes extrusion of carbon monoxide to afford a pyridine derivative (**IV**) with complete regioselectivity.

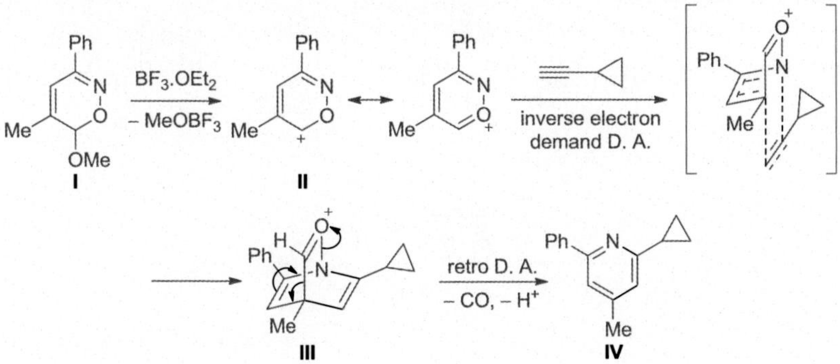

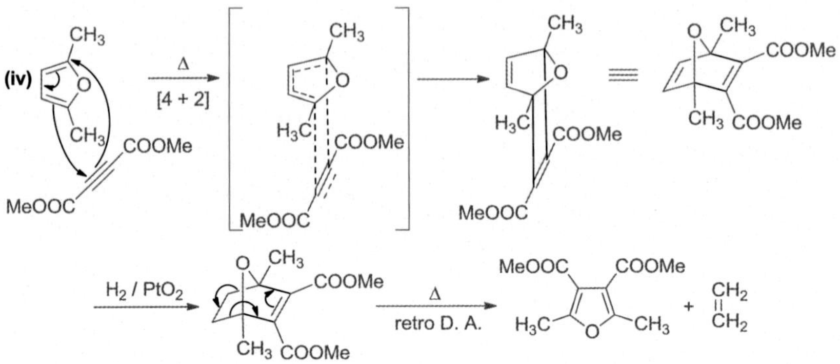

Sol 17. In the first example, the driving force for the rearrangement is relief in the steric strain. The steric effect arises by the steric repulsion between the two mesityl groups in **I**, which weakens the C—C bond that is cleaved in the rearrangement. However, in going from **I** to **II** a loss in imine conjugation is expected due to the fact that in the case of **I** the C=N bond can be planar with the phenyl group, whereas in **II** this bond cannot be fully planar or conjugated with the mesityl group owing to steric effects. But as the reaction goes to completion, it shows that the destabilization of **I**, as a result of the steric repulsion of the two mesityl groups, appears to be greater than the resonance stabilization of the imine group.

The second example shows that in contrast to the diaza-Cope rearrangement driven by steric forces, the hydrogen bond-directed reaction is more favored. The resonance-assisted hydrogen bond in **III** is more stable than the regular hydrogen bond in **IV**. Thus, the hydrogen-bond effect appears to be somewhat stronger than the steric effect for the rearrangement reaction.

Sol 18. Treatment of (*E*)-2-(phenyldiazenyl)pyridine (**I**) with trimethylsilyl iodide followed by reduction provides the corresponding pyridinium salt of 2-(2-phenylhydrazinyl)pyridine (**III**). The acid-catalyzed [5,5] sigmatropic shift (benzidine rearrangement) of **III** then gives the target compound 5-(4-aminophenyl)pyridin-2-amine (**II**). The Woodward—Hoffmann rules predict that [5,5] sigmatropic shifts would proceed suprafacially through a 10-membered transition state.

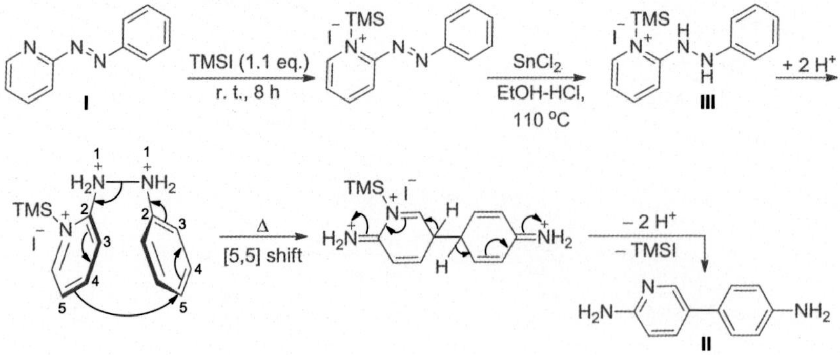

2-(2-Phenylhydrazinyl)pyridine (**IV**) obtained by the reduction of **I** may undergo tautomerization and disproportionation, which prevent correct orientation for [5,5] sigmatropic rearrangement of **IV** to provide the benzidine rearrangement product **II**. However, protection of nitrogen on the pyridine improves the yield of **II** by shifting the equilibrium from the thermodynamically stable hydrazono tautomer of **IV** to the less stable tautomer, and also by preventing disproportionation reaction.

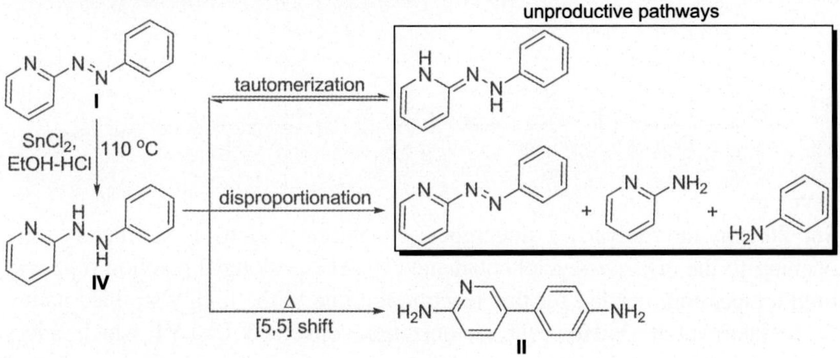

unproductive pathways

I

$SnCl_2$, EtOH-HCl, 110 °C

IV

tautomerization

disproportionation

Δ [5,5] shift

II

Sol 19. Thermal pericyclic rearrangement of **I** leads to the formation of *Z*-α,β,γ,δ-unsaturated amide **II**. The mechanism consists of initial double bond isomerization (a reaction that should be relatively facile, due to the relatively weak nature of the aldehyde π-bonds), 6π-electrocyclic ring closure, [1,5] sigmatropic shift of hydrogen, and 6π-electrocyclic ring opening. A second mechanism involving a change in the order of events has been also proposed. In this mechanism a [1,5] sigmatropic shift of hydrogen to give vinylketene intermediate precedes ring closure.

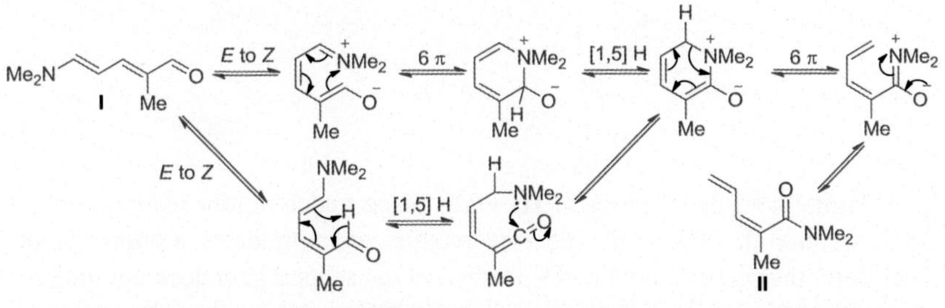

E to Z — 6 π — [1,5] H — 6 π

E to Z — [1,5] H

On heating, the aldehyde **III** bearing an *N*-allyl group gives the intermediate zwitterionic dienolate **VI** by the same mechanism as described. This intermediate undergoes usual 6π-electrocyclic ring opening reaction to afford the α,β,γ,δ-unsaturated amide, which then undergoes an intramolecular Diels–Alder reaction to form the bicyclic lactam **IV**. However, due to the presence of *N*-allyl groups, **VI** can also undergo a [3,3]-sigmatropic rearrangement with neutralization of charges to afford a substituted dihydropyridone. A further Cope rearrangement converts this into more conjugated dihydropyridone **V**.

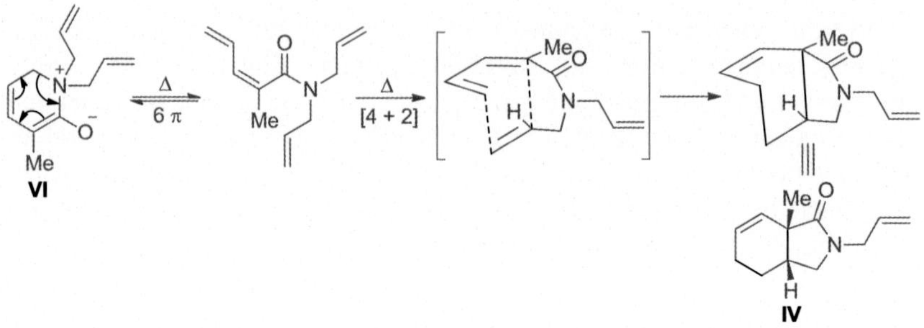

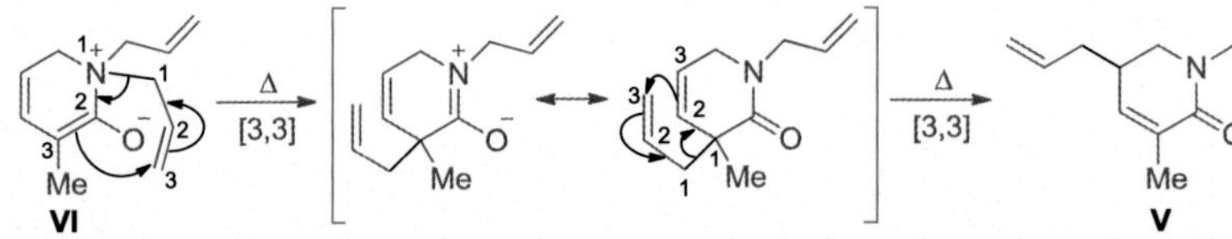

Sol 20. On thermolysis, **I** undergoes a symmetry-allowed conrotatory ring opening to the *cis,trans*-cycloheptadiene (**V**). The prolonged reaction time and high temperature needed for this reaction are due to the highly strained nature of the intermediate **V**. However, lactonization converts **V** into **VI**, which is less likely to undergo reversion to a cyclobutene because of strain. Two sequential 1,5-sigmatropic shifts then convert **VI** to *cis,cis*-diene **II**.

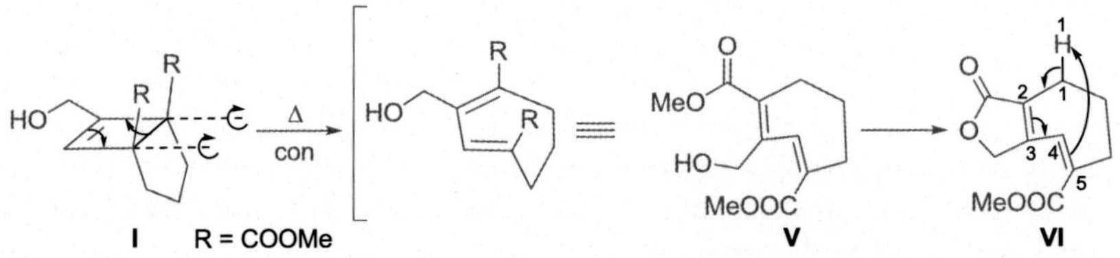

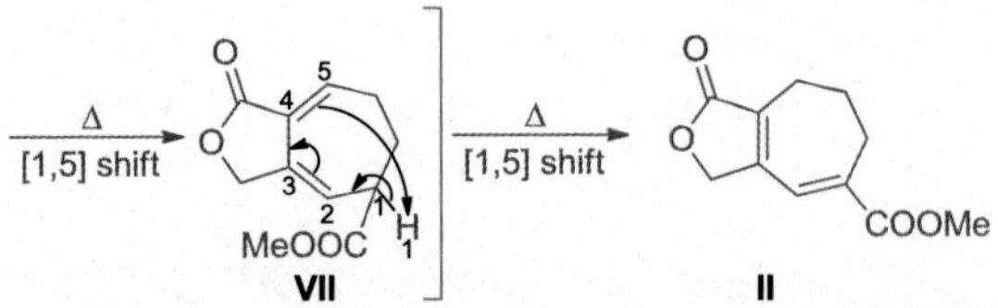

Conrotatory ring opening of **III** takes place under milder reaction conditions because the 10-membered ring readily accommodates a *trans*-alkene. Therefore, the bicyclic lactone **IV** formed in the second step does not undergo an isomerization and retains its *cis,trans* geometry.

Index

Note: Page numbers followed by "f" or "t" indicates figures and tables respectively.

D

E

F

G

H

I

J

K

L

Q

R

S

T

U

V

W

X

Z

Printed in the United States
By Bookmasters